Electricity for the Trades

THIRD EDITION

Electricity for the Trades

Frank D. Petruzella

Mc
Graw
Hill
Education

ELECTRICITY FOR THE TRADES

ISBN 978-1-260-54784-9
MHID 1-260-54784-1

Cover Image: ©*Chockchai Paralart/123RF*

mheducation.com/highered

BRIEF CONTENTS

CONTENTS

Section One

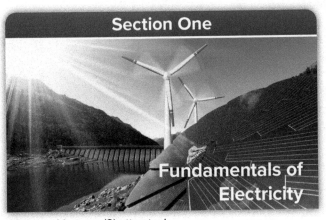

Fundamentals of Electricity

©Alberto Masnovo/Shutterstock

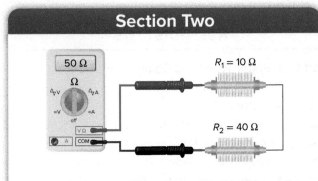
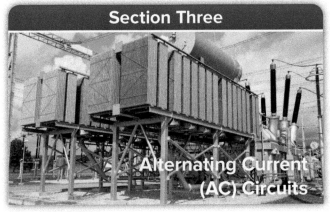

Chapter 16
Inductance and Capacitance   193

Chapter 17
Resistive, Inductive, Capacitive (*RLC*) Series Circuits   213

Chapter 18
Resistive, Inductive, Capacitive (*RLC*) Parallel Circuits   243

Chapter 19
Transformers   270

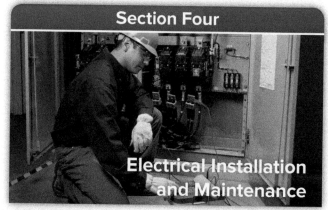

Section Four

Electrical Installation and Maintenance

©Fluke Corporation

Chapter 20
Circuit Conductors and Wire Sizes   294

The third edition of *Electricity for the Trades* focuses on the understanding of topics required for careers in the electrical industry and associated disciplines. The main objective of this text is to explain, **as simply as possible,** the electrical theory and its applications to related electrical circuits and products.

The textbook is divided into four sections:

Section 1 **Fundamentals of Electricity**
Section 2 **Direct Current (DC) Circuits**
Section 3 **Alternating Current (AC) Circuits**
Section 4 **Electrical Installation and Maintenance**

The organization of the text is designed to distinguish between the major areas of study in order to enhance the text's usability. Each section or combination of sections is available in both **eBook** and customized **McGraw-Hill Create** formats to meet more specific curriculum needs.

In this edition, **all chapters** have been enhanced, as requested by instructors, to provide up-to-date relevant subject matter. Most notable are the **new topics,** which include three-wire circuits, protection relays, DC motor control, and digital logic circuits. The text continues to use:

- **Boldface** fonts to highlight important points.
- **Bulleted** lists to summarize circuit operations.
- Schematic representations of circuits accompanied by **photos** of the devices being studied to increase the student's level of recognition of different electrical devices.

New to this edition is the **Simulation Lab Manual** that accompanies the text. Available through **McGraw-Hill Connect,** it features more than **250 Multisim** simulated lab exercises integrated into each chapter of the text. Features of this new concept of experimental labs assignments include:

- Each Multisim circuit file comes complete with **on-screen** detailed instructions for carrying out circuit simulation problems and exercises.
- All lab components have been **preselected** and with their required values requiring a **minimum of setup.** Students spend more time experimenting

rather than selecting, dragging, and assigning values to components.
- On-screen **formulas and graphics** from the text are designed to integrate the material covered in the text with the simulation assignment.

The lab assignments may require students to:

- Correctly connect and take measurements using a simulated voltmeter, ammeter, ohmmeter, digital multimeter, wattmeter, and oscilloscope.
- Properly record measured data.
- Calculate expected circuit values.
- Troubleshoot components and circuits.
- Modify circuit requirements.

For this edition, Multisim labs have been created to operate using National Instruments' **NI Multisim Student Edition Version 12** or higher.

Note: Multisim software must be purchased separately.

Chapter Changes to This Edition

Chapter 1
- Extended coverage of the National Electrical Code.
- Additional test bank questions.
- Additional chapter review questions.
- New and modified line diagrams and photos.

Chapter 2
- New section on the basic concepts of voltage, current, and resistance.
- Additional test bank questions.
- Additional chapter review questions.
- New and modified line diagrams and photos.

Chapter 3
- Coverage of static electric shock.
- Additional test bank questions.
- Additional chapter review questions.
- New and modified line diagrams and photos.

Chapter 4

- Extended coverage of relationship between the coulomb and the ampere.
- Additional test bank questions.
- Additional chapter review question.
- New and modified line diagrams and photos.

Chapter 5

- New section on short and open circuit faults.
- Additional test bank questions.
- Additional chapter review questions.
- New and modified line diagrams and photos.

Chapter 6

- Extended coverage of digital multimeter measurements.
- Additional test bank questions.
- Additional chapter review questions.
- New and modified line diagrams and photos.

Chapter 7

- Extended coverage of Ohm's law calculations.
- Additional test bank questions.
- Additional chapter review questions.
- New and modified line diagrams and photos.

Chapter 8

- Coverage of alphanumeric resistor code.
- Extended coverage of resistor connections.
- Additional test bank questions.
- Additional chapter review questions.
- New and modified line diagrams and photos.

Chapter 9

- New section on electromagnetic induction.
- Additional test bank questions.
- Additional chapter review questions.
- New and modified line diagrams and photos.

Chapter 10

- New coverage of pumped storage hydroelectric plant.
- Additional test bank questions.
- Additional chapter review questions.
- New and modified line diagrams and photos.

Chapter 11

- Extended coverage of series circuit troubleshooting problems.
- Additional test bank questions.

- Additional chapter review questions.
- New and modified line diagrams and photos.

Chapter 12

- Extended coverage of parallel circuit troubleshooting problems.
- Additional test bank questions.
- Additional chapter review questions.
- New and modified line diagrams and photos.

Chapter 13

- New section on three-wire circuits.
- Additional test bank questions.
- Additional chapter review questions.
- New and modified line diagrams and photos.

Chapter 14

- Additional test bank questions.
- New and modified line diagrams and photos.

Chapter 15

- New coverage of eddy currents and the skin effect.
- Additional coverage of alternator synchronization.
- Additional test bank questions.
- Additional chapter review questions.
- New and modified line diagrams and photos.

Chapter 16

- Additional coverage of Inductors.
- Additional coverage of AC resistive circuits.
- Additional coverage of energy storage versus energy dissipation.
- Additional test bank questions.
- Additional chapter review questions.
- New and modified line diagrams and photos.

Chapter 17

- Additional coverage of inductance.
- Additional coverage of capacitance.
- Additional coverage of series RLC resonance.
- Additional test bank questions.
- Additional chapter review questions.
- New and modified line diagrams and photos.

Chapter 18

- Additional coverage of parallel RLC circuits.
- Additional solved example problems.
- Additional test bank questions.

- Additional chapter review questions.
- New and modified line diagrams and photos.

Chapter 19
- New coverage of the open delta connection.
- Additional coverage of transformer basics.
- Additional test bank questions.
- Additional chapter review questions.
- New and modified line diagrams and photos.

Chapter 20
- Additional coverage of conductor ampacity.
- New coverage of teck cable.
- Additional test bank questions.
- Additional chapter review questions.
- New and modified line diagrams and photos.

Chapter 21
- New section on load centers and circuit breakers.
- Additional coverage of special application circuit breakers.
- Additional test bank questions.
- Additional chapter review questions.
- New and modified line diagrams and photos.

Chapter 22
- New section on protective relays.
- Additional coverage of relay types and applications.

- Additional test bank questions.
- Additional chapter review questions.
- New and modified line diagrams and photos.

Chapter 23
- New coverage of the split receptacle and lighting control devices.
- Additional coverage of LED color light control.
- Additional test bank questions.
- Additional chapter review questions.
- New and modified line diagrams and photos.

Chapter 24
- New chapter title.
- Additional coverage of two-wire and three-wire control.
- New Part 4 coverage of direct current motors.
- Additional test bank questions.
- Additional chapter review questions.
- New and modified line diagrams and photos.

Chapter 25
- New coverage of operational amplifiers.
- New Part 3 coverage of digital logic circuits.
- Additional test bank questions.
- Additional chapter review questions.
- New and modified line diagrams and photos.

ACKNOWLEDGMENTS

I would like to thank all the instructors who reviewed this manuscript and contributed their ideas to improve this edition. My sincerest appreciation goes to:

Max Rabiee
University of Cincinnati

I would like to thank Don Pelster of Nashville State Community College, who edited the manuscript, Multisim lab assignments, Test Bank, PowerPoint slides, and created the Instructor's Guide. This textbook would not be possible without the hard work, talent, and thoughtfulness of the editors and reviewers, and I am truly grateful to have worked with you all!

Electricity for the Trades is a practical text for a basic course in electricity. Through clearly written chapters, it focuses on the information students need to be successful in the field. The content is easy to read and is supported by helpful examples, colorful diagrams and illustrations, and review problems that evaluate students' understanding of the material.

LEARNING OUTCOMES give students an idea of what to expect in the following pages and what they should be able to accomplish by the end of the chapter.

LEARNING OUTCOMES

▶ Identify the factors that determine the severity of an electric shock.

▶ Be aware of general principles of electrical safety including wearing approved protective clothing and using protective equipment.

▶ Familiarity with arc flash hazards recognition and prevention.

▶ Explain the safety aspect of grounding an electrical installation.

▶ Outline the typical steps involved in lockout and tagout procedures.

▶ Be aware of the functions of the different organizations responsible for electrical codes and standards.

▶ Understand how the National Electrical Code is organized by chapters and article.

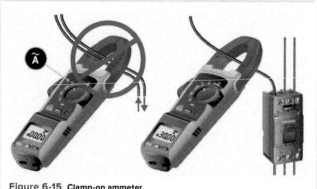

Figure 6-15 Clamp-on ammeter.
©Fluke Corporation

DIAGRAMS AND PHOTOS increase the students' recognition of key devices and processes. Dozens of new photos have been added to make this edition more up to date.

EXAMPLES emphasize the importance of using a systematic, step-by-step approach to problem solving.

EXAMPLE 13-5

Conductors carry current from the power supply to the loads. A conductor should have as little resistance as possible for it to carry this current with minimal voltage drop and power loss. All conductors have resistance that in certain instances must be taken into consideration.

Problem: Determine the voltage across the loads, total power of the circuit, wasted power dissipated in the conductors, and the total power delivered to the loads, for the electrical distribution system of Figure 13-25.

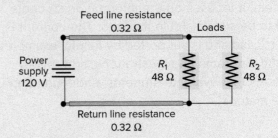

Figure 13-25 Circuit for example 13-5.

Step 1. Simplify the circuit, as shown in Figure 13-26, by representing the line wires as resistive loads.

REVIEW QUESTIONS appear at the end of each part within each chapter. They test students' knowledge of the material as they read, helping them identify areas that may need further study.

Part 2 Review Questions

1. What does the ampacity rating of a conductor specify?

2. List the factors taken into consideration when determining the ampacity rating of a conductor.

3. Why is a copper conductor rated at a higher ampacity than an aluminum conductor of equivalent gauge size or diameter?

4. State the effect (increase or decrease) of each of the following on the resistance value of a circuit conductor:
   a. Increasing the length of the conductor.
   b. Decreasing the diameter of the conductor.
   c. Increasing the operating temperature of the conductor.
   d. Using the same-size aluminum conductor in place of a copper one.

5. a. What causes line voltage drop in a circuit?
   b. Under what condition is the line voltage drop considered to be zero?
   c. In what type of electrical installation must the resistance of the conductors be taken into account?

The *Simulation Lab Manual for Electricity for the Trades,* third edition, contains 250 Multisim computer-simulated assignments. Each computer assignment comes complete with detailed instructions contained within the Multisim circuit file. Instructors can meaningfully integrate these assignments into each chapter of the text.

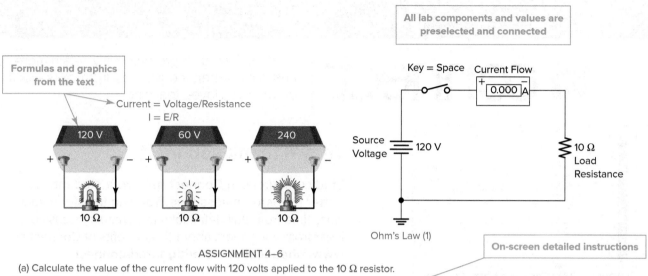

Formulas and graphics from the text

Current = Voltage/Resistance
I = E/R

All lab components and values are preselected and connected

Key = Space   Current Flow

On-screen detailed instructions

Ohm's Law (1)

ASSIGNMENT 4–6

(a) Calculate the value of the current flow with 120 volts applied to the 10 Ω resistor.
(b) Turn the simulation on and record the value of the current.
(c) Calculate the value of the current flow if the source voltage is decreased to 60 volts.
(d) Double-click on the source voltage icon and change its value to 60 volts. Turn the simulation on and record the value of the current.
(e) Calculate the value of the current flow if the source voltage is increased to 240 volts.
(f) Double-click on the source voltage icon and change its value to 240 volts. Turn the simulation on and record the value of the current.

Affordability & Outcomes = Academic Freedom!

You deserve choice, flexibility and control. You know what's best for your students and selecting the course materials that will help them succeed should be in your hands.

Thats why providing you with a wide range of options that lower costs and drive better outcomes is our highest priority.

Students—study more efficiently, retain more and achieve better outcomes. Instructors—focus on what you love—teaching.

They'll thank you for it.

Study resources in Connect help your students be better prepared in less time. You can transform your class time from dull definitions to dynamic discussion. Hear from your peers about the benefits of Connect at **www.mheducation.com/highered/connect**

Study anytime, anywhere.

Download the free ReadAnywhere app and access your online eBook when it's convenient, even if you're offline. And since the app automatically syncs with your eBook in Connect, all of your notes are available every time you open it. Find out more at **www.mheducation.com/readanywhere**

Learning for everyone.

McGraw-Hill works directly with Accessibility Services Departments and faculty to meet the learning needs of all students. Please contact your Accessibility Services office and ask them to email accessibility@mheducation.com, or visit **www.mheducation.com/about/accessibility.html** for more information.

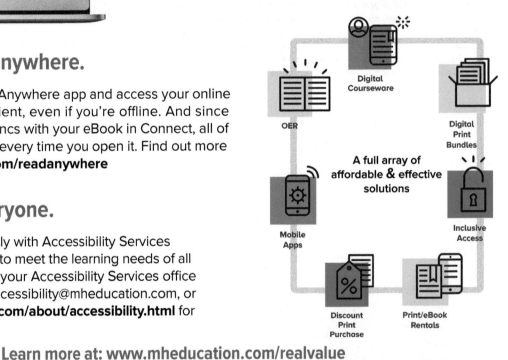

Digital Courseware

OER

Digital Print Bundles

A full array of affordable & effective solutions

Mobile Apps

Inclusive Access

Discount Print Purchase

Print/eBook Rentals

Learn more at: www.mheducation.com/realvalue

Rent It

Affordable print and digital rental options through our partnerships with leading textbook distributors including Amazon, Barnes & Noble, Chegg, Follett, and more.

Go Digital

A full and flexible range of affordable digital solutions ranging from Connect, ALEKS, inclusive access, mobile apps, OER and more.

Get Print

Students who purchase digital materials can get a loose-leaf print version at a significantly reduced rate to meet their individual preferences and budget.

Renewable Energy—Sunlight with solar panel. Wind with wind turbines. Rain with dam for hydropower
©Alberto Masnovo/Shutterstock

Fundamentals of Electricity

SECTION OUTLINE

CHAPTER ONE

Safety

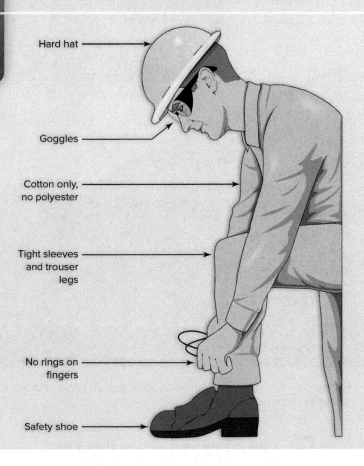

Hard hat

Goggles

Cotton only, no polyester

Tight sleeves and trouser legs

No rings on fingers

Safety shoe

Appropriate attire should be worn for each particular job site.

LEARNING OUTCOMES

▶ Identify the factors that determine the severity of an electric shock.

▶ Be aware of general principles of electrical safety including wearing approved protective clothing and using protective equipment.

▶ Familiarity with arc flash hazards recognition and prevention.

▶ Explain the safety aspect of grounding an electrical installation.

▶ Outline the typical steps involved in lockout and tagout procedures.

▶ Be aware of the functions of the different organizations responsible for electrical codes and standards.

▶ Understand how the National Electrical Code is organized by chapters and article.

Safety is the number one priority in any job. Every year, electrical accidents cause serious injury or death. Many of these casualties are young people just entering the workplace. They are involved in accidents that result from carelessness, from the pressures and distractions of a new job, or from a lack of understanding about electricity. This chapter is designed to develop an awareness of the dangers associated with electric power and the potential dangers that can exist on the job or at a training facility.

1.1 Electric Shock

Electric shock occurs when the **body** becomes a part of the electric circuit. The current must enter the body at one point and leave at another. The main factor for determining the severity of an electric shock is the amount of electric **current** that passes through the body. This current is dependent upon the voltage and the resistance of the path it follows through the body.

Resistance (R) is the opposition to the flow of current in a circuit and is measured in **ohms (Ω)**. The lower the body resistance, the greater the current flow and potential electric shock hazard. Body resistance can be divided into external (skin resistance) and internal (body tissues and bloodstream resistance). Dry skin is a good insulator; moisture lowers the resistance of skin, which explains why shock intensity is greater when hands are wet. Internal resistance is low owing to the salt and moisture content of the blood. There is a wide degree of variation in body resistance. Typical body resistance values are:

- Dry skin—100,000 to 600,000 Ω
- Wet skin—1,000 Ω
- Internal body (hand to foot)—400 to 600 Ω
- Ear to ear—100 Ω

Voltage (E) is the pressure that causes the flow of electric current in a circuit and is measured in units called **volts (V)**. The amount of voltage that is dangerous to life varies with each individual because of differences in body resistance and heart conditions. ***Generally, any voltage above 30 volts is considered dangerous.***

Current (I) is the rate of flow of electricity and is measured in **amperes (A)** or **milliamperes (mA)**. A milliampere is equal to one-thousandth (1/1000) of an ampere, or 0.001 A. In other words it takes 1,000 milliamperes to equal 1 ampere. The amount of current flowing through a person's body depends on the voltage and resistance. Current flow can be calculated using the following Ohm's law formula:

$$\text{Current } (I) = \frac{\text{Voltage } (E)}{\text{Resistance } (R)}$$

Voltage is not as reliable an indication of shock intensity because the body's resistance varies so widely that it is impossible to predict how much current will result from a given voltage. The three most reliable criteria of shock intensity are:

- Amount of current flowing through the body.
- Path of the current through the body.
- Length of time the body is in the circuit.

Figure 1-1 Pathways that can stop normal pumping of the heart.

Although it is not known the exact injuries that result from any given amperage, it doesn't take much current to cause a painful or even fatal shock. A current of 1 mA (1/1000 of an ampere) can be felt. A current of 10 mA will produce a shock of sufficient intensity to prevent voluntary control of muscles, which explains why, in some cases, the victim of electric shock is unable to release grip on the conductor while the current is flowing. A current of 100 mA passing through the body for a second or longer can be fatal. ***Generally, any current flow above 0.005 A, or 5 mA, is considered dangerous.***

A 1.5-V flashlight cell can deliver more than enough current to kill a human being, yet it is safe to handle. This is because the resistance of human skin is high enough to limit greatly the flow of electric current. In lower-voltage circuits, resistance restricts current flow to very low values. Therefore, there is little danger of an electric shock. Higher voltages, on the other hand, can force enough current though the skin to produce a shock. ***The danger of harmful shock increases as the voltage increases.***

Figure 1-1 illustrates electric current pathways that can stop normal pumping of the heart. For example, a current from hand to foot, which passes through the heart and part of the central nervous system, is far more dangerous than a shock between two points on the same arm.

1.2 Arc Flash Hazards

An **arc flash** is the ball of fire that explodes from an electrical **short circuit** resulting from one exposed live conductor to another conductor or to ground. The arc flash creates an enormous amount of energy, as shown in Figure 1-2, that can damage equipment and cause severe injury or loss of life. An arc flash can be caused by dropped tools, unintentional contact with electrical systems, or the buildup of conductive dust, dirt, corrosion, and particles.

Electrical short circuits are either bolted faults or arcing faults. A **bolted fault** is current flowing through bolted bus bars or other electric conductors. An **arcing fault** is current flowing through the air. Because air offers opposition to

Figure 1-2 **Arc flash.**
©2012 Coastal Training Technologies Corp. All Rights Reserved. Reprinted with permission.

⚠ DANGER
Arc Flash and Shock Hazard
Appropriate PPE Required

Figure 1-3 **An arc flash hazard exists when a person interacts with equipment.**
©Chemco Electrical Contractors Ltd.

electric current flow, the arc fault current is always lower than the bolted fault current. An **arc blast** is a flash that causes an explosion of air and metal that produces dangerous pressure waves, sound waves, and molten steel.

In order to understand the hazards associated with an arc flash incident, it's important to understand the difference between an arcing short circuit and a bolted short circuit. A bolted short circuit occurs when the normal circuit current bypasses the load through a very low conductive path, resulting in current flow that can be hundreds or thousands of times the normal load current. In this case, assuming all equipment remains intact, the fault energy is contained within the conductors and equipment, and the power of the fault is dissipated throughout the circuit from the source to the short. All equipment needs to have adequate interrupting ratings to safely contain and clear the high fault currents associated with bolted faults.

In contrast, an arcing fault is the flow of current through a higher-resistance medium, typically the air, between phase conductors or between phase conductors and neutral or ground. Arcing fault currents can be extremely high in current magnitude approaching the bolted short-circuit current but are typically between 38 and 89 percent of the bolted fault. The inverse characteristics of typical overcurrent protective devices generally results in substantially longer clearing times for an arcing fault due to the lower fault values.

Eighty percent of electrical workplace accidents are associated with arc flash and involve burns or injuries caused by intense heat or showers of molten metal or debris. In addition to toxic smoke, shrapnel, and shock waves, the creation of an arc flash produces an intense flash of blinding light. This flash is capable of causing immediate vision damage and can increase a worker's risk of future vision impairment.

An arc flash hazard exists when a person interacts with equipment in a way that could cause an electric arc. Such tasks may include testing or troubleshooting, application of temporary protective grounds, or the racking in or out of power circuit breakers as illustrated in Figure 1-3. *Arcs can produce temperature four times hotter than the surface of the sun.* To address this hazard, safety standards such as National Fire Protection Association (NFPA) 70E have been developed to minimize arc flash hazards. The NFPA standards require that any panel likely to be serviced by a worker be **surveyed** and **labeled.** Injuries can be avoided with training, proper work practices, and using protective face shields, hoods, and clothing that are NFPA-compliant.

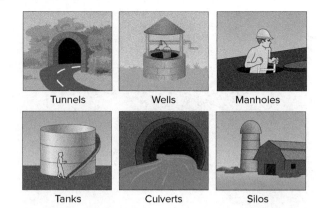

Figure 1-4 Confined spaces.

Figure 1-5 Typical safety signs.

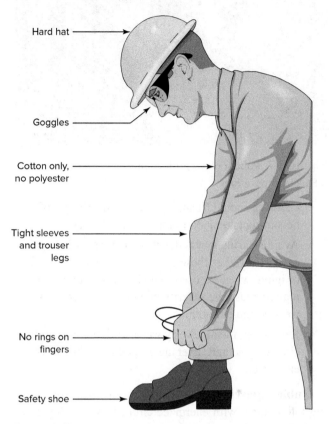

Figure 1-6 Appropriate attire should be worn for each particular job site.

1.3 Confined Spaces

Confined spaces can be found in almost any workplace. Figure 1-4 illustrates examples of typical confined spaces. In general, a **confined space** is an enclosed or partially enclosed space that:

- Is not primarily designed or intended for human occupancy.
- Has a restricted entrance or exit by way of location, size, or means.
- Can represent a risk for the health and safety of anyone who enters, because of its design, construction, location, or atmosphere; the materials or substances in it; work activities being carried out in it; or the mechanical, process, and safety hazards present.

All hazards found in a regular workspace can also be found in a confined space. However, they can be even more hazardous in a confined space than in a regular work site. Hazards in confined spaces can include poor air quality, fire hazard, noise, moving parts of equipment, temperature extremes, poor visibility, and barrier failure resulting in a flood or release of free-flowing solid.

A **permit-required** confined space is a confined space that has specific health and safety hazards associated with it. Permit-required confined spaces require assessment of procedures in compliance with Occupational Safety and Health Administration (OSHA) standards prior to entry.

1.4 Personal Protective Equipment

Construction and manufacturing work sites, by nature, are potentially hazardous places. For this reason, safety has become an increasingly large factor in the working environment. ***The electrical industry, in particular, regards safety to be unquestionably the most single important priority because of the hazardous nature of the business.***

A safe operation depends largely upon all personnel being informed and aware of potential hazards. Safety signs, such as those shown in Figure 1-5, indicate areas or tasks that can pose a hazard to personnel and/or equipment. Signs and tags may provide warnings specific to the hazard, or they may provide safety instructions.

Personal protective equipment (PPE) is equipment worn by a worker to minimize exposure to specific occupational hazards. Appropriate attire should be worn for each particular job site and work activity as illustrated in Figure 1-6 and summarized as follows:

- Hard hats, safety shoes, and goggles are normally required on almost any work site. In addition, nonconductive plastic hard hats must be of the approved type for the purpose of the electrical work being performed. ***Metal hats are not acceptable!***
- Safety earmuffs or earplugs must be worn in noisy areas.

Figure 1-7 Rubber glove protection.
©Fluke Corporation

- Clothing should fit snugly to avoid the danger of becoming entangled in moving machinery.
- Avoid wearing synthetic-fiber clothing such as polyester material as these types of materials may melt or ignite when exposed to high temperatures and may increase the severity of a burn. Instead always wear cotton clothing.
- Remove all metal jewelry when working on energized circuits; gold and silver are excellent conductors of electricity.

Rubber gloves are used to prevent the skin from coming into contact with energized circuits. A separate outer leather cover is used to protect the rubber glove from punctures and other damage, as shown in Figure 1-7.

Rubber blankets are used to prevent contact with energized conductors or circuit parts when working near exposed energized circuits. All rubber protective equipment must be marked with the appropriate voltage rating and the last inspection date. It is important that the insulating value of both rubber gloves and blankets have a voltage rating that matches that of the circuit or equipment they are to be used with. ***Insulating gloves must be given an air test, along with inspection.*** Twirl the glove around quickly, or roll it down to trap air inside. Squeeze the palm, fingers, and thumb to detect any escaping air. If the glove does not pass this inspection, you must dispose it.

The type of safety **eye protection** you should wear depends on the hazards in your workplace. If you are working in an area that has particles, flying objects, or dust, you must at least wear safety glasses with side protection (side shields). If you are working with chemicals, you should wear goggles. Approved listed **face shields,** such as shown in Figure 1-8, should be worn during all electrical switching operations where there is a possibility of

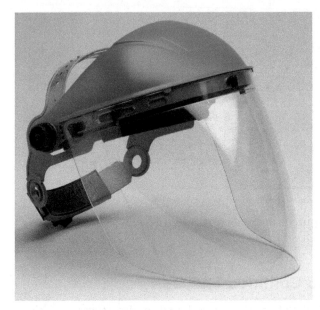

Figure 1-8 Typical face shield.
©Elvex Corporation

injury to the eyes or face from electric arcs or flashes, or from flying or falling objects that may result from an electrical explosion.

1.5 Fall Protection

Fall arrest systems are designed, not to necessarily prevent a fall, but to stop a fall once initiated. At a minimum, they must be rigged so that the workers will not free-fall more than **6 feet,** nor contact any lower level. They include personal fall arrest systems, such as the safety harness shown in Figure 1-9, and safety nets.

The misuse of ladders and scaffolds accounts for a high percentage of injuries in the workplace. Important rules for **all ladder** usage and safety include:

- Select the right ladder for the job; when performing electrical work, always use ladders made of **nonconductive material.**
- Inspect the ladder before you use it; inspect it for damaged rungs, steps, rails, or braces and traces of oil, grease, or other slippery substances.
- Never place the legs of a ladder on anything but a firm level surface.
- Never place a ladder in front of a door that swings open toward the ladder, unless the door is fastened open, locked, or guarded.
- Face the ladder when going up or down.
- Do not allow more than one person at a time on the ladder.
- Hold the ladder with both hands while climbing or descending. Use a tool belt or bucket attached to a

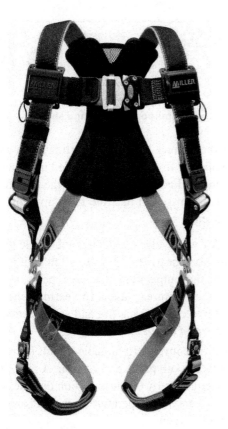

Figure 1-9 Typical safety harness.
Courtesy of Miller Fall Protection/Honeywell

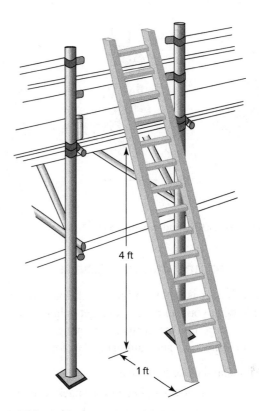

Figure 1-10 Ladder 4-to-1 placement rule.

hand line for raising and lowering the tools and materials you need.

- Make sure the ladder is clear of any power lines.
- Never climb higher than the second step from the top on a stepladder or the third from the top on a straight ladder.

Important rules for **stepladder** usage and safety include:

- Always open a stepladder to its fullest extent.
- Always lock both braces before climbing the ladder.
- Never use a stepladder as a straight ladder.
- Do not leave tools or materials on a stepladder.

Important rules for **extension ladder** usage and safety include:

- Always place a straight ladder at the proper angle; straight ladders should be placed at a **4-to-1** ratio. This means the base of the ladder should be 1 foot away from the wall or other vertical surface for every 4 feet of height to the point of support, as illustrated in Figure 1-10. When a person has to step off the ladder, it should extend about 3 feet above the roof, scaffold, or other kind of raised platform.
- Do not extend an extension ladder beyond the point where there is less than 3 feet of overlapping sections.

- When possible, secure the top of the ladder to the structure.
- When working on a ladder, hold a rung or rail with one hand at all times. Use a safety belt when it is absolutely necessary to work with both hands.
- Never add more extensions or fasten two ladders together to make a longer one.

At times it may be necessary to work in an elevated location. When this is the case, **scaffolds** provide the safest elevated working platforms. Important rules for scaffold usage and safety include:

- Scaffolds must be erected on rigid footing that can carry the maximum intended load using only materials designed and marked for this purpose.
- Guardrails and toe boards must be installed on the open sides and ends of platforms that are higher than 6 feet above the ground or floor.
- Work platforms must be completely decked with scaffold planks extended over their end supports not less than 6 inches nor more than 12 inches and must be properly blocked.
- Keep scaffold platforms clear of unnecessary material.

Figure 1-11 Lifting and moving loads.

1.6 Lifting and Moving Loads

When lifting, it is better to take small loads if possible. Lift only what you can handle, and get help if you need it. The basics steps for safe lifting and moving loads are illustrated in Figure 1-11 and summarized as follows:

- First stand close to the load.
- Then, squat down and keep your back straight.
- Get a firm grip on the load, and keep the load close to your body.
- Lift by straightening your legs. Make sure that you lift with your legs and not your back.
- Avoid lifting and twisting at the same time.
- Bend your knees rather than your back when putting a load down.
- If you bend from the waist to pick up a 50-pound object, you are applying *10 times the amount of pressure* (500 pounds) to your lower back.

1.7 Fire Prevention

Fire prevention is a very important part of any safety program. The fire triangle of Figure 1-12 illustrates the ingredients necessary for most fires, namely, **fuel, heat,** and **oxygen.** The fire is prevented or extinguished by removing any one of them.

Figure 1-12 Fire triangle.

Figure 1-13 Multipurpose fire extinguisher.

Using the correct type of fire extinguisher is vital to safely combating fires. Fires are divided into four classes: A, B, C, and D. Each class designates the fuel involved in the fire, as follows, and thus the most appropriate extinguishing agent:

Class A fires involve common combustible materials such as wood or paper. They are often extinguished by lowering the temperature of the fuel below the combustion temperature. Class A fire extinguishers often use water to extinguish a fire and as such *should never be used on an electrical fire.*

Class B fires involve flammable liquids such as gasoline, solvents, oil, paint, and varnish. A class B fire extinguisher generally employs carbon dioxide, which greatly lowers the temperature of the fuel and deprives the fire of oxygen.

Class C fires involve energized electrical equipment. A class C fire extinguisher uses a dry powder to smother the fire. *Under no circumstances use water, as the stream of water may conduct electricity through your body and give you a severe shock.*

Certain multipurpose dry-chemical extinguishers may be used on multiple types of fires. For example, an extinguisher labeled A/B/C, as shown in Figure 1-13, could be used on any of the three classes of fire listed.

Class D fires consist of burning metal. Class D extinguishers contain a sodium chloride or graphite metal–based powder. When discharged on a fire, the powder reacts to the heat, causing the powder to cake and form a crust which excludes air and dissipates the heat.

1.8 Hazardous Waste

Many products contain hazardous substances, which if not used and disposed of properly can result in the production of hazardous waste. A hazardous material is defined as any substance or material that could adversely affect the safety of the public, handlers, or carriers during transportation. Recognizing hazardous substances and the type of hazardous waste they produce is the first step in learning how to properly handle and dispose of them. One or more of the following dangerous properties

| Corrosive | Flammable | Toxic | Reactive |

Figure 1-14 Hazardous properties or characteristics.

or characteristics, illustrated in Figure 1-14, identify most common hazardous waste:

- **Corrosives** are materials that can attack and destroy human tissue, clothes, and other materials including metals on contact. For example, acids found in batteries are corrosive.
- A **flammable** material is one that is capable of bursting into flames. For example, gasoline and paint are flammable substances.
- **Toxic** materials can poison people and other life. Pesticides, weed killers, and many household cleaners are all examples of toxic materials.
- A **reactive** material can explode or create poisonous gas when mixed with another substance or chemical. For example, chlorine bleach and ammonia are reactive. When they come into contact with each other, they produce a poisonous gas.

Hazardous materials are required to be listed as such. A **material safety data sheet (MSDS)** is a form with data regarding the properties of a particular substance. MSDSs are a widely used system for cataloging information on chemicals, chemical compounds, and chemical mixtures. MSDS information may include instructions for the safe use and potential hazards associated with a particular material or product. These data sheets can be found anywhere where chemicals are being used.

Part 1 Review Questions

1. Does the severity of an electric shock increase or decrease with each of the following changes?
   a. A decrease in the source voltage.
   b. An increase in body current flow.
   c. An increase in body resistance.
   d. A decrease in the length of time of exposure.
2. In general, voltage levels above what value are considered dangerous?
3. In general, current levels above what value are considered dangerous?
4. What circuit fault can result in an arc flash?

5. Define each of the following terms associated with an arc flash:
   a. *Bolted fault*
   b. *Arcing fault*
   c. *Arc blast*
6. Explain why an arc flash is so potentially dangerous.
7. What is a permit-required confined space?
8. What does the term personal protective equipment (PPE) refer to?
9. What personal protective attire is required when taking measurements on energized circuits?
10. When should face shields be worn?
11. A fall arrest system must be rigged so that workers cannot free-fall more than how many feet?
12. A ladder is used to reach the top of a building 16 feet tall. According to the 4-to-1 ratio rule, what distance should the bottom of the ladder be placed from the side of the building?
13. What are the deck requirements for a scaffold work platform?
14. When lifting a load, why is it important to lift with your legs and not your back?
15. List the three ingredients required to sustain a fire.
16. Which classes of fire are multipurpose dry-chemical fire extinguishers approved for?
17. List four hazardous properties or characteristics.

PART 2 GROUNDING–FAULT PROTECTION–LOCKOUT–CODES

1.9 Grounding and Bonding

Proper grounding practices protect people from the hazards of electric shock and ensure the correct operation of overcurrent protection devices. Intentional grounding is required for the safe operation of electrical systems and equipment. Unintentional or accidental grounding is considered a fault in electrical wiring systems or circuits.

Grounding is the intentional connection of a current-carrying conductor to the earth. The prime reasons for grounding are:

- To limit the voltage surges caused by lightning, utility system operations, or accidental contact with higher-voltage lines.
- To provide a ground reference that stabilizes the voltage under normal operating conditions.

- To facilitate the operation of overcurrent devices such as circuit breakers and fuses under ground-fault conditions.

Bonding is the permanent joining together of metal parts that aren't intended to carry current during normal operation. Bonding creates an electrically conductive path that can safely carry current under ground-fault conditions. The prime reasons for bonding are:

- To establish an effective path for fault current that facilitates the operation of overcurrent protective devices.
- To minimize shock hazard to people by providing a low-conductive path to ground. Bonding limits the touch voltage when non-current-carrying metal parts are inadvertently energized by a ground fault.

The National Electrical Code (NEC) requires all metal used in the construction of a wiring system to be bonded to, or connected to, the ground system. The intent is to provide a low-impedance path back to the utility transformer in order to quickly clear faults. Figure 1-15 illustrates the ground-fault current path required to ensure that overcurrent devices operate to open the circuit as follows:

- The earth is not considered an effective ground-fault current path because its resistance is so high that

very little fault current returns to the electrical supply source through the earth.

- For this reason the main bonding jumper is used to provide the connection between the grounded service conductor and the equipment grounding conductor at the service.
- Bonding jumpers may be located throughout the electrical system, but a main bonding jumper is located only at the service.
- Grounding is accomplished by connecting the circuit to a metal underground water pipe, the metal frame of a building, a concrete-encased electrode, or a ground ring.

1.10 Ground-Fault and Arc-Fault Protection

A **ground fault** is defined as an unintentional, electrically conducting connection between an ungrounded conductor of an electric circuit and the normally non-current-carrying conductors, metallic enclosures, metallic raceways, metallic equipment, or earth. The **ground-fault circuit interrupter (GFCI)** is a device that can sense small ground-fault currents. The GFCI is fast-acting; the unit will shut off the current or interrupt the circuit within 1/40 second after its sensor detects a leakage as small as 5 milliamperes (mA). Most circuits are protected against overcurrent by 15-ampere or larger fuses or circuit breakers. While this protection is adequate against short circuits and overloads, leakage currents to ground may be much less than 15 amperes and still be hazardous.

Figure 1-16 shows the simplified circuit of a GFCI receptacle, the operation of which is summarized as follows:

- The device compares the amount of current in the ungrounded (hot) conductor with the amount of current in the grounded (neutral) conductor.
- Under normal operating conditions, the two will be equal in value.

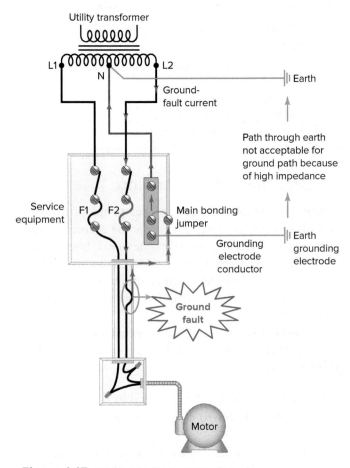

Figure 1-15 Ground-fault current path.

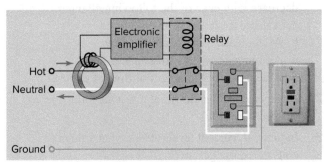

Figure 1-16 Simplified GFCI receptacle circuit.

- If the current in the neutral conductor becomes less than the current in the hot conductor, a ground-fault condition exists.
- The amount of current that is missing is returned to the source by the ground-fault path.
- Whenever the ground-fault current exceeds approximately 5 mA, the device automatically opens the circuit to the receptacle.

An **arcing fault** is an unintentional arcing condition in a circuit typically between the conductors in a frayed extension cord, arcing between a hot conductor and ground, or arcing in a loose connection in either the hot or neutral conductors. The **arc-fault circuit interrupter (AFCI)** is a device designed to protect against fires caused by arcing faults in residential electric wiring. The National Electrical Code requires arc-fault protection for all new installations of branch circuits supplying 125-V, single phase, 15- and 20-ampere outlets installed in dwelling unit bedrooms, dining rooms, living rooms, and other habitable areas.

The arc-fault circuit interrupter circuit breaker, shown in Figure 1-17, provides arc-fault protection by recognizing characteristics unique to arcing and operating to deenergize the circuit. Arc faults can occur as series or parallel arcs. A **series arc fault** can occur when the conductor in series with the load breaks. The series configuration means the arc current cannot be greater than the load current the conductor serves. More dangerous is the **parallel arc fault,** which can occur as a short circuit or a ground fault. One common example of a parallel arc fault is the insulation of a lamp cord or extension cord becoming damaged and allowing the two conductors to short together.

Conventional circuit breakers only respond to overloads and short circuits, while the AFCI circuit breaker provides both overcurrent and arc-fault protection. An AFCI is selective so that normal arcs, such as that which can occur when a switch is opened will not cause the breaker to trip.

When an unwanted arcing condition is detected, the control circuitry in the AFCI trips the internal contacts, thus deenergizing the circuit

An AFCI should not be confused with a ground fault circuit interrupter (GFCI). They are completely different and serve two totally different purposes. AFCIs are intended to reduce the likelihood of **fire** caused by electric arcing faults, whereas GFCIs are personnel protection intended to reduce the likelihood of electric **shock** hazard.

1.11 Lockout and Tagout

Electrical **lockout** is the process of removing the source of electric power and installing a lock, which prevents the power from being turned ON. Electrical **tagout** is the process of placing a danger tag on the source of electric power, which indicates that the equipment may not be operated until the danger tag is removed. Figure 1-18 contains an example of typical lockout-tagout devices.

Lockout and tagout procedures are necessary for the safety of personnel in that they ensure that no one will inadvertently energize the equipment while it is being worked on. This could apply to disconnect switches or circuit breakers. Most industries have their own lockout-tagout procedures that must be followed. Violating lockout and tagout procedures is considered to be an extremely serious offense and may result in immediate **termination of employment.** The following are the basic steps in a lockout procedure:

- **Prepare for machinery shutdown:** Management should have policies and procedures for safe lockout and should also educate and train everyone involved in locking out electrical or mechanical equipment.
- **Machinery or equipment shutdown:** Stop all running equipment by using the controls at or near the machine.

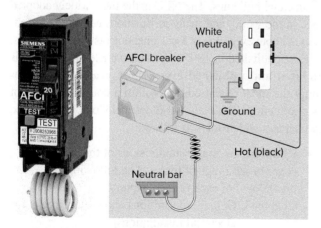

Figure 1-17 **AFCI circuit breaker.**
(*left*): ©Siemens Industry, Inc.

Figure 1-18 **Typical lockout-tagout devices.**
Courtesy of Honeywell Safety Products

- **Machinery or equipment isolation:** Open the main disconnect switch (do not operate if the switch is still under load). Stand clear of the box, and face away while operating the switch with the left hand (if the switch is on the right side of the box).
- **Lockout and tagout application:** Lock the main disconnect switch in the OFF position. Use a tamper-proof lock with one key, which is kept by the individual who owns the lock. Tag the lock with the signature of the individual performing the repair. There may be several locks and tags on the disconnect switch if more than one person is working on the machinery. When this is the case, a device that permits the use of multiple padlocks is used.
- **Release of stored energy:** All sources of energy that have the potential to unexpectedly start up, energize, or release must be identified and locked, blocked, or released.
- **Verification of isolation:** Use a voltage test to determine that voltage is present at the line side of the switch or breaker. When all phases of outlet are dead with the line side live, you can verify the isolation. Ensure that your voltmeter is working properly by performing the **live-dead-live** check before each use: First check your voltmeter on a known live voltage source of the same voltage range as the circuit you will be working on, as illustrated in Figure 1-19. Next check for the presence of voltage on the equipment you have locked out. Finally, to ensure that your voltmeter did not malfunction, check it again on the known live source.

1.12 Electrical Codes and Standards

Occupational Safety and Health Administration

In 1970, Congress created a regulatory agency known as the **Occupational Safety and Health Administration (OSHA).** OSHA's main goal is to ensure that employers provide each of their employees a place of employment that is free of recognized hazards. Employers have certain responsibilities under the act that require them to identify potential hazards and eliminate them, control them, or provide employees with suitable protection from them. Employees are responsible for following the safety procedures set up by the employer.

OSHA inspectors check on employers to make sure they are following prescribed safety regulations. OSHA also inspects and approves safety products. OSHA's electrical standards are designed to protect employees exposed to dangers such as electric shock, electrocution, fires, and explosions.

Figure 1-19 Testing for the presence of voltage.
©Fluke Corporation

National Electrical Code

The **National Electrical Code (NEC)** comprises a set of rules that, when properly applied, are intended to provide a safe installation of electrical wiring and equipment. This widely adopted minimum electrical safety standard has as its primary purpose *the practical safeguarding of persons and property from hazards arising from the use of electricity.* Standards contained in the NEC are enforced by being incorporated into the different city and community ordinances that deal with electrical installations in residences, industrial plants, and commercial buildings. The NEC is the most widely adopted code in the world, and many jurisdictions adopt it in its entirety without exception or local amendments or supplements.

First published in 1897, the National Electrical Code is **updated and published every three years.** Most states adopt the most recent edition within a couple of years of its publication. The National Electrical Code is organized by chapters and articles as follows:

- **Chapters** are major subdivisions of the NEC that cover broad range of topics. The following is a breakdown of each of the chapters:
  Chapter 1: General (definitions and requirements for electrical installations)
  Chapter 2: Wiring and Protection
  Chapter 3: Wiring Methods and Materials

Chapter 4: Equipment for General Use
Chapter 5: Special Occupancies
Chapter 6: Special Equipment
Chapter 7: Special Conditions
Chapter 8: Communications Systems
Chapter 9: Tables

- **Articles** are chapter subdivisions that cover a specific subject such as grounding, overcurrent protection, lighting fixtures, and so on. Articles are divided into **sections** and sometimes into parts. When an article is sufficiently large, or where necessary to logically group requirements, it is subdivided into **parts** that correspond to logical groupings of information.

National Fire Protection Association

The **National Fire Protection Association (NFPA)** develops codes governing construction practices in the building and electrical trades. It is the world's largest and most influential fire safety organization. NFPA has published almost 300 codes and standards, including the **National Electrical Code,** with the mission of preventing the loss of life and property.

Nationally Recognized Testing Laboratory

In accordance with OSHA safety standards, a **nationally recognized testing laboratory (NRTL)** must **test electrical products** for conformity to national codes and standards before they can be listed or labeled. The biggest and best-known testing laboratory is the Underwriters Laboratories, identified with the UL logo shown in Figure 1-20.

Article 100 of the NEC defines the terms **labeled** and **listed,** which are both related to product evaluation. Labeled or listed indicates the piece of electrical equipment or material has been tested and evaluated for the purpose for which it is intended to be used. Products that are big enough to carry a label are usually labeled. The smaller products are usually listed.

National Electrical Manufacturers Association

The **National Electrical Manufacturers Association (NEMA)** is a group that defines and recommends safety standards for electrical equipment. Standards established by NEMA assist users in proper selection of industrial control equipment, as illustrated in Figure 1-21. For example, NEMA standards provide practical information concerning the rating, testing, performance, and manufacture of devices such as enclosures, contactors, and starters.

Figure 1-20 Underwriters Laboratories logo.
Source: UL LLC

Figure 1-21 NEMA type 1 enclosure.
Courtesy of Schneider Electric

Part 2 Review Questions

1. List three reasons for employing proper grounding practices.
2. Compare the terms *grounding* and *bonding.*
3. Why is the earth not considered to be an effective path for ground-fault current?
4. Explain what is meant by a ground fault.
5. How does a ground-fault circuit interrupter (GFCI) sense ground-fault current?
6. What is the approximate current at which a GFCI will operate to open the circuit?
7. Explain what is meant by an arcing fault.
8. Compare the amount of current that results from series and parallel arc faults.
9. Compare what GFCIs and AFCIs protect against.
10. What does a lockout-tagout procedure refer to?
11. As part of a lockout-tagout procedure, you are required to open a disconnect switch. What is the safest way to proceed?
12. A voltmeter is used to verify that no voltage is present after the lockout-tagout procedure has been completed. Explain how you can ensure that the voltmeter is working properly.
13. What is the main goal of OSHA?
14. What is the primary purpose of the rules set forth in the National Electrical Code.
15. Assume a piece of electrical equipment has been listed by the Underwriters Laboratories. What does this indicate?
16. How often is the National Electrical Code updated and published?

CHAPTER TWO

Atoms and Electricity

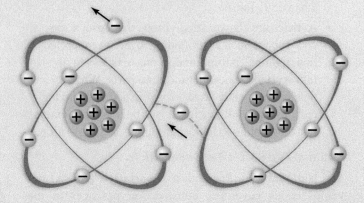

Movement of electrons between atoms.

LEARNING OUTCOMES

What is electricity? Where does it come from? How does it work? Before we understand all that, we need to learn the important relationship between atoms and electricity.

LEARNING OUTCOMES

▶ Describe the structure and electrical properties of an atom.

▶ Explain how the ionization process produces positive and negative charges.

▶ Define *electricity* in terms of electron flow.

▶ Compare the electrical properties of conductors, insulators, and semiconductors.

▶ Understand the basic concepts of voltage, current, and resistance.

PART 1   ATOMS AND IONS

2.1 Electron Theory of Matter

The **electron theory of matter** helps explain how electricity works. All matter—solid, liquid, or gas—is made up of tiny particles that are called molecules. A molecule is the smallest part of matter that has the physical and chemical properties of itself.

Molecules are made up of smaller particles, each of which is called an **atom.** Atoms, in turn, can be broken down into even smaller particles. Particles found within an atom are known as **electrons, protons,** and **neutrons** (Figure 2-1).

2.2 Bohr Model of Atomic Structure

The model of the atom as proposed by the physicist Niels Bohr gives a concept of its structure, which is helpful in understanding the fundamentals of electricity. According to Bohr, the atom is similar to a miniature solar system. As with the sun in the solar system, the **nucleus** is located in the center of the atom. As illustrated in Figure 2-2, tiny particles called electrons rotate in orbit around the nucleus, just as the planets rotate around the sun. The electrons are prevented from being pulled into the nucleus by the force of their momentum. They are prevented from flying off into space by an attraction between the electron and the nucleus. This attraction is due to the electric charge on the electron and the nucleus. The **electron has a negative (−)** charge, while the nucleus has a positive (+) charge. These unlike charges attract each other.

Most of the mass of an atom is found in its nucleus. The particles that can be found in the nucleus are called protons and neutrons. A **proton has a positive (+)** electric charge that is exactly equal in strength to that of the negative (−) charge of an electron. The electron is about three times larger than the proton but is extremely lighter. The mass or weight of the neutron is about the same as that of the proton, but it has no electric charge—hence its name, neutron. Neutrons, as far as is known, do not enter into ordinary electrical activity. Normally, every atom contains an **equal number of electrons and protons,** making its combined electric charge zero, or neutral.

Electrons are arranged in shells around the nucleus. A **shell** is an orbiting layer or energy level of one or more electrons. The major shell layers are identified by numbers or by letters, starting with K nearest the nucleus and continuing alphabetically outward. There are a maximum number of electrons that can be contained in each shell. Figure 2-3 illustrates the relationship between the energy shell level and the maximum number of electrons it can contain.

If the total number of electrons for a given atom is known, the placement of electrons in each shell can be determined easily. Each shell layer, beginning with the first and proceeding in sequence, is filled with the maximum number of electrons. For example, a normal copper atom that has 29 electrons would have the following arrangement of electrons (Figure 2-4):

Shell K (or 1) = 2 (full)
Shell L (or 2) = 8 (full)
Shell M (or 3) = 18 (full)
Shell N (or 4) = 1 (incomplete)

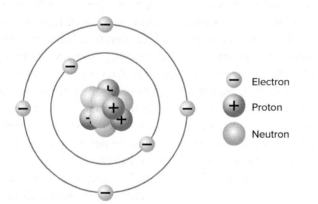

Figure 2-1   Structure of an atom.

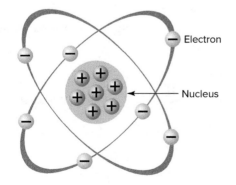

Figure 2-2   Bohr model of atomic structure.

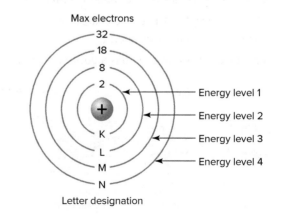

Figure 2-3   Electron shells.

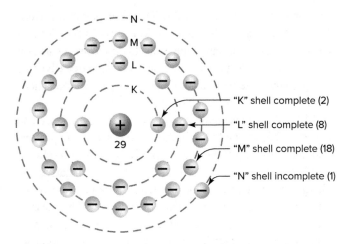

Figure 2-4 Placement of electrons in a copper atom.

The electrons in any shell of an atom are said to be located at certain energy levels. The farther away from the nucleus, the higher the energy level of the electrons. When outside energy such as heat, light, or electricity is applied to certain materials, the electrons within the atoms of these materials gain energy. This may cause the electrons to move to a higher energy level.

2.3 Ions

It is possible, through the action of some outside force, for an atom to lose or acquire electrons, as illustrated in Figure 2-5. The charged atom that results is called an ion. A **negatively charged ion** is an atom that has **acquired** electrons. It has more electrons than protons, and therefore it is negatively (−) charged. A **positively charged ion** is an atom that has **lost** electrons. It has fewer electrons than protons and therefore is positively (+) charged. The process by which atoms either gain or lose electrons is called **ionization.**

Part 1 Review Questions

1. What three particles can be found within an atom?

2. Which particle has a negative charge and rotates around the nucleus of the atom?

3. Which particle of an atom has a positive charge?

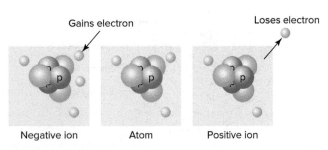

Figure 2-5 Positive and negative ions.

4. Compare the size and mass (weight) of the electron with that of the proton.

5. Which particle of an atom has no electric charge associated with it?

6. Normally every atom contains an _____ number of electrons and protons, making its combined electric charge _____.

7. Electrons can be considered to be arranged in _____ around the nucleus.

8. What is the process by which atoms gain or lose electrons called?

9. A negatively charged ion is an atom that has _____ electrons.

10. A positively charged ion is an atom that has _____ electrons.

PART 2   ELECTRON FLOW

2.4 Electricity Defined

The outermost shell or highest energy level of the atom is called the *valence shell,* and its electrons are called **valence electrons** (Figure 2-6). The number of valence electrons in a given atom determines its ability to gain or lose an electron, which in turn determines the electrical properties of the atom.

Whenever a valence electron is removed from its orbit, it becomes known as a **free electron.** Valence electrons can become free of their orbit by the application of some external force such as friction or voltage. A free electron leaves a void which can be filled by an electron forced out of orbit from another atom. As free electrons move from one atom to the next an **electron flow** is produced. Electricity can be defined as the flow of electrons through a conductor (Figure 2-7).

A material is said to conduct electricity when one electron of an atom is forced from its orbit path by another atom's electron (Figure 2-8). When one electron strikes another,

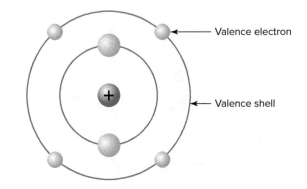

Figure 2-6 Valence shell and electrons.

Figure 2-7 Electricity is the flow of electrons through a conductor.

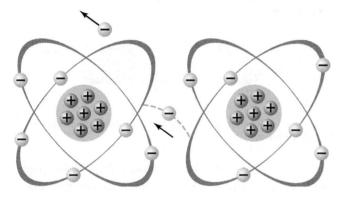

Figure 2-8 Movement of electrons between atoms.

the electron being hit takes energy from this action and jumps to a neighboring electron's orbit. This process is repeated so that the electron that impacted the first valence electron now takes the leaving electron's place as the new valence electron. Not all the energy is transferred when the electrons collide as some of the energy is lost in the form of **resistance.** This resistance comes from valence electrons not wanting to leave their orbits.

2.5 Electrical Conductors, Insulators, and Semiconductors

Conductors

Electrons can flow in all matter. However, this flow is much easier through some materials than others. Materials that permit free electrons to move easily from one atom to another are called **conductors.** Conductors offer very little resistance to the flow of electrons through them. Generally speaking, most metals are good conductors of electricity. **Copper** is the most common metal used as a conductor of electricity because of its relatively low cost and good conducting ability.

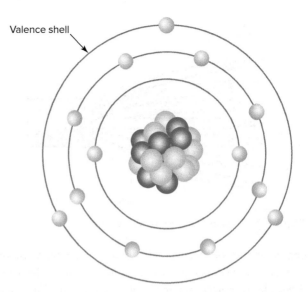

Valence shell

Figure 2-9 Placement of electrons in an aluminum atom.

Electrical conducting properties of various materials are determined by the number of electrons in the outer shell of their atoms. Conductors have very few electrons in their outer valence shell resulting in them being easily knocked out of the atom's orbit. Generally, a good conductor has an **incomplete valence shell** of one, two, or three electrons. The electrons are held loosely; there is room for more; and a low voltage will cause a flow of free electrons. Figure 2-9 shows the placement of electrons in an aluminum atom. Note that the valence shell is incomplete with only 3 out of a possible 18 electrons.

Insulators

Insulator is the name given to a material through which it is very difficult to produce a flow of electrons. **Insulators** have few, if any, free electrons and resist the flow of electrons. Generally, insulators have full valence shells of five to eight electrons. The electrons are held tightly, the shell is fairly full, and a very high voltage is needed to cause any electron flow. Some common insulators are air, glass, rubber, plastic, paper, and porcelain. An electrical cable is one example of how conductors and insulators are used together. Insulated copper conductors are used to keep electrons flowing along the intended path of the circuit (Figure 2-10).

No material has been found to be a **perfect insulator.** Every material can be forced to permit a small flow of electrons from atom to atom if enough energy in the form of voltage is applied. Whenever a material that is classified as an insulator is forced to pass an electric current, the insulator is said to have been broken down or ruptured.

Semiconductors

Semiconductors form the heart of modern electronics. A **semiconductor** is a material that has some of the

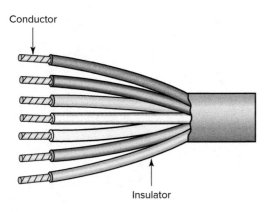

Figure 2-10 Insulated copper conductors.

characteristics of both a conductor and an insulator. Semiconductors have valence shells containing four electrons. Examples of pure semiconductor materials are silicon and germanium. They are not naturally semiconductors but can be made into semiconductors by melting them and adding very small amounts of other chemicals. This process is called **doping.** Whether a **p-type** semiconductor or an **n-type** semiconductor is produced depends on the type of doping chemical used.

Semiconductors are used to produce electronic components such as diodes, transistors, and integrated-circuit chips. A diode is one of the simplest semiconductor components. The main characteristic of a **diode** is its ability to pass electron current flow in one direction only. The operation of a diode circuit is shown in Figure 2-11. The diode acts like a conductor when a voltage is applied in one direction and like an insulator when the voltage is applied in the opposite direction. When connected in **forward-bias** to the battery voltage, the diode acts as a conductor and conducts current flow to the lamp. When connected in **reverse-bias** to the battery voltage, the diode acts as an insulator and blocks current flow to the lamp.

2.6 Basic Concepts of Voltage, Current, and Resistance

Voltage

Voltage is a form of electrical **pressure** that causes free electrons to move from one atom to another (Figure 2-12). Just as water needs some pressure to force it through a pipe, elec-

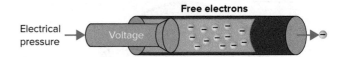

Figure 2-12 Voltage is a form of electrical pressure.

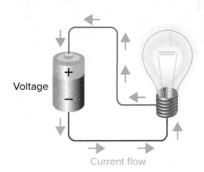

Figure 2-13 Current is a measure of the rate of electron flow.

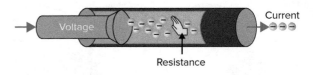

Figure 2-14 Resistance is a measure of the opposition to the electron flow.

trical current needs some force to make it flow. **Volts** refers to the measure of electrical pressure that causes current flow.

Current

Current is a measure of the rate of electron **flow** through a material (Figure 2-13). Electrical current is measured in units of **amperes.** This flow of electrical current develops when electrons are forced from one atom to another. Current cannot exist without voltage, but voltage can exist without current.

Resistance

Resistance is a measure of the **opposition** to the electron flow through a material (Figure 2-14). Electrical resistance is measured in units of **ohms.** Resistance elements essentially range somewhere between a conductor and an insulator. The lower the resistance of a material, the better the material acts as a conductor. For example, copper has a lower electrical resistance than aluminum; copper is a better conductor.

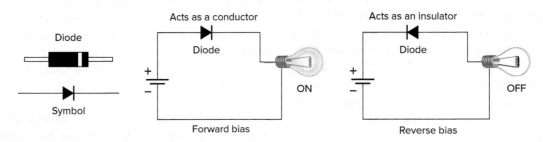

Figure 2-11 Operating circuit for a semiconductor diode.

2.7 Testing for Electrical Continuity

In electrical applications, when an electric circuit is capable of conducting current, it demonstrates **electrical continuity.** A simple circuit made up of a lamp, battery, and test leads, as shown in Figure 2-15, can be constructed to test for continuity. Connecting the two test leads across the two ends of the conductor path results in a flow of electrons that causes the lamp to come on at full brightness. An **open** in the conductor path produces no flow of electrons or light from the bulb.

The continuity test circuit can also be used to test electrical components out of circuit. In this case, the light will be activated if the probes detect a complete circuit between the leads. Practical applications for this type of testing include checking extension cords for open leads, checking for blown fuses, and checking for defective switches and pushbuttons. This tester is designed for checking electric components out of their normal circuits. *Under no circumstances should the continuity tester be connected to a circuit where other sources of voltage may be present!*

2.8 Electrical and Electronic Devices

The distinction between electrical and electronic devices is less clear than it used to be. Traditionally, electrical devices were those which used electric power but did not include electronic components such as diodes, transistors, integrated circuits, and microprocessors.

Electric or **electrical** are terms that can be used interchangeably because they relate to the concept of electricity. The word *electric* is used in the sense that it states the flow of electricity. Thus the term *electric* or *electrical* refers to the source and usage of power when it is conducted through a device.

On the other hand, **electronics** refer to the appliances and devices that are powered and run by electricity. The electronic devices are so made that they draw electricity from the source of power and manage the flow within them using electronic components. A **printed circuit board** assembly (Figure 2-16) is a self-contained module of interconnected electronic components found in most electronic devices.

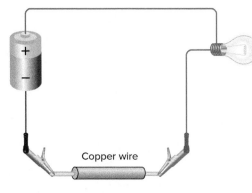

Figure 2-15 Continuity test circuit.

Copper wire

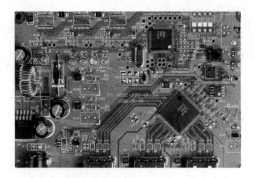

Figure 2-16 Printed circuit board assembly.
©Caspar Benson/Getty Images

The field of electronics is based on the ability of semiconductors to **generate, amplify, and control** an electric signal. Semiconductor devices are solid-state devices in which electrons flow in and through a semiconductor material. The adjective **solid-state** is sometimes used when referring to semiconductors because electrons flow through solid crystals of semiconductor materials rather than through a vacuum or a gas.

Part 2 Review Questions

1. _____ electrons are found in the outermost shell of atoms.

2. Whenever a valence electron is removed from its orbit, it is known as a(n) _____ electron.

3. Define *electricity*.

4. Conductors offer _____ resistance to current flow.

5. Conductors have incomplete valence shell of _____, _____, or _____ electrons.

6. Insulators have few, if any, _____ electrons.

7. Semiconductors have some of the characteristics of both a(n) _____ and a(n) _____.

8. Two examples of pure semiconductor materials are _____ and _____.

9. Explain the main characteristic of a semiconductor diode.

10. What is voltage similar to?

11. Electric current refers to the flow of _____ through a material.

12. Compare the resistance of a conductor with that of an insulator.

13. What does the circuit of a simple battery-operated continuity checker consist of?

14. In general, how would you classify a device as being electronic?

CHAPTER THREE

Sources and Characteristics of Electricity

Tesla coil
©Arthur S. Aubry/Getty Images

LEARNING OUTCOMES

▶ Define *static* and *current electricity*.

▶ Explain how static positive and negative charges are produced.

▶ State the law of electric charges.

▶ Recognize the difference between direct current (DC) and alternating current (AC) electricity.

▶ Be familiar with the basic sources of electricity and electrical devices used to convert the various energy forms.

Electricity is present in all matter in the form of electrons and protons. Any device that develops and maintains a voltage can be considered a voltage source. To accomplish this, the voltage source must remove electrons from one point and transfer electrons to a second point. In this chapter we explore the different methods used to produce voltage.

PART 1 STATIC ELECTRICITY

3.1 Static Electricity

Static electricity is the accumulation of electric charges on the surface of a material, usually an insulator or nonconductor of electricity. The term static means standing still or at rest, and **static electricity** refers to an electric charge at rest. The result of this buildup of static electricity is that objects may be attracted to each other or may even cause a spark to jump from one to the other.

One of the simplest ways to produce static electricity is by **friction.** For example, a static charge can be produced by rubbing a balloon with a piece of wool, as illustrated in Figure 3-1. The process causes electrons to be pulled from the wool to the balloon. As a result the balloon ends up with an excess of electrons and a negative charge. At the same time, the wool loses electrons, creating a shortage of electrons and a positive charge. Note that the charged atoms remain on the surface of the material. Static electricity is different from current electricity that flows through metal wires. Most often the materials involved in static electricity are nonconductors of electricity.

Static electricity is formed when we accumulate extra negatively charged electrons and they are discharged to an object or person. Take, for example, the rubber soles of your shoes and that wool carpet in the living room. When you walk across the carpet, your body builds up a negative charge of extra electrons it can't get rid of through the insulating soles of your shoes. Then when you reach for the doorknob, you may experience a **static electric shock** caused by electrons moving from your hand to the metal doorknob (Figure 3-2). Static electricity occurs more often during the colder seasons because the air is drier, and it's easier to build up electrons on the skin's surface. In warmer weather, the moisture in the air helps electrons move off of you more quickly so the static charge is not as great.

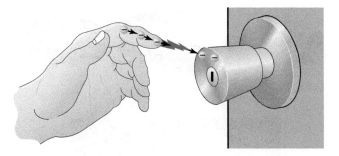

Figure 3-2 **Static electric shock.**

3.2 Charged Bodies

The first law of electric charges is illustrated in Figure 3-3 and states that *like charges repel and unlike charges attract.* An invisible **electrostatic field** exists in the space between and around charged balls. When two like-charged bodies are brought together, their electric fields repel one body from the other. When two unlike-charged bodies are brought together, their electric fields attract one body to the other.

The electrostatic field around two charged objects can be represented graphically by lines referred to as **electrostatic lines of force** (Figure 3-4). Lines are directed away from positively charged objects and toward negatively charged objects. *The attractive or repulsive force that is exerted between two charged particles is directly proportional to the strength of their charges and inversely proportional to the square of distance between the two charges.* This means that the bigger the charges, the more will be the force; the more distance they are apart, the less the force between them.

If two strongly charged bodies (one positive and one negative) are moved near to each other, before contact is

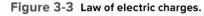

Like charges repel   Unlike charges attract

Figure 3-3 **Law of electric charges.**

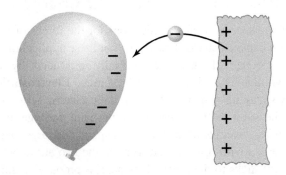

Figure 3-1 **Producing static electricity by friction.**

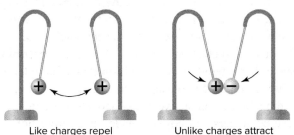
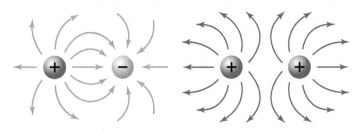

Figure 3-4 **Electrostatic field around two charged objects.**

Figure 3-5 **Methods of protection against ESD.**
Courtesy of Botron Company Inc.

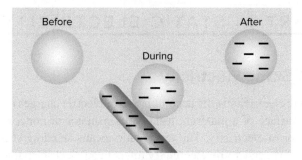

Figure 3-6 **Charging by conduction.**

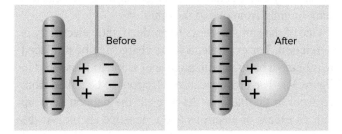

Figure 3-7 **Charging by induction.**

made you actually see the equalization of the charges take place in the form of an arc. **Lightning** is a perfect example. Cloud-to-ground lightning occurs when the electric charge travels between a negatively charged cloud base and the positively charged ground. A single bolt of lightning delivers about **1 trillion watts** of electricity.

Electrostatic discharge (ESD) is the rapid discharge of static electricity from one object to another of a different potential. This is a very rapid event that happens when two objects of different potentials come into direct contact with each other. Electrostatic discharge is one of the main causes of device failures in the **semiconductor** industry. Static electricity, so low that you can't feel it, can wreck havoc with today's large-scale microelectronic devices.

Methods of protection against ESD include **prevention** of static charge buildup and **safe dissipation** of any charge buildup (Figure 3-5). Packaging materials such as static shielding bags, conductive bags, and electrostatic discharge containers and boxes provide direct protection of devices from electrostatic discharge. Antistatic footwear and wrist straps when properly worn and grounded keep the human body near ground potential, thus preventing hazardous discharge between bodies and objects.

3.3 Charging by Conduction and Induction

The three common ways for a neutral object to become charged are by friction, conduction, or induction. Charging by **conduction** or contact occurs when a neutral object is placed in contact with an already-charged object. If the object is negatively charged, electric repulsion will push some of the excess electrons from the charged to the neutral object. If the object is positively charged, electric attraction will pull some electrons from the neutral object to the charged one. In the example shown in Figure 3-6, when a rod (that has an excess of electrons) touches a neutral ball, the charge distributes itself over both objects. When they are separated, the ball will now be electrically charged.

Charging by **induction** involves an already-charged object that is **brought close** to but does not touch the neutral object. In the example shown in Figure 3-7, the negatively charged rod is brought close to an electrically neutral ball. The electrons on the ball are repelled and move to the opposite side of the ball. Touching the negative right side of the ball drains electrons from the ball to ground, giving the ball a net positive charge. The rod is then removed, leaving a positively charged ball.

3.4 Practical Uses for Static Charges

There are several useful applications for the forces of attraction between charged particles. For these applications, static electric charges are normally produced by a high-voltage DC (direct current) source. **Electrostatic paint spraying** (Figure 3-8) uses static electricity to attract the paint to the target, reducing paint wastage and improving coverage of the target. The paint and target part are charged with **opposite charges** so the two attract and the paint sticks to the target. This process produces a uniform cover of paint with excellent adhesion.

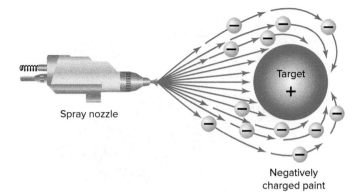

Figure 3-8 **Electrostatic paint spraying.**

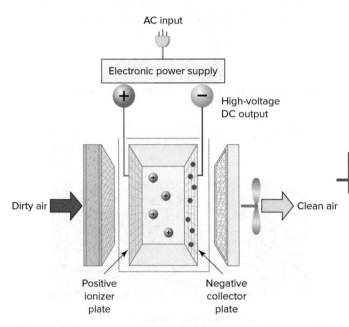

Figure 3-9 **Electrostatic air cleaner.**

Electrostatic **air cleaners or precipitators** use positively and negatively charged plates to remove dirt particles from the air. Figure 3-9 illustrates the operation an electronic air cleaner used in a home-heating system to clean the air as it circulates through the furnace. The dirty air passes through a paper filter that removes large dust and dirt particles from the air. The air then moves through an electrostatic precipitator consisting of two oppositely charged, high-voltage grids. The precipitator works by giving a **positive charge** to particles in the air and then attracting them with a **negatively charged** grid. Finally, the air passes through a carbon filter, which absorbs odors from the air.

Part 1 Review Questions

1. Define *static electricity*.
2. Explain why rubbing a balloon with a piece of wool results in a negative charge on the surface of the balloon.

3. State the law of electrostatic charges.
4. How is the force between two charged particles affected by the strength of their charges and the distance between them?
5. Define *electrostatic discharge*.
6. Electrostatic discharge is one of the main causes of device failure in the _____ industry.
7. Name three ways for a neutral object to become charged.
8. Charging by _____ involves an already-charged object that is brought close to but does not touch the neutral object.
9. Summarize the operation of an electrostatic paint spraying process.
10. Summarize the operation of an electrostatic air cleaner.
11. Why does static electricity occur more often during the colder seasons?

PART 2   CURRENT FLOW

3.5 Current Electricity

Current electricity is defined as an electric charge in motion. Current flow consists of a flow of negative electron charges from atom to atom, as illustrated in Figure 3-10. The external force that causes the electron flow is called the **electromotive force (emf) or voltage** which is supplied by the battery. The negative terminal of the battery has an excess of electrons, while the positive terminal has a deficiency of electrons. Since the positive terminal of the battery has a shortage of electrons, it attracts electrons from the conductor. Similarly, the negative terminal, with an excess of electrons, repels electrons into the conductor.

Current electricity is classified as being direct current (DC) or alternating current (AC) according to its voltage source. **Direct current voltage** produces a flow of electrons in one direction only. **Alternating current voltage**

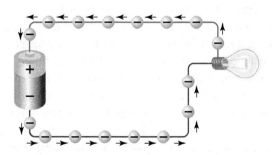

Figure 3-10 **Current electricity.**

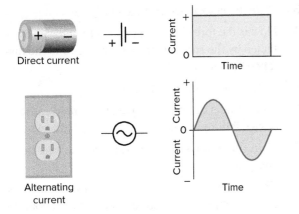

Figure 3-11 **DC and AC current electricity.**

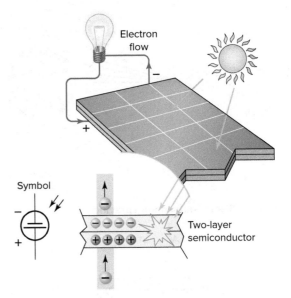

Figure 3-12 **Generating electricity from sunlight.**

produces a flow of electrons that changes both in direction and in magnitude. Typical symbols and waveforms for DC and AC voltage sources are shown in Figure 3-11. A battery is a common DC voltage source, while an electrical receptacle is the most common AC voltage source. All voltage sources share the characteristic of an excess of electrons at one terminal and a shortage at the other terminal. This results in a difference of electric potential between the two terminals.

Polarity identification (+ or −) is one way to distinguish a voltage source. Polarity can be identified on direct current circuits, but on alternating current circuits the current continuously reverses direction; therefore, the polarity **cannot** be identified.

3.6 Sources of Electromotive Force

For electrons to flow there must be a **source** of electromotive force (emf) or voltage. This voltage source can be produced from a variety of different primary energy sources. These primary sources supply energy in one form, which is then converted to electric energy. Primary sources of electromotive force include friction, light, chemical reaction, heat, pressure, and mechanical-magnetic action.

Light

A solar photovoltaic power system converts sunlight directly into electric energy using **solar or photovoltaic (PV) cells.** These are made from a semiconducting, light-sensitive material that makes electrons available when struck by the light energy (Figure 3-12). Solar cells operate on the photovoltaic effect, which occurs when light falling on a two-layer semiconductor material produces a DC voltage, between the two layers. The output voltage is directly proportional to the amount of light energy striking the surface of the cell. One of the best solar cells is the **silicon cell.** A single cell can produce up to **400 mV (millivolts)** with current in the milliampere range and can be used in constructing

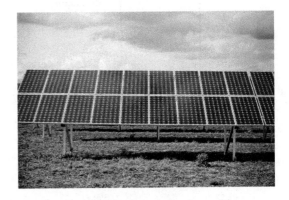

Figure 3-13 **Solar module or panel.**
©Rolfo Brenner/EyeEm/Getty Images

larger solar panels. Small current solar cells are often used as sensing devices in automatic control systems and to power electronic devices such as calculators.

A **solar module or panel** consists of solar cells electrically interconnected and encapsulated as shown in Figure 3-13. Solar panels typically have a sheet of glass, on the side facing the sun, and a translucent resin barrier, allowing light to pass through while protecting the semiconductor from the rain, snow, and hail. Solar panels can be grouped together to form an **array** capable of delivering large amounts of electric power.

A **grid-tie solar system** connects your solar power system to the electric power grid. This enables you to send any excess power you produce back to the electric company through a plan known as **net metering.** At night, or on cloudy days, you simply revert to buying power from the utility company. With this type of solar system installed, the electricity that you generate either offsets your usage or, if you are producing more than you are using, is fed

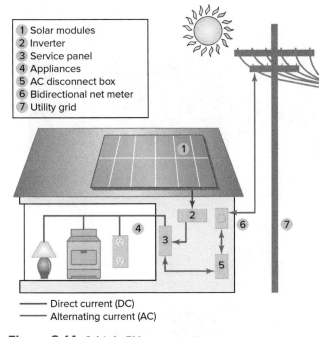

1 Solar modules
2 Inverter
3 Service panel
4 Appliances
5 AC disconnect box
6 Bidirectional net meter
7 Utility grid

—— Direct current (DC)
—— Alternating current (AC)

Figure 3-14 **Grid-tie PV power system.**

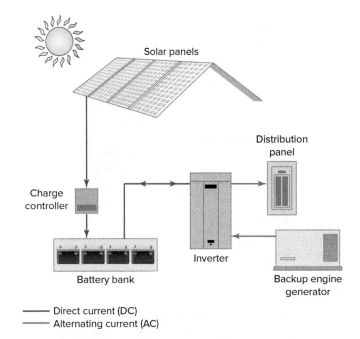

Solar panels

Distribution panel

Charge controller

Battery bank

Inverter

Backup engine generator

—— Direct current (DC)
—— Alternating current (AC)

Figure 3-15 **Off-grid PV power system.**

back to the electric grid, crediting your utility account. One key point to remember is that grid-tie PV system must be shut down when the power from the utility company is down. This is primarily a safety requirement to ensure that power is not being **fed back** into the grid while maintenance workers are restoring power.

Figure 3-14 shows the parts of a typical grid-tie photovoltaic power system. These systems use solar modules along with a DC to AC power inverter. The **inverter** changes **DC to AC** and synchronizes the power produced by solar modules with the electricity that comes from the utility company. The operation is straightforward. When the sun is shining, the solar array generates DC voltage. The inverter automatically connects to the electric utility grid and delivers AC power to the grid.

Off-grid PV solar systems are used for applications where utility lines are not available, not desired, or just too expensive to bring in. Off-grid solar systems use solar panels to produce DC electricity, which is then stored in a **battery bank** (Figure 3-15). An inverter converts the DC power stored in the batteries to AC power of the kind used in the residence or commercial establishment. Typically, off-grid systems will include a backup power generator to charge the batteries if they get too low and a charge controller to regulate the power flowing from a photovoltaic panel into the rechargeable battery bank.

Chemical Reaction

The **battery** or **voltaic cell** converts chemical energy directly into electric energy (Figure 3-16). Basically, a battery is made up of **two electrodes** and an **electrolyte**

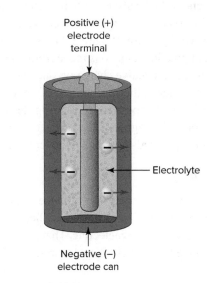

Positive (+) electrode terminal

Electrolyte

Negative (–) electrode can

Figure 3-16 **Battery converts chemical energy directly into electric energy.**

solution. One electrode connects to the (+) or positive terminal, and the other to the (−) or negative terminal.

When a battery is connected to a closed electric circuit, **chemical energy** is transformed into **electric energy.** The chemical action within the cell causes the electrolyte solution to react with the two electrodes. As a result, electrons are transferred from one electrode to the other. This produces a **positive charge** at the electrode that loses electrons and a **negative charge** at the electrode that gains electrons. Although the battery is a popular low-voltage, portable DC source, its relatively high energy cost limits its applications.

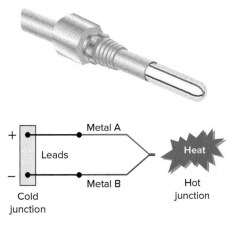

Figure 3-17 Thermocouple converts heat energy into electric energy.

Heat

Heat energy can be directly converted into electric energy by a device called a thermocouple. **Thermocouples** operate on the principle that when two dissimilar metals are joined, a predictable **DC voltage** will be generated that relates to the difference in **temperature** between the hot junction and the cold junction (Figure 3-17). When heat is applied to the hot junction, electrons move from one metal to the other creating a negative charge on one and a positive charge on the other. A thermocouple is often used as a temperature probe for temperature-measuring devices. A voltmeter, calibrated in degrees, is connected across the external thermocouple leads to indicate the temperature.

Piezoelectric Effect

A piezoelectric substance is one that produces an electric charge when a mechanical pressure is applied. Certain crystals such as quartz are **piezoelectric.** That means when they are **compressed** or **struck,** they generate an electric charge.

One common application of piezoelectricity is the piezo gas igniter shown in Figure 3-18. When you push the button, it makes a small, spring-powered hammer rise off the surface of the piezo crystal. When the hammer reaches the top, it releases and **strikes the crystal,** creating a high voltage. This voltage is high enough to make a spark which ignites the gas. Piezoelectric igniters are used on most gas furnaces and stoves.

Mechanical-Magnetic

Most of the electricity we use is produced using an **electric generator** that converts **mechanical-magnetic** energy into electric energy. The basic components and operation of an AC generator are shown in Figure 3-19. As the **armature rotates** through the **magnetic field,** a voltage is induced in the armature winding. Slip rings are attached

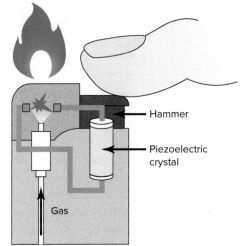

Figure 3-18 Piezo gas igniter.
Photo: ©Jo Ann Snover/Alamy Stock Photo

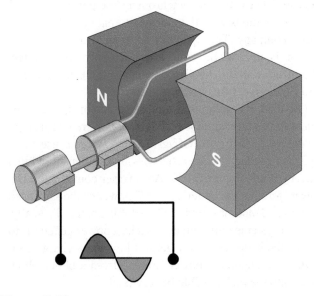

Figure 3-19 AC generator.

to the armature and rotate with it. Carbon brushes ride against the slip rings to conduct current from the armature. An armature is any number of conductive wires wound in

Figure 3-20 Wind generator.
©Ron_Thomas/Getty Images

loops which rotate through the magnetic field. For simplicity, one loop is shown. Although this generator produces AC electricity, it may be designed to produce AC or DC electricity.

Every generator must be driven by a turbine, a diesel engine, or some other machine that produces mechanical energy. **Prime mover** is a term used to identify the mechanical device that **drives** the generator. To obtain more electric energy from a generator, the prime mover must supply more mechanical energy. For example, **wind generators** are installed in locations with strong sustained winds (Figure 3-20). The wind pushes against the fan blades of the wind turbine, rotating the fan and a shaft that drives a generator to produce electricity. The electricity is either used or stored in batteries.

Part 2 Review Questions

1. Define *electric current*.
2. What is the external force that causes electron flow?
3. In which direction do electrons flow relative to the polarity of the applied voltage?
4. Compare electric current flow in a DC and an AC circuit.
5. Why is polarity normally identified on DC but not AC voltage sources?
6. How does a photovoltaic cell produce electricity?
7. Compare the operation of grid-tie and off-grid PV solar systems.
8. What is the function of an inverter as part of a solar energy system?
9. Name the three basic components of a battery.
10. How does a thermocouple produce electricity?
11. How does a piezoelectric substance produce electricity?
12. How does an electric generator produce electricity?
13. What type of prime mover is used as part of wind generator?

CHAPTER FOUR

Electrical Quantities and Ohm's Law

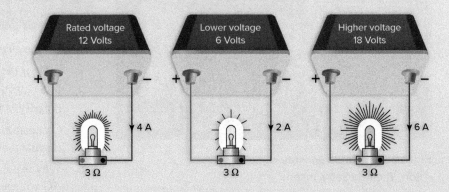

Effect of voltage on current flow.

In practical situations, you must be able to measure electricity if you want to work with it. This chapter deals with the standard units used for measuring of electric voltage, current, resistance, power, and energy. It also introduces Ohm's law which defines the relationship between voltage, current, and resistance.

PART 1 CURRENT, VOLTAGE, AND RESISTANCE

4.1 Current

The **coulomb** is a measurement for a quantity of electrons, and the practical unit for an electric *charge*. One coulomb of charge represents a total charge of 6.24×10^{18} (6,240,000,000,000,000,000) electrons. That appears to be a lot, but only because the charge on one electron is so small.

Charge in motion is referred to as current. **Current** is the rate of flow of electrons through a conductor. The letter I (which stands for *intensity*) is the symbol used to represent current. Current is measured in **amperes (A).** The word ampere (A) means coulombs per second (C/s). If we could count the individual electrons, we would discover that 1 ampere of current is equal to 1 coulomb of charge moving past a given point in 1 second, as illustrated in Figure 4-1.

The relationship between the coulomb and the ampere can be expressed by the mathematical equation:

$$Q = It$$

Where Q = quantity of coulombs (C)

I = current, in amperes (A)

t = time, in seconds (s)

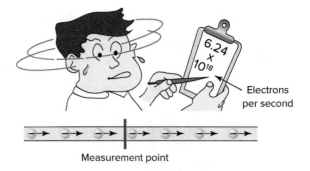

Electrons per second

Measurement point

Figure 4-1 **One ampere of current flow.**

EXAMPLE 4-1

Problem: What quantity of coulombs will flow through a circuit with a current flow 10 amperes over a 1-hour period?

Solution:

$$Q = It$$
$$= 10 \text{ A} \times 3,600 \text{ s}$$
$$= 36,000 \text{ C}$$

When the electrons begin to flow, the effect is felt instantly all along the conductor similar to the force that can be transmitted through a row of billiard balls, as illustrated in Figure 4-2. The electric current is actually the **impulse** of electric energy that one electron transmits to another as it changes orbit. The first electron repels the other out of orbit, transmitting its energy to it. The second electron repeats the action, and this process continues through the conductor.

Although the individual electrons travel less than an inch per second, the current effectively travels through the conductor nearly at the **speed of light** (186,000 miles per second). Because the atoms are close and the orbits overlap, the electron that is freed does not have to travel far to encounter a new orbit. This action is almost instantaneous so that even though the electrons are moving relatively slowly, the electric current travels at almost the speed of light.

An instrument called an **ammeter** is used to measure current flow in a circuit (Figure 4-3). The ammeter is inserted into the path of the current flow, or in **series,** to measure current. This means the circuit must be opened and the meter leads placed between the two open points.

Although the ammeter measures electron flow in coulombs per second, it is calibrated or marked in **amps or amperes.** For most practical applications, the term amps is used instead of coulombs per second when referring to the amount of current flow. Prefixes **micro** $\left(\frac{1}{1,000,000}\right)$ and **milli** $\left(\frac{1}{1,000}\right)$ are used to represent very small amounts of current and **kilo** (1,000) and **mega** (1,000,000) to represent very large amounts as follows.

Current	Base Unit	Units for Very Small Amounts		Units for Very Large Amounts	
Symbol	A	μA	mA	kA	MA
Pronounced As	Ampere (Amp)	Microampere	Milliampere	Kiloampere	Megampere
Multiplier	1	0.000001	0.001	1,000	1,000,000

Figure 4-2 **Transmission of electrons by impulse.**

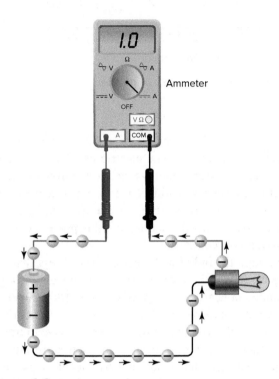

Ammeter

Figure 4-3 **Ammeter connected to measure current.**

4.2 Voltage

Voltage (*V, EMF,* or *E*) is **electric pressure,** a potential force or difference in electric charge between two points. Voltage pushes current through a wire similar to water pressure pushing water through a pipe. The voltage level or value is proportional to the difference in the electric potential energy between two points (Figure 4-4). Voltage is measured in **volts (V).** A voltage of 1 volt is required to force 1 ampere of current through 1 ohm of resistance. The letter *E,* which stands for emf (electromotive force), or the

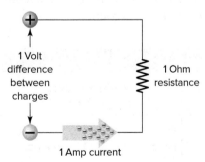

1 Volt difference between charges

1 Ohm resistance

1 Amp current

Figure 4-4 **Voltage is proportional to the difference between charges.**

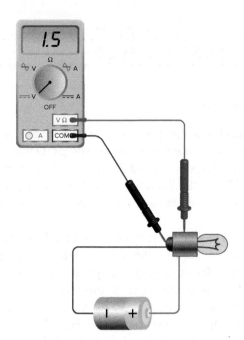

Figure 4-5 **Voltmeter connected to measure voltage.**

letter *V,* which stands for voltage, is commonly used to represent voltage in an algebraic formula.

Voltage, or a **difference in potential,** exists between any two charges that are not exactly equal to each other. Even an uncharged body has a potential difference with respect to a charged body; it is positive with respect to a negative charge and negative with respect to a positive charge. Voltage also exists between two unequal positive charges or between two unequal negative charges. Therefore, voltage is **purely relative** and is not used to express the actual amount of charge, but rather to compare one charge to another and indicate the electromotive force between the two charges being compared.

A **voltmeter** is used to measure the voltage, or potential energy difference of a load or source, as illustrated in Figure 4-5. Voltage exists between two points and does not flow through a circuit as current does. It is possible to have voltage without current, but current cannot flow without voltage. A voltmeter is connected **across,** or in **parallel,** with the two points.

Very small amounts of voltage are measured in **millivolts** $\left(\frac{1}{1,000}\right)$ and **microvolts** $\left(\frac{1}{1,000,000}\right)$, whereas high-voltage values are expressed in **kilovolts** (1,000) and **megavolts** (1,000,000). The typical units of measurement are as follows.

Voltage	Base Unit	Units for Very Small Amounts		Units for Very Large Amounts	
Symbol	V	µV	mV	kV	MV
Pronounced As	Volt	Microvolt	Millivolt	Kilovolt	Megavolt
Multiplier	1	0.000001	0.001	1,000	1,000,000

4.3 Resistance

Resistance (*R*) is the opposition to the flow of electrons or current. Basically, resistance is the measure of electric friction as electrons move through a conductor. Resistance is measured in **ohms.** The Greek letter Ω (omega) is used to represent ohms, and the typical units of measurement are as follows.

Resistance	Base Unit	Units for Very Small Amounts		Units for Very Large Amounts	
Symbol	Ω	$\mu\Omega$	$m\Omega$	$k\Omega$	$M\Omega$
Spelled as	Ohm	Microhm	Milliohm	Kilohm	Megohm
Multiplier	1	0.000001	0.001	1,000	1,000,000

Resistance is due, in part, to each atom resisting the removal of an electron by its attraction to the positive nucleus. Collisions of countless electrons and atoms as the electrons move through the conductor create additional resistance. The resistance created causes **heat** in the conductor when current flows through it; that is why a wire becomes warm when current flows through it. The **higher the resistance** of a conductor, the **greater its opposition** to current flow. The elements of an electric range become hot and the filament of an incandescent lamp becomes extremely hot because they have a much higher resistance than the conductors that carry the current to them.

Every electric component has resistance, and this resistance changes electric energy into another form of energy such as heat, light, or motion. An **ohmmeter** is used to measure **resistance,** as illustrated in Figure 4-6. Unlike the voltmeter and ammeter, which use energy in the current to make their measurements, the ohmmeter uses its **own**

power source. For example, a multimeter contains an ohmmeter that operates by a battery located inside the instrument. The ohmmeter applies a known voltage into a circuit, measures the resulting current, and then calculates the resistance. ***For this reason ohmmeters should never be connected to live circuits!***

Part 1 Review Questions

1. What is the practical unit of electric charge?
2. Define *electric current.*
3. What is the basic unit used to measure current?
4. Explain what a current flow of 1 ampere can be equated to.
5. Account for the fact that electric current travels at the speed of light, while the electron movement is relatively slow.
6. How must an ammeter be connected into a circuit to measure current?
7. Define *voltage.*
8. What is the basic unit used to measure voltage?
9. Explain what a voltage of 1 volt can be equated to.
10. How must a voltmeter be connected into a circuit to measure voltage?
11. Define *electric resistance.*
12. What is the basic unit used to measure resistance?
13. What is the cause and effect of resistance?
14. When using an ohmmeter to measure resistance, what precaution must be observed?
15. Convert the following data:
    a. 12,000 A to kA
    b. 0.04 V to mV
    c. 1.8 MΩ to Ω
    d. 40 μA to A
16. A current of 200 mA flows for 2 minutes. How much charge has passed?

PART 2   POWER, ENERGY, AND OHM'S LAW

Work can be defined as transfer of energy. Work involves the application of a force over a distance. Electrical work refers to the work done by the movement of electrons from one side of a voltage source to the other.

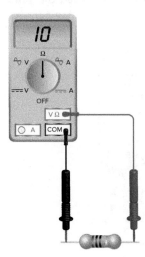

Figure 4-6 Ohmmeter connected to measure resistance.

4.4 Power

Electric power (P) is the amount of electric energy **converted** to another form of energy in a given length of time. Power represents the **work** performed by an electric circuit and is measured in **watts (W).** Power in an electric circuit is equal to

$$\text{Power} = \text{Voltage} \times \text{Current}$$
$$\text{Watts} = \text{Volts} \times \text{Amperes}$$
$$P = EI$$

where

P = the power in watts

E = the voltage in volts

I = the current in amperes

The electric power rating of an appliance is listed on its **nameplate** as illustrated in (Figure 4-7). This rating is not always specified simply in terms of watts. Power rating may be given in voltage and current, in which case, the power can be calculated using the equation $P = EI$. When the rating is given in watts and volts, the rated current can be calculated using the equation

$$I = \frac{P}{E}$$

EXAMPLE 4-2

Problem: An iron draws 10 A from an 120-V household supply. How much power is used?

Solution:

$$P = EI$$
$$= 120\ \text{V} \times 10\ \text{A}$$
$$= 1{,}200\ \text{W, or } 1.2\ \text{kW}$$

Figure 4-7 Appliance power rating.

Power can be measured using a **wattmeter.** The wattmeter is basically a **voltmeter** and **ammeter** combined into one instrument (Figure 4-8). The ammeter terminals are connected in series, and the voltmeter terminals are connected in parallel with the circuit in which the power is being measured. The wattage rating of a lamp indicates the **rate** at which the device can convert electric energy into light. The faster a lamp converts electric energy to light, the brighter the lamp will be.

4.5 Energy

Energy can be defined as the ability to do work. **Kinetic** energy is the energy that an object contains because of a particular motion. **Potential** energy is stored energy. For example, stretching a rubber band is a way of loading it with potential energy. When the rubber band is released, the potential energy is converted to kinetic energy, which can be used to do work.

Electric energy refers to the **energy of moving electrons.** In an operating electric circuit, the voltage pushes and pulls electrons into motion. When electrons are forced into motion, they have kinetic energy, or the energy of motion. The unit of energy, called the **joule (J),** is used in

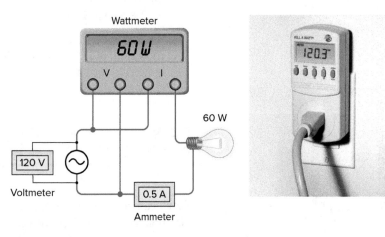

Figure 4-8 Wattmeter connected to measure power.
Photo: Courtesy of P3 International Corporation

Figure 4-9 Kilowatt-hour meter.
©Hydro One Networks Inc.

scientific work to measure electric energy. By definition, a joule is the amount of energy carried by 1 coulomb of charge propelled by an electromotive force of 1 volt.

The following are examples of mathematical relationships of joules, volts and watts:

1. What is the voltage required to move 100 coulombs of charge if 4000 Joules of energy is required?

$$\text{Volts} = \frac{\text{Joules}}{\text{Coulombs}} = \frac{4000\,\text{J}}{100\,\text{C}} = 40\,\text{volts}$$

2. If 100 joules of energy is used in 5 seconds. What is the power in watts?

$$\text{Watts} = \frac{\text{Joules}}{\text{Seconds}} = \frac{100\,\text{J}}{5\,\text{s}} = 20\,\text{watts}$$

The **watt-hour (Wh)** is the more practical unit of measurement of electric energy. **Power** and **time** are factors that must be considered in determining the amount of energy used. This is usually done by multiplying watts by hours. The result is watt-hours. If power is measured in kilowatts and multiplied by hours, the result is **kilowatt-hours**, abbreviated **kWh.** Energy measurements are used in calculating the cost of electric energy. A **kilowatt-hour meter** connected to a residential electrical system is used to monitor your daily power usage as illustrated in Figure 4-9.

EXAMPLE 4-3

Problem: Assuming that the cost of electricity is 5 cents ($0.05) per kWh, how much will it cost to operate a 100-W lamp continuously for 200 hours?

Solution:

$$\text{kWh} = \frac{100\,\text{W} \times 200\,\text{h}}{1{,}000}$$

$$= 20\,\text{kWh}$$

$$\text{Cost} = \text{kWh} \times \text{rate}$$

$$= 20 \times \$0.05$$

$$= \$1.00$$

4.6 Electric Circuit

The basic requirements for an electric circuit are summarized as follows:

*   **Power source**—provides the voltage (the force that pushes electrons through a conductor) and current (the rate of flow of electrons) to operate the load device connected to the circuit.
*   **Protection device**—automatically prevents dangerous or excess amounts of current that can occur in the event of circuit overloading or an electrical fault.
*   **Conductors**—wiring that provides the low resistance path of the circuit, through which the current flows. The conductor network interconnects all of the other components of the circuit.
*   **Control device**—provides the control that closes or opens the current path of the circuit.
*   **Load**—is the component or portion of a circuit that consumes electric power rather than provides it. Examples of loads are a light bulb, resistor, or motor.

The power source, conductors, switch, lamp, and fuse form a **closed-circuit** (Figure 4-10) conducting path from one side of the voltage source to the other. The lamp

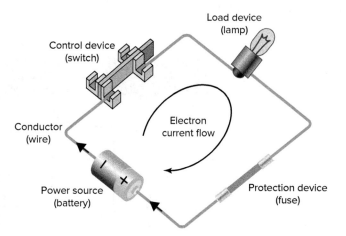

Figure 4-10 Closed electric circuit.

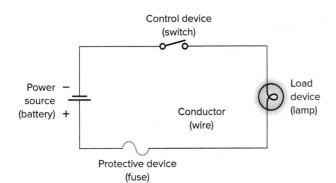

Figure 4-11 **Electric circuit schematic diagram.**

provides resistance to the circuit and limits the amount of current that can flow. If the switch is opened or the fuse blows open, the conducting path is opened so no current can flow. This is referred to as an **open circuit.**

Figure 4-11 shows the schematic diagram for an electric circuit. A **pictorial** circuit diagram uses simple images of components, while a **schematic** diagram shows the components of the circuit as simplified standard **symbols;** both types show the connections between the devices. The function of the parts can be summarized as follows:

- **Battery** power source supplies the electric pressure or voltage required to push current through the circuit.
- **Conductors** provide a low-resistance path from the source to the load.
- **Switch** control device opens and closes the circuit to switch the current flow OFF and ON.
- **Lamp** load converts the electric energy from the power source into light and heat energy.
- **Fuse** protection device melts to automatically open the circuit in the event of a higher than rated current flow. Too much current can cause damage to conductors and load devices.

4.7 Ohm's Law

Ohm's law defines the relationship between circuit current, voltage, and resistance and is stated as follows: ***The current flowing in a circuit is directly proportional to the applied voltage and inversely proportional to the resistance.*** Therefore, the **higher the voltage** applied to a circuit, **the higher the current** through the circuit. On the other hand, a decrease in the applied voltage will result in a decrease in circuit current. This assumes that the circuit resistance remains constant, as illustrated in Figure 4-12.

If the voltage is held constant, the current will change as the resistance changes, but in the opposite direction. Assuming that voltage is constant, **lowering the resistance** causes an **increase in current.** On the other hand, an increase in resistance results in a decrease in current flow, as illustrated in Figure 4-13.

Ohm's law can be stated as mathematical equations (Figure 4-14), all derived from the same principle as follows:

$I = \dfrac{E}{R}$ (current = voltage *divided by* resistance)

$E = I \times R$ (voltage = current *multiplied by* resistance)

$R = \dfrac{E}{I}$ (resistance = voltage *divided by* current)

EXAMPLE 4-4

Problem: A portable heater with a resistance of 10 Ω is plugged into a 120-volt electrical outlet. How much current will flow to the heater?

Solution:

$$I = \frac{E}{R}$$
$$= \frac{120 \text{ V}}{10 \text{ Ω}}$$
$$= 12 \text{ A}$$

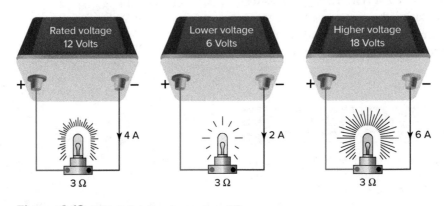

Figure 4-12 **Effect of voltage on current flow.**

Figure 4-13 **Effect of resistance on current flow.**

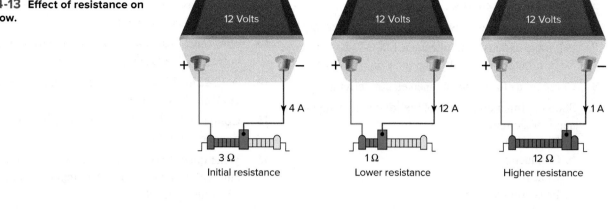

12 Volts	12 Volts	12 Volts
4 A	12 A	1 A
3 Ω	1 Ω	12 Ω
Initial resistance	Lower resistance	Higher resistance

$E = IR$ $I = E/R$ $R = E/I$

Figure 4-14 **Ohm's law equations.**

EXAMPLE 4-5

Problem: If a 25-Ω resistor has a current flow of 2 A through it, how much is the voltage across it?

Solution:

$$E = IR$$
$$= 2\,A \times 25\,\Omega$$
$$= 50\,V$$

EXAMPLE 4-6

Problem: What is the resistance of a toaster element that is rated for 120 V and 8 A?

Solution:

$$R = \frac{E}{I}$$
$$= \frac{120\,V}{8\,A}$$
$$= 15\,\Omega$$

4.8 Direction of Current Flow

The direction of current flow in a circuit can be designated either as electron flow or conventional current flow, as illustrated in Figure 4-15. **Electron flow** is based on the electron theory of matter and follows the motion of electrons in the circuit from **negative to positive. Conventional current flow** is based on an older theory of electricity and assumes a current flow in the opposite direction from **positive to negative.**

Both conventional current flow and electron flow are acceptable and are used for different applications. For purposes of circuit analysis it makes no difference which direction of current flow you use as long as the same direction is used consistently throughout the circuit. It is important to understand and be able to think in terms of both conventional current flow and in terms of electron flow.

Part 2 Review Questions

1. Define *electric power.*

2. What is the basic unit used to measure electric power?

3. A toaster draws 8 amperes when connected to a 120-volt source. What is the wattage rating of the toaster?

4. Explain why a 140-W soldering iron produces more heat than a 20-W soldering iron.

Figure 4-15 **Electron versus conventional current flow.**

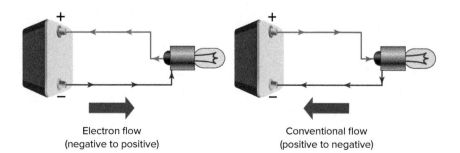

Electron flow
(negative to positive)

Conventional flow
(positive to negative)

5. Define *electric energy*.

6. What is the basic unit used to measure electric energy?

7. Is it more expensive to operate a 1,000-W hair dryer for 5 minutes or a 40-W lamp for 1 hour? Why?

8. Compare a closed and an open electric circuit.

9. State the function of each of the following components of a circuit:
   a. Power source
   b. Conductors
   c. Control device
   d. Protection device
   e. Load

10. What does Ohm's law tell us about current flow in a circuit?

11. Ohm's law is applied to a circuit using what three mathematical equations?

12. A load has a resistance of 20 Ω. What is its current when connected to 240 V?

13. How much resistance does a load have if it draws a current of 6 A from a 12-V battery?

14. A 12-Ω load is conducting a current of 2.5 amps. How much is its voltage?

15. Compare the direction of electron flow and conventional current flow.

16. What quantity of coulombs will flow through a circuit with a current flow of 8 amperes over a 15-minute period?

17. What is the voltage required to move 200 coulombs of charge if 500 joules of energy is required?

18. If 400 joules of energy is used in 20 seconds, what is the power in watts?

CHAPTER FIVE

Simple, Series, and Parallel Circuits

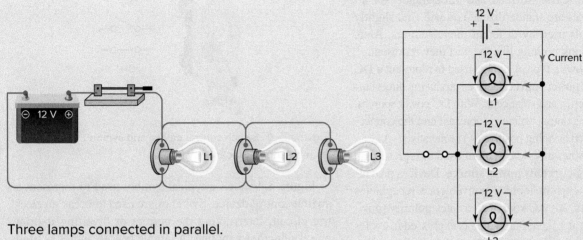

Three lamps connected in parallel.

Electric circuits can be as simple as a flashlight that uses but a single circuit or as complex as a large industrial or commercial installation that uses thousands of circuits working together. In this chapter you will study the three basic circuit types: **simple, series,** and **parallel.**

5.1 Circuit Symbols

Before you can read an electrical diagram, you must be able to identify the symbol used to represent each component. Use of **symbols** tends to make circuit diagrams less complicated and easier to read and understand. Not all electrical symbols are standardized. You will find slightly different symbols used by different manufacturers. Also, some symbols look nothing like the part they represent.

Figure 5-1 shows a typical symbol used to represent a **DC (direct current) power source.** Devices producing direct current include batteries and solar cells. With DC power sources, the polarity of the output voltage is constant and the connections are identified as being positive (+) or negative (−).

Figure 5-2 shows a typical symbol used to represent an **AC (alternating current) power source.** Devices providing alternating current include convenience receptacles and transformers. An AC source alternates polarity (positive to negative) at a fixed rate which is expressed in cycles per second, or hertz. Therefore, the output voltage connections are not identified as being positive or negative.

Figure 5-3 shows a typical symbol used to represent a **circuit breaker** protection device. Protection devices include circuit breakers and fuses. A circuit breaker's function is similar to that of a fuse in that they both open the circuit path whenever too much current is passed.

Figure 5-1 DC power supply and symbol.
©Ingram Publishing/Alamy Stock Photo

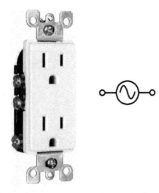

Figure 5-2 AC power supply and symbol.
©Joebelanger/iStock/Getty Images

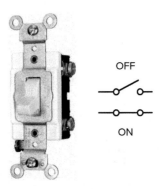

Figure 5-3 Circuit breaker protection device and symbol.
©Aleks_G/iStock/Getty Images

Figure 5-4 Switch control device and symbol.
©Joebelanger/iStock/Getty Images

Figure 5-4 shows a typical symbol used to represent a **switch** control device. Switches are used to break an electric circuit, interrupting the current or diverting it from one conductor to another. A simple ON/OFF switch is used to either start or stop the current flow through a circuit.

Figure 5-5 shows a typical symbol used to represent a **pushbutton** control device. Pushbuttons are spring-loaded switches designed to operate by finger pressure. The push-to-make normally open pushbutton returns to its normally open (OFF) position when you release the button. A push-to-break normally closed switch returns to its normally closed (ON) position when you release the button.

Figure 5-6 shows a typical symbol used to represent an incandescent lamp **load** device. Load devices convert the

Figure 5-5 Pushbutton control device and symbol.

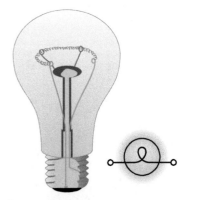

Figure 5-6 Incandescent lamp load and symbol.

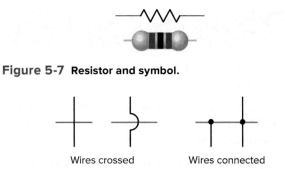

Figure 5-7 Resistor and symbol.

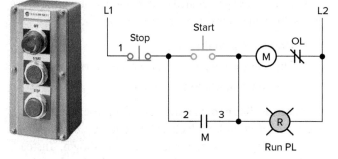

Figure 5-8 Crossed and connected wiring symbols.

energy of the source to another form of energy in order to produce the desired function or useful work of the circuit. Electrical load devices in buildings include equipment for space heating, cooling, water heating, and lighting.

Figure 5-7 shows a typical symbol used to represent a **resistor.** Resistors are used to limit and control the flow of current in a circuit. They are used in conjunction with other electric and electronic components and can be found in virtually all electronic circuits. A fixed resistor offers a specified ohmic amount of resistance.

Figure 5-8 shows typical types of lines used throughout electrical diagrams to show the wire connections between components. Wires or conductors are shown as connected together or crossed without connecting.

5.2 Circuit Diagrams

Different types of diagrams are used to show the layout of circuits. A **pictorial diagram** is used to show the physical details of the circuit visually. Figure 5-9 shows a pictorial diagram for the wiring of a single lamp fluorescent fixture. The advantage here is that a person can simply take a group of parts, compare them with the pictures in the diagram, and

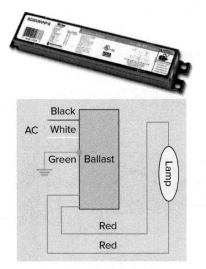

Figure 5-9 Pictorial diagram for the wiring of a fluorescent fixture.

Photo: Courtesy of Universal Lighting Technologies

Figure 5-10 Ladder-type schematic diagram of a motor start/stop circuit.

Photo: Courtesy of Rockwell Automation, Inc.

wire the circuit as shown. The main disadvantage is that many circuits are so complex that this method is impractical.

A **schematic diagram** uses symbols to represent the various components and, as a result, is not as cluttered as a pictorial diagram. The components are arranged in a manner that makes it **easier to read** and understand the operation of the circuit. This type of diagram is most often used to explain the sequence of operation of a circuit.

Figure 5-10 shows a **ladder-type** schematic diagram for a motor start/stop control circuit. The two vertical power lines (*L1* and *L2*) connect to the power source, and the control circuit connects horizontally across them like rungs in a ladder. This type of diagram is not intended to show the physical relationship of the various components of the circuit; rather, it leans toward simplicity, emphasizing only the operation of the circuit.

A **wiring diagram** is intended to show, as closely as possible, the actual connections and placements of all component parts in a circuit. Unlike the schematic, the components are shown in their **relative physical position.** Figure 5-11 shows a wiring diagram for the motor start/stop circuit.

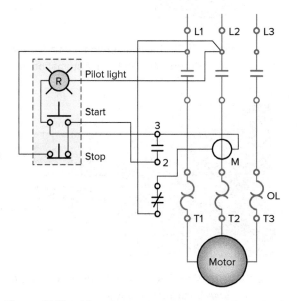

Figure 5-11 Wiring diagram for the motor start/stop circuit.

All connections are included to show the actual routing of the wires. Such diagrams show the necessary information for:

- Doing the actual wiring
- Physically tracing wires in the event of troubleshooting
- Making changes to the circuit

A **block diagram** is a method of representing the major parts of complex electrical systems by blocks. Individual components and wires are not shown. Instead, each block represents electric circuits that perform **specific functions** in the system. Figure 5-12 shows the block diagram for a variable-frequency motor drive. The functions the circuits perform are written in each block. Lines connecting the blocks may indicate the general direction of current paths, useful in creating an understanding of the overall system.

The **one-line or single-line diagram** permits **greater simplification** for representing three-phase AC power systems. One-line diagrams are most often used where a considerable amount of information needs to be shown and yet the main object is to make the layout as simple as possible. Figure 5-13 shows a one-line diagram of the routing of power for a grid-tied solar system.

Part 1 Review Questions

1. Why are symbols used in diagrams to represent components?

2. Polarity markings are shown on symbols representing DC voltage sources but not shown on AC voltage sources. Why?

3. Compare the type of voltage produced by a solar panel with that available from a convenience wall receptacle.

4. Explain how circuit breakers and fuses protect a circuit.

5. What are the two basic functions of a switch?

6. Compare the operation of a normally open and normally closed pushbutton.

7. What type of circuit component is a light bulb classified as?

8. In general, what is the function of resistors found in electronic equipment?

9. Two wires are shown on an electrical diagram joined by a dot. What does this indicate?

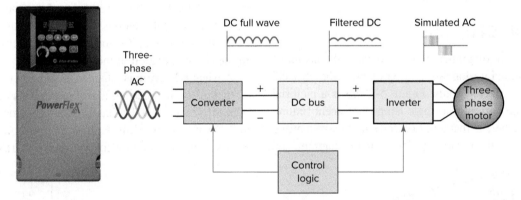

Figure 5-12 Block diagram for a variable-frequency motor drive.
Photo: Courtesy of Rockwell Automation, Inc.

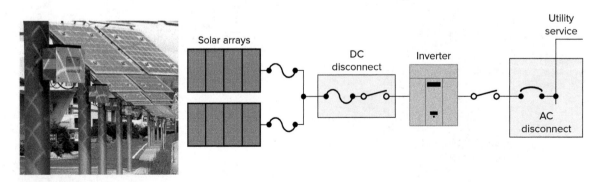

Figure 5-13 One-line diagram of the routing of power for a grid-tied solar system.
Photo: ©SMA Solar Technology AG

10. Give one advantage and one limitation of the pictorial type of electrical diagram.

11. Explain why schematic diagrams are easier to read.

12. What is a wiring diagram intended to show?

13. What are block diagrams normally used to represent?

14. The one-line diagram is often used with a power distribution system. What type of AC power is usually associated with these systems?

PART 2 SIMPLE, SERIES, AND PARALLEL CIRCUITS

5.3 Simple Circuit

A simple circuit can be classified as one that has only one control, one load device, and one voltage source. Figure 5-14 is an example of a **simple circuit.** Each component is connected or wired **end-to-end** in a continuous loop. The lamp is turned ON and OFF by closing and opening the switch. When the switch is closed, the voltage across the lamp is the same as the voltage of the source.

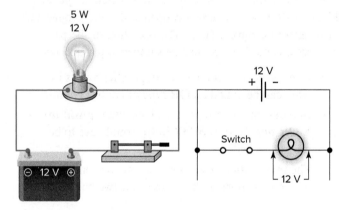

Figure 5-14 Simple circuit.

5.4 Series Circuit

Series-Connected Loads

If two or more loads are connected end to end, they are said to be connected in **series.** Figure 5-15 shows two lamps connected in series to a 12-volt source. The characteristics of series connected loads can be summarized as follows:

- The **same amount of current** flows through each of the loads.
- There is only **one path** through which the current flows.
- The current stops flowing if the path is opened or the circuit is broken at any point. For example, if two lamps are connected in series and one burns out, both lamps will go out.
- Each load receives part of the applied source voltage. For example, if the two lamps have the same ratings, each will receive one-half of the applied source voltage.
- The amount of voltage each load in series receives is directly proportional to its electric resistance. The **higher the resistance** of the load connected in series, the **more voltage** it receives.

Series-Connected Control Devices

Two or more control devices may also be connected in series. The connection is the same as that used for loads, which is **end to end.** Figure 5-16 shows two switches connected in series to control a lamp. The operation of these series-connected control devices can be summarized as follows:

- Connecting control devices in series results in what can be called an **AND** type of control circuit.
- In order for the lamp to turn ON both **switch A AND switch B** would have to be closed.

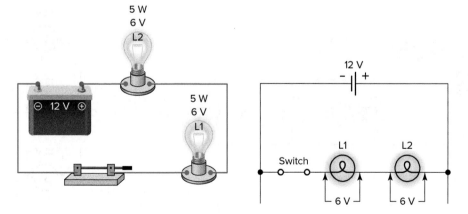

Figure 5-15 Two lamps connected in series.

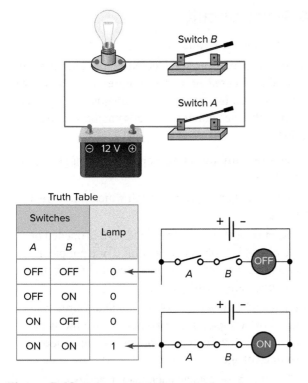

Truth Table

Switches		Lamp
A	B	
OFF	OFF	0
OFF	ON	0
ON	OFF	0
ON	ON	1

Figure 5-16 Two switches connected in series.

- A **truth table** is a way of showing how the switches in the circuit operate the lamp: 0 indicates the lamp is OFF; 1 indicates the lamp is ON.

5.5 Parallel Circuit

Parallel-Connected Loads

If two or more loads are connected **across** the two voltage source leads, they are said to be connected in **parallel.** Figure 5-17 shows three lamps connected in parallel to a 12-volt source. The characteristics of parallel-connected loads can be summarized as follows:

- The **same amount of voltage** is applied to each load.
- The value of the voltage across each load is the same as that of the **voltage source.**
- The current to each load will vary with the resistance of the device. The amount of current that each in-parallel load passes is inversely proportional to its resistance value. The **higher the resistance** of the load connected in parallel, the **less current** it passes.
- Each load operates **independently** of the others. This is because there are as many current paths as there are loads. For example, when three lamps are connected in parallel, there will be three paths created. If one lamp burns out, the remaining two will not be affected.

Parallel circuits are the most widely used circuits in residential, commercial, and industrial installations. Receptacles, lights, appliances, and power tools are all wired using parallel circuits. Parallel circuits deliver the full system voltage to each and every load connected to the circuit. Parallel circuits allow for the **standardization of voltages,** in the case of residential loads 120 volts, as illustrated in Figure 5-18.

Parallel-Connected Control Devices

When two or more control devices are connected **across each other,** they are said to be connected in **parallel.** Figure 5-19 shows two normally open pushbuttons connected in parallel to control a lamp. The operation of the parallel-connected control devices can be summarized as follows:

- Connecting control devices in parallel results in what can be called an **OR** type of control circuit.
- In order for the lamp to turn ON, either **pushbutton A OR pushbutton B OR both** would have to be pressed closed.
- The **truth table** shows how the lamp responds to various combinations of the states of the two pushbuttons.

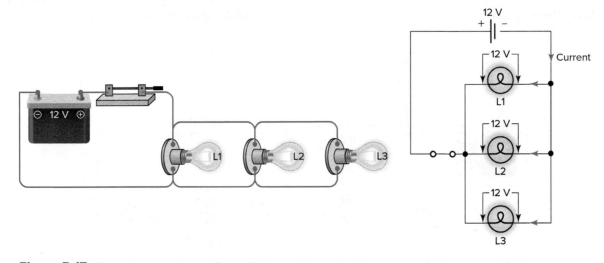

Figure 5-17 Three lamps connected in parallel.

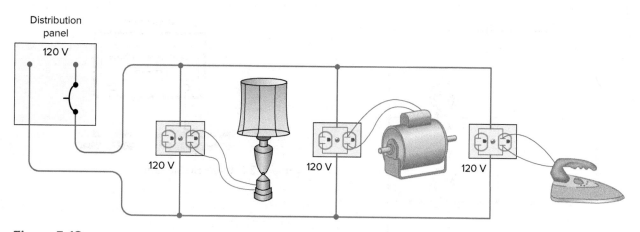

Figure 5-18 Parallel circuits allow standardization of voltages.

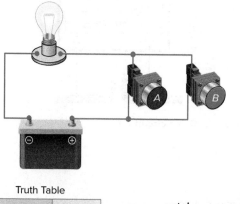

Truth Table

Switches		Lamp
A	B	
OFF	OFF	0
ON	OFF	1
OFF	ON	1
ON	ON	1

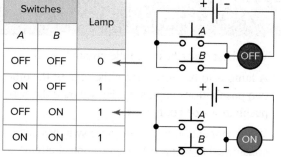

Figure 5-19 Two normally open pushbuttons connected in parallel.

5.6 Breadboarding and Computer Simulation of Circuits

Breadboarding refers to the building of a **temporary version** of a circuit. This technique is used extensively for experimenting with electric and electronic circuits. Figure 5-20 shows a typical breadboard used for building and testing circuits. No soldering is required so it is easy to change connections and replace components. Breadboards have many tiny sockets or holes arranged on a grid. The leads of most components can be pushed straight into the holes with wire connections made using 0.6-mm diameter solid wire. Integrated circuits (ICs) are inserted across the central gap. Power is supplied to the top and bottom rows which are connected horizontally all the way across. The other holes are linked vertically in blocks of five with no link across the center.

The computer has become a powerful learning tool. **Computer simulation software** allows you to experiment with electrical and electronic circuits without using laboratory facilities and materials. Programs such as **Multisim**

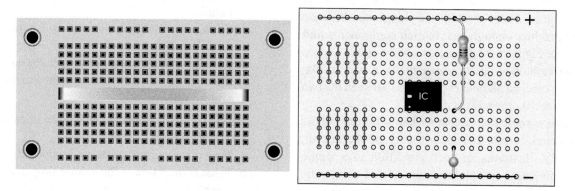

Figure 5-20 Typical breadboard.

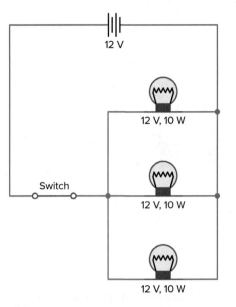

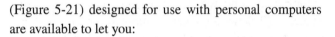

Figure 5-21 Typical Multisim circuit simulation.

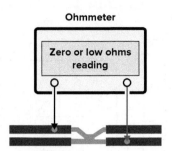

Figure 5-22 Short circuit fault.

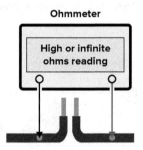

Figure 5-23 Open circuit fault.

(Figure 5-21) designed for use with personal computers are available to let you:

- Construct a schematic for a circuit on a computer.
- Simulate the operation of that circuit.
- Display its activity on test instruments contained within the program.
- Print a copy of the circuit and instrument readings.
- Practice troubleshooting circuits.

5.7 Short and Open Circuit Faults

Electrical circuits are subjected to various types of faults. When a fault occurs, there is a deviation of voltages and currents from normal values. The two basic types of faults are short and open circuit faults.

A **short circuit** is a faulty low-resistance connection that causes current to flow around a normal load instead of through it. Because the resistance is much less, the current is much higher. **Short circuits are dangerous because they can produce enough heat to melt components and start a fire.** Figure 5-22 illustrates a short circuit caused by two wires shorted due to a breakdown of wire insulation. In this case, an ohmmeter connected between the two wires will read 0 or low resistance.

An **open circuit** fault occurs when a failure occurs in the normal current conduction path of the circuit. Figure 5-23 illustrates an open condition in a single wire. In this case, an ohmmeter connected across the open ends of the wire will read high or infinite ohms.

Part 2 Review Questions

1. What best describes a simple circuit?

2. A lamp is to turn ON when either one or the other of two pushbuttons is pressed. How would the two pushbuttons be connected?

3. What connection of two or more switches can be considered to be an AND-type control circuit?

4. Three 120-V, 40-W incandescent lamps are connected in series to a 120-VAC source.
   a. How many current paths are produced?
   b. What is the value of the voltage drop across each lamp?
   c. Comment on the brightness of each lamp (full or dim). Why?
   d. Assume that one lamp burns open (or out). What affect, if any, does this have on the other two lamps? Why?

5. Two 12-V, 10-W lamps are connected in parallel to a 12-VDC source.
   a. How many current paths are produced?
   b. What is the value of the voltage drop across each lamp?

c. Comment on the brightness of each lamp (full or dim). Why?

d. Assume that one lamp burns open (or out). What affect, if any, does this have on the other lamp? Why?

6. A control system calls for a light to come on when switch A and either switch B or switch C is closed. State the connection of the switches that will accomplish this. Draw a schematic diagram of the circuit.

7. One bulb in a 20-string series-connected holiday tree lamp set needs to be replaced. If the lamp set is rated for an input voltage of 120-VAC, what voltage rating of bulb would be required?

8. Define the term *breadboarding*.

9. Why are short circuit faults potentially more dangerous than open circuit types?

10. An open in a single strand of a wire housed in a wiring harness is suspected. Explain how an ohmmeter test would be applied to confirm this.

CHAPTER SIX

Measuring Voltage, Current, and Resistance

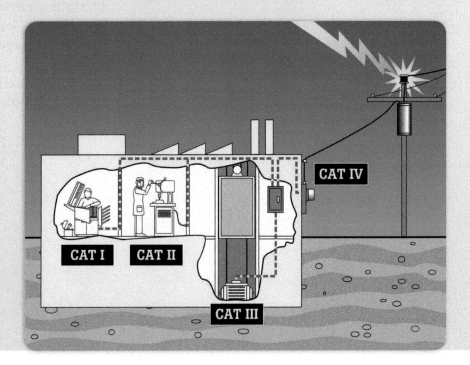

Overvoltage energy-level categories.

LEARNING OUTCOMES

► Compare the operation of analog and digital meters.

► Correctly read an analog scale and digital meter display.

► Use a multimeter to measure voltage, current, and resistance.

► Be aware of safety precautions to follow when taking circuit measurements.

► Understand multimeter specifications and special features.

Electricity cannot be seen. However, making circuit measurements of voltage, current, or resistance, using a meter, can help in understanding an electric circuit and determining if it is operating properly. This chapter helps you understand how meters work and how to use them to make basic electrical measurements.

6.1 Analog and Digital Meters

Analog meters use a needle and **mechanical type** of meter movement to indicate the measurement. The permanent magnet, moving-coil galvanometer, illustrated in Figure 6-1, is the basis of most analog meters. It consists of a moving coil suspended between the poles of a horseshoe magnet. With no current in the coil, no magnetic field is set up, and the needle balances, so there is no tension on the spring. This gives a reading of zero. With current applied, the coil becomes an electromagnet. Magnetic forces cause the coil to turn until the forces are balanced with the spring force. This results in a reading other than zero.

The major difference between analog and digital meters is the type of display used. In a digital meter the meter movement is replaced by an **electronic digital display,** as shown in Figure 6-2. The basis for digital meter operations is an electronic **analog-to-digital** converter circuit. This circuit converts analog values at the input to an equivalent digital form. Values are processed by electronic circuits and shown on the display as decimals.

6.2 Multimeter

Most often, the voltmeter, ammeter, and ohmmeter are combined in a single instrument called a **multimeter,** which is the basic measuring tool of the electrical industry. Ultimately the diagnosis of an electrical system comes down to using a multimeter to pinpoint the exact location of a problem.

An analog multimeter consists of a single meter movement and the associated meter circuitry required of a volt-

Figure 6-2 **Digital meter circuit board.**
©Juan Peña, Sparkfun Electronics

meter, ammeter, and ohmmeter. The **analog multimeter,** such as that shown in Figure 6-3, is often referred to as a **volt-ohm-milliammeter (VOM).** The operating controls of an analog multimeter can be summarized as follows:

- **Function** selects the type of measurement: voltage, current, or resistance.
- **Range** selects the full-scale range of the measurement. This permits proper selection of the internal circuits

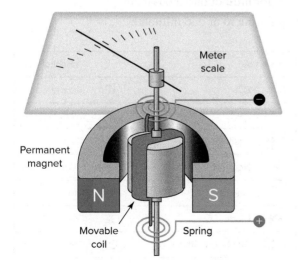

Figure 6-1 **Analog meter movement.**

Figure 6-3 **Analog (VOM) multimeter.**
©DonNichols/iStock/Getty Images

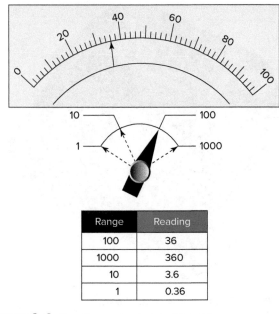

Range	Reading
100	36
1000	360
10	3.6
1	0.36

Figure 6-4 **Reading an analog meter scale.**

so that only one range of one type of measurement is selected at any one time.

- The **test leads and input jacks** connect the multimeter to the circuit or component you want to measure. Test leads are usually red and black. Normally you put the **red test lead in the POS (+)** input jack and the **black test lead in the NEG (−)** or common input jack.

- Multimeters usually have more than these two input jacks. You use other jacks primarily for measuring higher voltages and current and other special functions.

- To take an accurate reading from an analog scale, you must have your eye **in line** with the pointer.

Reading analog meters may require simple mental calculations. For example, a meter might have four voltage ranges—1 V, 10 V, 100 V and 1,000 V—but only one scale—100 V, as illustrated in Figure 6-4. Therefore when using the:

- 1,000-V range you need to multiply the needle reading by 10.
- 10-V range you need to divide the needle reading by 10.
- 1-V range you need to divide the needle reading by 100.

A **digital multimeter (DMM)** also has function and range switches and jacks to accept test leads, as illustrated in Figure 6-5. Since digital multimeters contain electronic circuits to produce their measurements, they need internal **batteries** to supply power for **all measurements.** Therefore, unlike analog multimeters, digital multimeters require an ON/OFF power switch that connects the power supply to the electronic circuits. Digital meter values can be read directly from digital displays so they are easy to read accurately.

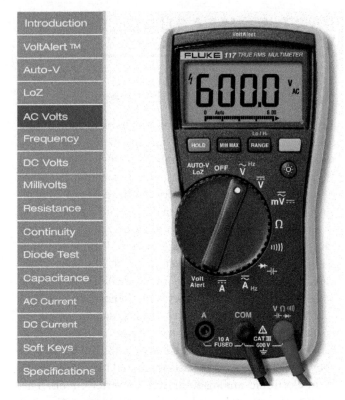

Figure 6-5 **Digital multimeter (DMM).**
Photo: ©Fluke Corporation

Many digital meters are **auto-ranging;** that is, the meter itself **adjusts to the range** needed for the circuit or device it is connected to. Other digital meters require the operator to select the proper range. It is important to understand the symbols used with digital multimeters for purposes of circuit connections and interpretation of readings, as illustrated in Figure 6-6. The electrical units of measure used are:

- M for mega or million
- K for kilo or thousand
- m for milli or one-thousandth
- μ for micro or one-millionth

∼	AC	⊣⊢	Capacitor
⎓	DC	μF	Microfarad
Hz	Hertz	μ	Micro
+	Positive	m	Milli
−	Negative	M	Mega
Ω	Ohms	K	Kilo
⊣▷⊢	Diode	OL	Overload
)))))	Audible continuity	COM	Common

Figure 6-6 **Symbols used with digital multimeters.**

6.3 Measuring Voltage

One of the most basic tasks of a multimeter is measuring voltage. Voltmeters are connected in **parallel** to points in circuit across which voltage needs to be measured. All digital meters contain a battery to power the display, so they use virtually no power from the circuit under test. This means that on their voltage ranges they have a very high resistance (usually called input impedance) of 1 MΩ or more, usually 10 MΩ, and they are very unlikely to affect the loading of the circuit under test. Due to their lower-input impedance, analog multimeters tend to **load the circuit under test** when used for voltage measurements. For example, taking a voltage reading across a high-impedance load can affect the circuit and display a lower-than-actual voltage reading.

Figure 6-7 shows a digital multimeter set to measure an AC voltage. In general, the procedure to follow when making voltage measurements is:

- Select V~ (AC) or V (DC), as required for the circuit.
- Plug the black test probe into the COM input jack.
- Plug the red test probe into the V input jack.
- If the DMM has manual ranging only, select the highest range so as not to overload the input.
- Touch the probe tips to the circuit across a load or power source (in parallel to the circuit).
- View the reading, being sure to note the unit of measurement.
- **Note:** AC voltage does not have polarity.
- **Caution:** Do not let fingers touch the lead tips or allow the tips to contact one another.

Digital multimeters will **automatically** indicate the **correct polarity** of a DC voltage measurement, as illustrated in Figure 6-8. When the positive lead of the meter

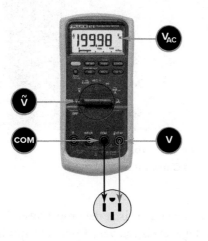

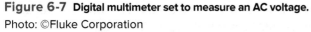

Figure 6-7 Digital multimeter set to measure an AC voltage.
Photo: ©Fluke Corporation

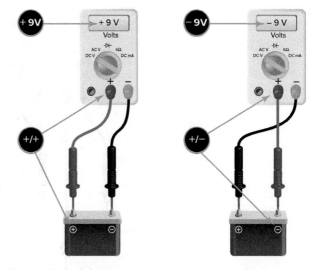

Figure 6-8 Digital multimeter polarity identification.

(V input jack) is connected to the positive point of the circuit, the meter will indicate a positive (+) polarity on the digital display. When the positive lead of the meter is connected to the negative point of the circuit, the meter will indicate a negative (−) polarity on the digital display.

Analog voltmeters must be **connected with the correct polarity**. The negative (−) voltmeter lead connects to the negative (−) side of the circuit, and the positive (+) voltmeter lead connects to the positive (+) side of the circuit. If the leads are reversed, the needle will move off scale to the left of zero and the movement may be damaged.

Voltage drop is the loss of voltage caused by the flow of current through a resistance. The greater the resistance, the greater the voltage drop. To check the voltage drop, use a voltmeter connected between the points where the voltage drop is to be measured. In DC circuits and AC **resistive** circuits the total of all the voltage drops across series-connected loads **should add up** to the voltage applied to the circuit, as shown in Figure 6-9. Each load device must receive its rated voltage to operate properly. If not enough or too much voltage is available, the device may not operate as it should.

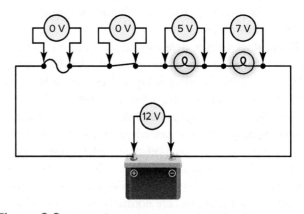

Figure 6-9 Measuring voltage drops in a circuit.

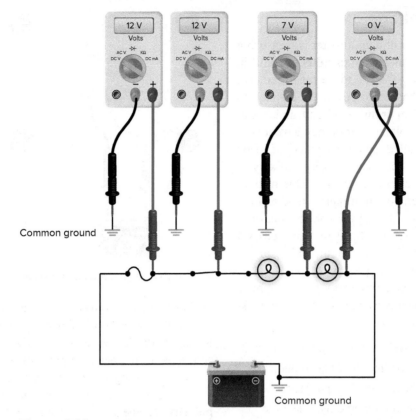

Figure 6-10 Ground-referenced voltage measurements.

Measurement of voltage drops is useful in determining the voltage that exists across individual components of a circuit. Another method of measuring voltage is to measure voltage with respect to a common, or a ground, point, as illustrated in Figure 6-10. **Ground-referenced** voltage measurement is done by first connecting the black common test probe of the multimeter to the circuit ground or common point of the circuit. Next, touch the red probe (V input jack) to the point in the circuit you want to measure. This will measure the voltage of this point **relative** to the circuit's ground potential.

The **voltage tester,** shown in Figure 6-11, is a **ruggedly** constructed voltmeter ideally suited for on-the-job voltage measurements. A voltage tester indicates the **approximate** level of voltage present and not the exact value. It is primarily used to test for the presence or absence of a voltage.

Noncontact voltage detectors are designed for live and not live voltage detection of voltage on **AC circuits.** Voltage detectors provide a critical safety function by telling you whether or not a circuit is live before you work on it, as illustrated in Figure 6-12. **Always verify that the voltage detector is working properly** before you rely on it. Use the detector to test a known live circuit both before and after you test an unknown circuit, and make sure it gives you the proper response. The same practice applies to voltage testers and multimeters.

Voltage measurements are normally taken to establish that voltage exists at a given point in the circuit and to

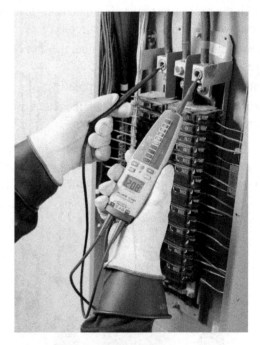

Figure 6-11 Voltage tester.
©Fluke Corporation

confirm that the voltage is at the proper level. In many instances, voltage levels can vary from their specified value yet cause no problems in a circuit. The measured voltage value should be within the specified range.

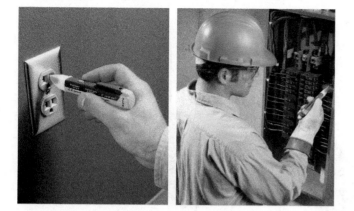

Figure 6-12 Noncontact voltage detector.
©Fluke Corporation

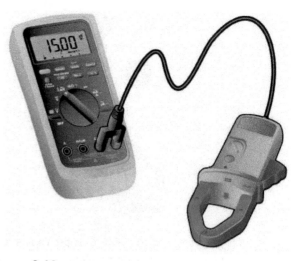

Figure 6-14 Digital multimeter current probe.
©Fluke Corporation

6.4 Measuring Current

Current measurements taken with a multimeter require setting the meter to measure current and placing the meter in **series** with the circuit being measured. This requires **opening the circuit** and using the multimeter test leads to complete the circuit. As a result, all the circuit current flows through the multimeter circuitry, and the resistance of the circuit limits the current flow through the meter. The meter itself has a very **low internal resistance** value which allows normal current to flow during the metering process.

Figure 6-13 shows a digital multimeter set to measure a DC current. In general, the procedure to follow when making current measurements is:

- Turn off power to the circuit.
- Open the circuit, creating a place where the meter probes can be inserted in series.
- Select A~ (AC) or A (DC) as desired.
- Plug the black test probe into the COM input jack.
- Plug the red test probe into the amp (A) input jack.

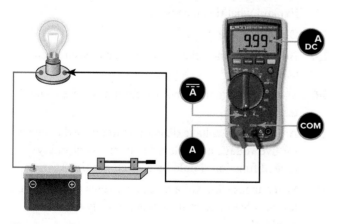

Figure 6-13 Digital multimeter set to measure a DC current.
Photo: ©Fluke Corporation

- Connect the probe tips to the circuit across the open in the circuit so that all current will flow through the multimeter (a series connection).
- Turn the circuit power back on.
- View the reading, being sure to note the unit of measurement.
- If the test leads are reversed for a DC measurement, a – (minus sign) will show in the display.
- *When finished with the measurement, it is best to place the test probe back to the voltage measurement position. Failure to do so, and inadvertent measuring of a voltage, will create a short circuit because of the low resistance of the multimeter when set to measure current.*

An indirect method of measuring current on a digital multimeter can be performed using a **current probe,** as shown in Figure 6-14. The probe clamps around the outside of the conductor, thus avoiding opening the circuit and connecting the meter in series. It measures current indirectly by measuring the **strength of the magnetic field** generated by a current conductor. The **clamp-on ammeter,** shown in Figure 6-15, works on the same principle and is often used for higher-current applications. **Note:** Current flowing in opposite directions cancels each other. If current is moving in opposite directions, place just one conductor within the clamp during measurement.

The remote display digital multimeter, shown in Figure 6-16, contains a **wireless display.** Special features of this type of multimeter include:

- Low-power wireless technology allows the display to be carried up to 10 meters (33 ft) away from the point of measurement for added flexibility.
- The removable magnetic display can be conveniently mounted where it is easily seen.

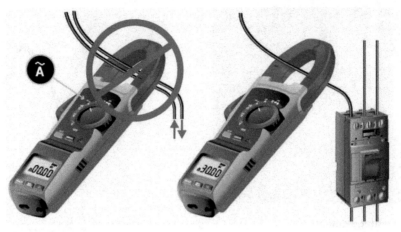

Figure 6-15 Clamp-on ammeter.
©Fluke Corporation

Figure 6-16 Remote display digital multimeter.
©Fluke Corporation

- Make measurements without holding the meter to improve visual focus on probes and augment safe electrical measurements.
- Take measurements in hard-to-reach places or in user-prohibited areas such as clean rooms and hazardous areas.
- Use as a conventional multimeter when the display is connected.

Part 1 Review Questions

1. Explain how measurements are made using an analog meter movement.

2. Explain how measurements are made using a digital meter.

3. Name the three basic metering functions that can be performed using a multimeter.

4. What is the purpose of the multimeter function switch?

5. What is the purpose of the multimeter range switch?

6. An analog voltmeter uses a single scale calibrated from 1 to 10 for several different ranges of voltage. What is the correct voltage reading if the needle of the scale points to 5 and the range switch is set to 100?

7. Unlike analog multimeters, digital multimeters require an ON/OFF power switch. Why?

8. In what way does an auto-ranging digital multimeter differ from the more traditional type?

9. How are voltmeters connected relative to the circuit voltage that needs to be measured?

10. Digital voltmeters have little or no loading effect on the circuit under test. Why?

11. A digital multimeter is used to measure a DC supply voltage and displays a minus voltage reading. Explain what this means.

12. Define *voltage drop*.

13. How are ground-referenced voltage measurements carried out?

14. State one advantage and one limitation of a voltage tester.

15. A noncontact voltage detector is to be used to check for live voltage at an electric outlet. What safety-related function should be carried out first?

16. When measuring voltages and currents of unknown levels, what meter range-switch setting should be used? Why?

17. How must ammeters be connected relative to the circuit current that needs to be measured?

18. Ammeters are required to have very low resistance values. Why?

19. What is the advantage of taking a current reading with a DMM current probe?

20. Explain how a clamp-on ammeter measures current flow through a conductor.

21. While attempting a voltage measurement across a power supply, the two test lead tips accidently contact one another. As a result, what type of circuit fault would be created?

22. A clamp-on ammeter is to be used to determine the current flow to a 120-V portable electric space heater. What problem would you encounter if you attempt to measure the current by placing the cord jacket containing both power supply conductors within the clamp?

PART 2 METER RESISTANCE MEASUREMENT, SAFETY, AND SPECIFICATIONS

6.5 Measuring Resistance

The **ohmmeter** is used to measure the amount of **electric resistance** offered by a complete circuit or a circuit component. A series analog ohmmeter circuit is shown in Figure 6-17. The internal circuit is made up of an analog meter movement, battery, fixed resistor, and variable

zero-adjust resistor, all connected in series. The scale ranges form zero (input leads short-circuited) to infinity (input leads open).

The principle of operation of the **analog ohmmeter** is based on Ohm's law and can be summarized as follows:

- Current is made to flow through an unknown resistance.
- According to Ohm's law, the amount of current flow will be **inversely** proportional to the resistance value.
- The amount of current measured by the meter is an indication of the unknown resistance so the scale can be calibrated in ohms.
- When there is infinite resistance (open between test leads), there is zero current through the meter movement, and the needle points toward the far left of the scale.
- When the test leads are directly shorted together (**measuring zero Ω**), the meter movement will have a maximum amount of current through it, limited only by the battery voltage and the movement's internal resistance.
- Series analog ohmmeters also have **nonlinear** scales, expanded at the low end of the scale and compressed at the high end to be able to provide for a range from zero to infinite resistance.
- It is important that you have good contact between the test leads and circuit you're testing. Dirt, oil, bodily contact, and poor test lead connection can significantly increase resistance readings.

Before you take a reading with an analog ohmmeter, the meter scale must be **set for zero** as follows:

- Touch the ohmmeter's probe leads together.
- Zero the meter by turning the resistance zero adjustment knob until the needle points to zero, at the right extreme of the meter scale.

If the battery is dead or weak the ohmmeter scale won't adjust to zero, and the battery will have to be replaced.

Digital multimeters can measure resistance down to 0.1 Ω, and as high as 300 MΩ (300,000,000 ohms). Infinite resistance (open circuit) is indicated as **OL** on some digital meter displays, which means the resistance is greater than the meter can measure. Both digital and analog ohmmeters are self-powered by a battery located within the meter itself and can *be damaged if connected to a live circuit.*

Figure 6-18 shows a digital multimeter set to measure resistance. In general, the procedure to follow when making resistance measurements is:

- Check that all power is OFF.
- Set the function switch to resistance (Ω).
- Plug the black test probe into the COM input jack.

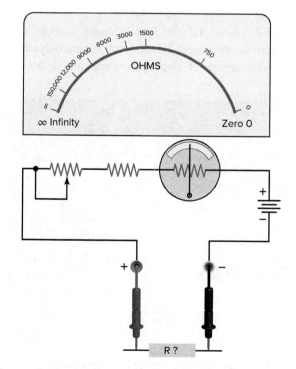

Figure 6-17 Series analog ohmmeter circuit.

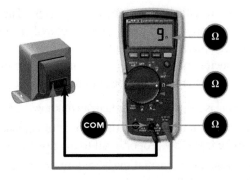

Figure 6-18 Digital multimeter set to measure resistance.
Photo: ©Fluke Corporation

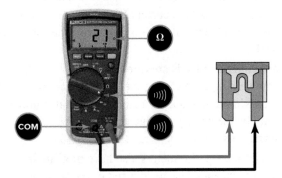

Figure 6-19 Digital multimeter set to test the continuity of a fuse.
Photo: ©Fluke Corporation

- Plug the red test probe into the Ω input jack.
- For in-circuit measurements, disconnect, if possible, one lead end of the component to open any parallel paths so only the resistance of the single component is measured.
- Connect the probe tips across the component or portion of the circuit for which you want to determine resistance.
- View the reading, being sure to note the unit of measurement—ohms (Ω), kilohms (kΩ), or megohms (MΩ).
- After completing all measurements, turn the meter OFF to prevent possible battery drain.
- For accurate, low-resistance measurements, resistance in the test leads must be subtracted from the total resistance measured. Typical test lead resistance should be between 0.2 Ω and 0.5 Ω.

In addition to measuring resistance, an ohmmeter is used to make continuity tests. **Continuity** is a quick resistance test that distinguishes between **an open and a closed circuit.** In Figure 6-19 a digital multimeter set to test the continuity of a fuse. The continuity beeper allows you to complete many continuity tests easily and quickly. The meter beeps when it detects a closed circuit, so you don't have to look at the meter

as you test. A continuous tone sounds if the resistance between the terminals is less than a specified level.

6.6 Multimeter Safety

The occurrence and levels of **transient overvoltages** on power systems have given rise to more stringent safety standards for electrical measurement equipment. As an example, lightning strikes on outdoor transmission lines also cause extremely hazardous high-energy transients. The **International Electromechanical Commission (IEC) 1010** standard defines four energy-level categories (Figure 6-20) as follows:

- *CAT IV,* the primary supply level, is the highest level of power and covers utility lines, transformers, and lines that come into a building. This is the most dangerous level of transient overvoltage that electrical workers are likely to encounter while working with utility service to a facility. For this category, the IEC calls for overvoltage protection of up to **12,000 V.**
- *CAT III* includes feeders, short branch circuits, distribution panel devices, or heavy appliances with short connections to the service entrance. For this category, the IEC calls for overvoltage protection of up to **8,000 V.** *For the electrical trade all test instruments should carry a Category III (or higher) rating.*
- *CAT II* is the local level for fixed and nonfixed power devices. Outlets that are some distance away from a Measurement Category IV or III source are rated Measurement Category II. The required overvoltage protection is up to **6,000 V.**
- *CAT I* is the signal level for telecommunications and electronic equipment. Instruments that are only rated for Measurement Category I are not suitable for

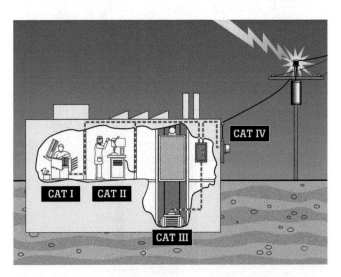

Figure 6-20 Overvoltage energy-level categories.

measurements on an electrical distribution system. This category is for measurements of voltages from specially protected secondary circuits. Such voltage measurements include signal levels, special equipment, limited-energy parts of equipment, circuits powered by regulated low-voltage sources, and electronics.

Poorly made, worn, or underrated **test leads** can cause inaccurate readings and may pose a serious shock or electrocution hazard if you touch live wires that the meter has read as being deenergized. Procedures to verify the condition and rating of your leads include:

- Visual inspection of the insulation, probe handles, and connections of the leads. Check to make sure the insulation is not nicked or cracked.
- Conduct a simple resistance measurement of the test probes to confirm they are electrically reliable. Place your DMM in the resistance (ohms) function, plug the leads into the DMM, and touch the probe tips together—red to black, as illustrated in Figure 6-21. The meter should read about 0.5 ohms or less for good-quality test leads.

Tips to make electrical measurements safely include:

- Never use the ohmmeter on a live circuit.
- Never connect an ammeter in parallel with a voltage source.
- Never overload an ammeter or voltmeter by attempting to measure currents or voltages in excess of the range-switch setting.
- Start out with high ranges when the value of the measurement is unknown.

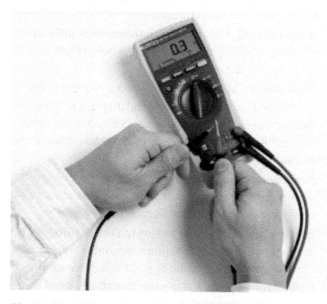

Figure 6-21 Test probe resistance measurement.
©Fluke Corporation

Figure 6-22 Personal protection equipment (PPE).
©Fluke Corporation

- Make certain that any terminals you are measuring across are not accidentally shorted together, or to ground, by the test leads.
- Never measure unknown voltages in *high-voltage* circuits. Refer to the technical information or the manufacturer for a reference voltage before proceeding.
- Avoid touching the bare metal clips or tips of test probes.
- Whenever possible, remove voltage before connecting meter test leads into the circuit.
- To reduce the danger of accidental shock, disconnect meter test leads immediately after the test is completed.
- Use equipment rated for the test environment.
- NFPA standard 70E provides detailed guidelines about when and where approved personal protection equipment (PPE) should be used when making electrical measurements. This may include eye and hearing protection, insulated hand tools, insulated gloves, and fire resistant clothing, as shown in Figure 6-22.

6.7 Multimeter Specifications

Most **digital multimeters** feature other functions beyond the basic voltage, current, and resistance measurement. When selecting a multimeter, it is important to be sure that the meter's capabilities will cover the type of test procedures you usually do. Some important specifications and features to consider are as follows:

Input impedance Impedance refers to the combined opposition to current created by the resistance, capacitance, and inductance. A meter with **low impedance** can draw enough current to cause an inaccurate measurement of voltage drop. A **high-impedance** meter will draw little current, ensuring accurate readings. Typical analog voltmeters have impedance values of 20,000 to 30,000 Ω per volt. Some electronic systems and components may be damaged or give inaccurate results when such meters are used. Most digital voltmeters have a 10-MΩ (or 10 million ohms) of input impedance, which results in **little or no loading effect** on a circuit while measurements are being taken.

Accuracy and resolution The **accuracy** of a multimeter represents the maximum amount of error that occurs when it takes a measurement. Simple accuracy specifications are given as a plus/minus percentage of the full scale. Digital meters have another type of rating, this one more commonly called resolution. **Resolution** refers to the smallest numerical value that can be read on the display of a digital meter. Factors that determine resolution are the number of digits displayed and the number of ranges available for each function. One of the most common DMM displays is called a 3½-digit display, because it shows three full digits (0–9), preceded by a so-called half-digit. At this position, the meter can only display 0 or 1. Thus, the largest number that can be displayed on a 3½-digit meter is 1,999. This is also called a 2,000-count display, because the actual count may be from 0,000 to 1,999.

Overload Protection Fuses protect against overcurrent. High-input impedance of the volts/ohms terminals of digital multimeters ensures that an overcurrent condition is unlikely, so fuses aren't necessary. A **fuse** connected in series with the input lead usually protects the resistance and current-measuring circuitry. Safety precautions, involving the possible misuse of meters, have lead to the use of a second **high-energy fuse** within the meter circuit. High-energy fuses provide greater protection from high-voltage transients (Figure 6-23). In high-voltage environments, a regular fuse (typically in glass casing) will not be able to withstand a surge of high energy and thus will explode. The filler material contained in a high-energy fuse is designed to prevent explosions from happening. High-energy fuses are usually longer, and the greater length also helps in preventing arc flashes from forming.

Combination digital and analog display Analog meters are particularly suited for trend observation, as in slowly changing voltage levels. Some digital multimeters use a combination display that includes a **bar graph,** similar to that shown in Figure 6-24, which

Figure 6-23 High-energy fuses installed in a multimeter.
©Fluke Corporation

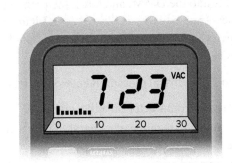

Figure 6-24 Bar graph display.

simulates an analog needle for watching changing signals or for circuit adjustments.

Auto-ranging Auto-ranging **automatically adjusts** the meter's measuring circuits to the correct voltage, current, or resistance ranges.

Hold feature Many digital multimeters have a HOLD **button** that captures a reading and displays it from **memory** even after the probe has been removed from the circuit. This is particularly useful when making measurements in a confined area.

Response time Response time is the number of seconds a digital multimeter requires for its electronic circuits to settle to their rated accuracy.

Battery life Whereas the analog multimeter simply draws power from the circuit under test for voltage and current measurements, the digital multimeter requires a battery to operate. Battery life is normally rated in hours. With some multimeters, if the meter is ON, but inactive for a specified period of time, the display goes blank to preserve battery life.

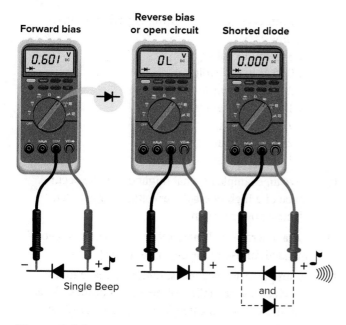

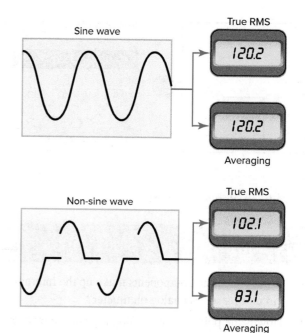

Figure 6-25 Diode test function.

Figure 6-26 True RMS and averaging meters.

Auto polarity With the automatic polarity feature, a + or − activated on the digital display indicates the polarity of DC measurements and eliminates the need for reversing leads.

Diode test This test is used to check the forward and reverse bias of a **semiconductor diode,** as illustrated in Figure 6-25. Typically, when the diode is connected in **forward bias,** the meter displays the forward voltage drop and beeps briefly. When connected in **reverse bias** or open circuit, the meter displays OL. If the diode is shorted, the meter displays zero and emits a continuous tone when connected in forward or reverse bias.

Averaging or true RMS A **true RMS meter** responds to the effective heating value of an AC waveform. Meters having rectifier-type circuits have scales that are calibrated in RMS values for AC measurements, but actually are measuring the average value of the input voltage and are depending on the voltage to be a sine wave. Figure 6-26 shows a comparison between **true RMS and averaging** digital meter voltage readings. Where the AC signal approximates a pure sine wave, there will be little or no difference in the two readings. For a non-sine wave voltage the true-RMS meter reads higher than the averaging meter.

Multifunction multimeter The integrated-circuit (IC) chip revolution has helped combine the capabilities of other test instruments into a **multifunction multimeter.** Volts, ohms, and amps are the most often used functions; however, the multifunction digital meter allows the reading of **frequency** and **capacitance** measurements as well (Figure 6-27).

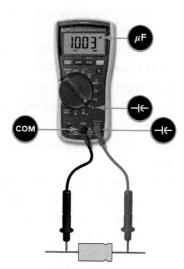

Figure 6-27 Multifunction multimeter set to measure capacitance.
Photo: ©Fluke Corporation

6.8 Virtual Multimeter

Virtual instrumentation is based on using the computer and associated software to create the measuring instrument. In principle, this allows users to create instruments at lower costs that are capable of higher performance.

Figure 6-28 shows a typical virtual multimeter used as one of the simulated instruments in the **Electronics Workbench Multisim** simulation package. This multimeter can be used to measure AC or DC voltage or current, or resistance, between any two points in the circuit. The multimeter is auto-ranging so a measurement range does not need to be specified.

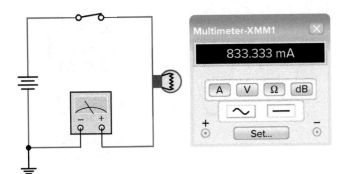

Figure 6-28 Virtual multimeter.

Part 2 Review Questions

1. What four basic components make up the internal circuit of a series analog ohmmeter?

2. How is the pointer of an analog-type ohmmeter set for zero?

3. Series analog ohmmeters have nonlinear scales. Why?

4. What does a resistance reading of OL on a digital multimeter indicate?

5. A multimeter set to measure resistance is inadvertently connected to a live circuit. What can this result in?

6. When doing in-circuit component resistance measurements, why is it advisable to disconnect one end lead of the component?

7. Explain the purpose of the multimeter continuity beeper test function.

8. An ohmmeter is connected across the two leads of a single-pole toggle switch to check its operation out of circuit. What reading you would expect to get if the:
   a. Switch is operating properly?
   b. Switch has an open circuit fault?
   c. Switch has a short circuit fault?

9. Convert each of the following digital multimeter readings:
   a. 340 mV to volts
   b. 0.75 V to millivolts
   c. 2 A to milliamps
   d. 1,950 mA to amps
   e. 7.5 Ω to kilohms
   f. 2.2 kΩ to ohms
   g. 1.5 MΩ to ohms

10. Give an example of what might cause the occurrence of a high-energy overvoltage surge on an electric power system.

11. What energy-level category of multimeter is recommended for use in the electrical trade?

12. A simple resistance measurement is made to test the electrical reliability of meter test probes. What reading would indicate that the probes are of good quality?

13. What type of hand protection may be required when making electrical measurements?

14. Compare the input impedance of analog and digital multimeters.

15. Define *meter accuracy.*

16. Define *meter resolution.*

17. What protection is afforded by the fuse connected in series with the input lead compared with that afforded by the high-energy fuse?

18. Explain the auto-ranging feature of a DMM.

19. The diode test function of a DMM is to be used to test a diode out of circuit. What reading on the meter display would indicate that the diode is shorted?

20. Under what condition will true RMS and averaging meters register the same reading of an AC voltage measurement?

CHAPTER SEVEN

Ohm's Law

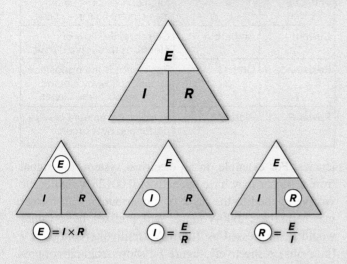

Ohm's law triangle.

Ohm's law is easily the most important formula for understanding and solving electric circuits. A simple equation summarizes the relationship between the values of current, voltage, and resistance in an electric circuit. Before George Simon Ohm discovered his now famous law in 1827, work with electric circuits was a hit-and-miss affair. This chapter also introduces you to the concept of electric power.

7.1 Metric Prefixes

Most often measurements made on electric circuits are that of current, voltage, resistance, and power. The base units—ampere, volt, ohm, and watts—are the values most commonly used to measure them. Table 7-1 lists these basic electrical quantities and the symbols that identify them.

In certain circuit applications the basic units—volt, ampere, ohm, and watt—are either too small or too big to express conveniently. In such cases metric prefixes are often used. Recognizing the meaning of a prefix reduces the possibility of confusion in interpreting data. Common metric prefixes are shown in Table 7-2.

A metric prefix precedes a unit of measure or its symbol to form decimal multiples or submultiples. **Prefixes** are used to reduce the quantity of zeros in numerical equiva-

lencies. For example, in an electrical system, the signal from a sensor may have strength of 0.00125 V, while the voltage applied to the input of a distribution transformer may be in the 27,000-V range. With prefixes, these values would be expressed as 1.25 mV (millivolts) and 27 kV (kilovolts), respectively. Figure 7-1 shows examples of prefixes used in the rating of electric components.

Table 7-1 Current, Voltage, and Resistance

Quantity	Measurement	Function
Voltage V, emf, or E	Volt V	Voltage is the electromotive force or pressure that creates current flow in a circuit.
Current I	Ampere A	Current is the flow of electrons through a circuit.
Resistance R	Ohm Ω	Resistance is the opposition to current flow by electric components in the circuit.
Power P	Watt W	Power is the amount of work performed by a circuit.

Table 7-2 Common Metric Prefixes

Number		Power	Prefix	Symbol
One billion	1,000,000,000	10^9	giga	G
One million	1,000,000	10^6	mega	M
One thousand	1,000	10^3	kilo	k
One	1	10^0	None	None
One-thousandth	0.001	10^{-3}	milli	m
One-millionth	0.000001	10^{-6}	micro	μ
One-billionth	0.000000001	10^{-9}	nano	n
One-trillionth	0.000000000001	10^{-12}	pico	p

4-GB flash memory
(4,294,967,296 bytes)

250-kcmil cable
(250,000 circular mil)

47-kΩ resistor
(47,000 ohms)

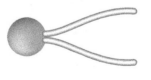

12 mV @ 300°C
thermocouple
(0.012 volts)

Smoke detector
standby current 45 μA
(0.000045 amperes)

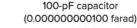

100-pF capacitor
(0.000000000100 farad)

Figure 7-1 Prefixes used in the rating of electric components.

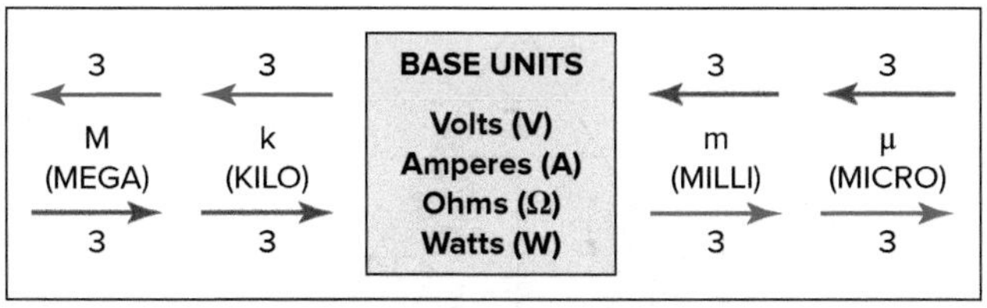

Figure 7-2 Movement of the decimal point to and from base units.

Knowing how to convert metric prefixes back to base units is needed when reading digital multimeters or using electric circuit formulas. Figure 7-2 and the following examples illustrate how many positions the decimal point is moved to get from a base unit to a multiple or a submultiple of the base unit.

EXAMPLE 7-1

To convert amperes (A) to milliamperes (mA), it is necessary to move the decimal point **three places** to the **right** (this is the same as multiplying the number by 1,000).

$$0.012 \text{ A} = ? \text{ mA}$$
$$0.012 \text{ A} = 0.012$$
$$0.012 \text{ A} = 12 \text{ mA}$$

EXAMPLE 7-2

To convert milliamperes (mA) to amperes (A), it is necessary to move the decimal point **three places** to the **left** (this is the same as multiplying by 0.001).

$$450.0 \text{ mA} = ? \text{ A}$$
$$450.0 \text{ mA} = 450.0$$
$$450.0 \text{ mA} = 0.45 \text{ A}$$

EXAMPLE 7-3

To convert volts (V) to kilovolts (kV), it is necessary to move the decimal point **three places** to the **left**.

$$47,000.0 \text{ V} = ? \text{ kV}$$
$$47,000.0 \text{ V} = 47,000.0$$
$$47,000.0 \text{ V} = 47.0 \text{ kV}$$

EXAMPLE 7-4

To convert from megohms (MΩ) to ohms (Ω), it is necessary to move the decimal point **six places** to the **right**.

$$2.2 \text{ MΩ} = ? \text{ Ω}$$
$$2.2 \text{ MΩ} = 2.200000$$
$$2.2 \text{ MΩ} = 2,200,000 \text{ Ω}$$

EXAMPLE 7-5

To convert from microamperes (μA) to amperes (A), it is necessary to move the decimal point **six places** to the **left**.

$$500 \text{ μA} = ? \text{ A}$$
$$500 \text{ μA} = 000500.$$
$$500 \text{ μA} = 0.0005 \text{ A}$$

Part 1 Review Questions

1. What is the base unit and symbol used for electric:
   a. Current
   b. Voltage
   c. Resistance
   d. Power

2. Write the metric prefix and symbol used to represent each of the following:
   a. One-thousandth
   b. One million
   c. One millionth
   d. One thousand

3. Convert each of the following:
   a. 2,500 Ω to kilohms
   b. 120 kΩ to ohms
   c. 1,500,000 Ω to megohms
   d. 2.03 MΩ to ohms
   e. 0.000466 A to microamps
   f. 0.000466 A to milliamps
   g. 378 mV to volts
   h. 475 Ω to kilohms
   i. 28 μA to amps
   j. 5 kΩ + 850 Ω to kilohms
   k. 40,000 kV to megavolts
   l. 4,600,000 μA to amps
   m. 2.2 kΩ to ohms

PART 2 MATHEMATICAL RELATIONSHIPS BETWEEN CURRENT, VOLTAGE, RESISTANCE, AND POWER

7.2 Ohm's Law

Electricity always acts in a predictable manner. By using different laws for electric circuits, we can predict what should happen in a circuit or diagnose why things are not operating as they should.

Ohm's law expresses the relationship between the voltage (E), the current (I), and the resistance (R) in a circuit. Ohm's law can be stated as follows:

The *current* (*I*) in a circuit is *directly proportional* to the applied *voltage* (*E*) and *inversely proportional* to the circuit *resistance* (*R*).

Mathematically, Ohm's law can be expressed in the form of three formulas: one basic formula and two others derived from it as follows:

$$\text{Current } (I) = \frac{\text{Voltage } (E)}{\text{Resistance } (R)} \quad \text{or} \quad I = \frac{E}{R}$$

$$\text{Voltage } (E) = \text{Current } (I) \times \text{Resistance } (R) \quad \text{or} \quad E = I \times R$$

$$\text{Resistance} = \frac{\text{Voltage } (E)}{\text{Current } (I)} \quad \text{or} \quad R = \frac{E}{I}$$

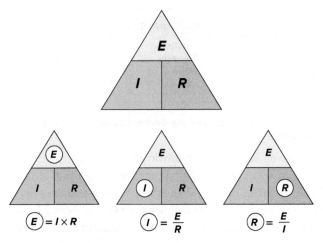

Figure 7-3 Ohm's law triangle.

The three variations of the Ohm's law formulas can be easily remembered by arranging the three quantities within a triangle, as shown in Figure 7-3. This arrangement represents the actual position of each quantity in the Ohm's law formulas. Using these three formulas and knowing any two of the values for voltage, current, or resistance, it is possible to find the third value.

When applying the mathematical equations of Ohm's law to a circuit, it is important to use the correct set of units. **Improperly mixing units will result in incorrect answers.** The Ohm's law equations work if you use the following *sets of units*:

- Volts (V), amperes (A), and ohms (Ω)
- Volts (V), milliamperes (mA), and kilohms (kΩ)
- Volts (V), microamperes (μA), and megohms (MΩ)

Electric Circuits	Electronic Circuits	Microelectronic Circuits
If	If	If
$E = 240$ V $R = 10\ \Omega$	$I = 8$ mA $R = 5$ kΩ	$I = 20\ \mu$A $E = 24$ V
Then, $I = \dfrac{E}{R}$ $= \dfrac{240\text{ V}}{10\Omega}$ $= 24$ A	Then, $E = I \times R$ $= 8\text{ mA} \times 5\text{ k}\Omega$ $= 40$ V	Then, $R = \dfrac{E}{I}$ $= \dfrac{24\text{ V}}{20\ \mu\text{A}}$ $= 1.2$ MΩ

7.3 Applying Ohm's Law to Calculate Current

By using Ohm's law, you are able to find the current of a circuit while knowing only the voltage and the resistance of the circuit.

EXAMPLE 7-6

A portable electric heater with a resistance of 15 Ω is directly connected to a 120 VAC electric outlet, as shown in Figure 7-4. The expected current flow in this circuit can be calculated as follows:

$$\text{Current} = \text{Voltage} \div \text{Resistance}$$
$$I = E \div R$$
$$= 120\text{ V} \div 15\ \Omega$$
$$= 8\text{ A}$$

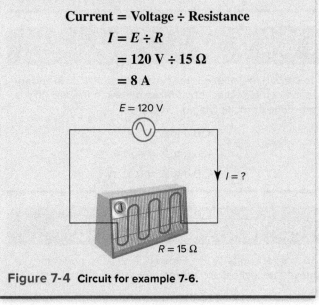

Figure 7-4 Circuit for example 7-6.

Electronic and microelectronic circuits operate at much lower current values. This is mainly because they usually contain much higher resistance values. If the resistance of these circuits is expressed in kilohms or megohms, and the voltage in volts, then the current can be calculated directly in milliamperes or microamperes as follows:

$$\text{mA} = \text{Volts} \div \text{k}\Omega$$
$$\text{or}$$
$$\mu\text{A} = \text{Volts} \div \text{M}\Omega$$

EXAMPLE 7-7

A 10-kΩ resistor is connected to a 12-VDC battery, as shown in Figure 7-5. The expected current flow in this circuit can be calculated as follows:

$$I = E \div R$$
$$= 12 \text{ V} \div 10 \text{ k}\Omega$$
$$= 1.2 \text{ mA}$$

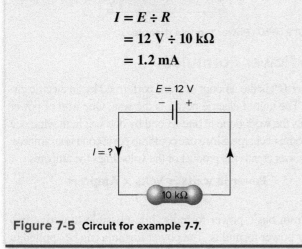

Figure 7-5 Circuit for example 7-7.

7.4 Applying Ohm's Law to Calculate Voltage

When the current flow and resistance of a circuit are known, the voltage being applied can be calculated using the formula $E = I \times R$.

EXAMPLE 7-8

A solar panel is delivering a current of 2.5 A to a load that has a resistance of 50 Ω, as shown Figure 7-6. The voltage output of the solar panel can be calculated as follows:

$$\text{Voltage} = \text{Current} \times \text{Resistance}$$
$$E = I \times R$$
$$= 2.5 \text{ A} \times 50 \text{ } \Omega$$
$$= 125 \text{ V}$$

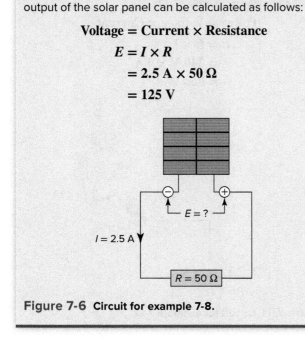

Figure 7-6 Circuit for example 7-8.

For low-current electronic and microelectronic circuits, if the resistance of these circuits is expressed in kilohms or megohms, and the current in milliamperes or microamperes, then the voltage can be calculated directly as follows:

$$\text{Volts} = \text{mA} \times \text{k}\Omega$$

or

$$\text{Volts} = \mu\text{A} \times \text{M}\Omega$$

EXAMPLE 7-9

A thermopile consists of an array of thermocouples connected in series to provide a higher output voltage. The thermopile of Figure 7-7 provides a current of 2.5 mA to a 0.5-kΩ gas solenoid valve. The voltage output of the thermopile is

$$E = I \times R$$
$$= 2.5 \text{ mA} \times 0.5 \text{ k}\Omega$$
$$= 1.25 \text{ V}$$

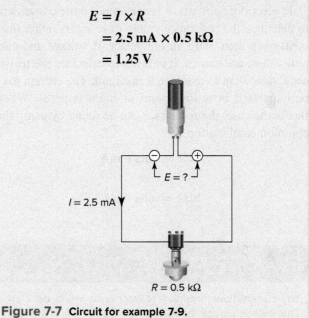

Figure 7-7 Circuit for example 7-9.

7.5 Applying Ohm's Law to Calculate Resistance

The resistance of a circuit can be calculated when the applied and the current flow through it are known and using the formula $R = E \div I$.

EXAMPLE 7-10

An electric toaster draws a current of 8 amperes when connected to 120-VAC outlet, as shown in Figure 7-8. The resistance of the toaster heating element would be

$$\text{Resistance} = \text{Voltage} \div \text{Current}$$
$$R = E \div I$$
$$= 120 \text{ V} \div 8 \text{ A}$$
$$= 15 \text{ } \Omega$$

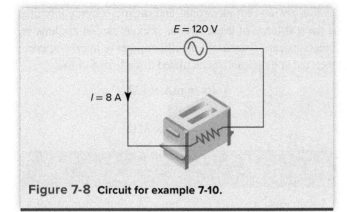

Figure 7-8 Circuit for example 7-10.

In electronic circuits, it is sometimes more convenient to calculate the resistance values of resistors rather than measuring them with an ohmmeter. If voltage and current values are known, it is faster to calculate the resistance value than to measure it in circuit. The current may be expressed in milliamperes or microamperes. When this is the case, the resistance can be found by using the common combination of

$$k\Omega = volts \div mA$$

or

$$M\Omega = volts \div \mu A$$

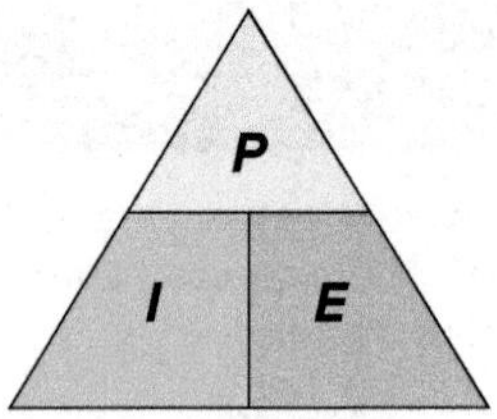

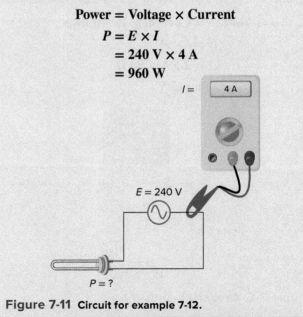

Figure 7-10 Power formula triangle.

7.6 Power Formulas

Power (P) is the amount of work performed by an electric circuit. The unit of electric power is the watt. One *watt* of power equals the work done in one second by one volt in moving one coulomb of charge. Since one coulomb per second is an ampere, the power equals the product of the volts times the amperes.

Power in watts = Volts × Amperes
$$P = E \times I$$

From basic power formula, it is possible to derive two other power formulas. The power formula can be applied in any of the following three ways depending on whether you want to calculate *P, I,* or *E*. The three variations of the power formulas can be easily remembered by arranging the three quantities within a triangle as shown in Figure 7-10.

$$P = E \times I$$
$$I = P \div E$$
$$E = P \div I$$

EXAMPLE 7-11

The current flow through a resistor is known to be 2 μA. The voltage across the resistor is measured and found to be 9 volts, as shown in Figure 7-9. The resistance value of the resistor would be

$$R = E \div I$$
$$= 9\,V \div 2\,\mu A$$
$$= 4.5\,M\Omega$$

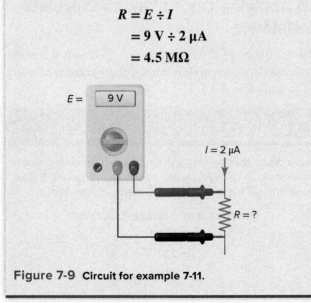

Figure 7-9 Circuit for example 7-11.

EXAMPLE 7-12

An immersion electric water heater element draws a current of 4 amperes when connected to its rated voltage of 240 volts, as shown in Figure 7-11. The rated power of the heater element would be

Power = Voltage × Current
$$P = E \times I$$
$$= 240\,V \times 4\,A$$
$$= 960\,W$$

Figure 7-11 Circuit for example 7-12.

EXAMPLE 7-13

An incandescent lamp is rated for 250 watts when operated at 120 volts, as shown in Figure 7-12. The operating current of the lamp would be

$$\textbf{Current} = \textbf{Power} \div \textbf{Voltage}$$
$$I = 250 \text{ W} \div 120 \text{ V}$$
$$= 2.08 \text{ A}$$

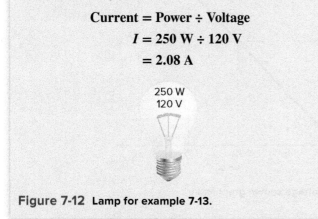

Figure 7-12 Lamp for example 7-13.

Figure 7-13 shows an expanded formula chart for finding values of voltage, current, resistance, and power. To use this chart, from the center circle, select the value you need to find (P, R, I or E). Then select the formula containing the values you know from the corresponding chart quadrant.

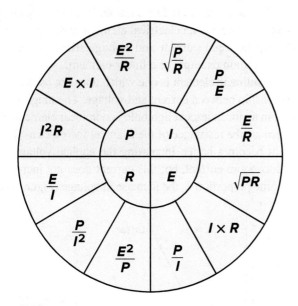

Figure 7-13 Formula chart for finding values of *P, I, E,* and *R.*

EXAMPLE 7-14

A current of 30 A is being supplied to an electric range, as shown in Figure 7-14. If the total resistance of the conductors used to supply this current is 0.1 Ω, the power that is lost in the wire would be

$$\textbf{Power} = \textbf{Current}^2 \times \textbf{Resistance}$$
$$P = I^2 R$$
$$= 30 \text{ A} \times 30 \text{ A} \times 0.1 \text{ }\Omega$$
$$= 90 \text{ W}$$

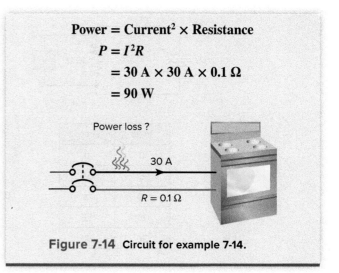

Figure 7-14 Circuit for example 7-14.

The **wattage** rating of a resistor indicates the maximum amount of power a resistor can dissipate as heat before drastically changing value or burning up. To keep a resistor from becoming overheated, its wattage rating should be about twice the wattage computed from a power formula.

EXAMPLE 7-15

A 100-Ω resistor required to be wired in circuit across a 6-V source, as shown Figure 7-15. The calculated wattage required to be dissipated by the resistor would be

$$\textbf{Power} = \textbf{Voltage}^2 \div \textbf{Resistance}$$
$$P = E^2 \div R$$
$$= 6 \text{ V} \times 6 \text{ V} \div 100 \text{ }\Omega$$
$$= 0.36 \text{ W}$$

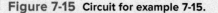

Figure 7-15 Circuit for example 7-15.

To keep a resistor from becoming overheated, its wattage rating should be about twice the wattage rating computed from a power formula. Thus, the resistor used in this circuit should have a wattage rating of 1 W.

Ohm's law can be used to diagnose why circuits may not be working as they should. For example, if the specified operating voltage to a load is changed both the current and wattage will change but the resistance will remain the same.

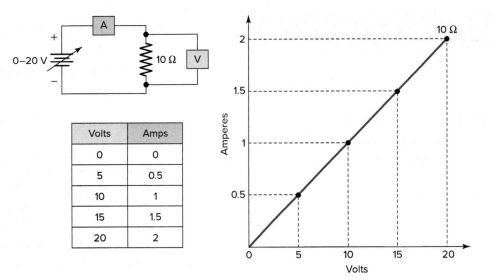

Volts	Amps
0	0
5	0.5
10	1
15	1.5
20	2

Figure 7-16 Relationship between current and voltage shown graphically.

EXAMPLE 7-16

An electric water heater element is rated for 1440 W and 240 V.

a. Calculate the current rating of the element.

$$\text{Current} = \text{Power} \div \text{Voltage}$$
$$I = 1440\ \text{W} \div 240\ \text{V}$$
$$= 6\ \text{A}$$

b. Calculate the resistance of the element.

$$\text{Resistance} = \text{Voltage} \div \text{Current}$$
$$R = E \div I$$
$$= 240\ \text{V} \div 6\ \text{A}$$
$$= 40\ \Omega$$

c. Calculate the current of the element if it is operated at 200 V.

$$\text{Current} = \text{Voltage} \div \text{Resistance}$$
$$I = E \div R$$
$$= 200\ \text{V} \div 40\ \Omega$$
$$= 5\ \text{A}$$

d. Calculate the wattage of the element if it operated at the 200-V level.

$$\text{Power} = \text{Voltage} \times \text{Current}$$
$$P = E \times I$$
$$= 200\ \text{V} \times 5\ \text{A}$$
$$= 1000\ \text{W}$$

7.7 Ohm's Law in Graphical Form

The relationship between current and voltage is shown graphically in Figure 7-16. The current increase is directly proportional to the voltage increase. Each time the voltage increases by 5 volts, the current increases by 0.5 amperes.

A linear function in mathematics is one that tracks a straight line when plotted on a graph. In an electric circuit, a **linear** element is an electrical element with a linear relationship between current and voltage. Resistors are the most common example of a linear element.

A **nonlinear** element is one which does not have a linear relationship between current and voltage. The tungsten filament in an incandescent lightbulb is nonlinear element. This is because the resistance of the filament increases as the filament becomes hotter. Increasing the applied voltage does produce more current, but this current does not increase in the same proportion as the increase in voltage (Figure 7-17).

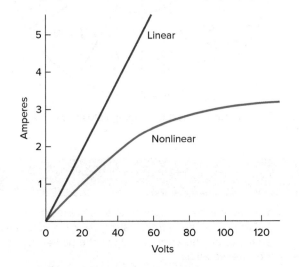

Figure 7-17 Linear and nonlinear elements.

1. State Ohm's law.

2. List the three formulas associated with Ohm's law.

3. Calculate the unknown value for each circuit (a) through (l) listed in the following table.

	Current	Resistance	Voltage
(a)	$I = ?$	$R = 6\ \Omega$	$E = 12\ V$
(b)	$I = 0.1\ A$	$R = 120\ \Omega$	$E = ?$
(c)	$I = ?$	$R = 10\ \Omega$	$E = 120\ V$
(d)	$I = 20\ A$	$R = ?$	$E = 24\ V$
(e)	$I = 0.001\ A$	$R = 30\ \Omega$	$E = ?$
(f)	$I = 0.0005\ A$	$R = ?$	$E = 40\ V$
(g)	$I = ?$	$R = 1.5\ k\Omega$	$E = 12\ V$
(h)	$I = 24\ mA$	$R = ?$	$E = 12\ V$
(i)	$I = 1.2\ mA$	$R = 12\ k\Omega$	$E = ?$
(j)	$I = ?$	$R = \tfrac{1}{2}\ \Omega$	$E = 12\ V$
(k)	$I = 40\ \mu A$	$R = ?$	$E = 3\ V$
(l)	$I = 0.02\ mA$	$R = 0.5\ M\Omega$	$E = ?$

4. A pilot lamp with 12 volts applied to it draws a current of 3 amperes. What is the hot resistance of the bulb filament? (*Hot resistance* refers to the resistance of the bulb filament under normal operating condition and is much higher than its cold or out-of-circuit resistance.)

5. A resistor has a resistance of 220 Ω. The current flow through the resistor is measured and found to be 30 mA. What is the value of the voltage across the resistor?

6. An electric soldering iron with a 40-Ω heating element is plugged into a 120-V outlet. How much current will be drawn by the iron?

7. A baseboard heater draws a current of 8 A when connected to a 240-V source. What is the resistance value of the heater element?

8. A sensor module has an internal resistance of 50,000 ohms. What is its normal operating current, in microamperes, if the applied working voltage is 6 VDC?

9. The current through and the voltage drop across a precision resistor are measured and found to be 8 mA and 1.5 V, respectively. What is the resistance value of the resistor?

10. The heating element for an electric floor heating system draws 2 A when connected to a 120-VAC source. What is the amount of power dissipated by the system?

11. How much power is lost in the form of heat when 25 A of current flows through a conductor that has 0.02 Ω of resistance?

12. The voltage drop across a 330-Ω resistor is measured and found to be 9 V. How much wattage is being dissipated by the resistor?

13. A digital multimeter display indicates a reading of 150.0 kΩ. Express this reading in ohms and megohms.

14. Calculate the unknown value for each circuit (a) through (c) listed in the following table.

	Current	Resistance	Voltage	Power
(a)	100 mA	?	250 V	?
(b)	?	$6\ \Omega$	120 V	?
(c)	3 A	$40\ \Omega$	?	?

15. Assume that applying 120 volts to a resistive load results in a current flow of 10 amperes. What will the value of the current flow be if:
    a. The voltage is increased to 240 V?
    b. The voltage is decreased to 60 V?
    c. The load resistance is doubled?
    d. The load resistance is halved?

16. The voltage drop across a 10-W, 330-Ω resistor is measured and found to be 48 V. Is the wattage being dissipated by the resistor within its normal operating range? Why?

17. The cold resistance of a tungsten lamp filament is measured using an ohmmeter and found to be much lower than the hot resistance calculated using Ohm's law and the measured voltage and current values. How can this be possible?

18. An electric baseboard heater is rated for 1200 W and 120 V.
    a. Calculate the current rating of the heater element.
    b. Calculate the resistance of the heater element.
    c. Calculate the current of the heater element if it is operated at 105 V.
    d. Calculate the wattage of the element if it operated at the 105-V level.

19. A lamp dimmer switch is rated for 5 A at 120 V. What is the maximum wattage lamp that can be safely connected to the dimmer?

20. A 400-W 4-Ω public address speaker is to be fused for overload protection. What maximum current rating of a fuse should be used? (**Hint:** $P = I^2 R$.)

CHAPTER EIGHT

Resistors

Close up electronic circuit board background.
R&P: ©Rueangsin Phuthawil/123RF

LEARNING OUTCOMES

▶ Identify the different types of resistors.

▶ Explain the different ways in which resistors are used.

▶ State the ways in which resistors are rated.

▶ Use resistor color codes to determine resistance value.

▶ Calculate the total resistance of different resistor configurations.

▶ Show how resistors are used as voltage and current dividers.

▶ Calculate the resistance and wattage value for a series voltage dropping resistor.

Resistors are components that are specifically designed to have a certain amount of resistance. The principal applications of resistors are to generate heat, limit current, and divide voltage. Resistors can be used to represent any type of resistive load, and the electrical theory can be applied in such formulas as Ohm's law. This chapter discusses the different types of resistors and the ways in which they are used in circuits.

8.1 Resistance Wire

A **resistance wire** (Figure 8-1) is used to produce heat for heating with electricity. The most popular type of resistance wire is made of a high-resistance alloy of nickel and chromium and is referred to by the trade name **nichrome** wire. This wire is used for the heating elements in stoves, dryers, toasters, and other heating appliances.

When a voltage is applied to the heating element, the high resistance of the wire converts most of the electric energy into **heat** energy. The tubular heating element, shown in Figure 8-2, consists of the current-carrying nichrome wire enclosed in tubing with powder insulation. The powder insulation insulates the wire from the tubing as well as seals the wire off from contact with the air. This seal prevents oxidation and prolongs the life of the element.

Figure 8-1 Resistance wire.
©Tutco, Inc.

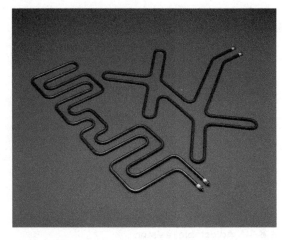

Figure 8-2 Tubular heating element.
©Durex Industries

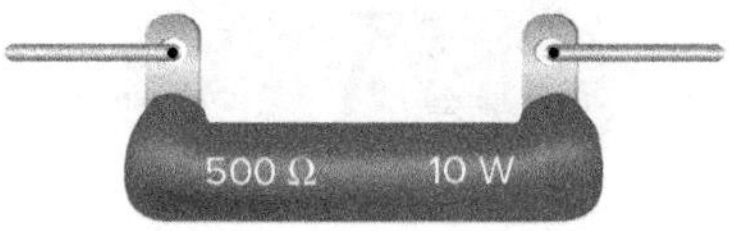

Figure 8-3 Resistor rating.

8.2 Resistors

Resistors are commonly used to adjust and set voltage and current levels. **Resistors** (Figure 8-3) are rated according to **resistance, tolerance,** and **wattage** as follows:

- The resistance value given in ohms is the key parameter.
- It is hard to manufacture a resistor with an exact number of ohms of resistance. Resistor tolerance is a measure of the resistor's variation from the specified resistance value and is expressed as a percentage of its nominal value.
- Electric current passing through a resistor causes it to heat up. If the temperature rises too high, the material of the resistor may burn out. Resistors are rated by the value of their resistance and the power in watts that they can safely dissipate based mainly upon their size. The larger the physical size of the resistor, the more heat it can safely dissipate, thus, the greater the power rating.

All electrical devices are classed into either being passive devices or active devices. An **active device** (such as a transistor) is any type of circuit component with the ability to electrically control electron flow (electricity controlling electricity). Components incapable of controlling current by means of another electric signal are called **passive devices.** All resistors are classified as passive devices.

8.3 Types of Resistors

Resistors can be classified into different types according to their construction. **Wire-wound** resistors are made by wrapping high-resistance wire around an insulated cylinder, as illustrated in Figure 8-4. This type of resistor is generally used in circuits that carry high currents. Large wire-wound resistors are called *power resistors* and range in size from ½ watt to tens or even hundreds of watts.

Figure 8-4 Wire-wound resistor.
Courtesy of Post Glover

Figure 8-5 **Fusible resistor.**

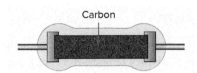

Figure 8-6 **Carbon-composition resistor.**

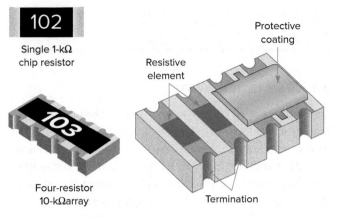

Figure 8-8 **Chip resistors.**

Special wire-wound **fusible resistors** (Figure 8-5) are designed to burn open when their power rating is exceeded. They serve the dual functions of a fuse and resistor to limit the current. To prevent flames or explosion when fusing, the device has an inflammable construction with high dielectric strength.

Carbon-composition resistors are made from a paste consisting of carbon graphite and a resin bonding material, as illustrated in Figure 8-6. The resistance of a carbon resistor is determined by the amount of carbon graphite used in making the resistor. The resistor element is enclosed in a plastic case for insulation and mechanical strength. Joined to the two ends of the carbon resistance element are metal caps with leads for soldering the connections into a circuit. Carbon-composition resistors at one time were the most common type used. Generally, they cannot handle large currents, and their actual value of resistance can vary as much as 20 percent from their rated value.

Presently, the most popular type of resistor is the **film-type resistor** shown in Figure 8-7. In these resistors, a resistance film is deposited on a nonconductive rod. Then the value of resistance is set by cutting a spiral groove through the film. The length and width of the groove determine the resistance value. This method of manufacture allows for much closer tolerance resistors (1 percent or less) as compared to the simpler carbon-composition types.

A **chip** resistor is a surface-mounted printed circuit board mini-resistor. Chip resistors differ from conventional axial lead resistors in that they are soldered directly onto the conductive tracks of printed circuit boards. There are two basic configurations for chip resistors: single resistor

and resistor chip array (Figure 8-8). **Single-chip** resistors are standard, passive resistors with a single resistance value. **Resistor chip arrays** contain several resistors in a single package. The chip resistors shown are marked with a three-digit code (similar to the resistor color code). The first two numbers indicate the significant digits, and the third will be the multiplier, telling you the power of 10 to which the two significant digits must be multiplied.

Resistors are also classified as being fixed, adjustable, or variable. The **fixed resistor,** shown in Figure 8-9, has a specified value that cannot be changed. The sliding contact **adjustable resistor,** shown in Figure 8-10, has an

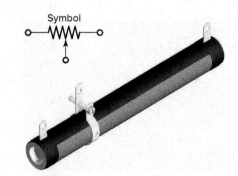

Figure 8-9 **Fixed resistor.**
©Ohmite Manufacturing

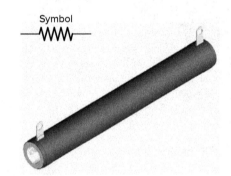

Figure 8-10 **Adjustable resistor.**
©Ohmite Manufacturing

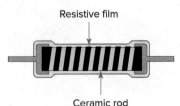

Figure 8-7 **Film-type resistor.**

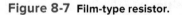

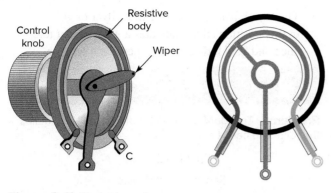

Figure 8-11 Variable resistor.

adjustable collar that can be moved to tap off any resistance within the ohmic value range of the resistor.

The **variable resistor,** shown in Figure 8-11, is designed to provide continuous adjustment of resistance. Variable resistors have a resistive body and a wiper. The wiper slides on the resistive body, changing the length of the resistive material between one end of the device and the wiper. Since resistance depends directly on length, increasing the length of the resistive material between the end of the resistor and the wiper makes the resistance higher.

8.4 Rheostats and Potentiometers

Variable resistors are of two different types: rheostat and potentiometer. A **rheostat** is a variable resistor connected using only two of its terminals. The rheostat is used to control current by varying the resistance in a circuit. Figure 8-12 shows a rheostat in a low-power lamp dimmer circuit. By changing the resistance of the rheostat, the current flow through it is varied.

A **potentiometer** (or *pot,* as it is commonly known) is a variable resistor that makes use of all three of its terminals. Potentiometers are generally used to adjust the level of AC or DC voltage. Figure 8-13 shows a variable DC voltage control potentiometer circuit. The two fixed maximum-resistance leads are connected across the voltage source, and the variable wiper arm lead provides a voltage that varies from 0- to the 9-V maximum.

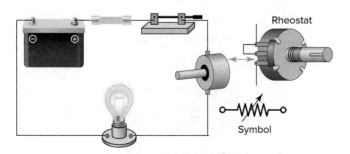

Figure 8-12 Rheostat lamp dimmer circuit.

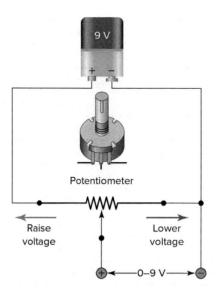

Figure 8-13 Voltage control potentiometer circuit.

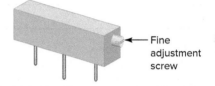

Figure 8-14 Multiturn trimmer potentiometer.

Trimmer (trim) potentiometers are used when the ohmic value of a resistor is set at the time a circuit is manufactured and tested. They are generally a miniature size and mounted on a printed circuit board. Often they are used to fine-tune or calibrate a circuit. Unlike a typical potentiometer that rotates just short of one complete revolution, some trim pots, such as the one shown in Figure 8-14, are what we call *multiturn pots.* A 10-turn trim pot, for example, must be rotated completely 10 times in order for the wiper to move from one end of the resistive element to the other. This allows for very precise adjustments.

Whether rheostat or potentiometer, the resistive track can be classified as either a linear or a tapered (nonlinear) resistance. With a **linear** rheostat or potentiometer the resistance changes in direct proportion to rotation. The resistance for a **nonlinear** rheostat or potentiometer changes more gradually at one end, with bigger changes at the opposite end. The effect is accomplished by different densities of the resistance element in one-half than in the other. Audio volume controls are of the nonlinear type, allowing greater control of loudness at normal or low listening levels.

8.5 Resistor Color Code

Some resistors are large enough to have their resistance value, tolerance, and power rating stamped on them. For small fixed resistors, a system of **color coding** is often

Color	Digit	Multiplier	Tolerance
Black	0	$10^0 = 1$	
Brown	1	$10^1 = 10$	±1%
Red	2	$10^2 = 100$	±2%
Orange	3	$10^3 = 1,000$	
Yellow	4	$10^4 = 10,000$	
Green	5	$10^5 = 100,000$	±0.5%
Blue	6	$10^6 = 1,000,000$	±0.25%
Violet	7	$10^7 = 10,000,000$	±0.1%
Gray	8	$10^8 = 100,000,000$	
White	9	$10^9 = 1,000,000,000$	
Silver		$10^{-2} = 0.01$	±10%
Gold		$10^{-1} = 0.1$	±5%
None			±20%

Figure 8-15 Resistor color code.

used to identify the resistance value and tolerance. The basis of this system is the use of colors for numerical values, as shown in Figure 8-15.

General-purpose resistors use a **four-band color code** (Figure 8-16), which is read as follows:

- The first three bands will show the value of the resistor (that is, its resistance) in ohms.
- Bands are read from left to right starting from the end that has the band closest to it.
- The first two bands identify the first and second digits of the resistance value.
- The third band is the multiplier exponent (power-of-10) band and indicates the number of zeros to be inserted after the second digit. An exception to this is when the resistor is under 10 Ω. In this case the third band is either silver or gold, which indicates a 0.01 or 0.1 multiplier, respectively.

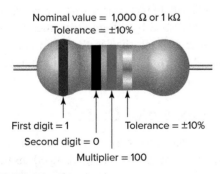

Nominal value = 1,000 Ω or 1 kΩ
Tolerance = ±10%

First digit = 1
Second digit = 0
Multiplier = 100
Tolerance = ±10%

Figure 8-16 Reading a four-band resistor.

- The fourth band is always either silver or gold, and in this position silver indicates a ±10 percent tolerance and gold indicates a ±5 percent tolerance.

EXAMPLE 8-1

Problem: The four-band resistor of Figure 8-17 contains the following bands of color:

First band = red
Second band = blue
Third band = orange
Fourth band = Gold

Determine its nominal resistance value and permissible ohmic range.

Figure 8-17 Resistor for example 8-1.

Solution:

Red = 2
Blue = 6
Orange = × 1,000
Gold = ±5% tolerance

Therefore, its nominal resistance value is 26,000 Ω ± 5 percent. The permissible ohmic range is calculated as follows:

Tolerance = 26,000 Ω × 0.05
= 1300 Ω
Range = (26,000 Ω + 1,300 Ω) to (26,000 Ω − 1,300 Ω)
= 27,300 Ω to 24,700 Ω

EXAMPLE 8-2

Problem: The four-band resistor of Figure 8-18 contains the following bands of color:

First band = orange
Second band = black
Third band = gold
Fourth band = silver

Determine its nominal resistance value.

Figure 8-18 Resistor for example 8-2.

Solution:

$$Orange = 3$$
$$Black = 0$$
$$Gold = \times 0.1$$
$$Silver = \pm 10\% \text{ tolerance}$$

Therefore, its nominal resistance value is 3 Ω ±10 percent.

EXAMPLE 8-3

Problem: What would be the color code for a four-band 500-Ω resistor with a tolerance of ±5 percent?

Solution:

$$5 = green$$
$$0 = black$$
$$\times 10 = brown$$
$$\pm 5\% \text{ tolerance} = Gold$$

Therefore, the color code would be green, black, brown, and gold.

Precision resistors are labeled with a **five-band color code.** The five-band resistor (Figure 8-19) is read as follows:

- Color and multiplier values are the same as that used for the four-band color code.
- The first three bands indicate the first three significant digits.
- The fourth band is the multiplier.
- The fifth band indicates the percentage of tolerance.

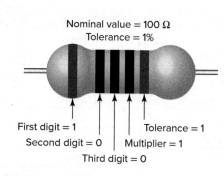

Nominal value = 100 Ω
Tolerance = 1%

First digit = 1
Second digit = 0
Third digit = 0
Multiplier = 1
Tolerance = 1

Figure 8-19 Reading a five-band resistor.

EXAMPLE 8-4

Problem: The five-band resistor of Figure 8-20 contains the following bands of color:

$$First band = red$$
$$Second band = orange$$
$$Third band = violet$$
$$Fourth band = orange$$
$$Fifth band = brown$$

Determine its nominal resistance value and permissible ohmic range.

Figure 8-20 Resistor for example 8-4.

Solution:

$$Red = 2$$
$$Orange = 3$$
$$Violet = 7$$
$$Orange = \times 1,000$$
$$Brown = \pm 1\% \text{ tolerance}$$

Therefore, its nominal resistance value is 237,000 Ω ±1 percent.
The permissible ohmic range is calculated as follows:

$$Tolerance = 237,000 \ \Omega \times 0.01$$
$$= 2,370 \ \Omega$$
$$Range = (237,000 \ \Omega + 2,370 \ \Omega) \text{ to } (237,000 \ \Omega - 2,370 \ \Omega)$$
$$= 239,370 \ \Omega \text{ to } 234,630 \ \Omega$$

Resistors values can change with temperature. A **six-band color-coded** resistor has a sixth band added to it, which represents its **temperature coefficient** or *tempco.* This color represents the amount the resistance value will change with temperature.

The physical size of a resistor has nothing to do with its resistance value. A very small resistor can have a very low or a very high value of resistance. Physical resistor size is an indication of the resistor's **power or wattage** rating. For a given value of resistance, the physical size of a resistor increases as the wattage rating increases, as illustrated in Figure 8-21.

In addition to the required ohmic value, a resistor must have a wattage rating high enough to dissipate the heat produced by the current flow through it. If a resistor's wattage rating is exceeded, the resistor will become overheated and permanently damaged. The amount of heat that must be

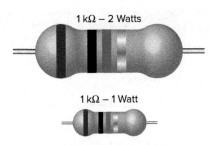

1 kΩ – 2 Watts

1 kΩ – 1 Watt

Figure 8-21 For a given value of resistance, the physical size of a resistor increases as the wattage rating increases.

dissipated by the resistor in a given circuit can be determined by applying any one of the following power formulas:

$$P = E \times I$$
$$P = I^2 \times R$$
$$P = E^2 \div R$$

EXAMPLE 8-5

Problem: Calculate the amount of watts of power a 33-Ω resistor must dissipate when connected across a 5-volt power source.

Solution:

$$P = E^2 \div R$$
$$= 5^2 \div 33$$
$$= 0.76 \text{ W}$$

Resistor values are sometimes written on circuit diagrams using an **alphanumeric code** system. This avoids using a decimal point because it is easy to miss the small dot. Instead, the letters R, K, and M are used in place of the decimal point. To read this code, you replace the letter with a decimal point, then multiply the value by 1,000 if the letter was K, or 1,000,000 if the letter was M. The letter R means multiply by 1. For example:

560R means 560 × 1 = 560 Ω

2K7 means 2.7 × 1,000 = 2,700 Ω = 2.7 kΩ

39K means 39 × 1,000 = 39,000 Ω = 39 kΩ

1M0 means 1.0 × 1,000,000 = 1,000 kΩ = 1 MΩ

Part 1 Review Questions

1. In what way is resistance wire different from the wire used for circuit conductors?

2. What is the trade name for the most popular type of resistance wire?

3. What circuit functions are resistors commonly used for?

4. Name the three ways by which resistors are rated.

5. Resistors are classified as passive devices. Why?

6. Identify the type of circuit that generally requires wire-wound resistors.

7. State two functions served by a fusible resistor.

8. What advantage do film resistors have over carbon-composition types?

9. What are the two configurations for chip resistors?

10. In what way is the construction of an adjustable resistor different from that of a fixed resistor?

11. Compare the connection and control function of a rheostat with that of a potentiometer.

12. If the wiper arm of a linear potentiometer is one-quarter of the way around the contact surface, what is the resistance between the wiper arm and each terminal if the total resistance is 25 kΩ?

13. Explain how a 10-turn trim pot is adjusted.

14. Compare the way the resistance varies in a linear and nonlinear potentiometer.

15. Identify the color bands for each of the following four-band color-coded resistors:
    a. 100 Ω ±10%
    b. 2,200 Ω ±5%
    c. 47,000 Ω ±20%
    d. 1,000,000 Ω ±10%
    e. 2.5 Ω ±5%

16. A 680-Ω resistor has a rated tolerance of ±10 percent. Determine the permissible ohmic range for this resistor.

17. What is the color code for a 365-Ω five-band precision resistor with a tolerance of ±5 percent?

18. Determine the resistance value and percentage of tolerance for each of the four-band color-coded resistors shown in the following table:

	1st Band	2nd Band	3rd Band	4th Band
a.	Red	Green	Yellow	Silver
b.	Orange	Blue	Brown	Gold
c.	White	Brown	Red	None
d.	Gray	Black	Blue	Gold
e.	Violet	Green	Gold	Silver
f.	Blue	Red	Black	Gold

19. Determine the resistance value and percentage of tolerance for each of the five-band color-coded resistors shown in the following table:

	1st Band	2nd Band	3rd Band	4th Band	5th Band
a.	Green	Blue	Red	Red	Brown
b.	Violet	Gray	Violet	Silver	Red
c.	Orange	Blue	Green	Black	Brown
d.	Brown	Black	Green	Brown	Red
e.	Red	Red	Blue	Brown	Green
f.	Brown	Black	Green	Gold	Violet

20. What information is given by the sixth band of a six-band color-coded resistor?

21. What resistor rating is determined by the physical size of the resistor?

22. Calculate the amount of watts of power a 100-Ω resistor must dissipate when connected across a 10-volt power source?

PART 2 RESISTOR CONNECTIONS

8.6 Series Connection of Resistors

Individual resistors can be connected together in a series connection, a parallel connection, or combinations of both series and parallel together. This results in a more complex circuit whose total circuit resistance is a combination of the individual resistors.

To connect resistors in **series,** they are joined **end to end** together in a single line as illustrated if Figure 8-22. The characteristics of series connected resistors can be summarized as follows:

- The total resistance of the circuit (R_T) increases if additional resistors are connected in series and decreases if resistors are removed.
- To determine the total resistance of the circuit, simply find the sum of the individual resistance loads.

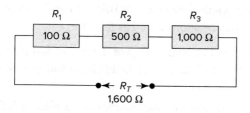

Figure 8-22 Resistors connected in series.

- In this example, if the resistors are labeled R_1, R_2, and R_3, then the total resistance R_T is calculated using the formula

$$R_T = R_1 + R_2 + R_3$$
$$= 100\ \Omega + 500\ \Omega + 1{,}000\ \Omega$$
$$= 1{,}600\ \Omega$$

EXAMPLE 8-6

Problem: Three resistors, R_1 (25 Ω), R_2 (50 Ω), and R_3 (75 Ω) are connected in series as shown in Figure 8-23. Determine the value of the total combined circuit resistance.

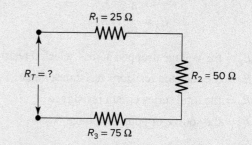

Figure 8-23 Circuit for example 8-6.

Solution:

$$R_T = R_1 + R_2 + R_3$$
$$= 25\ \Omega + 50\ \Omega + 75\ \Omega$$
$$= 150\ \Omega$$

Resistors connected in series are used as **voltage dividers,** as illustrated in the circuit of Figure 8-24. Voltage dividers are widely used in circuits where a single voltage source must supply several different voltage values for different parts of a circuit. The characteristics of a series voltage divider circuit can be summarized as follows:

- The same amount of current flows through each resistor.
- The input voltage is divided proportionally across the series-connected resistors.

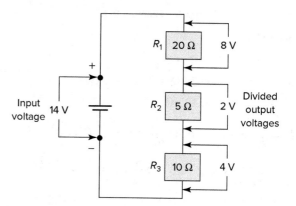

Figure 8-24 Voltage divider circuit.

- The voltage drop across a resistor in a series circuit is directly proportional to the ohmic value of the resistor.
- The higher the resistance value, the greater the voltage drop.

For a voltage divider circuit, the voltage dropped across each resistor is normally a factor that needs to be determined. The voltage drop across any one resistor is proportional to the ratio of its resistance value to that of the total circuit resistance. The **voltage divider formula** allows you to calculate the voltage drop across any one of the resistors connected in series without having to first calculate the value of circuit current. Stated as a formula:

$$E_X = \frac{R_X}{R_T} \times E_S$$

where E_X = the voltage dropped across selected resistor

R_X = the selected resistor's resistance value

R_T = the total series circuit resistance

E_S = the source or applied voltage

EXAMPLE 8-7

Problem: Resistors R_1 (5 kΩ), R_2 (3 kΩ), and R_3 (2 kΩ) are connected in series to form a voltage divider as shown in Figure 8-25. If an input voltage of 9 volts is applied to the circuit, calculate the value of the voltage drop across each of the resistors, using the voltage divider formula.

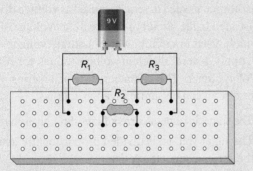

Figure 8-25 Circuit for example 8-7.

Solution:

$$R_T = R_1 + R_2 + R_3$$
$$= 5\ k\Omega + 3\ k\Omega + 2\ k\Omega$$
$$= 10\ k\Omega$$

$$E_1 = \frac{R_1}{R_T} \times E_S \qquad E_2 = \frac{R_2}{R_T} \times E_S \qquad E_3 = \frac{R_3}{R_T} \times E_S$$

$$= \frac{5\ k\Omega}{10\ k\Omega} \times 9\ V \quad = \frac{3\ k\Omega}{10\ k\Omega} \times 9\ V \quad = \frac{2\ k\Omega}{10\ k\Omega} \times 9\ V$$

$$= 4.5\ V \qquad\qquad = 2.7\ V \qquad\qquad = 1.8\ V$$

EXAMPLE 8-8

Problem: You have a 120-V source and want to use a series dropping resistor in conjunction with a 6 V @ 150 mA pilot light to indicate when power is applied (Figure 8-26). Determine the series dropping resistance value and wattage required.

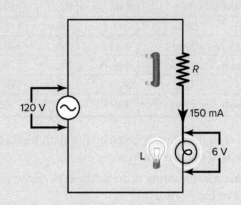

Figure 8-26 Circuit for example 8-8.

Solution:

Voltage across the resistor = 120 V − 6 V

$$= 114\ V$$

$$\text{Resistance of the resistor} = \frac{E_R}{I_R}$$

$$= \frac{114\ V}{0.150\ A\ (150\text{mA})}$$

$$= 760\ \Omega$$

Wattage of the resistor = $E_R \times I_R$

$$= 114\ V \times 0.150\ A$$

$$= 17.1\ W$$

8.7 Parallel Connection of Resistors

Resistors are connected in **parallel** by connecting them **side by side** across one another, as illustrated in Figure 8-27. Note that the two ends of the resistors are connected to the same two points. The characteristics of parallel-connected resistors can be summarized as follows:

- The total resistance (R_T) of the circuit formed is **less** than that of the lowest value of resistance present in any of the branches.
- Each resistor provides a separate parallel path for the current to flow.

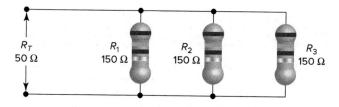

Figure 8-27 **Resistors connected in parallel.**

- If you have several resistors of the same value in parallel, then the total resistance is found most easily by dividing the common resistance value by the number of resistors connected. For three 150-Ω resistors in parallel the total resistance is

$$R_T = \frac{R_{\text{common value}}}{\text{Number of resistors}}$$

$$= \frac{150\ \Omega}{3}$$

$$= 50\ \Omega$$

To find the total resistance of two unequal values of resistors connected in parallel (a very common use) the product over sum formula is used. This formula is

$$R_T = \frac{R_1 \times R_2}{R_1 + R_2}$$

EXAMPLE 8-9

Problem: A 60-Ω resistor is connected in parallel with one of 40 Ω, as shown in Figure 8-28. Determine the value of the total combined resistance of the two using the product over sum formula.

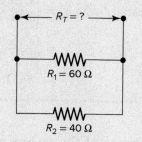

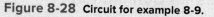

Figure 8-28 **Circuit for example 8-9.**

Solution:

$$R_T = \frac{R_1 \times R_2}{R_1 + R_2}$$

$$= \frac{60\ \Omega \times 40\ \Omega}{60\ \Omega + 40\ \Omega}$$

$$= \frac{2{,}400\ \Omega}{100\ \Omega}$$

$$= 24\ \Omega$$

The product over sum formula works best for two resistors in parallel. Where more than two resistors are in parallel, it becomes more difficult, and less practical, to use. For more than two unequal resistor values connected in parallel, the general formula for total resistance of a parallel circuit is used. This formula is

$$R_T = \frac{1}{\dfrac{1}{R_1} + \dfrac{1}{R_2} + \dfrac{1}{R_3}}$$

EXAMPLE 8-10

Problem: Three resistors, R_1 (120 Ω), R_2 (60 Ω), and R_3 (40 Ω) are connected in parallel, as shown in Figure 8-29. Determine the value of the total combined circuit resistance.

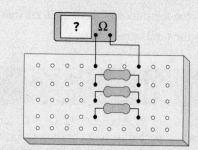

Figure 8-29 **Circuit for example 8-10.**

Solution:

$$R_T = \frac{1}{\dfrac{1}{120\ \Omega} + \dfrac{1}{60\ \Omega} + \dfrac{1}{40\ \Omega}}$$

$$= \frac{1}{0.0083 + 0.0167 + 0.025}$$

$$= \frac{1}{0.050}$$

$$= 20\ \Omega$$

Parallel resistor circuits can be considered as **current dividers** because the current splits or divides between the various resistors, as illustrated in Figure 8-30. The characteristics of a parallel current divider circuit can be summarized as follows:

- Current flow through each branch resistor is inversely proportional to its resistance value.
- The smaller the resistance value, the greater the current flow and vice versa.
- Resistors of the same ohmic value will have the same amount of current through them.

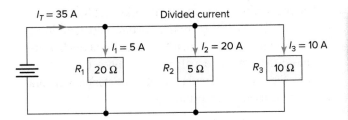

Figure 8-30 Current divider circuit.

- The formula describing a current divider is similar in form to that for the voltage divider and can be expressed as follows:

$$I_X = \frac{R_T \times I_T}{R_X}$$

where I_X = the branch current desired
R_T = the total resistance value
R_X = the resistance of the branch current desired
I_T = the total current flow

EXAMPLE 8-11

Problem: Resistors R_1, R_2 and R_3 (2 Ω, 3 Ω, and 6 Ω, respectively) are connected in parallel, as shown in Figure 8-31. Use the current divider formula to calculate the value of the current flow through each of the load resistors if the total current flow to the circuit is 10 amperes.

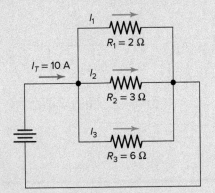
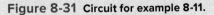

Figure 8-31 Circuit for example 8-11.

Solution:

$$R_T = \frac{1}{\frac{1}{2} + \frac{1}{3} + \frac{1}{6}}$$

$$= 1\ \Omega$$

$$I_1 = \frac{R_T \times I_T}{R_1} \qquad I_2 = \frac{R_T \times I_T}{R_2} \qquad I_3 = \frac{R_T \times I_T}{R_3}$$

$$= \frac{1 \times 10}{2} \qquad = \frac{1 \times 10}{3} \qquad = \frac{1 \times 10}{6}$$

$$= 5\,\text{A} \qquad\qquad = 3.33\ \text{A} \qquad = 1.67\ \text{A}$$

EXAMPLE 8-12

Problem: As additional loads are connected in parallel the total resistance of the circuit **decreases.** For the circuit shown in Figure 8-32, determine the total circuit resistance under each of the following operating conditions:

a. Switches 1 and 2 closed.

b. Switches 1, 2, and 3 closed.

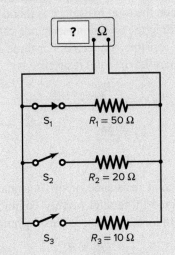

Figure 8-32 Circuit for example 8-12.

Solution:

a. R_1 and R_2 (in parallel) $= \dfrac{R_1 \times R_2}{R_1 + R_2}$

$$= \frac{50 \times 20}{50 + 20}$$

$$= 14.3\ \Omega$$

b. R_1 and R_2 in parallel with $R_3 = \dfrac{14.3 \times R_3}{14.3 + R_3}$

$$= \frac{14.3 \times 10}{14.3 + 10}$$

$$= 5.88\ \Omega$$

8.8 Series-Parallel Connection of Resistors

Combination resistive circuits, otherwise known as **series-parallel** resistive circuits, combine resistors in series with resistors in parallel, as shown in the Figure 8-33. The rules governing these circuits are the same as those developed for series circuits and for parallel circuits. First, the resistance of the combined total resistance of the parallel portion is found. Then the total resistance of the parallel section is added to any series resistance to find the total resistance of the series-parallel combination circuit.

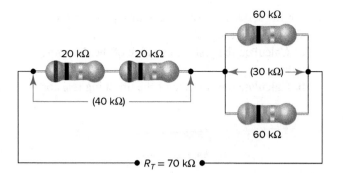

Figure 8-33 Series-parallel connection of resistors.

EXAMPLE 8-13

Problem: A 30-Ω resistor, R_1, and a 60-Ω resistor, R_2, are connected in parallel with each other and in series with a 40-Ω resistor, R_3, as shown in Figure 8-34. Determine the total resistance of this series-parallel combination of resistors.

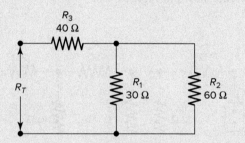

Figure 8-34 Circuit for example 8-11.

Solution:

$$R_1 \text{ and } R_2 \text{ (in parallel)} = \frac{R_1 \times R_2}{R_1 + R_2}$$

$$= \frac{30\ \Omega \times 60\ \Omega}{30\ \Omega + 60\ \Omega}$$

$$= \frac{1,800}{90}$$

$$= 20\ \Omega$$

$$R_T = R_1 \text{ and } R_2 \text{ (in parallel)} + R_3$$

$$= 20\ \Omega + 40\ \Omega$$

$$= 60\ \Omega$$

EXAMPLE 8-14

Problem: Resistance readings can be used to check circuits for fault conditions. As determined in the previous example, the normal total resistance of this series-parallel circuit arrangement of Figure 8-35 is 60 ohms.

a. Find what the new value of R_T would be should resistor R_1 be faulted **open** while the resistance values of R_2 and R_3 remain the same.

b. Similarly, find what the new value of R_T would be should resistor R_3 be faulted **shorted** while the resistance values of R_1 and R_2 remain the same.

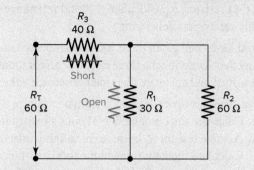

Figure 8-35 Circuit for example 8-14.

Solution:

a. With R_1 faulted open, the circuit would consist of R_3 in series with R_2 and the total resistance would be:

$$R_T = R_3 + R_2$$

$$= 40 + 60$$

$$= 100\ \Omega$$

b. With R_3 faulted shorted, the circuit would consist of R_1 in parallel with R_2 and the total resistance would be:

$$R_T = \frac{R_1 \times R_2}{R_1 + R_2}$$

$$= \frac{30 \times 60}{30 + 60}$$

$$= 20\ \Omega$$

Part 2 Review Questions

1. Calculate the total resistance for each of the following resistor circuits:
   a. Series circuit: $R_1 = 40\ \Omega$, $R_2 = 75\ \Omega$
   b. Parallel circuit: $R_1 = 200\ \Omega$, $R_2 = 200\ \Omega$, $R_3 = 200\ \Omega$
   c. Series circuit: $R_1 = 2,000\ \Omega$, $R_2 = 6,000\ \Omega$, $R_3 = 2,200\ \Omega$
   d. Parallel circuit: $R_1 = 14\ \Omega$, $R_2 = 32\ \Omega$
   e. Series circuit: $R_1 = 4,700\ \Omega$, $R_2 = 800\ \Omega$, $R_3 = 200\ \Omega$
   f. Parallel circuit: $R_1 = 60\ \Omega$, $R_2 = 30\ \Omega$, $R_3 = 15\ \Omega$

2. Resistors R_1, R_2, and R_3 (50 Ω, 30 Ω, and 20 Ω, respectively) are connected in series across an applied voltage of 200 V to form a voltage divider. Using the voltage divider formula, calculate voltages E_1, E_2, and E_3.

3. The total current to two parallel-connected resistors is 3 A. The resistance of R_1 is 10 Ω and the resistance of R_2 is 40 Ω. Using the current divider formula, calculate currents I_1 and I_2.

4. A 5-Ω resistor, R_1, and a 20-Ω resistor, R_2, are connected in parallel with each other and in series with a 6-Ω resistor, R_3. Calculate the total resistance of this series-parallel circuit.

5. You have been given three 100-Ω resistors to connect together. Describe the three circuit configurations possible, and calculate their total resistance values.

6. For the series-parallel circuit shown in Figure 8-36:
   a. Calculate the combined total circuit resistance.
   b. Assume resistor R_1 burnt open (infinite resistance). Calculate the new value of the total resistance.
   c. Assume resistor R_2 of the original circuit is short-circuited (zero resistance across it). Calculate the new value of the total resistance.

7. For the series resistor voltage dropping circuit shown in Figure 8-37:
   a. Calculate the resistance value of the dropping resistor.
   b. Calculate the wattage of the dropping resistor.

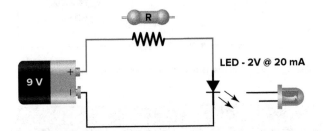

Figure 8-37 Circuit for question 7.

8. Calculate the total resistance for the series-parallel circuit shown in Figure 8-38.

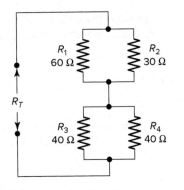

Figure 8-36 Circuit for question 6.

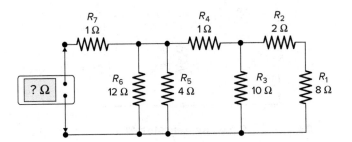

Figure 8-38 Circuit for question 8.

CHAPTER NINE

Electricity and Magnetism

Crane electromagnet
©Marmaduke St. John/Alamy Stock Photo

LEARNING OUTCOMES

▶ Define common magnetic terms.

▶ State the law of magnetic poles.

▶ Describe the characteristics of magnetic lines of force.

▶ Determine the magnetic polarity of an electromagnet.

▶ State the factors that determine the strength of an electromagnet.

▶ Explain how solenoids, solenoid valves, transformers, relays, generators, and motors operate.

▶ Explain the principle of electromagnetic induction.

Although electricity and magnetism may appear to be two apparently distinct topics, we will see that there actually is an important connection between them. A magnet is a piece of iron oxide or special alloy that exerts an invisible force of attraction on objects made of iron, nickel, or cobalt. The invisible force itself is called magnetism or magnetic force. Electromagnetism is the magnetism produced around a conductor whenever current flows through the conductor. In this chapter we discuss phenomena associated with electricity and magnetism.

9.1 Properties of Magnets

The ability of certain materials to attract objects made of iron or iron alloys is the most familiar of all magnetic effects. This property of a material to attract pieces of iron or steel is called **magnetism** (Figure 9-1). In addition, all magnets if free to move will assume a north-south orientation.

Magnetic substances are those that are attracted to a magnet while **nonmagnetic substances** are not attracted to a magnet. Magnetic substances include iron, steel, nickel, and cobalt. Examples of nonmagnetic substances are copper, aluminum, lead, silver, brass, wood, glass, liquids, and gases.

9.2 Types of Magnets

Certain ores of iron have long been known to posses the magnetic property of attracting iron. This type of mineral in olden days was called lodestone. These **natural magnets** have very little practical use because it is possible to produce much stronger magnets by artificial means. **Artificial magnets** are those made from ordinary unmagnetized magnetic materials. The bar magnet, horseshoe magnet, and compass needle (Figure 9-2) are all examples of artificial magnets.

Most artificial magnets are produced electrically. The process used is illustrated in Figure 9-3. To magnetize a magnetic material using electricity, the material to be magnetized is first placed in a coil of insulated wire. A **direct current** voltage source is then momentarily applied to the coil leads. To demagnetize an artificial magnet, the same process is repeated but the voltage source used is **alternating current.**

The two basic types of magnets are permanent and temporary. A **permanent magnet** retains its magnetic

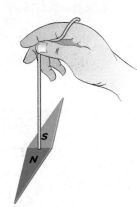

Figure 9-2 A compass needle is an example of an artificial magnet.

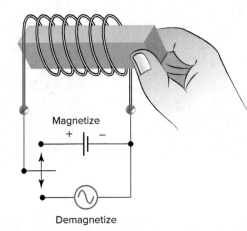

Figure 9-3 Magnetizing and demagnetizing process.

properties for a long period of time. A **temporary magnet** acts as a magnet only as long as it is in the magnetic field produced by a permanent magnet or an electric current. Temporary magnets are made from soft irons which are easily magnetized. Permanent magnets are made from hard iron and steel and require more energy to magnetize. Today all magnets that are used commercially are made from synthetic magnetic materials such as alloys containing aluminum, nickel, and cobalt combined with iron. Magnetic materials containing such rare-earth elements form very strong permanent magnets.

One special category of permanent magnets is that of **ceramic magnets,** often referred to as **ferrites.** These permanent magnets are charcoal gray in color and usually appear in the forms of discs, rings, blocks, cylinders, and sometimes arcs for generators and motors (Figure 9-4). Ferrite magnets are made by combining iron oxide particles with a ceramic compound. They can be molded to any shape and are nonconductive in that they have a very high electric resistance.

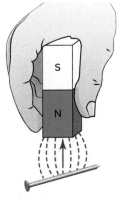

Figure 9-1 Magnetism is the property to attract iron or steel.

Figure 9-4　Ceramic ferrite permanent magnet.
©Sergei Kornilev/Shutterstock

9.3 Laws of Magnetic Poles

Every magnet has two poles, one **north pole** and one **south pole.** Invisible magnetic **lines of force** are assumed to leave the north pole and enter the south pole of a magnet, as illustrated in Figure 9-5. The number of lines of force per unit area is called the **flux density,** and the greater the flux density, the stronger the magnetic field. Flux density is greatest inside the magnet and where the lines of flux enter and leave the magnet. Therefore, the effects of magnetism are strong at the ends of the magnet and weak in the middle.

The law of magnetic poles states that *like poles repel and unlike poles attract* (Figure 9-6). When two magnets are brought together, the magnetic field around the magnets causes some form of interaction. Placing the north pole of a magnet near the south pole of a second magnet causes the magnets to attract each other and move together. Repeating this using the two south pole ends (or two north pole ends) will cause the magnets to repel each other and move apart. The attracting or repelling force will increase as the distance between the poles of the magnet decreases.

The **horseshoe-shaped** magnet (Figure 9-7) is actually a bar magnet that is bent around in the shape of a horseshoe.

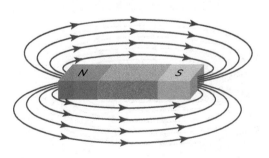

Figure 9-5　Magnetic poles.

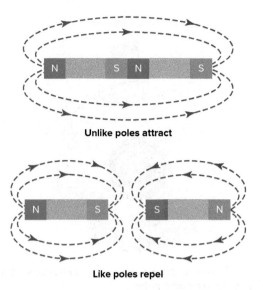

Unlike poles attract

Like poles repel

Figure 9-6　Law of magnetic poles.

Figure 9-7　Horseshoe-shaped magnet.

This brings the two poles of the magnet closer together than they are in a straight bar magnet. A more direct path exists for the lines of force to travel between the poles, resulting in a much stronger magnetic force.

9.4 Polarity of Magnets

Every magnet has a north and a south pole which represents its magnetic polarity. The **north pole** of a magnet was established as being the pole which is attracted to the **geographic north pole** as illustrated in Figure 9-8. The earth itself is a natural magnet with magnetic poles located near the north and south geographic poles. The compass is a permanent magnet pivoted at its midpoint so that it is free to move in a horizontal plane. Due to the magnetic attraction between poles, the compass will always come to rest with the same end pointing toward the north. The end of the compass that points to the geographic north was established as being the north-seeking pole of the compass. Thus, the north-seeking end of the compass is considered

North geographic pole

Figure 9-8 The north pole of a compass needle is attracted to the geographic north pole.

North

?

Figure 9-9 Using a compass to determine the polarity of a magnet.

to be the north pole of the compass. The opposite end is the south pole of the compass.

The polarity of a magnet can be determined using a compass, as illustrated in Figure 9-9. The steps are as follows:

- First, identify the north and south poles of the compass. Remember the north pole of the compass points to the geographic north pole.

- Next, place the compass at a short distance from one of the pole ends of the magnet.
- Apply the law of magnetic poles to identify the unmarked pole of the magnet.
- If the north pole of the compass is attracted to the pole, then the unknown pole is a south pole.
- If the south pole of the compass is attracted to the pole, then the unknown pole is a north pole.

9.5 Magnetic Field

The area surrounding a magnet, in which the invisible magnetic force is evident, is called the **magnetic field** of the magnet. Although the lines of force or flux are invisible, the effects of magnetic fields can be made visible. When a sheet of paper is placed on a magnet and iron filings loosely scattered over it, the filings arrange themselves along the invisible lines of flux (Figure 9-10).

Lines of force are assumed to have certain characteristics which are summarized as follows:

- Lines of force never cross one another.
- Lines of force form closed loops.
- Lines of force travel from the north pole to the south pole outside the magnet and from the south pole to north pole inside the magnet.
- Lines of force follow the easiest path, passing most easily through soft iron.
- The stronger the magnet, the greater the flux density (lines of force per square inch).
- Lines of force repel each other.
- There is no known insulator for lines of force.

Mapping of magnetic field can be done using a compass. When a compass is placed in a magnetic field, the north pole of the compass will point in the direction of the lines of force (Figure 9-11).

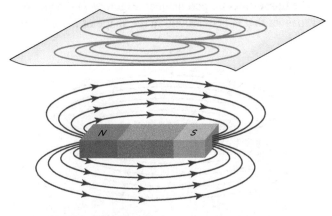

Figure 9-10 Visualizing magnetic field patterns.

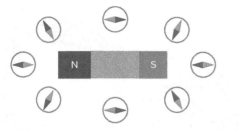

Figure 9-11 Mapping of the magnetic field using a compass.

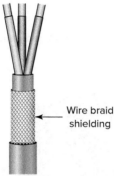

Figure 9-12 Shielded cable.

The purpose of **magnetic shielding** is to prevent magnetic fields from interfering with electrical devices or signals. Magnetic shielding limits the coupling of a magnetic field by providing an alternative good conducting magnetic path. Figure 9-12 shows a shielded cable in which the conductors are surrounded by a protective wire braid shield. The shield material will vary and may be composed of braided strands of copper (or other metal), a nonbraided spiral winding of copper tape, or a layer of conducting polymer. This shielding material is normally terminated to ground. The same principle of design is applied in motors and transformers to minimize the radiation of lines of force from the magnetic fields of these devices.

9.6 Theories of Magnetism

Different theories have been developed over the years in an attempt to explain what causes magnetism. The **molecular theory of magnetism** (Figure 9-13) assumes that each molecule (group of atoms) of a substance is, in fact, a small magnet. When a material is unmagnetized, its mo-

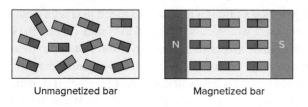

Figure 9-13 Molecular theory of magnetism.

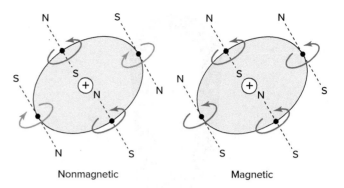

Figure 9-14 Electron theory of magnetism.

lecular magnets are arranged in a random fashion. The net result is a cancellation of the magnetic effect. In a magnetized bar the molecular magnets are arranged so that their magnetic fields are aligned in the same direction. If a magnet is split in half, the molecular theory also applies, since each half will possess both a north and south pole.

The **electron theory of magnetism** is a more modern theory of magnetism. Electrons are believed to spin on their axes, in the same way that the earth turns on its axis, as they orbit around the nucleus. The spinning effect of the electron creates a magnetic field. The polarity of this magnetic field is determined by the direction the electron is spinning, as illustrated in Figure 9-14. Nonmagnetic materials have electrons spinning in different directions causing the cancellation of the magnetic effect. Magnetic materials tend to have most or all of their electrons spinning in the same direction.

There is a definite limit to the amount of magnetism a material can have. This limit is reached when all the molecular magnets are aligned or all the electrons are spinning in the same direction. When maximum magnetic strength is reached, the material is said to be **magnetically saturated.**

The magnetism that remains in a magnetic material, once the magnetizing force is removed, is called **residual magnetism.** This term is usually only applied to temporary magnets. Residual magnetism is of importance in certain types of generators because it provides the initial voltage required for the generator to build up to its rated voltage.

Proper handling of permanent magnets is important. A piece of steel that has been magnetized can lose much of its magnetism by improper handling. If it is jarred or heated, there will be a rearrangement of its molecules resulting in the loss of some of its effective magnetism. A magnet may also become weakened from loss of flux. Thus when storing magnets, one should always try to avoid excess leakage of magnetic flux. For example, a horseshoe magnet should always be stored with a **keeper,** a soft-iron bar used to join the magnetic poles, as illustrated in Figure 9-15. By your using the keeper while the magnet is being stored, the magnetic flux will

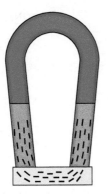

Figure 9-15 **Horseshoe magnet and keeper.**

continuously circulate through the magnet and not leak off into space.

Permeability (μ) is a measure of the ease with which magnetic lines of force pass through a material. Materials that are easy to magnetize have higher magnetic permeability. For example, iron and steel have a much greater permeability than air and other nonmagnetic materials.

9.7 Uses for Permanent Magnets

Permanent magnets of various shapes are used extensively in electrical and electronic equipment. Horseshoe magnets are often used in the construction of **analog-type measuring devices,** such as illustrated in Figure 9-16. Such meter movements were at the heart of the moving coil meters, such as voltmeters and ammeters, until they were largely replaced with solid-state digital meters.

Permanent-magnet generators, such as the one shown in Figure 9-17, are used in wind turbines as part of the generation process. Wind power is used to rotate the shaft of the generator and provides the mechanical force or motion required for generator action. The rotor structure can consist of a ring of magnetic iron with magnets mounted on its surface to provide the required magnetism.

Permanent-magnet motors are used to convert electric energy into mechanical energy. The operation of an electric motor depends on the interaction between two magnetic fields. One magnetic field is produced by a fixed permanent magnet and the other by an electromagnet wound on a movable armature (Figure 9-18).

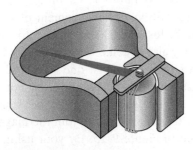

Figure 9-16 **Analog meter movement.**

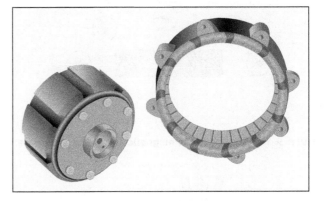

Figure 9-17 **Permanent-magnet generator.**

Figure 9-18 **Permanent magnet motor.**
©American Honda Motor Co., Inc.

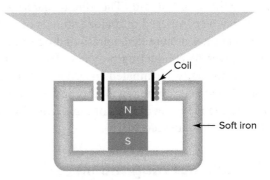

Figure 9-19 **Permanent-magnet speaker.**

Permanent-magnet loudspeakers are the most common of all speakers. They are designed to convert electric energy into sound energy. The voice coil is suspended in the air gap of a permanent-magnet arrangement. When current flows through the coil, a second magnetic field is established, which causes the coil to vibrate (Figure 9-19).

Magnetic switches are used in alarm systems to detect the opening of a door or window. A permanent magnet is mounted on the window or door, and the switch is mounted on the frame, as illustrated in Figure 9-20. When the window or door is closed, the two units are aligned, and the

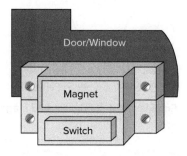

Figure 9-20 **Permanent-magnet switch.**

strength of the magnetic field closes the switch's normally open contacts to signal a closed condition. When the window or door is opened, the reduced strength of the magnetic field opens the switch's contacts to signal an open condition.

Part 1 Review Questions

1. Define *magnetism*.
2. Classify each of the following as being a magnetic or nonmagnetic material:
   a. Copper.
   b. Aluminum.
   c. Iron.
   d. Brass.
   e. Nickel.
   f. Steel.
3. Explain how electricity is used to magnetize and demagnetize an iron bar.
4. What is the difference between permanent and temporary magnets?
5. How are permanent ferrite magnets made?
6. Define the term *flux density*.
7. State the law of magnetic poles.
8. What is the relationship between the distance between two unlike poles and the amount of magnetic attraction between them?
9. The magnetic polarity of one end of an unmarked bar magnet is to be determined using a compass. If the north pole of the compass needle is attracted to this end, what is its magnetic polarity?
10. List six characteristics of magnetic lines of force.
11. What is the purpose of magnetic shielding?
12. Compare the way the magnetic effect is explained according to the molecular theory of magnetism and the electron theory of magnetism.

13. Explain each of the following magnetic terms:
    a. Magnetic saturation.
    b. Residual magnetism.
    c. Magnetic permeability.
14. When storing a permanent horseshoe magnet, it is recommended that a soft-iron bar be placed across the space between its poles. Why?
15. List five practical electrical applications for permanent magnets.

PART 2   ELECTROMAGNETISM

9.8 Magnetic Field around a Current-Carrying Conductor

Whenever electrons flow through a conductor, a magnetic field is created around the conductor. This important relationship between electricity and magnetism is known as **electromagnetism,** or the magnetic effect of current. When DC current flows, the magnetic field will act in one direction, as illustrated in Figure 9-21. AC current flow will produce a magnetic field that varies in direction with the direction of the electron current flow.

The amount of current flowing through a single conductor determines the strength of the magnetic field it produces. Increasing the current flow increases the strength of the magnetic field. Normally the strength of the magnetic field around a single current-carrying conductor is comparatively weak and therefore goes undetected.

A compass can be used to reveal both the presence and direction of this magnetic field, as illustrated in Figure 9-22. When the compass is brought close to a conductor carrying DC current, the north-seeking pole of the compass needle will point in the direction the magnetic lines of force are traveling. As the compass is rotated around the conductor, a definite circular pattern will be indicated.

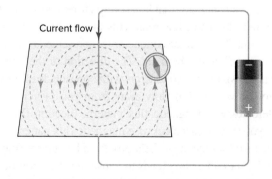

Figure 9-21 **Magnetic field around a current-carrying conductor.**

Current flowing

No current

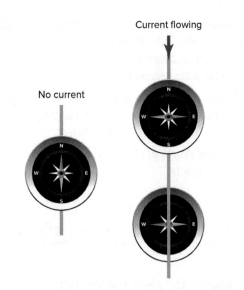

Figure 9-22 Using a compass to trace the magnetic field.

Lines of force

Electron flow

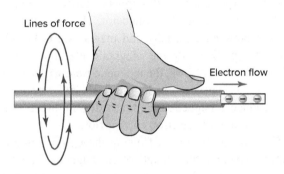

Figure 9-23 Left-hand conductor rule.

A relationship exists between the direction of current flow through a conductor and the direction of the magnetic field. The **left-hand rule** for conductors, illustrated in Figure 9-23, demonstrates this relationship. *If a current-carrying conductor is grasped with the left hand with the thumb pointing in the direction of electron flow (− to +), the fingers point in the direction of the magnetic lines of force or flux.*

An end view of the wire is sometimes used to simplify the drawing of a conductor that is carrying current, as illustrated in Figure 9-24. The circle represents the end of the conductor. Current flow into the conductor is represented by a cross and current flow out by a dot. The direction of the lines of force is shown for two conductors with opposite directions of electron flow. Electron flow away from you (+) creates a counterclockwise magnetic flux pattern. Electron flow towards you (·) causes a clockwise magnetic flux pattern.

The resulting magnetic field produced by current flow in two adjacent conductors tends to cause the attraction or repulsion of the two conductors. If the two parallel conductors are carrying current in opposite directions, the direction of

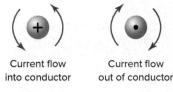

Current flow
into conductor

Current flow
out of conductor

Figure 9-24 End view of a conductor and magnetic field.

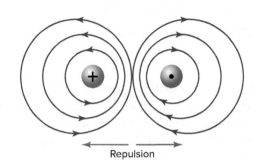

Repulsion

Figure 9-25 Parallel conductors with current flow in opposite directions.

the magnetic field is clockwise around the one conductor and counterclockwise around the other. This sets up a repelling action between the two individual magnetic fields, and the conductors would tend to move apart, as illustrated in Figure 9-25.

When parallel conductors are carrying current in the same direction, the direction of the magnetic field is the same around each field. The lines of force act in the same direction and link together around both conductors, as illustrated in Figure 9-26. As a result the two conductors will be attracted to each other and tend to move together. The two conductors under this condition will create a magnetic field equivalent to one conductor carrying twice the current.

When a conductor carries a current, it creates a magnetic field which interacts with any other magnetic field present to produce a force. This fact must be taken into consideration when designing large pieces of electrical equipment that handle very high current flows. For example, bus

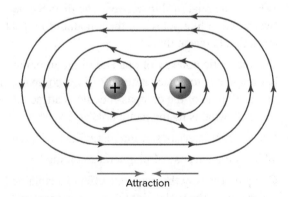

Attraction

Figure 9-26 Parallel conductors with current flow in the same direction.

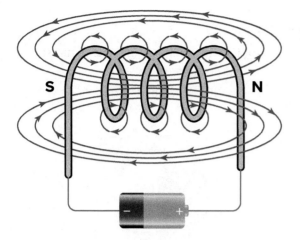

Figure 9-27 Magnetic field produced by a coil.

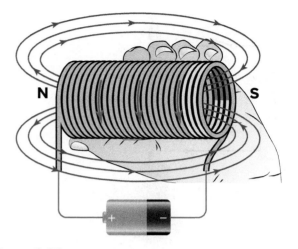

Figure 9-28 Left-hand coil rule.

bars carrying high currents must be anchored securely lest they attract each other and short out. Also, short circuit fault currents will increase stress that can result in damage to conductors and raceways if they are not properly secured and protected.

9.9 Electromagnets

If we take a single piece of wire and wind it into a number of loops to form a **coil,** we create the equivalent of several parallel conductors carrying current in the same direction. The total resulting magnetic field is the sum of all the single-loop magnetic fields. A coil so formed will have a magnetic field pattern similar to that of a bar magnet with a definite north and south pole, as illustrated in Figure 9-27.

A definite relationship exists between the direction of current flow through a coil, the direction in which the wire is wound to form the coil, and the location of the north and south poles. This relationship can be shown by the left-hand coil rule illustrated in Figure 9-28 and stated as follows: *If the coil is grasped in the left hand, with fingers pointing in the direction of the current flow, the thumb will point in the direction of the north pole of the magnet.*

The **magnetic polarity** of a coil depends on the direction of current flow and the direction in which the coil is wound (Figure 9-29). If the electromagnet is operated with

direct current, the polarity of its magnetic poles remains fixed. However, when operated with alternating current, its magnetic polarity reverses with each reversal of the direction of the current.

An **electromagnet** can be defined as a soft-iron core that is magnetized temporarily by passing a current through a coil of wire wound on the core. The operating circuit for a simple electromagnet is shown in Figure 9-30. The strength of the magnetic field is greatly increased by the addition of the soft-iron core. When current flows through the coil, the core becomes magnetized by induction. The magnetic lines of force produced by the magnetized core align themselves with those of the coil to produce a much stronger magnetic field. If current flow through the coil is interrupted, both the coil and the soft-iron core lose their magnetism.

Several factors affect the magnetic field strength of an electromagnet formed by a coil. These include:

- Core material, length, and area. For example, the greater the area of the core, the stronger the magnetic field.

- Number of turns on the coil and the spacing of the turns. The more turns and the closer they are spaced together, the stronger the magnetic field.

- Amount of current flowing through each turn. The greater the current flow, the stronger the magnetic field.

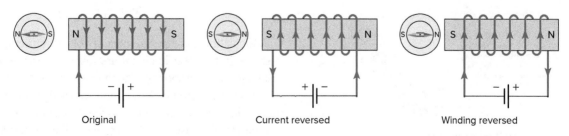

Figure 9-29 Magnetic polarity of a coil.

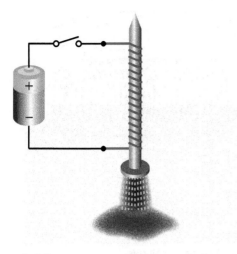

Figure 9-30 **Operating circuit for a simple electromagnet.**

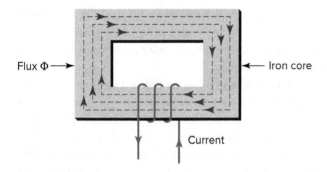

Flux Φ →   ← Iron core

Current

Figure 9-31 **Magnetic circuit.**

9.10 The Magnetic Circuit

A **magnetic circuit** (Figure 9-31) is a closed-loop path for magnetic lines of force. The total number of lines of force in the magnetic circuit is called **magnetic flux (Φ),** and the unit commonly used to measure flux is the weber (Wb). The source of a magnetic circuit's magnetomotive force (mmf) is a current-carrying coil. In an electric circuit the electron current flow is the result of an electromotive force (emf), acting

EXAMPLE 9-1

Problem: What is the magnetomotive force produced by the circuit of Figure 9-32 when a 50-ampere of current flows through the 4 turns of the coil?

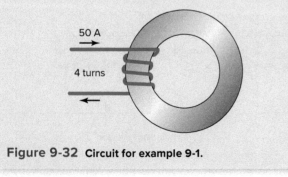

50 A

4 turns

Figure 9-32 **Circuit for example 9-1.**

Solution:

$$\text{mmf} = I \text{ (current)} \times N \text{ (number of turns)}$$
$$= 50 \text{ A} \times 4 \text{ t}$$
$$= 200 \text{ At (ampere-turns)}$$

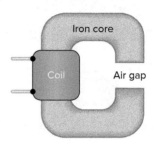

Iron core

Coil   Air gap

Figure 9-33 **Magnetic reluctance.**

on the circuit. Similarly, in a magnetic circuit the magnetic flux (Φ) is the result of a **magnetomotive force (mmf)** acting on the circuit. The magnetomotive force produced is the product of the current in amperes (A) and the number of turns (N) on the coil. The unit commonly used to measure mmf is the **ampere-turn (A · t).**

Magnetic reluctance (R) is the opposition offered by a magnetic circuit to the setting up of flux (Figure 9-33). Magnetic flux is reluctant to travel through air. It is much easier for it to travel through iron. We say that air has a high reluctance and iron has a low reluctance. This is similar to resistance in an electric circuit. The current in an electric circuit can be confined very effectively to the desired path by insulating the conducting parts from each other. Lines of force of a magnetic circuit cannot be isolated nearly as effectively.

The similarity between magnetic and electric circuits extends to Ohm's law. Just as electromotive force (E) must work against resistance (R) to produce current (I) in the electric circuit, magnetomotive force (mmf) must work against reluctance (R) to produce flux (Φ) in the magnetic circuit. The magnetic circuit Ohm's law formula states: *The flux produced by a magnetic circuit is directly proportional to the magnetomotive force and inversely proportional to the reluctance or*

$$\Phi = \frac{\text{mmf}}{R}$$

Calculation and measurement of voltage, current, and resistance in the electric circuit are relatively easy to do and useful for troubleshooting. The same is not true for quantities in the magnetic circuit. A thorough knowledge of the laws of magnetic circuits is important to the designer of equipment, but for practical use in fieldwork, we will consider this only to better understand the operation of equipment.

9.11 Electromagnetic Induction Principles

Electromagnetic induction involves the **induction** of an electromotive force by the **motion** of a conductor across a magnetic field or by a **change in magnetic flux** in a magnetic field. Figure 9-34 illustrates the electromagnetic induction of a voltage by the relative motion of a conductor across a magnetic field.

- When the permanent magnet is moved in and out of the coil, it induces a voltage and resulting current in the coil.
- Moving the magnet into the coil will deflect the DC voltmeter away from its center position in one direction only. When the magnet stops moving, the inductive effect ceases, and the needle returns back to zero because there is no physical movement of the magnetic field.
- In a similar fashion, when the magnet is moved out of the coil in the other direction, the needle deflects in the opposite direction, indicating a change in polarity. Continuous movement of the magnet back and forth within the coil will deflect the needle to the left or right, positive or negative, relative to the directional motion of the magnet.
- The amount of induced voltage produced is proportional to the speed or velocity of the movement. The faster the movement of the magnetic field, the greater will be the induced voltage in the coil.

Figure 9-35 illustrates the electromagnetic induction of a voltage by a change in magnetic flux in a magnetic field.

- When AC is applied to a coil, a varying magnetic field will be produced around it.
- When a second coil is placed within that changing magnetic field, the changing magnetic field will induce a voltage and resulting current flow in the second coil. This principle is called **mutual inductance.**

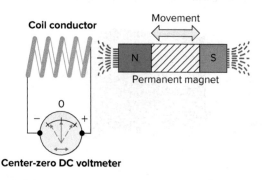

Figure 9-34 Induction of a voltage by motion.

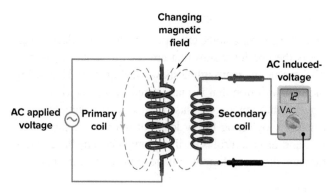

Figure 9-35 Induction by a change in the magnetic field.

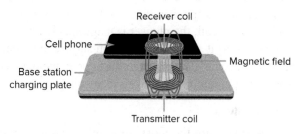

Figure 9-36 Wireless cell phone charger.

- The coil in which current is changed is called the **primary** coil, and the coil in which voltage is induced is called the **secondary** coil.

The wireless cell phone charger, shown in Figure 9-36, operates on the principle of electromagnetic induction.

- Alternating current is supplied from an electrical outlet to the base station.
- Inside the base station, a transformer steps down this voltage, which is then applied to the transmitter coil.
- Current flow in the transmitter coil creates a changing magnetic field as the current passes through it.
- The magnetic field of the transmitter coil then induces a low-voltage current in the receiver coil.
- The low-voltage current from receiver coil then flows into the cell phone battery to charge it.

9.12 Uses for Electromagnets

Electromagnets can be made much more powerful than permanent magnets. In addition, the strength of the electromagnet can be easily controlled from zero to maximum by controlling the current flowing through the coil. For these reasons, electromagnets have many more practical applications than do permanent magnets.

Crane Electromagnets

One of the most graphic examples of a working electromagnet is the one for cranes that are used to move scrap iron. The **crane electromagnet** shown in Figure 9-37 is a big block of soft iron that is magnetized by an electric current flowing through a coil. This type of electromagnet has the capability of lifting heavy loads of magnetic scrap metal. Lift-and-drop control is easily accomplished by the connection and disconnection of voltage to the electromagnet.

Solenoids

A **solenoid** is an electromagnet with a movable iron core or plunger. Upon applying power or energizing the coil, the magnetic field that is produced pulls or pushes the plunger into the coil, as illustrated in Figure 9-38. Whenever power is applied to the coil, the plunger is magnetically attracted into the stator as it slides along a nylon surface in a linear fashion. By attaching a nonferrous push rod to the plunger, a pushing or pulling motion can be accomplished.

Solenoid Valves

Solenoid valves are the most frequently used device to control liquid or gas flow. A solenoid valve has two main parts: the solenoid and the valve. Figure 9-39

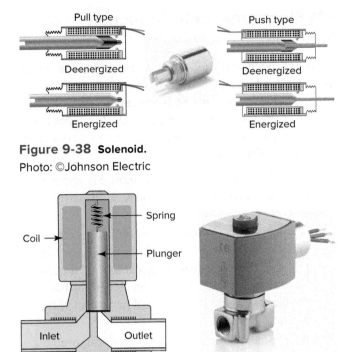

Figure 9-38 Solenoid.
Photo: ©Johnson Electric

Figure 9-39 Solenoid valve.
Photo: ©ASCO Valve, Inc.

shows a typical solenoid valve. One inlet and one outlet are used to permit and shut off fluid flow. A spring is used to hold the valve closed when the solenoid coil is deenergized. When the coil is energized, the magnetic field created pulls the plunger upward to open the valve, permitting liquid flow from inlet to outlet.

Transformers

Most home heating/cooling system's thermostats operate on low-voltage (typically 24 volts AC) control circuits. The source of the 24-volt AC power is a control transformer installed as part of the heating/cooling equipment. The advantage of the low-voltage control system is the ability to operate system load devices using inherently safe voltage and current levels.

Transformers are electrical devices that are used to raise or lower AC voltages. Figure 9-40 is a typical

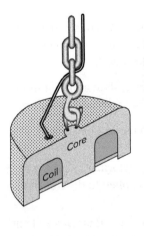

Figure 9-37 Crane electromagnet.
Photo: ©Photolibrary/Age Fotostock

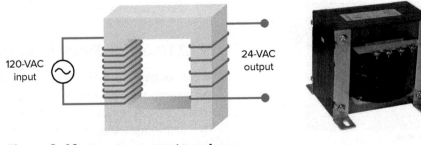

Figure 9-40 Step-down control transformer.
Photo: ©Hammond Power Solutions Inc.

step-down control transformer. The transformer uses two electromagnetic coils to transform or change the AC input voltage level. An input voltage of 120 VAC is applied to the primary coil wound around an iron core. An output voltage of 24 VAC emerges from a secondary coil also wound around the core. The AC voltage input current produces a magnetic field that continually varies in magnitude. The core transfers this field to the secondary coil where it induces an output voltage. The change in voltage depends on the ratio of turns in the primary and secondary coils.

Electricity is generated at a comparatively low voltage. In order to transport this energy great distances, the voltage has to be increased, or stepped up, to values as high as 765,000 volts. This is accomplished through the use of large step-up transformers located near the generating station and provides an efficient and economical method of transmission. A series of transformers then step-down the voltage to levels that are safe in businesses and residences.

Control Relays

An **electromagnetic control relay** is a switch that is operated by an electromagnet. The relay generates electromagnetic force when input voltage is applied to the coil. The electromagnetic force moves the armature that switches the contacts. A relay is made up of two circuits: the coil input or control circuit and the contact output or load circuit, as illustrated in Figure 9-41. Relays are used to control small loads of 15 A or less, which include solenoids, pilot lights, alarms, and small fan motors.

Generators

An **electric generator** is a machine that uses magnetism to convert mechanical energy into electric energy. Generators can be subdivided into two major categories depending on whether the electric current produced is alternating cur-

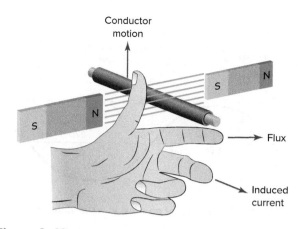

Figure 9-42 Generator principle.

rent (AC) or direct current (DC). The basic principle on which both types of generator work is the same, although the details of construction of the two may differ somewhat.

The generator principle states that *a voltage is induced in a conductor whenever the conductor is moved through a magnetic field so as to cut lines of force.* Figure 9-42 illustrates the generator principle and relationships between the direction the conductor is moving, the direction of the magnetic field, and the resultant direction of the induced current flow.

In its basic form the AC generator consists of a magnetic field, an armature, slip rings, and brushes. For most applications the magnetic field is created by an electromagnet, but for the simple generator shown in Figure 9-43, permanent magnets are used. The armature is rotated through the magnetic field and may contain any number of conductors wound in loops. As the armature is rotated, a voltage is generated in the single loop conductor which causes current to flow. Slip rings are attached to the armature and rotate with it. Carbon brushes ride against the slip rings and conduct the current from the armature to the output load.

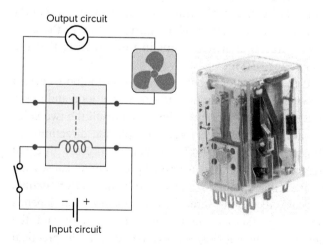

Figure 9-41 Electromagnetic control relay.

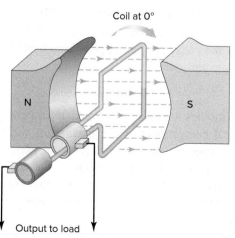

Figure 9-43 Basic AC generator.

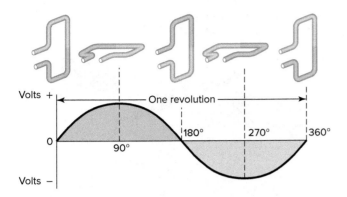

Figure 9-44 **Generation of an AC sine wave voltage.**

Figure 9-45 **AC electric motor.**
Courtesy of Baldor Electric Co.

As the armature is rotated through one complete revolution, an **AC sine wave** voltage is produced, as illustrated in Figure 9-44. This generated voltage varies in both voltage value and polarity as follows:

- At 0 degrees the coil is moving parallel to the magnetic field. It cuts no lines of force, so no voltage is generated.
- When the coil rotates from 0 to 90 degrees, it cuts more and more lines of flux. As the lines of flux are cut, voltage is generated in the positive direction and reaches a maximum value at 90 degrees.
- As the coil continues to rotate from 90 to 180 degrees, it cuts fewer and fewer lines of flux. Therefore, the voltage generated goes from maximum back to zero.
- When the coil continues to rotate past 180 to 270 degrees, each side of the coils moves through the magnetic field in the opposite direction. More and more lines of flux are cut, and a voltage is generated in the negative direction and reaches a maximum value at 270 degrees
- As the coil continues to rotate from 270 to 360 degrees, it cuts fewer and fewer lines of flux. Therefore, the voltage generated goes from maximum back to zero.

Motors

Electric motors are used to convert electric energy into mechanical energy. Motors use magnetism and electric currents to operate. There are two basic categories of motors, AC (Figure 9-45) and DC. Both use the same fundamental parts but with variations to allow them to operate from two different kinds of electric power supply, alternating current or direct current.

The operation of an electric motor depends upon the interaction of two magnetic fields. Whenever a

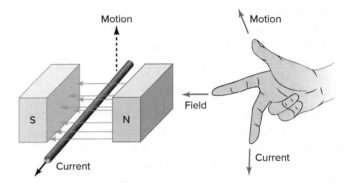

Figure 9-46 **Current-carrying conductor in a magnetic field.**

conductor carrying current is placed in a magnetic field, it will experience a force which tends to push it out of the magnetic field. The **right-hand rule for motors** (Figure 9-46) indicates the direction that a current-carrying conductor will be moved in a magnetic field. When the forefinger is pointed in the direction of the magnetic lines of force, and the center finger in the direction of the current flow in the conductor, the thumb will point in the direction that the conductor will move.

If the conductor is bent in the form of a loop, current flows in one direction on one side of the loop and in the opposite direction on the other. As a result the two sides of the coil experience forces in opposite directions, as illustrated in Figure 9-47. The pair of forces with equal magnitude but acting in opposite directions causes the loop to rotate around its axis, developing **motor torque.** The overall turning force on the armature loop depends on several factors, including field strength and the amount of current flow through the loop. In the practical motor, if the motor does not develop enough torque to pull its load, it stalls.

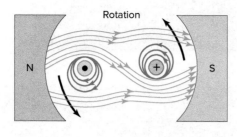
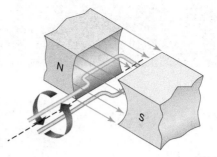

Figure 9-47 Developing motor torque.

Part 2 Review Questions

1. What is the relationship between electricity and magnetism?

2. When a compass is used to track a magnetic field, which pole of the compass will point in the direction of the lines of force?

3. What determines the strength of the magnetic field that is produced around a single conductor?

4. Assume two parallel conductors are carrying the same amount of current in the same direction:
   a. In what direction will the magnetic forces tend to move the conductors?
   b. What is the equivalent strength of the magnetic field created?

5. How are practical electromagnets constructed?

6. What two factors determine the location of the north and south poles of an electromagnet?

7. List three factors that determine the strength of an electromagnet.

8. A similarity exists between magnetic and electric circuits. Give the electric circuit equivalent for each of the following magnetic circuit quantities:
   a. Magnetic flux.
   b. Magnetomotive force (mmf).
   c. Reluctance.

9. What is the Ohm's law formula for a magnetic circuit?

10. List two advantages that electromagnets have over permanent magnets for many practical applications.

11. Explain how the lift-and-drop control of a crane-operated lifting magnet is accomplished.

12. Explain how solenoids operate.

13. Explain how solenoid valves operate.

14. In the operation of a transformer, what determines the change in voltage between the primary and secondary coils?

15. Explain how an electromagnetic control relay operates.

16. State the principle of operation of a generator.

17. In the operation of an AC generator the output voltage varies in both voltage value and polarity.
    a. What causes the output voltage to vary in value?
    b. What causes the output voltage to vary in polarity?

18. Compare the function of electric generators and motors.

19. What is the basis for the operation of an electric motor?

20. Describe the two methods used to create electromagnetic induction.

21. When does mutual inductance between two coils occur?

22. Which side of a transformer is designated as the primary and what side as the secondary?

CHAPTER TEN

Electric Power and Energy

Hydroelectric power.
©Maxim Burkovskiy/Shutterstock

LEARNING OUTCOMES

▶ Describe the various methods of power generation.

▶ Outline the method used in transmitting power over long distances.

▶ Define *electric power*.

▶ Perform calculations using the electric power formula.

▶ Properly connect a wattmeter into a circuit.

▶ Define *electric energy*.

▶ Calculate energy consumption and costs.

Early electric power plants were small and generated direct current. Today's plants are huge by comparison and use alternating current generators or alternators. The main purpose of all electrical installations is to deliver electric energy to a load. The load, in turn, uses this energy to develop power and do work for us. In this chapter we explain a variety of electric power and energy concepts and measurements.

PART 1 GENERATION AND TRANSMISSION OF ELECTRICITY

10.1 Electric Generating Stations

Most of the electric energy supplied to industries, commercial establishments, and homes is produced at a central location called a **generating station (or power plant).** This station is usually equipped with several large generators, each being driven separately. Most of the electric power generated today is three-phase, 60-cycle alternating current (AC). The circle chart of Figure 10-1 shows the proportional amount of electricity generated annually in the United States by the major energy sources.

The generation of electricity is the process of generating electric power from sources of primary energy. In general, it involves converting **chemical energy** in fuels or the **flowing energy** of wind, water, or steam into electrical energy, using a mechanical **turbine** connected to an electric **generator.**

Thermal Generation

Thermal electric generating stations are the major source of power generation. In a thermal power plant, the chemical energy stored in fossil fuels such as **coal** and **natural gas** is converted successively into thermal energy, mechanical energy, and finally electric energy. The thermal electric generating process is illustrated in Figure 10-2 and operates as follows:

- Thermal electric stations use heat to convert water to steam, which then forces a turbine wheel to turn.
- After the water in the boiler turns to steam, it is fed to a steam turbine.

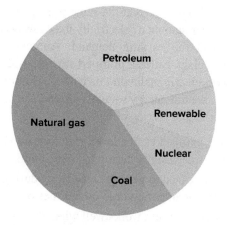

Figure 10-1 Amount of electricity generated by the major energy sources.

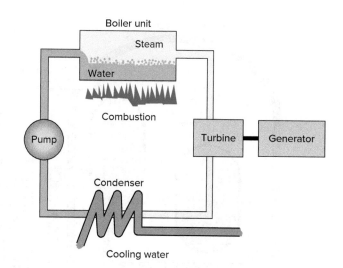

Figure 10-2 Thermal electric generating process.

- The steam rushes into the turbine under great pressure and it forces the blades around very rapidly.
- As the shaft holding the blades revolves, it spins the generator and electric energy is produced.
- In addition to its fuel requirements, a thermal plant must have a supply of water to cool and condense the exhaust steam from the turbine so that it can be reused in the steam cycle.

Nuclear Power

Nuclear fission is the process where the nucleus of a fissionable atom is split. Enormous amounts of energy are released in this process. This energy, in the form of heat, is transferred to steam turbines to generate electricity. So, apart from the source of energy, a **nuclear energy plant** is in essence no different from any fossil fuel power plant. The heart of a nuclear power generating station is its nuclear reactor. The nuclear generating process is illustrated in Figure 10-3 and operates as follows:

- The splitting of uranium atoms or nuclear fission generates immense heat.
- The heat is transferred from the reactor to a boiler where it boils water to steam.
- The turbine-generator part of a nuclear plant is basically the same as a thermal electric plant.

Hydroelectric Generation

Renewable energy is energy that is generated from natural processes that are continuously replenished. Hydroelectricity uses renewable energy derived from falling or flowing water.

Hydroelectric generating stations use water power in the form of waterfalls and water under pressure from giant dams or reservoirs to drive the generators. The **hydroelectric**

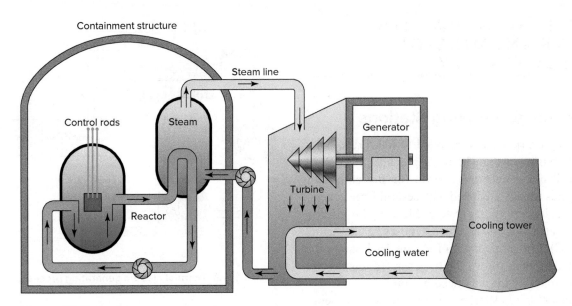

Figure 10-3 **Nuclear power generation.**

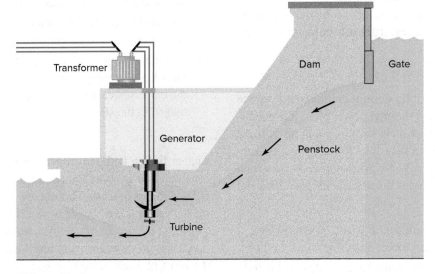

Figure 10-4 **Hydroelectric generating process.**

generating process is illustrated in Figure 10-4 and operates as follows:

- The generating station is a building located at the base of the dam.
- A large pipe, the penstock, carries water from the reservoir to the turbines. The water flows into the penstock and races down the long slope.
- At the bottom, it meets the blades of the turbine with great force and makes the turbine spin.
- A generator is attached to the turbine, and it also begins turning.
- The flow of water is adjustable to match power demand. The turbine must run at a constant speed in order to generate electricity at a constant frequency.

Electricity is consumed at the same time as it is generated. The proper amount of electricity must always be provided to meet the varying demand. Another type of hydroelectric plant, called a **pumped storage system,** can be used to efficiently maintain this balance (Figure 10-5). These plants are **not** energy sources; instead, they are storage devices that operate as follows:

- Water is pumped from a lower reservoir into an upper reservoir, usually during off-peak hours.
- Flow is reversed to generate electricity during the daily peak load period or at other times of need.
- This type of system works well when the water is pumped up into the reservoir at night, when electricity costs are low, and the water is used to generate

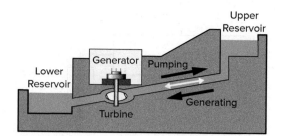

Figure 10-5 **Pumped storage hydroelecric plant.**

electricity during the day, when the need for electricity is highest.

- Although the losses of the pumping process make such a plant a net energy consumer, the plant provides large-scale energy storage system benefits.

10.2 Alternative Ways of Generating Electricity

At the present time, we are seeing a growing number of small energy suppliers using a variety of renewable energy sources to produce electricity.

Electricity from Wind Power

Wind power systems convert the kinetic energy in wind into electric energy. **Wind turbines** are installed in locations

with strong, sustained winds. Figure 10-6 shows a wind turbine used to generate electricity. Its operation can be summarized as follows:

- The energy in the wind turns two or three propeller-like blades around a rotor.
- The rotor is connected to the main shaft, which spins a generator to create electricity.
- Wind turbines are typically mounted on a tower at an elevation high above grade away from ground obstructions where wind currents are strong and consistent.
- To extract as much energy from the wind as possible, the wind turbine blades are huge—up to 330 feet tip to tip.
- Wind sensors enable the turbine's computer to control the movement of the rotor and to produce optimum power in all wind conditions.
- Wind turbines are usually grouped together in what are called **wind farms.**
- The electrical power from the generator is typical 60-Hz, AC power with 600-volt output for large wind turbines. A transformer may be required to increase or decrease the voltage so it is compatible with the end use, distribution, or transmission voltage, depending on the type of interconnection.

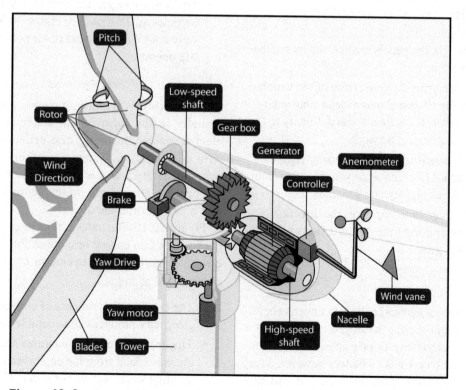

Figure 10-6 **Wind turbine used to generate electricity.**

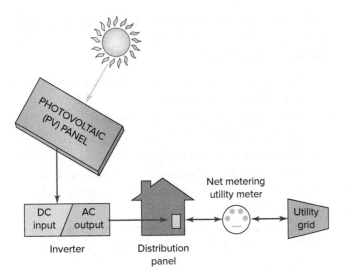

Figure 10-7 **Grid-tied solar system.**

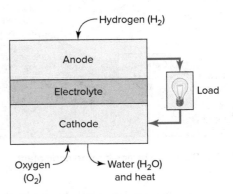

Figure 10-8 **Hydrogen-oxygen fuel cell.**

Electricity from Solar Energy

Solar energy is energy harnessed from the sun. Solar electric or **photovoltaic (PV)** technology converts sunlight directly into electricity. Figure 10-7 illustrates a grid-tied solar system. Its operation can be summarized as follows:

- Grid-tied solar electric systems generate electricity for your home or business and route the excess power into the electric utility grid for compensation from the utility company.
- Photovoltaic (PV) panels gather solar energy in the form of sunlight and convert it into direct current (DC) electricity.
- The more sunlight the panels receive, the more electricity they produce.
- The inverter performs the conversion of the variable DC output of the PV panel into a clean sinusoidal 60-Hz AC current that is then applied directly to the commercial electric grid network.
- **Net metering** means the utility company charges you the difference between what you consume from the grid and the electricity you generate.
- Current from solar panels reverses the direction of the electric meter during the day as it flows back into the electric grid.

Fuel Cell

A **fuel cell** is an electrochemical cell that converts chemical energy from a fuel such as hydrogen and an oxidant (air or oxygen) into electricity. In principle, a fuel cell operates much like a battery. Unlike a battery however, a fuel cell does not run down or require recharging. It will produce electricity as long as fuel and an oxidizer are sup-

plied. The block diagram of a hydrogen-oxygen fuel cell is shown in Figure 10-8 and operates as follows:

- The fuel cell device uses hydrogen fuel and oxygen to create electricity.
- The hydrogen can come from a variety of sources (hydrogen, natural gas, methanol, and gasoline).
- The cell contains an anode electrode, a cathode electrode and an ion-conducting material called an **electrolyte.**
- Hydrogen passes over one electrode and oxygen over the other, generating electricity, water, and heat.
- Fuel cells are classified by their electrolyte material.
- Hydrogen-oxygen fuel cells can be used to power vehicles as illustrated in Figure 10-9. These vehicles only emit water, instead of air pollutants, while they are operating.

Cogeneration Power Systems

Cogeneration is the simultaneous production of heat (usually in the form of hot water and/or steam) and electric power, utilizing one primary fuel. Facilities with cogeneration systems use them to produce their own electricity and use the unused excess (waste) heat for process steam, hot water heating, space heating, and other thermal needs.

Figure 10-10 illustrates a typical cogeneration system consisting of an engine or combustion turbine driving an electric generator. The operation of the system can be summarized as follows:

- The cogeneration process of converting fuel into electricity produces considerable amounts of heat.
- The heat recovery unit captures and makes use of the waste heat from the engine and exhaust gases by creating steam and/or hot water. This heat is then routed to the facility's heat distribution system.

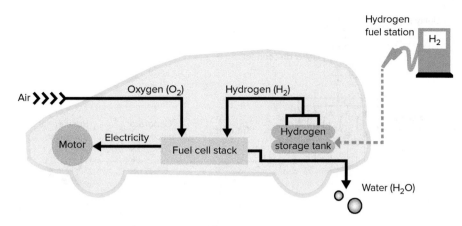

Figure 10-9 **Fuel cell vehicle.**

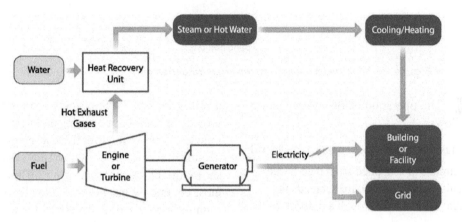

Figure 10-10 **Typical cogeneration system.**

- The electricity created is tied into the main switch gear and operates in parallel with the local utility, reducing the facility's electrical load on the local grid.

Emergency Power Supply Systems

When power in an electrical system is disrupted for an extended period, life-threatening conditions may occur. Continuous lighting and power are essential in places of public assembly, theaters, hotels, sports arenas, health care facilities, and the like. **Engine-driven alternators** are commonly used to provide **emergency power** in the event of a power failure. Figure 10-11 shows a residential standby generator unit that allows ongoing operation of essential loads during a utility outage. Such units include a small engine and generator mounted together and called a **generator set.** Natural gas or liquid propane gas is a commonly used fuel source available for powering the engine. Normally the generator set is mounted on a cement pad, preferably near the fuel source.

Residential generator installations must be coordinated with the local electric utility, including obtaining any necessary approvals and taking all safety precautions

Figure 10-11 **Residential standby generator.**
©Aleksandr Rado/123RF

to prevent feedbacks onto the utility system. A common approach in the design of a standby system is to divide the dwelling's circuits into essential and nonessential loads, as illustrated in Figure 10-12. The **essential loads** design approach requires a smaller generator with a

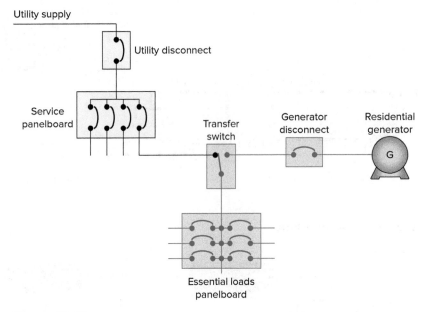

Figure 10-12 Essential loads standby generator design.

lower-power rating. The operation of the system can be summarized as follows:

- Essential circuits have their own separate panelboard.
- Nonessential circuits are supplied from the service panelboard. This service panelboard subfeeds the smaller essential loads panel through a transfer switch.
- If a utility power outage occurs, the transfer switch (either manual or automatic) changes from utility power to generator power.
- The nonessential loads in the service equipment are shed, and the generator supplies electricity only to the essential loads.

Even a brief power outage can result in loss of unsaved data on a personal computer (PC) that was running. An **uninterruptable power supply (UPS)** is designed to supply electricity to a load for a certain period of time during

a utility failure. UPS systems are normally designed to provide operating time during an interruption from 5 to 20 minutes. There are several basic methodologies used by manufacturers in the design of UPS systems. Figure 10-13 shows the block diagram of an **offline** UPS system, the most commonly used type for small applications such as running a single desktop computer or workstation. The operation of the system can be summarized as follows:

- In normal operation, the main AC power input is supplied directly to the load.
- Should a voltage drop occur, the UPS then switches on instantaneously by means of a static switch to a battery-powered inverter, providing backup power to the load.
- The rectifier/charge regulator converts mains supply AC to DC, keeping the battery charged to full capacity. The charge regulator protects the battery bank from overcharge and prevents excessive discharge.

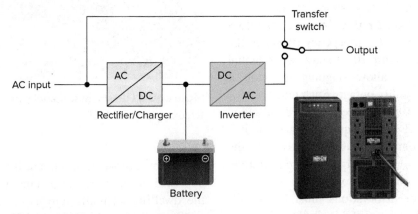

Figure 10-13 Typical offline UPS system.
Photo: Courtesy of Tripp Lite

10.3 Transmitting Electricity

Electric power transmission refers to the large-scale transfer of electric energy, from generating power plants to substations located near population centers. To minimize electricity losses in the transmission system, power is transmitted across lines with **high voltage and low current.**

The power transmitted in a system is proportional to the voltage multiplied by the current. If the voltage is raised, the current can be reduced to a small value while still transmitting the same amount of power. For example, transmitting 10,000 watts of power at 100 volts would require a current of 100 amperes, as illustrated in Figure 10-14:

$$\text{Power} = \text{volts} \times \text{amperes}$$
$$P = E \times I$$
$$= 100\ \text{V} \times 100\ \text{A}$$
$$= 10,000\ \text{W}$$

If the transmission voltage is stepped up to 10,000 volts (Figure 10-15), a current flow of only 1 ampere is needed to transmit the same 10,000 watts of power:

$$\text{Power} = \text{volts} \times \text{amperes}$$
$$P = E \times I$$
$$= 10,000\ \text{V} \times 1\ \text{A}$$
$$= 10,000\ \text{W}$$

Because of the reduction of current flow at high voltage, the **size and cost** of wiring are greatly reduced. Reducing the current also minimizes the **voltage and power losses** in the lines.

Without transformers the widespread distribution of electric power would be impractical. **Transformers**

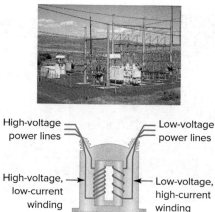

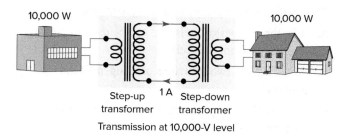

Figure 10-14 **Transmission of 10,000 watts of power at 100 volts.**

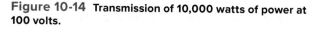

Figure 10-15 **Transmission of 10,000 watts of power at 10,000 volts.**

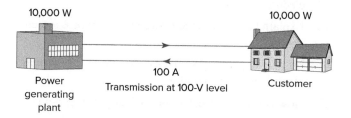
Figure 10-16 Distribution transformer.
Photo: ©KingWu/E+/Getty Images

(Figure 10-16) are electrical devices that transfer energy from one electric circuit to another by magnetic coupling. Their purpose in a power distribution system is to convert AC power at one voltage level to AC power of the same frequency at another voltage level.

There are certain **limitations to the use of high voltage** in power transmission and distribution systems. The higher the voltage, the more difficult and expensive it becomes to safely insulate between line wires, as well as from line wires to ground. The use of transformers in power systems allows generation of electricity at the most suitable voltage level for generation and at the same time allows this voltage to be changed to a higher and more economical voltage for transmission. At the load centers transformers allow the voltage to be lowered to a safer voltage and more suitable voltage for a particular load, as illustrated in Figure 10-17.

The place where the conversion from transmission to distribution occurs is in a power **substation.** It has transformers that step transmission voltage levels down to distribution voltage levels. Basically a **power substation** consists of equipment installed for switching, changing, or regulating line voltages. The power needs of some

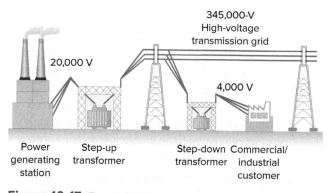

Figure 10-17 **Transmission stages of a power distribution system.**

Figure 10-18 Typical onsite unit substation.
Courtesy of Schneider Electric

users are so great that they are fed through individual substations dedicated to them. Most large commercial and industrial customers will have their own onsite **unit substation** that receives power at the transmission voltage level (Figure 10-18).

There are several different types of electrical disturbances that can affect the **power quality** of a distribution system and have an adverse affect on the operation of sensitive equipment. These can be summarized as follows:

- **Momentary power interruption.** Momentary power interruptions are brief disruptions in electric service, usually lasting no longer than a few seconds. These interruptions are the result of temporary faults such as lightning strikes, fallen branches, or animals (for example, squirrels) coming into contact with power lines. Power is automatically restored once the fault is cleared.

- **Power outage.** A power outage (also **blackout or power failure**) is a short- or long-term loss of the electric power to an area. Causes include faults at power stations; damage to electric transmission lines, substations, or other parts of the distribution system; a short circuit; or the overloading of electricity mains.

- **Sag or brownout.** A decrease in voltage levels of more than 10 percent, which is usually of short duration but can last from fractions of a second to hours. Sags or brownouts can be caused by heavy equipment coming on line, short circuits, or undersized electric circuitry. Also, the utility may deliberately decrease voltage levels in an attempt to cope with peak load times by forcing a reduction in power.

- **Voltage spike.** An instantaneous (usually no longer than one-millionth of a second) and tremendous increase in voltage levels. Spikes are generally caused by a nearby lightning strike, or if utility power lines are downed in a storm or as a result of an accident.

- **Voltage surge.** An increase in voltage of more than 10 percent above the rated nominal line voltage lasting from 0.5 cycle up to 1 minute. Voltage surges can occur when a large load is turned off. As an example, a surge in voltage commonly occurs in office areas of a plant when production lines with large loads are shut down.

- **Electrical noise.** A disruption in the normal smooth AC sine wave waveform caused by electromagnetic interference (EMI) and radio frequency interference (RFI). Lightning, load switching, generators, transmitters, transformers, and industrial equipment commonly cause noise.

- **Harmonic distortion.** A distortion of the normal AC current waveform caused by the presence of frequencies other than the standard 60-cycle fundamental frequency. Nonlinear loads that draw current in short pulses are common sources of harmonic distortion. These include variable-speed motor drives, electronic ballasts used in lighting circuits, and power supplies used in personal computers.

Part 1 Review Questions

1. How is electricity generated using the thermal process?

2. How is electricity generated using the hydroelectric electric process?

3. Explain how a pumped hydroelectric storage system operates.

4. Name two primary sources of heat used for thermal electric power generation.

5. How is the heat created for nuclear power generation?

6. Explain how electricity is generated by a wind turbine.

7. Explain how a grid-tied solar system works.

8. List three types of fuels that can be used to create electricity from fuel cells.

9. What are the two functions of a cogeneration system?

10. How is electricity generated in a residential standby generator unit?

11. Explain the essential loads approach to the design of an emergency power supply system.

12. What is the main application for an uninterruptable power supply?

13. Assume 100 kW of power is to be transmitted. Determine the amount of line current when transmission voltages of 1,000 volts and 100,000 volts are used.

14. What is the function of a transformer as part of a power distribution system?

15. What are the limitations to the use of high-voltage levels in power transmission and distribution systems?

16. At what point in a typical power grid system does the conversion from transmission to distribution occur?

17. Define each of the following types of electrical disturbances that can occur on a power distribution system:
   a. Momentary power interruption.
   b. Power outage.
   c. Brownout.
   d. Electrical noise.
   e. Harmonic distortion.

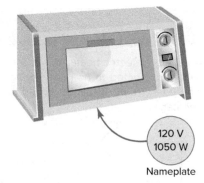

Figure 10-19 Nameplate power rating.

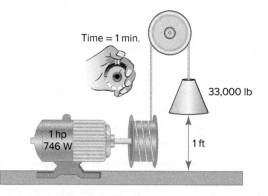

Figure 10-20 One horsepower.

PART 2 POWER METERING AND COSTS

10.4 Energy, Work, and Power

The terms energy, work, and power can be defined as follows:

Energy is the capacity for doing work.

Work is the result of conversion of energy from one form to another.

Power is the rate at which work is done.

Electric energy is the work performed by an electric current. Whenever current exists in a circuit, there is a conversion of electric energy into other forms of energy. For example, current flow through a lamp filament converts electric energy into heat and light by heating the filament to a very high temperature, causing it to glow.

Electric power is the rate of doing work using electricity. The watt (W) is the base unit used to measure electric power. Electrical devices normally carry an electrical wattage or power rating, as illustrated in Figure 10-19. The wattage rating indicates the rate at which the device can convert electric energy into another form of energy, such as light, heat, or motion. The faster a device converts electric energy, the higher its wattage rating. The power rating of a device can be read directly from its nameplate or calculated from other values given.

If the normal wattage rating is exceeded, the equipment or device will overheat and perhaps be damaged. For example, if a lamp is rated for 100 watts at 120 volts and is connected to a source of 240 volts, the current through the lamp will double. The lamp will then use four times the wattage for which it is rated, and it will overheat and burn out quickly.

There is a direct relationship between measurements of electric power and mechanical power. **One horsepower (hp)** is a unit of mechanical power equal to **746 watts** of electric power. One horsepower is also defined as the amount of energy or work required to raise a weight of 33,000 pounds a height of 1 foot in 1 minute of time (Figure 10-20).

10.5 Calculating Electric Power

In any DC circuit, or AC circuit that contains resistive type loads, power is equal to the voltage multiplied by the current. Electric power can be calculated directly in watts using the measurements of voltage and current as follows:

$$P = E \times I$$

where P = power in watts (W)
E = voltage in volts (V)
I = current in amperes (A)

EXAMPLE 10-1

Problem: A portable electric heater draws a current of 8 amperes when connected to its rated voltage of 120 volts, as illustrated in Figure 10-21. Determine the power rating of the heater.

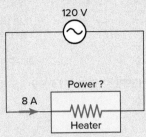

Figure 10-21 Circuit for example 10-1.

Solution:

$$\text{Power} = \text{Voltage} \times \text{Current}$$
$$P = E \times I$$
$$= 120 \text{ V} \times 8 \text{ A}$$
$$= 960 \text{ W}$$

Electrical load devices such as lights and appliances are designed to operate at a specific voltage. They may be marked with both their operating voltage and power output at that voltage. If the current rating is not marked, it can be calculated by using the rated voltage and power values. To calculate the current, the power is divided by the rated voltage:

$$I = \frac{P}{E}$$

EXAMPLE 10-2

Problem: The lightbulb shown in Figure 10-22 is rated for 150 watts and 120 volts. Determine the amount of current flowing through the lamp when it is operated at 120 volts.

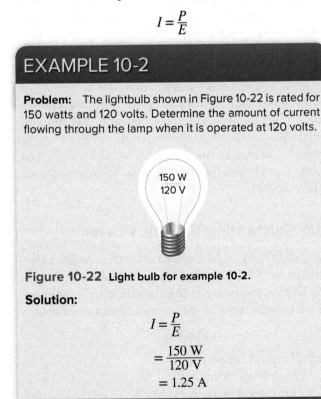

Figure 10-22 Light bulb for example 10-2.

Solution:

$$I = \frac{P}{E}$$
$$= \frac{150 \text{ W}}{120 \text{ V}}$$
$$= 1.25 \text{ A}$$

In some cases it is more convenient to calculate power by using the resistance of the component rather than the voltage applied to it. This occurs at times when calculating the power dissipated in a resistor or lost as heat in a conductor. By substituting the product IR for E in the basic power formula, we get

$$P = E \times I$$
$$P = IR \times I$$
$$= I^2 \times R$$

EXAMPLE 10-3

Problem: The current flowing through a 100-ohm resistor is measured and found to be 0.5 ampere, as illustrated in Figure 10-23. Determine:

  a. How much power is being dissipated in the form of heat?

  b. How much power would be required to be dissipated if the current was doubled to 1 ampere?

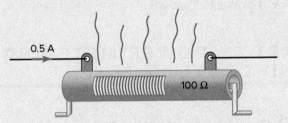

Figure 10-23 Resistor for example 10-3.

Solution:

  a. $P = I^2 \times R$
$$= 0.5 \text{ A} \times 0.5 \text{ A} \times 100 \text{ }\Omega$$
$$= 25 \text{ W}$$

  b. $P = I^2 \times R$
$$= 1 \text{ A} \times 1 \text{ A} \times 100 \text{ }\Omega$$
$$= 100 \text{ W (4 times the original power)}$$

When working with very larger amounts of power, it is more convenient to express the power in terms of kilowatt (kW) and megawatt (MW). A **megawatt** (MW) is 1 million watts and a **kilowatt** (kW) is 1 thousand watts. For instance, a 100-MW-rated wind farm may be capable of producing 100 MW during peak winds.

Equipment nameplate ratings (generally power or current along with rated voltage) are accurate only when the equipment is operated at the manufacture's **rated voltage.** Fixed resistance is determined from the nameplate information, and this resistance does not change regardless of the applied operating voltage.

EXAMPLE 10-4

Problem: The hot water heater element, shown in Figure 10-24, is rated by the manufacturer for 1.15 kW, 230 V, and 5 A. Assuming the heater element is operating at 215 volts, determine:

  a. The resistance of the element.
  b. The current drawn by the element.
  c. The amount of wattage dissipated by the element.

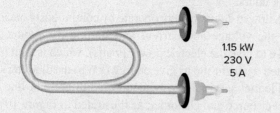

1.15 kW
230 V
5 A

Figure 10-24 Element for example 10-4.

Solution:

  a. $R = \dfrac{E}{I}$

$= \dfrac{230\ \text{V}}{5\ \text{A}}$

$= 46\ \Omega$

  b. $I = \dfrac{E}{R}$

$= \dfrac{215\ \text{V}}{46\ \Omega}$

$= 4.67\ \text{A}$

  c. $P = E \times I$

$= 215\ \text{V} \times 4.67\ \text{A}$

$= 1{,}004\ \text{W}$

$= 1\ \text{kW}$

10.6 Measuring Electric Power

Electric power is measured by means of a wattmeter. The **wattmeter** is a combination of both a voltmeter and an ammeter. It measures both voltage and current at the same time and indicates the resultant power value. The analog wattmeter contains two coils—one for voltage and the other for current, as illustrated in Figure 10-25. The **voltmeter section** is connected the same as a regular voltmeter is, in parallel or across the load. As the voltage coil responds to voltage, it has a multiplier resistor in series with it. The **ammeter section** is connected like a regular ammeter is, in series with the load. Wattmeters are rated for maximum-measured current and voltage as well as power.

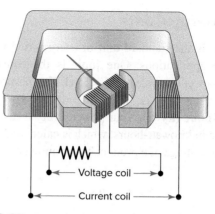

Figure 10-25 Analog wattmeter.

A wattmeter operates on DC as well as AC since when the current reverses, the magnetic polarity of both coils reverses and the turning force is still in the same direction. However, wattmeters are more useful in AC circuit measurements than in DC measurements. With **DC circuits,** watts always **equal volts times amps.** In **nonresistive AC circuits,** there are situations in which the watts **do not equal volts times amps,** but the wattmeter still indicates the true or real power consumed in the circuit. It automatically compensates for any phase difference between the circuit voltage and current.

Figure 10-26 illustrates a typical in-circuit connection of a wattmeter. When connecting the wattmeter into either a DC or AC circuit the relative polarity of the voltage and current coil must be observed in order to ensure a forward movement of the pointer. Connecting the plus-minus (±) terminals of both the voltage coil and the current coil to one side of the load provides the correct polarity. The other end of the current coil is connected to one side of the line, ensuring that the entire load current flows through the current coil. The other end of the voltage coil is connected to the other side of the line, which puts the full line voltage across the voltage coil circuit.

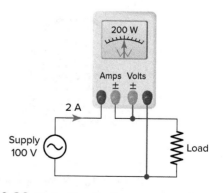

Figure 10-26 Typical in-circuit connection of a wattmeter.

10.7 Energy Costs

The joule is the standard unit of energy used in general scientific applications. One **joule** is the equivalent of 1 watt of power dissipated for 1 second. The most commonly used practical unit for measuring electric energy is the **kilowatt-hour (kWh).** Your power utility measures electricity use in kilowatt-hours, which is calculated by multiplying the wattage of a product by the number of hours it's in use:

$$E = P \times t$$

where E = energy in kilowatt-hours (kWh)

P = power in kilowatts (kW)

t = time in hours (h)

If P and t are not specified in kilowatts and hours, respectively, then they must be converted to those units before determining E in kilowatt-hours.

Kilowatt-hour meters (Figure 10-28) measure the electric energy used in an electric installation. The meter records how many watts of power are used over a period of time. Periodic readings of electric meters establish billing cycles and energy used during a cycle. The meters fall into two basic categories, **electromechanical** and **electronic.** Mechanical-type kWh meters will either have dials or a register similar to the odometer on your car. Electronic meters will have a digital display that flashes metered quantities. Electronic meters can also transmit readings to remote places. In addition to measuring energy used, they can also support time-of-day billing, for example, recording the amount of energy used during on-peak and off-peak hours.

Kilowatt-hours meters operate by continuously measuring the voltage and current and finding the product of these to give the electric power (watts), which is then integrated against time to give energy (kWh) used. Meters for residential services can be connected directly in line between source and customer, as illustrated in Figure 10-29.

The electric utility charges for the total amount of energy we use during a billing period. Electric energy is sold by the kilowatt-hour. The price per kWh varies widely, and your bill might have multiple charges per kWh. The bill

---

EXAMPLE 10-5

Problem: The electric coffeemaker shown in Figure 10-27 is rated for 900 W of power. Assuming it is used an average of 6 hours per month, calculate the monthly energy consumption in kilowatt-hours.

Solution:

$$\text{Energy} = \text{Power} \times \text{Time}$$

$$E = \frac{900\ \text{W}}{1,000} \times 6\text{h}$$

$$= 5.4\ \text{kWh}$$

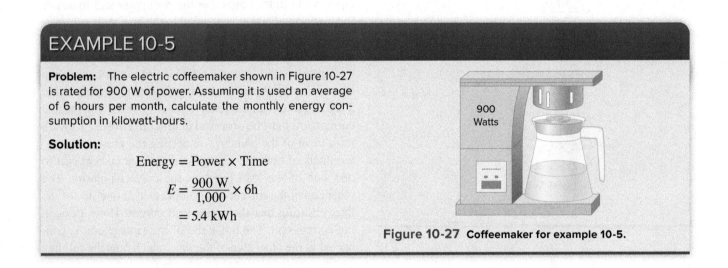

Figure 10-27 Coffeemaker for example 10-5.

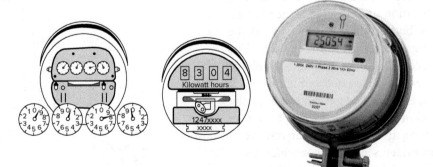

Figure 10-28 Types of kWh meters.
Photo: ©kenneth-cheung/Getty Images

Figure 10-29 Residential service kWh meter connection.

from the utility company always gives the previous and present meter readings. The difference is the amount used. The company also issues a rate card with the charge per kilowatt-hour. To calculate your bill, you simply multiply the kWh used by the cost per kWh.

EXAMPLE 10-6

Problem: The billing from an electric utility company indicates a present meter reading of 5,060 kWh and a previous reading of 3,140 kWh. Determine the energy cost for this billing period based on a rate 15 cents per kWh.

Solution:

$$\text{Present reading} = 5{,}060 \text{ kWh}$$
$$\text{Previous reading} = 3{,}140 \text{ kWh}$$
$$\text{kWh used} = 5{,}060 \text{ kWh} - 3{,}140 \text{ kWh}$$
$$= 1{,}920 \text{ kWh}$$
$$\text{Cost} = \text{kWh} \times \text{Rate}$$
$$= 1920 \text{ kWh} \times \$0.15$$
$$= \$288.00$$

The amount of electric energy used by load devices depends on two factors. One is the length of **time** they are used. The other factor is the amount of electric operating **power** required by each load device. The kWh meter keeps a record of the total electric energy used, but does not provide any information on the energy usage of the individual load devices. Energy costs for individual load devices can be calculated from their rated power consumption and normal operating cycle.

EXAMPLE 10-7

Problem: An electric clothes dryer rated for 4.2 kW is used an average of 20 hours per month. Assuming the rate for the energy used is 18 cents per kWh, calculate the cost of operating the dryer for one month.

Solution:

$$\text{Energy} = \text{Power} \times \text{Time}$$
$$E = 4.2 \text{ kW} \times 20 \text{ h}$$
$$= 84 \text{ kWh}$$
$$\text{Cost} = \text{Energy} \times \text{Rate}$$
$$= 84 \text{ kWh} \times \$0.18$$
$$= \$15.12$$

Electricity is not a natural resource. It is difficult to store large amounts of electric energy until it is wanted. For large-scale use, it must be manufactured as it is needed. This characteristic creates many problems for the power-supply authority. It must be able to generate electric energy to meet varying demands. **Time-of-day metering** allows customers to be charged at different rates depending on the time of day that they are consuming power. A higher **on-peak** rate is charged during the daylight hours of Monday to Friday, and a lower **off-peak** rate is charged overnight and on the weekends. With this incentive, medium- and large-energy users need to monitor their total power needs closely and to know exactly where and when their energy is used.

One way companies are realizing energy savings is through **building automation systems** (Figure 10-30).

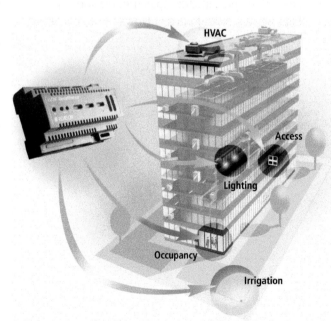

Figure 10-30 Building automation system.
Graphic was provided on behalf of Echelon Corporation

The objective of a building automation system is to achieve an optimal level of control of occupant comfort while minimizing energy use. The earliest forms of energy management involved simple time clock- and thermostat-based systems. Indeed, many of these systems are still being used. Typically, these systems are wired directly to the end-use equipment and mostly function autonomously from other system components. With today's increasing availability of **microprocessor-based systems,** energy management has quickly moved to its current state of a computer-based, digitally controlled system.

Part 2 Review Questions

1. Define the terms *energy, work,* and *power.*
2. What is the basic unit used to measure electric power?
3. Explain the relationship between the wattage rating of a device and the rate at which it converts electric energy into another form.
4. What is 1 horsepower (hp) of mechanical power equal to in terms of watts of electric power?
5. Calculate the power rating of a range element that draws 7.5 amperes of current when connected to its rated voltage of 230 volts.
6. How many amperes will a 1,250-W toaster draw when connected to its rated voltage of 120 V?
7. How much power is dissipated in the form of heat when 0.5 A of current flows through a 50-Ω resistor?

8. A water heater has a nameplate rating of 7,400 watts at 250 volts. How much current will the heater draw when operated at 208 volts?
9. A wattmeter is a combination of what two types of meters?
10. Wattmeter measurements are more useful in AC circuit measurement than in DC measurement. Why?
11. What is the practical unit for measuring electric energy?
12. A 150-W lightbulb is on for a total of 16 hours. How many kilowatt-hours of energy are converted?
13. What two metering functions are possible only with electronic kWh energy meters?
14. A 2.5-kW water heater is operated for a total of 42 hours over a 1-month period. Calculate the cost of operating the heater for this time period at a rate of 10 cents per kWh.
15. The billing from an electric utility indicates a present reading of 31,450 kWh and a previous reading of 30,000 kWh. Determine the energy cost for this period based on a rate of 12 cents per kWh.
16. How does time-of-day metering work?
17. What is the main object of a building automation system?
18. a. How much power loss is there in a wire carrying 20 A of current if the resistance of the wire is 0.05 Ω?
    b. How much voltage would be dropped by the wire's resistance?
19. A portable generator is to be purchased to operate a 2.5-hp motor load. What maximum power rating (in watts) of this generator would be required?

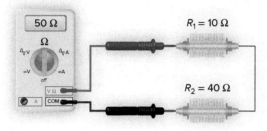

$R_1 = 10 \, \Omega$

$R_2 = 40 \, \Omega$

$50 \, \Omega$

Series circuit resistance.

Direct Current (DC) Circuits

SECTION OUTLINE

Solving the DC Series Circuit

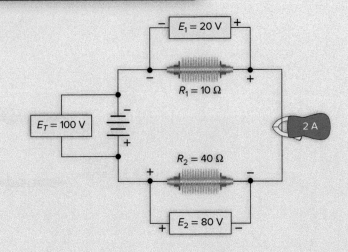

$E_1 = 20$ V

$R_1 = 10$ Ω

$E_T = 100$ V

2 A

$R_2 = 40$ Ω

$E_2 = 80$ V

Series circuit voltage and current.

LEARNING OUTCOMES

▶ Understand the voltage, current, resistance, and power characteristics of a DC series circuit.

▶ Solve for unknown circuit values in a DC series circuit.

▶ Measure current, voltage drops, and resistance values in a series DC circuit.

▶ Apply the concepts of relative polarity, series-aiding, and series-opposing voltages.

▶ Troubleshoot a series DC resistive circuit.

The main feature of a DC series circuit is the way its component parts are connected: They form a single loop, beginning and ending at the power supply. As a result only one path for current is established between the negative (−) and positive (+) terminals of the power supply. Series circuits obey a specific set of rules that apply only to them. This chapter examines these special characteristics of a series circuit.

PART 1 SERIES CIRCUIT CHARACTERISTICS

11.1 Series Circuit Connection

A **series circuit** contains one or more loads but only one path for current to flow from the source voltage through the loads and back to the source. All the components in the circuit are connected in a single line **end to end,** as illustrated in Figure 11-1. Only one pathway can be traced for the flow of current from one side of the voltage source to the other. If this path is broken at any point, all current flow in the circuit stops.

The symbols used when referring to voltage, current, resistance, and power are E, I, R, and P, respectively. In circuits that contain more than one load resistor, it is necessary to use a system of *letter and number subscripts* to correctly identify the different circuit quantities.

Figure 11-2 shows an example of how quantities are identified in a series circuit. The total resistance of the circuit is represented by the symbol R_T and the individual resistors by the symbols R_1, R_2, R_3, etc. The applied source voltage is represented by the symbol E_T and the voltage drop across the individual resistors by the symbols E_1, E_2, E_3, etc. The total or source current is represented by the symbol I_T and the current flow through the individual resistors is represented by the symbols I_1, I_2, I_3, etc. The total power dissipated is represented by the symbol P_T while the power dissipated by the individual resistors is given by the symbols P_1, P_2, P_3, etc.

11.2 Determining Current Flow

The current is the **same value** throughout a series circuit. This is because there is only one current path. The same amount of current must flow through each component of the circuit. Although the load resistors may be different in value, they all carry the same amount of current when connected in series:

$$I_T = I_1 = I_2 = I_3 \cdots$$

To determine the current in a series circuit, only the current through one of the components need be known. Since there is only one path for current, the current reading will be the same regardless of what point the ammeter is connected into the circuit. The fact that the same current flows through each component of a series circuit can be verified by inserting ammeters into the circuit at various points, as shown in Figure 11-3. If this were done, each meter would be found to indicate a current value of 2 amperes.

The total resistance of a series circuit is equal to the **sum** of the individual load resistances, as illustrated in Figure 11-4. Since there is only one path for electrons to follow, they must flow through each of the load resistors in their journey from one side of the voltage source to the other. Therefore, the total resistance in a series circuit is equal to the sum of all resistive load components of the circuit:

$$R_T = R_1 + R_2 + R_3 \cdots$$

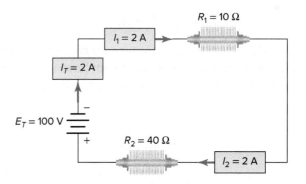

Figure 11-3 **Current is the same value at all points in a series circuit.**

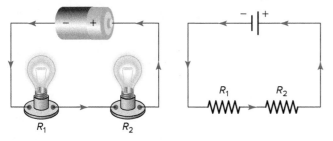

Figure 11-1 **DC series circuit.**

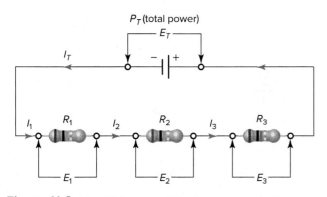

Figure 11-2 **Identifying quantities in a series circuit.**

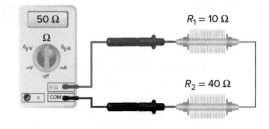

Figure 11-4 **Series circuit resistance.**

The current that flows in a series circuit depends on the applied voltage and the total circuit resistance. When the supply voltage (E_T) and the total resistance (R_T) are known, Ohm's law for current can be applied as follows to find the total current (I_T):

$$I_T = \frac{E_T}{R_T}$$
$$= \frac{100 \text{ V}}{50 \text{ }\Omega}$$
$$= 2 \text{ A}$$

Changes in either the total series circuit resistance or the total applied voltage will affect the total current. For example, if the applied voltage doubles and the circuit resistance remains the same, current will double. Similarly, if the circuit resistance doubles and the applied voltage remains the same, the current will decrease to one-half its original value.

11.3 Determining Voltage

Total voltage applied to a series circuit is divided between each of the loads. The voltage across each load resistor is known as a **voltage drop.** Voltage drop is directly proportional to the resistance value of the load component. The higher the resistance value, the greater the voltage drop. Components in series having the same resistance value will have the same amount of voltage drop. Total source voltage is then the sum of the voltage drops across each of the individual load resistors:

$$E_T = E_1 + E_2 + E_3 \cdots$$

The value of the voltage drop across a component load in a series circuit depends on the current flowing through the load and its resistance, as illustrated in Figure 11-5.

When the current flow and the resistance of a series-connected load are known, Ohm's law for voltage can be applied to determine its voltage drop:

$$E_1 = I_1 \times R_1$$
$$= 2 \text{ A} \times 10 \text{ }\Omega$$
$$= 20 \text{ V}$$
$$E_2 = I_2 \times R_2$$
$$= 2 \text{ A} \times 40 \text{ }\Omega$$
$$= 80 \text{ V}$$

Changes in either the total series circuit resistance or the total applied voltage will affect the total current. This in turn will affect the voltage drops across the load resistors. An increase in the applied voltage causes the voltage drop across each component to increase by the same factor: If the voltage of the source is doubled, the component voltage drops will also double.

11.4 Determining Individual Resistance Values

The resistance value of the individual resistive loads of a series circuit depends on the current flowing through a particular load and the voltage drop across it. When the current flow and the voltage of a series-connected load is known, Ohm's law for resistance can be applied to determine its resistance, as illustrated in Figure 11-6.

$$R_1 = \frac{E_1}{I_1}$$
$$= \frac{20 \text{ V}}{2 \text{ A}}$$
$$= 10 \text{ }\Omega$$

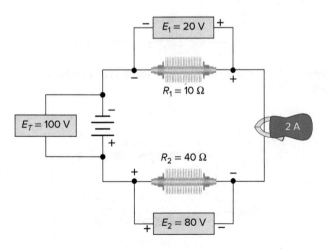

Figure 11-5 Determining voltage drops.

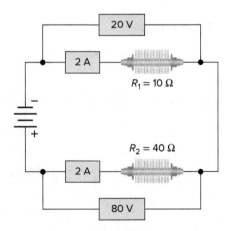

Figure 11-6 Determining individual resistance values.

$$R_2 = \frac{E_2}{I_2}$$
$$= \frac{80 \text{ V}}{2 \text{ A}}$$
$$= 40 \ \Omega$$

11.5 Determining Power

The power dissipated by the individual resistive loads in a DC series circuit is determined by the amount of the current flowing through the load and the value of the voltage drop across it, as illustrated in Figure 11-7. To determine the power requirements of the individual series circuit components, two parameters have to be known about each load and one of the following formulas applied to each load:

$$P_1 = E_1 \times I_1$$
$$P_1 = I_1^2 \times R_1$$
$$P_1 = \frac{E_1^2}{R_1}$$

The total power dissipated in a series circuit is always the **sum** of the power dissipated by the individual resistive loads or the product of the source voltage and total current (Figure 11-8):

$$P_T = P_1 + P_2 + P_3 \cdots$$
$$P_T = E_T \times I_T$$

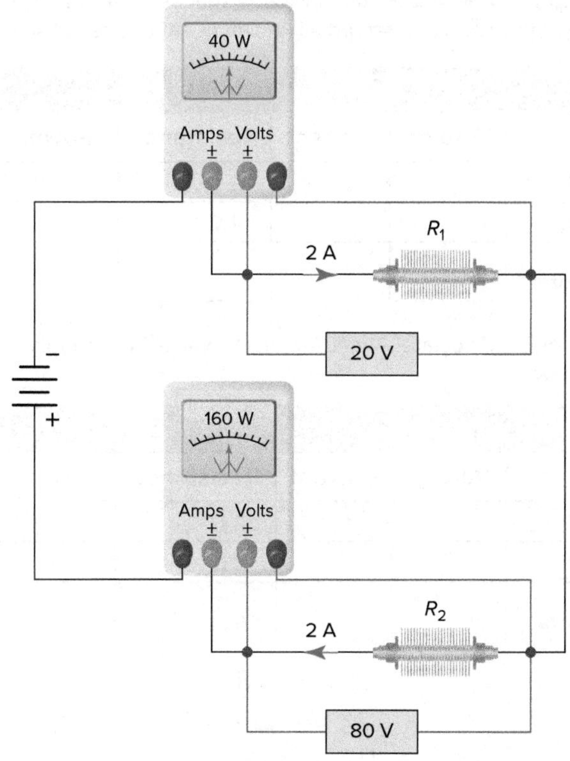

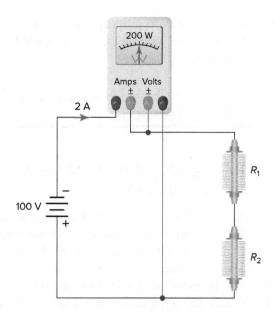

Figure 11-8 **Total power of a DC series circuit.**

Figure 11-7 Power dissipated by the individual resistive loads.

Part 1 Review Questions

1. Describe how the component parts of a series circuit are connected.

2. **a.** State the current characteristic of a series circuit.
   b. Express this characteristic in the form of an equation.

3. **a.** State the voltage characteristic of a series circuit.
   b. Express this characteristic in the form of an equation.

4. **a.** State the resistance characteristic of a series circuit.
   b. Express this characteristic in the form of an equation.

5. **a.** State the power dissipation characteristic of a series circuit.
   b. Express this characteristic in the form of an equation.

6. The total power consumed in a series circuit is 100 watts. If there are two loads and one consumes 65 watts, what does the other consume?

7. Three loads (R_1, R_2, and R_3) are connected in series to a 120-VDC source. The values of E_1 and E_3 are measured and found to be 32 V and 8 V, respectively. What is the value of E_2?

8. Resistors R_1 (8 Ω), R_2 (4 Ω), and R_3 (6 Ω), are connected in series to a voltage source. Which resistor will dissipate the most power? Why?

9. What is the total current of a series circuit that has a DC source voltage of 48 V and the following resistors: 1 Ω, 5 Ω, and 2 Ω?

10. Two 100-Ω resistors are connected in series to a 24-VDC source. What is the value of the voltage across each?

PART 2 SOLVING THE DC SERIES CIRCUIT

11.6 Solving for Current, Voltage, Resistance, and Power

Given DC series circuit values of voltage, current, and resistance, it is possible to calculate any unknown values of voltage, current, resistance, and power. The ability to make these calculations is important if you are truly to understand the operation of a DC series circuit.

In solving a DC series circuit problem, you must know the voltage, current, resistance, and power characteristics (as previously presented) of the circuit. They are most easily remembered by recalling the following equations for a series DC circuit:

$$I_T = I_1 = I_2 = I_3 \cdots$$
$$E_T = E_1 + E_2 + E_3 \cdots$$
$$R_T = R_1 + R_2 + R_3 \cdots$$
$$P_T = P_1 + P_2 + P_3 \cdots$$

Solving the DC series circuit will also require the use of Ohm's law. Ohm's law is most easily remembered by recalling the different forms of the Ohm's law equation:

$$E = I \times R$$
$$I = \frac{E}{R}$$
$$R = \frac{E}{I}$$
$$P = E \times I$$

Ohm's law is applied to the total circuit or to the individual loads. When applying it to the total circuit, it becomes:

$$E_T = I_T \times R_T$$
$$I_T = \frac{E_T}{R_T}$$
$$R_T = \frac{E_T}{I_T}$$
$$P_T = E_T \times I_T$$

When applying Ohm's law to the individual load, it becomes

$$E_1 = I_1 \times R_1$$
$$I_1 = \frac{E_1}{R_1}$$

$$R_1 = \frac{E_1}{I_1}$$
$$P_1 = E_1 \times I_1$$

One helpful method of solving circuits is to *use a table* to help organize the steps for solving the problem. Start by recording all given values of voltage, current, resistance, and power. Next, calculate the unknown values, and record each in the table as they are determined. We demonstrate this method in solving the following typical series circuit problems.

EXAMPLE 11-1

Problem: Find all the unknown quantities of *E, I, R,* and *P,* for the series circuit of Figure 11-9.

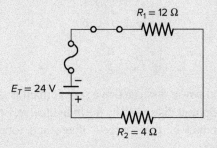

Figure 11-9 Circuit for example 11-1.

Solution:

Step 1. Draw a table that includes spaces for all values to be determined and record all known quantities as follows:

	Voltage	Current	Resistance	Power
R_1			12 Ω	
R_2			4 Ω	
Total	24 V			

Step 2. Calculate R_T and enter the value in the table as follows:

	Voltage	Current	Resistance	Power
R_1			12 Ω	
R_2			4 Ω	
Total	24 V		16 Ω	

$$R_T = R_1 + R_2$$
$$= 12\ \Omega + 4\ \Omega$$
$$= 16\ \Omega$$

Step 3. Calculate I_T, I_1, and I_2 and enter their values in the table as follows:

	Voltage	Current	Resistance	Power
R_1		1.5 A	12 Ω	
R_2		1.5 A	4 Ω	
Total	24 V	1.5 A	16 Ω	

$$I_T = \frac{E_T}{R_T}$$
$$= \frac{24\ \text{V}}{16\ \Omega}$$
$$= 1.5\ \text{A}$$
$$I_T = I_1 = I_2 = 1.5\ \text{A}$$

Step 4. Calculate E_1 and E_2 and enter their values in the table as follows:

	Voltage	Current	Resistance	Power
R_1	18 V	1.5 A	12 Ω	
R_2	6 V	1.5 A	4 Ω	
Total	24 V	1.5 A	16 Ω	

$$E_1 = I_1 \times R_1$$
$$= 1.5\ \text{A} \times 12\ \Omega$$
$$= 18\ \text{V}$$
$$E_2 = I_2 \times R_2$$
$$= 1.5\ \text{A} \times 4\ \Omega$$
$$= 6\ \text{V}$$

Step 5. Calculate P_T, P_1, and P_2 and enter their values in the table as follows:

	Voltage	Current	Resistance	Power
R_1	18 V	1.5 A	12 Ω	27 W
R_2	6 V	1.5 A	4 Ω	9 W
Total	24 V	1.5 A	16 Ω	36 W

$$P_T = E_T \times I_T$$
$$= 24\ \text{V} \times 1.5\ \text{A}$$
$$= 36\ \text{W}$$
$$P_1 = E_1 \times I_1$$
$$= 18\ \text{V} \times 1.5\ \text{A}$$
$$= 27\ \text{W}$$
$$P_2 = E_2 \times I_2$$
$$= 6\ \text{V} \times 1.5\ \text{A}$$
$$= 9\ \text{W}$$

EXAMPLE 11-2

Problem: Find all the unknown quantities of *E*, *I*, *R*, and *P* for the series circuit of Figure 11-10.

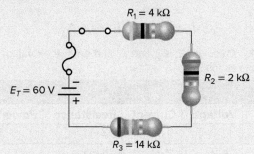

Figure 11-10 Circuit for example 11-2.

Solution:

Step 1. Draw a table that includes spaces for all values to be determined and record all known quantities as follows:

	Voltage	Current	Resistance	Power
R_1			4 kΩ	
R_2			2 kΩ	
R_3			14 kΩ	
Total	60 V			

Step 2. Calculate R_T and enter the value in the table as follows:

	Voltage	Current	Resistance	Power
R_1			4 kΩ	
R_2			2 kΩ	
R_3			14 kΩ	
Total	60 V		20 kΩ	

$$R_T = R_1 + R_2 + R_3$$
$$= 4\ \text{k}\Omega + 2\ \text{k}\Omega + 14\ \text{k}\Omega$$
$$= 20\ \text{k}\Omega$$

Step 3. Calculate I_T, I_1, I_2, and I_3, and enter their values in the table as follows:

	Voltage	Current	Resistance	Power
R_1		3 mA	4 kΩ	
R_2		3 mA	2 kΩ	
R_3		3 mA	14 kΩ	
Total	60 V	3 mA	20 kΩ	

$$I_T = \frac{E_T}{R_T}$$
$$= \frac{60 \text{ V}}{20 \text{ k}\Omega}$$
$$= 3 \text{ mA}$$
$$I_T = I_1 = I_2 = I_3 = 3 \text{ mA}$$

Step 4. Calculate E_1, E_2, and E_3 and enter their values in the table as follows:

	Voltage	Current	Resistance	Power
R_1	12 V	3 mA	4 kΩ	
R_2	6 V	3 mA	2 kΩ	
R_3	42 V	3 mA	14 kΩ	
Total	60 V	3 mA	20 kΩ	

$$E_1 = I_1 \times R_1$$
$$= 3 \text{ mA} \times 4 \text{ k}\Omega$$
$$= 12 \text{ V}$$

$$E_2 = I_2 \times R_2$$
$$= 3 \text{ mA} \times 2 \text{ k}\Omega$$
$$= 6 \text{ V}$$

$$E_3 = I_3 \times R_3$$
$$= 3 \text{ mA} \times 14 \text{ k}\Omega = 42 \text{ V}$$

Step 5. Calculate P_T, P_1, and P_2 and enter their values in the table as follows:

	Voltage	Current	Resistance	Power
R_1	12 V	3 mA	4 kΩ	36 mW
R_2	6 V	3 mA	2 kΩ	18 mW
R_3	42 V	3 mA	14 kΩ	126 mW
Total	60 V	3 mA	20 kΩ	180 mW

$$P_T = E_T \times I_T$$
$$= 60 \text{ V} \times 3 \text{ mA}$$
$$= 180 \text{ mW}$$

$$P_1 = E_1 \times I_1$$
$$= 12 \text{ V} \times 3 \text{ mA}$$
$$= 36 \text{ mW}$$

$$P_2 = E_2 \times I_2$$
$$= 6 \text{ V} \times 3 \text{ mA} = 18 \text{ mW}$$

$$P_3 = E_3 \times I_3$$
$$= 42 \text{ V} \times 3 \text{ mA} = 126 \text{ mW}$$

EXAMPLE 11-3

Problem: Find all the unknown quantities of E, I, R, and P, for the series circuit of Figure 11-11.

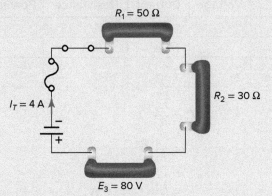

Figure 11-11 Circuit for example 11-3.

Solution:

Step 1. Draw a table that includes spaces for all values to be determined and record all known quantities as follows:

	Voltage	Current	Resistance	Power
R_1			50 Ω	
R_2			30 Ω	
R_3	80 V			
Total		4 A		

Step 2. Calculate I_1, I_2, and I_3, and enter their values in the table as follows:

	Voltage	Current	Resistance	Power
R_1		4 A	50 Ω	
R_2		4 A	30 Ω	
R_3	80 V	4 A		
Total		4 A		

$$I_T = I_1 = I_2 = I_3 = 4 \text{ A}$$

Step 3. Calculate E_1, E_2, and E_T and enter their values in the table as follows:

	Voltage	Current	Resistance	Power
R_1	200 V	4 A	50 Ω	
R_2	120 V	4 A	30 Ω	
R_3	80 V	4 A		
Total	400 V	4 A		

$$E_1 = I_1 \times R_1$$
$$= 4\text{ A} \times 50\ \Omega$$
$$= 200\text{ V}$$
$$E_2 = I_2 \times R_2$$
$$= 4\text{ A} \times 30\ \Omega$$
$$= 120\text{ V}$$
$$E_T = E_1 + E_2 + E_3$$
$$= 200\text{ V} + 120\text{ V} + 80\text{ V}$$
$$= 400\text{ V}$$

Step 4. Calculate R_3, and R_T and enter their values in the table as follows:

	Voltage	Current	Resistance	Power
R_1	200 V	4 A	50 Ω	
R_2	120 V	4 A	30 Ω	
R_3	80 V	4 A	20 Ω	
Total	400 V	4 A	100 Ω	

$$R_3 = \frac{E_3}{I_3}$$
$$= \frac{80\text{ V}}{4\text{ A}} = 20\ \Omega$$
$$R_T = \frac{E_T}{I_T}$$
$$= \frac{400\text{ V}}{4\text{ A}} = 100\ \Omega$$

Step 5. Calculate P_T, P_1, P_2, and P_3 and enter their values in the table as follows:

	Voltage	Current	Resistance	Power
R_1	200 V	4 A	50 Ω	800 W
R_2	120 V	4 A	30 Ω	480 W
R_3	80 V	4 A	20 Ω	320 W
Total	400 V	4 A	100 Ω	1,600 W

$$P_T = E_T \times I_T$$
$$= 400\text{ V} \times 4\text{ A}$$
$$= 1,600\text{ W}$$
$$P_1 = E_1 \times I_1$$
$$= 200\text{ V} \times 4\text{ A}$$
$$= 800\text{ W}$$

$$P_2 = E_2 \times I_2$$
$$= 120\text{ V} \times 4\text{ A}$$
$$= 480\text{ W}$$
$$P_3 = E_3 \times I_3$$
$$= 80\text{ V} \times 4\text{ A} = 320\text{ W}$$

11.7 Series Circuit Polarity

Every DC voltage source has a positive and negative terminal that establishes polarity in a circuit. Each component in a DC series circuit (fuse, switch, load, etc.) will have a positive (+) polarity side and a negative (−) polarity side. The side of the component closest to the positive voltage terminal is the positive (+) side, and the side closest to the negative terminal is the negative (−) side, as illustrated in Figure 11-12.

Voltage drops across all resistive loads of a series circuit have polarities. The easiest way to find these polarities is to use the direction of the electron current as a basis. Current enters at the negative polarity and leaves at the positive polarity. Figure 11-13 illustrates the relationship between current direction and polarity of the voltage drop. Any point in the circuit is positive or negative only in relation to another point. For the circuit shown:

- Point *A* is **negative** relative to Point *B*.
- Point *B* is **negative** relative to Point *C*.
- Point *C* is **negative** relative to Point *D*.

In some DC circuits, one point in the circuit is designated as the **common point,** and all voltages are referenced relative to this

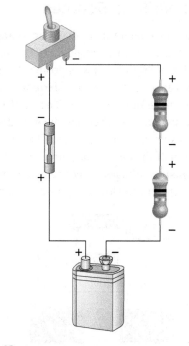

Figure 11-12 **Series circuit polarity.**

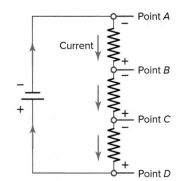

Figure 11-13 Relationship between current direction and polarity.

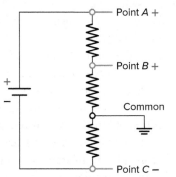

Figure 11-14 Circuit with common reference point.

common point. Using a common reference point, both negative and positive output voltages can be obtained, depending on what point in the circuit is the common reference point. In the voltage divider circuit of Figure 11-14 points A and B are positive relative to the common point, while point C is negative.

11.8 Series-Aiding and Series-Opposing Voltage Sources

Some circuits may contain more than one voltage source. When this is the case, the voltage sources may be connected as series aiding or series opposing. **Series-aiding** voltage sources are connected with polarities that allow the current to flow in the same direction. The **positive** terminal of one voltage source is connected to the **negative** terminal of the other, as shown in Figure 11-15. The total voltage is equal to the sum of the individual source voltages.

Series-opposing voltages are connected with polarities that allow the current to flow in opposite directions. The

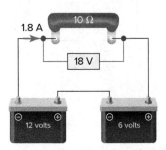

Figure 11-15 Series-aiding voltages.

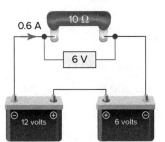

Figure 11-16 Series-opposing voltages.

positive terminal of one voltage source is connected to the **positive** terminal of the other, as shown in Figure 11-16. Series-opposing voltages are subtracted to obtain the total equivalent voltage. Subtract the smaller voltage value from the larger value, and give the equivalent voltage the polarity of the larger voltage. If the two voltages are equal, the net voltage and current will be zero.

11.9 Voltage Source Resistance

The **ideal** voltage source maintains a constant output voltage regardless of the resistance of the load or the amount of current drawn from it. In reality all voltage sources have some **internal resistance,** such as the conductors in the coils of a generator or the chemicals in a battery.

The internal resistance of a voltage source can be represented as a resistance in series with an ideal voltage source, as illustrated in Figure 11-17. Normally this resistance is very small compared to that of the load and has little effect on circuit operation. However when the internal resistance becomes a significant part of the total circuit resistance, we must take it into account. For example, when a battery fails, it is typically because it has built up enough internal resistance that it can no longer supply a useful amount of power to an external load. If you measure the voltage of a failed battery which is disconnected, you will usually find that it has a nearly normal voltage, so that a voltmeter is not a useful tool to judge the degree of life left in a battery. If you connect the battery to an external resistance, then you will find that the terminal voltage of the battery drops; this can be interpreted as dropping most of its voltage across its internal resistance so that it is not available for external service.

Figure 11-17 Voltage source internal resistance.

EXAMPLE 11-4

Problem: The voltage source of Figure 11-18 has a no-load voltage of 100 volts. Its internal resistance is 0.5 Ω. Calculate the terminal voltage when this source is connected to a 10-Ω load.

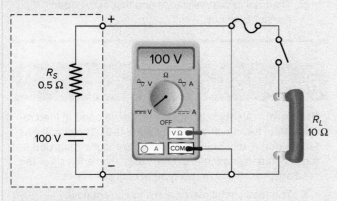

Figure 11-18 Circuit for example 11-4.

Solution:

Step 1. Draw a table that includes spaces for all values to be determined, and record all known quantities as follows:

	Voltage	Current	Resistance
R_S			0.5 Ω
R_L			10 Ω
Total	100 V		

Step 2. Calculate R_T and enter the value in the table as follows:

	Voltage	Current	Resistance
R_S			0.5 Ω
R_L			10 Ω
Total	100 V		10.5 Ω

$$R_T = R_S + R_L$$
$$= 0.5\ \Omega + 10\ \Omega$$
$$= 10.5\ \Omega$$

Step 3. Calculate I_T, I_S, and I_L and enter their values in the table as follows:

	Voltage	Current	Resistance
R_S		9.52 A	0.5Ω
R_L		9.52 A	10 Ω
Total	100 V	9.52 A	10.5 Ω

$$I_T = \frac{E_T}{R_T}$$
$$= \frac{100\ \text{V}}{10.5\ \Omega}$$
$$= 9.52\ \text{A}$$
$$I_T = I_S = I_L$$

Step 4. Calculate E_S and E_L and enter their values in the table as follows:

	Voltage	Current	Resistance
R_S	4.76 V	9.52 A	0.5 Ω
R_L	95.2 V	9.52 A	10 Ω
Total	100 V	9.52 A	10.5 Ω

$$E_S = I_S \times R_S$$
$$= 9.52\ \text{A} \times 0.5\ \Omega$$
$$= 4.76\ \text{V}$$
$$E_L = I_L \times R_L$$
$$= 9.52\ \text{A} \times 10\ \Omega = 95.2\ \text{V}$$

11.10 Series Conductor Resistance

Although the resistance of electric conductors is very low, a long length of wire could cause a substantial voltage drop. The voltage dropped across the conductors of a circuit is directly proportional to their resistance and the magnitude of the current they are carrying. When conductor resistance needs to be taken into account, the wire resistance is assumed to be connected in series with the load. The following table shows all given and calculated values of voltage, current, resistance, and power for the circuit of Figure 11-19 which takes into account the resistance of the two conductors feeding power to the load.

	Voltage	Current	Resistance	Power
R_1	20 V	20 A	1 Ω	400 W
R_2	20 V	20 A	1 Ω	400 W
R_3	200 V	20 A	10 Ω	4,000 W
Total	240 V	20 A	12 Ω	4,800 W

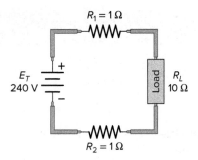

Figure 11-19 Series conductor resistance.

11.11 Troubleshooting a Series Circuit

Troubleshooting a circuit involves locating and repairing one or more faults in a circuit. Using Ohm's law and series circuit characteristics, you can **predict** what changes will occur in a circuit under different fault conditions. This information often can be very useful in pinpointing the cause of a problem. As an example, we will use one of the original series circuits studied and examine how different faults can affect voltage, current, resistance, and power.

EXAMPLE 11-5

Problem: An **open** circuit fault has developed in load resistor R_1 of the circuit shown in Figure 11-20. The circuit was tested under this fault condition and results recorded in the accompanying table. On the basis of an analysis of the readings, surmise how an open in one of the loads of a series circuit will effect:

a. Current flow and power dissipation.

b. Voltage across the open load resistor.

c. Voltage across all remaining load resistors.

d. Resistance of the circuit as a whole and that of the open load resistor.

	Voltage	Current	Resistance	Power
R_1	24 V	Zero A	Infinite	Zero
R_2	0 V	Zero A	4 Ω	Zero
Total	24 V	Zero A	Infinite	Zero

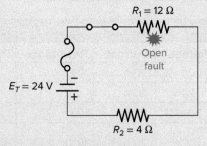
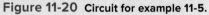

Figure 11-20 Circuit for example 11-5.

Solution:

a. No current flows, so no power is dissipated.

b. The full source voltage can be measured across the open load resistor.

c. All voltages across other load resistors drop to zero.

d. The total circuit resistance and that of the open resistor is infinite.

EXAMPLE 11-6

Problem: A **short** circuit fault has developed in load resistor R_3 of the circuit shown in Figure 11-21. The circuit was tested under this fault condition and results recorded in the accompanying table. On the basis of an analysis of the readings, surmise how this type of fault will effect:

a. The total circuit resistance and current flow.

b. Voltage drops across the loads.

c. Power dissipation in the circuit.

	Voltage	Current	Resistance	Power
R_1	40 V	10 mA	4 kΩ	400 mW
R_2	20 V	10 mA	2 kΩ	200 mW
R_3	Zero	10 mA	Zero	Zero
Total	60 V	10 mA	6 kΩ	600 mW

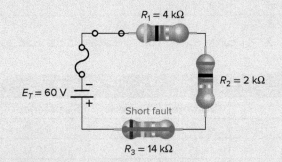

Figure 11-21 Circuit for example 11-6.

Solution:

a. The total circuit resistance decreases, causing a proportional increase in the total circuit current.

b. The voltage across the shorted component drops to zero, and the voltage across the other load components increases.

c. The total power dissipated by the circuit and the other load components increases.

It should be noted that a short in any load component of a series circuit could result in other components being damaged from excessive current. For example, the increase in circuit current could cause another component to burn open if power dissipation exceeds the power rating of the component. If a faulty series circuit contains a burnt component or an open fuse, there's a good chance that a shorted component is somewhere in the circuit. When either of these conditions exists, the other components in the circuit should be tested before the circuit is assumed to be repaired.

Not all components fail due to shorts or opens. Because of aging or overheating, original component **values** may change. When a component has increased in resistance, then the effects would be similar to an open circuit. Similarly, if a component has decreased in resistance, then the effects would be similar to a short circuit.

Part 2 Review Questions

1. Find all unknown values for *E, I, R,* and *P* for the series circuit drawn in Figure 11-22. Construct a table and record the given and calculated values. Show all steps and equations used to arrive at your answers.

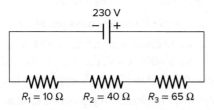

Figure 11-22 Circuit for review question 1.

2. Find all unknown values for *E, I, R,* and *P* for the series circuit drawn in Figure 11-23. Construct a table and record the given and calculated values. Show all steps and equations used to arrive at your answers.

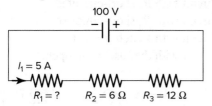

Figure 11-23 Circuit for review question 2.

3. Find all unknown values for *E, I, R,* and *P* for the series circuit drawn in Figure 11-24. Construct a table and record the given and calculated values. Show all steps and equations used to arrive at your answers.

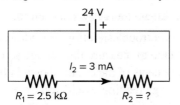

Figure 11-24 Circuit for review question 3.

4. For the series circuit shown in Figure 11-25, what is the polarity of:
   a. Point *A* relative to the common reference point?

b. Point *B* relative to the common reference point?
c. Point *C* relative to the common reference point?

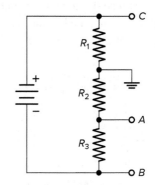

Figure 11-25 Circuit for review question 4.

5. Answer the following with reference to the dual-voltage circuit drawn in Figure 11-26:
   a. Are the two voltage sources connected series-aiding or series-opposing?
   b. What is the value of E_T applied to the circuit?
   c. What is the polarity of point *A* relative to point *B*?
   d. What is the value of current flow I_1?
   e. What is the value of current flow I_2?

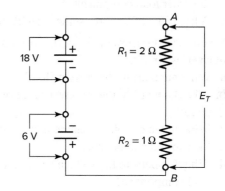

Figure 11-26 Circuit for review question 5.

6. The no-load output voltage of the battery source drawn in Figure 11-27 is 12 volts, and its internal series resistance (R_s) is 5 Ω. Find all unknown values for *E, I,* and *R* for the series circuit with the 100-Ω load connected. Construct a table and record the given and calculated values. Show all steps and equations used to arrive at your answers.

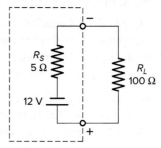

Figure 11-27 Circuit for review question 6.

7. For the circuit shown in Figure 11-28, the voltage at the load is to be determined taking into consideration the resistance of the two conductors feeding power to it. Construct a table and record the given and calculated circuit values for *E, I, R,* and *P.*

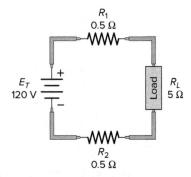

Figure 11-28 Circuit for review question 7.

8. When an open fault occurs in one of the load components of a series circuit:
   a. What happens to the current flow?
   b. How much power is dissipated by the circuit?
   c. What value of voltage can be measured across the open component?
   d. What value of voltage can be measured across the other load components?
   e. What value of total resistance can be measured?

9. When a short occurs in one of the load components of a series circuit:
   a. What happens to the current flow?
   b. What happens to the total resistance of the circuit?
   c. What value of voltage can be measured across the shorted component?
   d. What happens to the voltage across the other load components?
   e. What happens to the total power dissipated by the circuit and the other load components?

10. A circuit has three resistors connected in series. Resistor R_2 has a resistance of 22 Ω and a voltage drop of 88 V. What is the value of the current flow through resistor R_3?

11. Assume one of three identical series-connected lightbulbs is shorted. Explain what affect, if any, this will have on the other two light bulbs.

12. While putting four 1.5-V batteries in a flashlight, you accidentally put one of them in backward. What will be the value of the resultant voltage available to operate the flashlight bulb?

13. Three resistors with values of 1 kΩ, 5 kΩ, and 3 kΩ, respectively, are connected in series with a fourth resistor of unknown value. The applied

voltage is 120 V, and the current flow is 5 mA. What is the resistance value of the unknown resistor?

14. As part of a troubleshooting procedure, an ohmmeter is connected to measure the total resistance of the circuit shown in Figure 11-29. What resistance reading you would expect to see for each of the following circuit conditions?
   a. Circuit operating normally as drawn.
   b. Circuit with R_1 short-circuited.
   c. Circuit with R_2 burnt open.
   d. Circuit with R_3 short-circuited.
   e. Circuit with switch opened.
   f. Circuit with the fuse burnt open.

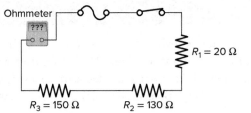

Figure 11-29 Circuit for review question 14.

15. How much resistance R_1 must be added in series with a 400-Ω resistance R_2 to limit the current to 0.15 A with 120 V applied to the circuit?

16. As part of a troubleshooting procedure, an ammeter is connected to measure the current of the circuit shown in Figure 11-30. What current reading would you expect to see for each of the following circuit conditions?
   a. Circuit operating normally as drawn.
   b. Circuit with R_1 burnt open.
   c. Circuit with R_2 short-circuited.
   d. Circuit with R_3 short-circuited.
   e. Circuit with switch opened.
   f. Circuit with the fuse burnt open.

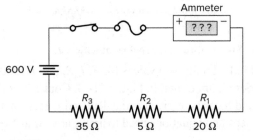

Figure 11-30 Circuit for review question 16.

17. As part of a troubleshooting procedure, a voltmeter is connected to measure the voltage across the various components of the circuit shown in Figure 11-31. What voltage reading would you expect to see for each of the following circuit conditions?
   a. With the circuit operating normally, the voltage drop across: (I) the switch, (II) the fuse, (III) R_1 (IV) R_2, and (V) R_3.

b. With R_1 short circuited, the voltage drop across: (I) the switch, (II) the fuse, (III) R_1 (IV) R_2, and (V) R_3.

c. With R_2 burnt open, the voltage drop across: (I) the switch, (II) the fuse, (III) R_1 (IV) R_2, and (V) R_3.

d. With R_3 short circuited, the voltage drop across: (I) the switch, (II) the fuse, (III) R_1 (IV) R_2, and (V) R_3.

e. With the fuse burnt open, the voltage drop across: (I) the switch, (II) the fuse, (III) R_1 (IV) R_2, and (V) R_3.

f. With the switch manually opened, the voltage drop across: (I) the switch, (II) the fuse, (III) R_1 (IV) R_2, and (V) R_3.

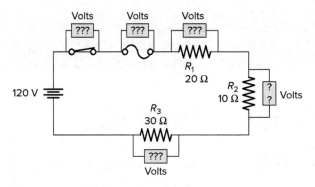

Figure 11-31 Circuit for review question 17.

18. a. How many 1.5-V batteries would be required for the standby power supply of a 12-V alarm system?
   b. How would the batteries be connected?

CHAPTER TWELVE

Solving the DC Parallel Circuit

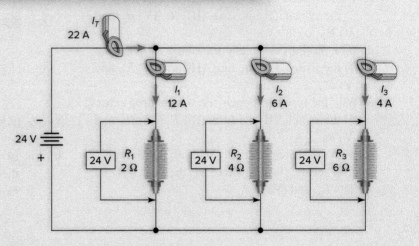

Parallel circuit voltage and current.

LEARNING OUTCOMES

▶ Understand the voltage, current, resistance, and power characteristics of a DC parallel circuit.

▶ Solve for unknown circuit values in a DC parallel circuit.

▶ Measure current paths, voltages, and resistance values in a DC parallel circuit.

▶ Apply the concepts of relative polarity to a DC parallel circuit.

▶ Troubleshoot a parallel DC resistive circuit.

Parallel circuits are used more frequently than all other circuits. By tracing the path of current, we can determine whether a circuit has series-connected or parallel-connected components. In a series circuit, there is only one path for current, whereas in parallel circuits the current has two or more paths. Components connect across each other to form parallel circuits. In this chapter, we study the unique characteristics of the parallel circuit.

PART 1 PARALLEL CIRCUIT CHARACTERISTICS

12.1 Parallel Circuit Connection

A **parallel** circuit is a circuit that has two or more paths for current flow. In parallel circuits, the loads connect across each other, like rungs on a ladder. The voltage applied to each parallel component is the same. In the parallel circuit of Figure 12-1, the total source current splits to travel through each separate load resistance. Each load acts independently of the other, and this arrangement results in as many pathways for the current as there are load components connected in parallel.

The main reason for the popularity of parallel circuits is the fact that all loads can be designed for a **common supply voltage** such as 120 volts for household circuits, as illustrated in Figure 12-2. In addition, devices like lights and receptacles are normally connected in parallel allowing them to operate **independently** of each other. If lights in a home were wired in series and one were to burn out, all other lights on the circuit would go out. Similarly, if receptacles were connected in series, a device would have to be connected into each receptacle before power could be supplied to any other device.

12.2 Determining Current Flow

In the parallel circuit, each load provides a separate path for current flow. The separate paths are called **branches** and the current flowing in each branch is called the **branch current.**

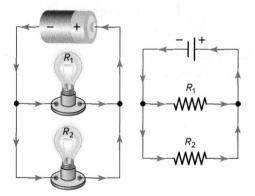

Figure 12-1 DC parallel circuit.

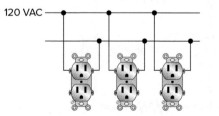

Figure 12-2 Receptacles are connected in parallel in all residential wiring.

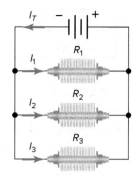

Figure 12-3 Total current is equal to the sum of the branch currents.

All the branches have the same voltage applied to them. Total current in a circuit containing parallel-connected loads equals the sum of the current through all the loads, as illustrated in Figure 12-3. To find the total current in a parallel circuit, apply the formula

$$I_T = I_1 + I_2 + I_3 \cdots$$

Current flowing through each branch of a parallel circuit depends on the resistance of that branch and the value of the voltage applied to it. Current flow through the parallel branches is inversely proportional to the amount of resistance in the branch: The higher the resistance, the lower the current. Figure 12-4 illustrates how branch current flow can be calculated when the voltage and the resistance of a branch are known. Ohm's law for current can be applied as follows to determine the current flow through each branch:

$$I_1 = \frac{E_1}{R_1}$$
$$= \frac{100 \text{ V}}{10 \text{ }\Omega}$$
$$= 10 \text{ A}$$
$$I_2 = \frac{E_2}{R_2}$$
$$= \frac{100 \text{ V}}{40 \text{ }\Omega}$$
$$= 2.5 \text{ A}$$
$$I_T = I_1 + I_2$$
$$= 10 \text{ A} + 2.5 \text{ A}$$
$$= 12.5 \text{ A}$$

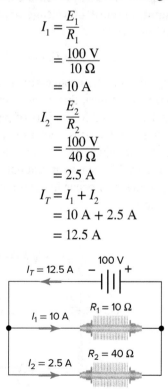

Figure 12-4 Calculating branch current flow.

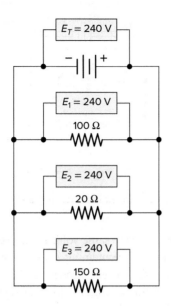

Figure 12-5 Voltage is the same in a parallel circuit.

12.3 Determining Voltage

The voltage across each branch of a parallel circuit is the same and is equal in value to that of the applied source voltage. Figure 12-5 shows that voltmeters connected across the branch circuit loads read the same voltage as the one connected across the power supply. This is true, regardless of the number of branches connected in parallel. Although the load resistors may be different in value, they all have the **same amount of voltage** applied to them when connected in parallel:

$$E_T = E_1 = E_2 = E_3 \cdots$$

Series voltage drop is the **progressive** loss of voltage, due to **conductor resistance,** that occurs when feeding a string of parallel-connected loads. It is of particular concern when feeding parallel loads, such as lights on a walkway or parking lot, in which each lamp is located a long distance apart. Figure 12-6 shows an example of the differences in voltages that can occur in such a circuit due to conductor resistance.

- In this example, all conductors are of the same gauge size, and all lights have the same resistance.
- Lamp L3, which is furthest from the power source, receives the lowest voltage because it encounters the greatest amount of series line resistance.
- The first segment of the string carries the most current and the last segment carries the least.
- The voltage difference between the lights can be minimized by using larger size wire near the voltage source.

12.4 Determining Total Resistance

In a series circuit the total resistance is the sum of the individual load resistances. In a parallel circuit, the relationship between the total resistance and the individual resistances is completely opposite to that of the series circuit. Adding more loads in parallel increases the number of current paths, making it easier for current to flow. This **reduces** the total circuit resistance, as illustrated in Figure 12-7. Because increasing the number of resistors in parallel reduces the circuit resistance, the total resistance cannot be determined from the sum of the individual resistances.

The total resistance of a parallel circuit is always **lower** than the smallest value resistor in any of the branches. There are several formulas for calculating total resistance in parallel circuits. The number of resistors in the circuit and their values determine the one that is used. Parallel circuits with resistors of **equal value** are the easiest to calculate. The total resistance is simply the value of one of the resistors divided by the number of resistances:

$$R_T = \frac{R_X \text{ (value of one resistor)}}{R_N \text{ (number of resistors)}}$$

For example, for a parallel circuit with five 100-Ω resistors, the total resistance is

$$R_T = \frac{R_X}{R_N}$$
$$= \frac{100 \ \Omega}{5}$$
$$= 20 \ \Omega$$

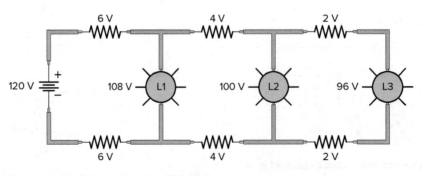

Figure 12-6 Series voltage drop due to conductor resistance

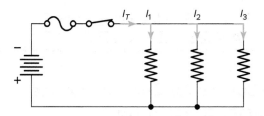

Figure 12-7 Additional loads in parallel reduces the total circuit resistance.

The recommended method to determine the total resistance of parallel circuits with **two resistors** that have different values is to use the **product-over-sum** method. The formula is as follows:

$$R_T = \frac{R_1 \times R_2}{R_1 + R_2}$$

For example, for a parallel circuit with a 40-Ω and a 10-Ω resistors (Figure 12-8), the total resistance is

$$R_T = \frac{R_1 \times R_2}{R_1 + R_2}$$
$$= \frac{40\ \Omega \times 10\ \Omega}{40\ \Omega + 10\ \Omega}$$
$$= \frac{400\ \Omega}{50\ \Omega}$$
$$= 8\ \Omega$$

The recommended method to determine the total resistance of parallel circuits with three or more resistors that have different values is the **reciprocal** method. The formula is as follows:

$$R_T = \frac{1}{\dfrac{1}{R_1} + \dfrac{1}{R_2} + \dfrac{1}{R_3}} \cdots$$

For example, for a parallel circuit with 80-kΩ, 20-kΩ, and 4-kΩ resistors, the total resistance can be calculated as follows:

$$R_T = \frac{1}{\dfrac{1}{R_1} + \dfrac{1}{R_2} + \dfrac{1}{R_3}}$$
$$= \frac{1}{\dfrac{1}{80\ \text{k}\Omega} + \dfrac{1}{20\ \text{k}\Omega} + \dfrac{1}{4\ \text{k}\Omega}}$$
$$= \frac{1}{0.0125 + 0.05 + 0.25}$$
$$= \frac{1}{0.3125}$$
$$= 3.2\ \text{k}\Omega$$

You can also use Ohm's law to find the total resistance of a parallel circuit directly. If the total current and the applied voltage are known, the total resistance can easily be found using Ohm's law in the form

$$R_T = \frac{E_T}{I_T}$$

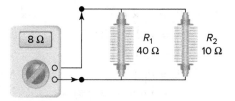

Figure 12-8 Product-over-sum method.

The following formula is used to determine the resistance of an unknown resistor (R_2) when the total resistance and the resistance of all other resistors (R_1 and R_2) is known.

$$R_2 = \frac{1}{\dfrac{1}{R_T} - \dfrac{1}{R_1} - \dfrac{1}{R_3}}$$

For example, three resistors—R_1, 25 Ω; R_2, unknown; R_3, 50 Ω—are connected in parallel, and the total resistance of this combination is 6.25 Ω. The resistance of R_2 is

$$R_2 = \frac{1}{\dfrac{1}{R_T} - \dfrac{1}{R_1} - \dfrac{1}{R_3}}$$
$$= \frac{1}{\dfrac{1}{6.25} - \dfrac{1}{25} - \dfrac{1}{50}}$$
$$= 10\ \Omega$$

12.5 Determining Power

Power consumption in a parallel circuit is calculated in the same manner as that of a series circuit. Total power can be calculated directly from total current and source voltage, or it can be found indirectly by taking the sum of the individual power consumptions of the loads. The power for the circuit of Figure 12-9 can be calculated as follows:

$$P_1 = E_1 \times I_1$$
$$= 100\ \text{V} \times 10\ \text{A}$$
$$= 1{,}000\ \text{W}$$

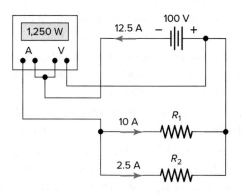

Figure 12-9 Power consumption in a parallel circuit.

$$P_2 = E_2 \times I_2$$
$$= 100 \text{ V} \times 2.5 \text{ A}$$
$$= 250 \text{ W}$$
$$P_T = P_1 + P_2$$
$$= 1{,}000 \text{ W} + 250 \text{ W}$$
$$= 1{,}250 \text{ W}$$

Part 1 Review Questions

1. Describe how the component parts of a parallel circuit are connected.

2. Give two reasons why multiple electrical receptacles in homes are always connected in parallel.

3. **a.** State the current characteristic of a parallel circuit.
   b. Express this characteristic in the form of an equation.

4. **a.** State the voltage characteristic of a parallel circuit.
   b. Express this characteristic in the form of an equation.

5. Give the recommended formula used to find the total resistance for each of the following:
   a. Several parallel-connected resistors with the same resistance value.
   b. Two parallel-connected resistors with different resistance values.
   c. Three or more parallel-connected resistors with different resistance values.
   d. Parallel circuit when the voltage and total current are known.

6. The total power consumed in a parallel circuit is 1,000 watts. If there are three loads, and two consume 300 watts, what does the other consume?

7. Two identical loads (R_1 and R_2) are connected in parallel to a 120-volt source. What is the value of the voltage across each of the loads?

8. Resistors R_1 (8 Ω), R_2 (4 Ω), and R_3 (6 Ω) are connected in parallel to a voltage source. Which resistor will dissipate the most power? Why?

9. What is the total current of a parallel circuit that has a DC source voltage of 48 V and the following resistors: 6 Ω, 4 Ω, and 2 Ω?

10. Two 100-Ω resistors are connected in parallel to a 24-VDC source. What is the total circuit resistance?

11. The resistance of a parallel circuit consisting of two branches is 4 Ω. If the resistance of one of the branches is 14 Ω, what is the resistance of the other branch?

PART 2 SOLVING THE DC PARALLEL CIRCUIT

12.6 Solving for Current, Voltage, Resistance, and Power

The procedure for solving parallel circuit values of voltage, current, resistance, and power is similar to that used for solving values for series circuits. Ohm's law as it applies to the circuit as a whole and to individual loads as well is used. In addition, the parallel circuit characteristics of voltage, current, resistance, and power must also be used. The parallel circuit characteristics expressed in the form of equations are as follows:

$$I_T = I_1 + I_2 + I_3 \cdots$$
$$E_T = E_1 = E_2 = E_3 \cdots$$
$$R_T = \frac{R_1 \times R_2}{R_1 + R_2}$$
$$R_T = \frac{1}{\dfrac{1}{R_1} + \dfrac{1}{R_2} + \dfrac{1}{R_3} \cdots}$$
$$P_T = P_1 + P_2 + P_3 \cdots$$

EXAMPLE 12-1

Problem: Find all the unknown quantities of E, I, R, and P for the parallel circuit of Figure 12-10.

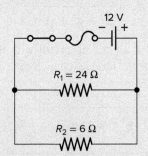

Figure 12-10 Circuit for example 12-1.

Solution:

Step 1. Draw a table that includes spaces for all values to be determined and record all known quantities as follows:

	Voltage	Current	Resistance	Power
R_1			24 Ω	
R_2			6 Ω	
Total	12 V			

Step 2. Calculate E_1 and E_2 and enter their values in the table as follows:

	Voltage	Current	Resistance	Power
R_1	12 V		24 Ω	
R_2	12 V		6 Ω	
Total	12 V			

$$E_T = E_1 = E_2 = 12 \text{ V}$$

Step 3. Calculate I_1, I_2, and I_T and enter their values in the table as follows:

	Voltage	Current	Resistance	Power
R_1	12 V	0.5 A	24 Ω	
R_2	12 V	2 A	6 Ω	
Total	12 V	2.5 A		

$$I_1 = \frac{E_1}{R_1}$$
$$= \frac{12 \text{ V}}{24 \text{ Ω}}$$
$$= 0.5 \text{ A}$$

$$I_2 = \frac{E_2}{R_2}$$
$$= \frac{12 \text{ V}}{6 \text{ Ω}}$$
$$= 2 \text{ A}$$

$$I_T = I_1 + I_2$$
$$= 0.5 \text{ A} + 2 \text{ A}$$
$$= 2.5 \text{ A}$$

Step 4. Calculate R_T and enter the value in the table as follows:

	Voltage	Current	Resistance	Power
R_1	12 V	0.5 A	24 Ω	
R_2	12 V	2 A	6 Ω	
Total	12 V	2.5 A	4.8 Ω	

$$R_T = \frac{E_T}{I_T}$$
$$= \frac{12 \text{ V}}{2.5 \text{ A}}$$
$$= 4.8 \text{ Ω}$$
$$\text{or}$$

$$R_T = \frac{R_1 \times R_2}{R_1 + R_2}$$
$$= \frac{24 \text{ Ω} \times 6 \text{ Ω}}{24 \text{ Ω} + 6 \text{ Ω}}$$
$$= \frac{144}{30}$$
$$= 4.8 \text{ Ω}$$

	Voltage	Current	Resistance	Power
R_1	12 V	0.5 A	24 Ω	6 W
R_2	12 V	2 A	6 Ω	24 W
Total	12 V	2.5 A	4.8 Ω	30 W

Step 5. Calculate P_T, P_1, and P_2 and enter their values in the table as follows:

$$P_T = E_T \times I_T$$
$$= 12 \text{ V} \times 2.5 \text{ A}$$
$$= 30 \text{ W}$$

$$P_1 = E_1 \times I_1$$
$$= 12 \text{ V} \times 0.5 \text{ A}$$
$$= 6 \text{ W}$$

$$P_2 = E_2 \times I_2$$
$$= 12 \text{ V} \times 2 \text{ A}$$
$$= 24 \text{ W}$$

EXAMPLE 12-2

Problem: Find all the unknown quantities of E, I, R, and P for the parallel circuit of Figure 12-11.

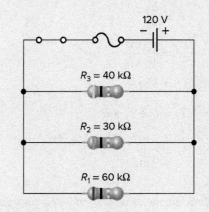

Figure 12-11 Circuit for example 12-2.

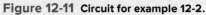

Solution:

Step 1. Draw a table that includes spaces for all values to be determined and record all known quantities as follows:

	Voltage	Current	Resistance	Power
R_1	120 V		60 kΩ	
R_2	120 V		30 kΩ	
R_3	120 V		40 kΩ	
Total	120 V			

Step 2. Calculate I_1, I_2, I_3, and I_T and enter their values in the table as follows:

	Voltage	Current	Resistance	Power
R_1	120 V	2 mA	60 kΩ	
R_2	120 V	4 mA	30 kΩ	
R_3	120 V	3 mA	40 kΩ	
Total	120 V	9 mA		

$$I_1 = \frac{E_1}{R_1}$$
$$= \frac{120 \text{ V}}{60 \text{ k}\Omega}$$
$$= 2 \text{ mA}$$

$$I_2 = \frac{E_2}{R_2}$$
$$= \frac{120 \text{ V}}{30 \text{ k}\Omega}$$
$$= 4 \text{ mA}$$

$$I_3 = \frac{E_3}{R_3}$$
$$= \frac{120 \text{ V}}{40 \text{ k}\Omega}$$
$$= 3 \text{ mA}$$

$$I_T = I_1 + I_2 + I_3$$
$$= 2 \text{ mA} + 4 \text{ mA} + 3 \text{ mA}$$
$$= 9 \text{ mA}$$

Step 3. Calculate the value of R_T and enter the value in the table as follows:

	Voltage	Current	Resistance	Power
R_1	120 V	2 mA	60 kΩ	
R_2	120 V	4 mA	30 kΩ	
R_3	120 V	3 mA	40 kΩ	
Total	120 V	9 mA	13.3 kΩ	

$$R_T = \frac{E_T}{I_T}$$
$$= \frac{120 \text{ V}}{9 \text{ mA}}$$
$$= 13.3 \text{ k}\Omega$$

Step 4. Calculate P_T, P_1, P_2, and P_3 and enter their values in the table as follows:

	Voltage	Current	Resistance	Power
R_1	120 V	2 mA	60 kΩ	240 mW
R_2	120 V	4 mA	30 kΩ	480 mW
R_3	120 V	3 mA	40 kΩ	360 mW
Total	120 V	9 mA	13.3 kΩ	1,080 mW

$$P_T = E_T \times I_T$$
$$= 120 \text{ V} \times 9 \text{ mA}$$
$$= 1,080 \text{ mW} \ (1.08 \text{ W})$$

$$P_1 = E_1 \times I_1$$
$$= 120 \text{ V} \times 2 \text{ mA} = 240 \text{ mW}$$

$$P_2 = E_2 \times I_2$$
$$= 120 \text{ V} \times 4 \text{ mA}$$
$$= 480 \text{ mW}$$

$$P_3 = E_3 \times I_3$$
$$= 120 \text{ V} \times 3 \text{ mA} = 360 \text{ mW}$$

EXAMPLE 12-3

Problem: Find all the unknown quantities of E, I, R, and P for the parallel circuit of Figure 12-12.

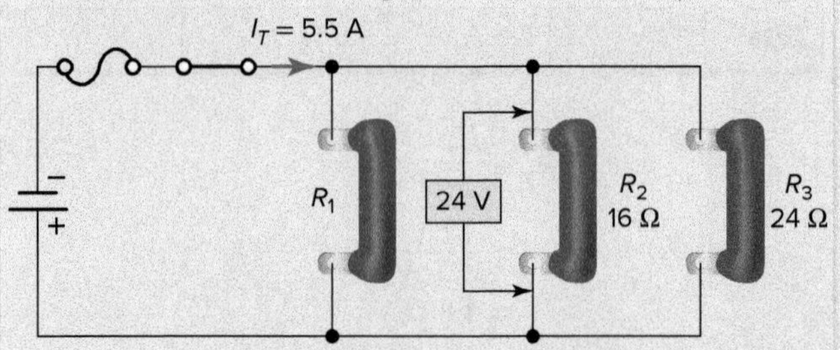

Figure 12-12 Circuit for example 12-3.

Solution:

Step 1. Draw a table that includes spaces for all values to be determined and record all known quantities as follows:

	Voltage	Current	Resistance	Power
R_1	24 V			
R_2	24 V		16 Ω	
R_3	24 V		24 Ω	
Total	24 V	5.5 A		

Step 2. Calculate I_2, I_3, and I_1 and enter their values in the table as follows:

	Voltage	Current	Resistance	Power
R_1	24 V	3 A		
R_2	24 V	1.5 A	16 Ω	
R_3	24 V	1 A	24 Ω	
Total	24 V	5.5 A		

$$I_2 = \frac{E_2}{R_2}$$
$$= \frac{24\ V}{16\ \Omega}$$
$$= 1.5 A$$
$$I_3 = \frac{E_3}{R_3}$$
$$= \frac{24\ V}{24\ \Omega}$$
$$= 1 A$$
$$I_1 = I_T - (I_2 + I_3)$$
$$= 5.5\ A - 2.5\ A$$
$$= 3\ A$$

Step 3. Calculate R_1 and R_T and enter their values in the table as follows:

	Voltage	Current	Resistance	Power
R_1	24 V	3 A	8 Ω	
R_2	24 V	1.5 A	16 Ω	
R_3	24 V	1 A	24 Ω	
Total	24 V	5.5 A	4.36 Ω	

$$R_1 = \frac{E_1}{I_1}$$
$$= \frac{24\ V}{3\ A}$$
$$= 8\ \Omega$$
$$R_T = \frac{E_T}{I_T}$$
$$= \frac{24\ V}{5.5\ A}$$
$$= 4.36\ \Omega$$

Step 4. Calculate P_T, P_1, P_2, and P_3 and enter their values in the table as follows:

	Voltage	Current	Resistance	Power
R_1	24 V	3 A	8 Ω	72 W
R_2	24 V	1.5 A	16 Ω	36 W
R_3	24 V	1 A	24 Ω	24 W
Total	24 V	5.5 A	4.36 Ω	132 W

$$P_T = E_T \times I_T$$
$$= 24\ V \times 5.5\ A$$
$$= 132\ W$$
$$P_1 = E_1 \times I_1$$
$$= 24\ V \times 3\ A$$
$$= 72\ W$$
$$P_2 = E_2 \times I_2$$
$$= 24\ V \times 1.5\ A$$
$$= 36\ W$$
$$P_3 = E_3 \times I_3$$
$$= 24\ V \times 1\ A$$
$$= 24\ W$$

12.7 Parallel Circuit Polarity

In any DC circuit the positive and negative terminals of the voltage source establish the polarity relationships in the circuit. Polarity markings for a DC parallel circuit are similar to that used for the DC series circuit. The side of the component closest to the positive voltage terminal is the positive (+) side and the side closest to the negative terminal is the negative (–) side, as illustrated in Figure 12-13.

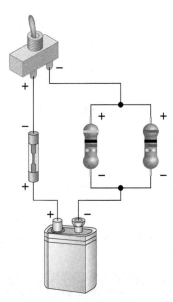

Figure 12-13 Parallel circuit polarity.

12.8 Parallel Voltage Sources

Connecting batteries in parallel has the effect of increasing the current or ampere-hour capacity of the power supply while keeping the voltage the same. To connect batteries in parallel, the positive (+) terminals are connected together and the negative (–) terminals are connected together as illustrated in Figure 12-14.

A **deep-cycle battery** is designed to provide a steady amount of current over a long period of time in contrast to automotive batteries designed to deliver short, high-current bursts for cranking the engine. Deep-cycle batteries are commonly used in solar power systems where the sun produces power during the day and the batteries store some of the power for use at night. All deep-cycle batteries are rated in **amp-hours (Ah).** Simply stated, amp-hours refer to the amount of current, in amperes, which can be supplied by the battery over the

period of hours. As an example, a 200-Ah (20-h rate) battery can sustain a 10-amp draw for 20 hours before reaching its state of discharge.

Voltage sources of **different voltage values** should never be connected in parallel. If sources with different voltages were put in parallel, some current would flow from the source with the higher voltage into the one with the lower voltage. Also, if by error, a battery connection is reversed in a parallel group, it will act as a **short circuit.** All other batteries in the group will discharge their energy through this short-circuit path. ***Both scenarios are unsafe practices that cause excessive currents capable of damaging the batteries and causing personal injury from battery casing ruptures.***

12.9 Troubleshooting a Parallel Circuit

Troubleshooting parallel DC circuits for open or shorted loads is not much different from troubleshooting a DC series circuit. Again, using Ohm's law, you can predict what changes will occur under different fault conditions. Using one of the original parallel circuits studied, we examine how different faults can affect voltage, current, resistance, and power.

EXAMPLE 12-4

Problem: An *open* circuit fault has developed in load resistor R_1 of the circuit shown in Figure 12-15. The circuit was tested under this fault condition and results recorded in the accompanying table. Based on an analysis of the readings, surmise how an open in one of the branches of a parallel circuit will effect:

a. The total resistance, current, and power dissipation.

b. The open branch voltage, resistance, current, and power dissipation.

c. The remaining branches voltage, resistance, current, and power dissipation.

	Voltage	Current	Resistance	Power
R_1	120 V	Zero	Infinite	Zero
R_2	120 V	4 mA	30 kΩ	480 mW
R_3	120 V	3 mA	40 kΩ	360 mW
Total	120 V	7 mA	17.1 kΩ	840 mW

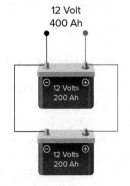

12 Volt
400 Ah

Figure 12-14 Parallel connection of batteries.

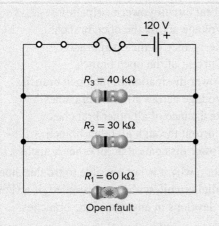

Figure 12-15 Circuit for example 12-4.

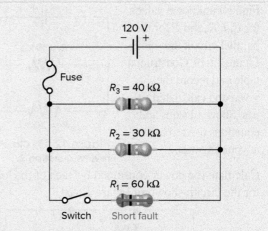

Figure 12-16 Short-circuit fault.

Solution:

a. The total resistance increases, resulting in less total current flow and less total power dissipation.

b. The full open branch will have normal voltage across it, but infinite resistance, the result being zero current flow and power dissipation.

c. All other branches will remain unchanged with normal voltage, resistance, current, and power dissipation.

The first step in troubleshooting a parallel circuit is to determine whether you are looking for a shorted branch or open branch. In most cases, the circuit protection device or the power supply identifies the type of problem you are looking for. When a branch is open, the source voltage is normal and the fuse is intact. A blown fuse indicates a shorted branch. If the shorted branch burns open, the circuit will take on the characteristics of an open circuit.

The circuit of Figure 12-16 illustrates the effects of a **short circuit** in one of the branches of a parallel circuit. The best way to find the source of the short is to disconnect each branch in the circuit and measure its resistance. The resistance of the shorted branch will be near zero ohms. When you close the switch of the shorted resistor (R_1), the following sequence of events occurs:

- The power supply voltage drops to an extremely low value.

- Current jumps to the maximum value the power supply can provide.

- The power supply shorts out through the shorted branch, and the fuse opens to prevent burnout of the power supply and wiring.

- With the power supply removed from the circuit, an ohmmeter will measure near zero ohms across all branches.

Part 2 Review Questions

1. Find all unknown values for *E, I, R,* and *P* for the parallel circuit drawn in Figure 12-17. Construct a table and record the given and calculated values. Show all steps and equations used to arrive at your answers.

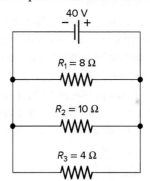

Figure 12-17 Circuit for review question 1.

2. Find all unknown values for *E, I, R,* and *P* for the parallel circuit drawn in Figure 12-18. Construct a table and record the given and calculated values. Show all steps and equations used to arrive at your answers.

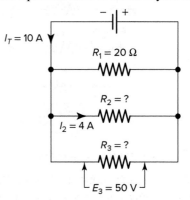

Figure 12-18 Circuit for review question 2.

3. Find all unknown values for E, I, R, and P for the parallel circuit drawn in Figure 12-19. Construct a table and record the given and calculated values. Show all steps and equations used to arrive at your answers.

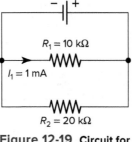

Figure 12-19 Circuit for review question 3.

4. Calculate the power consumed by each of the loads for the circuit shown in Figure 12-20.

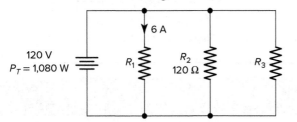

Figure 12-20 Circuit for review question 4.

5. A circuit contains twelve 60-W, 120-V lamps connected to a 120-VDC circuit. How much power does the circuit consume?

6. Determine how many 40-W, 120-V lamps can be connected in parallel on a circuit rated for 20 A and 120 V without exceeding 80 percent of its current rating.

7. Find all unknown values for E, I, R, and P for the parallel circuit drawn in Figure 12-21. Construct a table and record the given and calculated values. Show all steps and equations used to arrive at your answers.

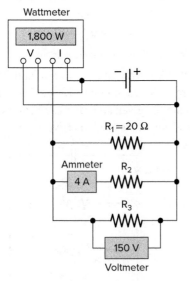

Figure 12-21 Circuit for review question 7.

8. A bank of lamps are connected in parallel. If one of the branch lamps burns open, what change, if any, will occur in each of the following?
   a. Total circuit resistance.
   b. Total circuit current flow.

c. Total circuit power dissipation.
d. Voltage across the open branch.
e. Resistance of the open branch.
f. Current of the open branch.
g. Power dissipation of the open branch.
h. Voltage across all other branches.
i. Resistance of all other branches.
j. Current through all other branches.
k. Power dissipation of all other branches.

9. Explain why it is not possible to troubleshoot a parallel circuit, with power applied, if a short circuit develops in any one of the branches.

10. Two identical baseboard heaters are connected in parallel to a 230-volt supply source. If each heater draws 6 amperes of current, how much power is delivered by the circuit?

11. For the circuit shown in Figure 12-22, determine how many 50-Ω resistive loads can be connected in parallel with the 12-volt supply before the maximum 3-ampere current rating of the fuse is exceeded.

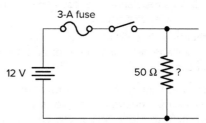

Figure 12-22 Circuit for review question 11.

12. Construct a schematic circuit according to the following requirements. Indicate what value of current would be read on each meter.
   - 48-VDC power supply.
   - A switch to operate the circuit and a fuse for circuit protection.
   - Four resistive (R_1, R_2, R_3, and R_4) loads connected in parallel.
   - R_1 to have a resistance of 10 Ω.
   - R_2 to have a resistance of 12 Ω.
   - R_3 to have a resistance of 6 Ω.
   - R_4 to have a resistance of 8 Ω.
   - Five ammeters to meter the total and branch current flows.

13. For the circuit shown in Figure 12-23, determine the total circuit resistance as seen by the power supply for each of the following switching scenarios.
   a. S_1 open, S_2 closed, and S_3 closed.
   b. S_1 closed, S_2 open, and S_3 open.
   c. S_1 closed, S_2 closed, and S_3 open.
   d. S_1 closed, S_2 open, and S_3 closed.
   e. S_1 closed, S_2 closed, and S_3 closed.

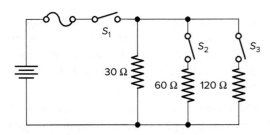

Figure 12-23 Circuit for review question 13.

14. The resistance directly across one of three loads connected in parallel is measured and found to be zero. Can you tell from this reading which one of the three is shorted? Why?

15. A circuit consists of a 10-V battery connected to 10 equal-value resistors connected in parallel. If the total power dissipated by the circuit is 40 watts, what is the value of each resistor?

16. Deep-cycle batteries provide energy storage for solar, wind, and other renewable energy systems. What special design feature makes them suited for this type of application?

17. A common application for parallel connected loads can be found in residential wiring. Assume that a table lamp is connected to one outlet of a wall receptacle and a hair dryer to the other. Compare the voltage and operation of both.

18. Determine the current flow through the resistor of the circuit shown of Figure 12-24.

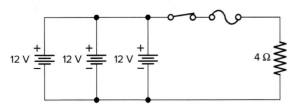

Figure 12-24 Circuit for review question 18.

19. **a.** For the circuit of Figure 12-25, assuming that the fuse and switches have zero resistance, what is the reading for each voltmeter?
    b. What would the readings be with switch S_2 open?

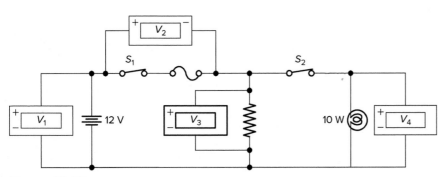

Figure 12-25 Circuit for review question 19.

20. You are taking a voltage measurement across the branch of a parallel circuit when you accidentally short the two meter test leads together. The circuit fuse blows. Why?

21. If you know the voltage across one branch of a parallel circuit, how would this help in solving values in other components?

22. Why do tail-end loads in long-distance parallel circuits get a lower voltage than loads closer to the source?

23. For the circuit shown in Figure 12-26, determine how many 50-Ω resistors can be connected in parallel before the maximum current rating of the fuse is exceeded.

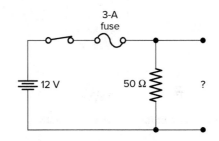

Figure 12-26 Circuit for review question 23.

24. For the circuit shown in Figure 12-27 determine:
    a. The current flow for each lamp branch (L1, L2, L3).
    b. The minimum size of fuse that could be used to protect the circuit.

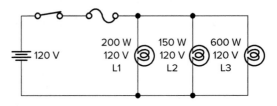

Figure 12-27 Circuit for review question 24.

25. Given the partial circuit shown in Figure 12-28, find the resistance of R_2.

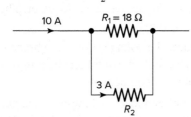

Figure 12-28 Circuit for review question 25.

26. Given the partial circuit shown in Figure 12-29, find the current in each branch.

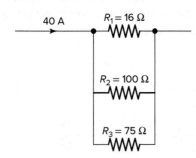

Figure 12-29 Circuit for review question 26.

27. You have been assigned the task of installing 40 light fixtures in an office. Each fixture is rated for 200 W and 120 V. If the maximum wattage allowed on any one circuit is 1800 W, how many 120-V branch circuits do you need?

28. A 120-V branch circuit to a kitchen is rated for 20 A. The following appliances are on at the same time:

$-$1000-W, 120-V toaster

$-$500-W, 120-V coffeepot

$-$1200-W, 120-V microwave

Does this load condition exceed the rating of the circuit? Why or why not?

Solving the DC Series-Parallel Circuit

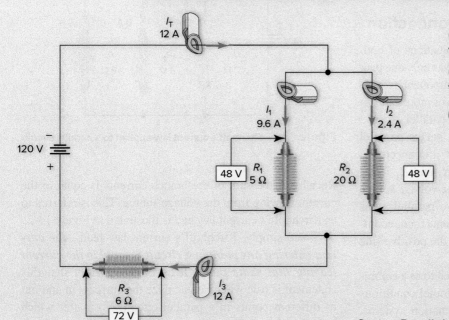

I_T
12 A

I_1
9.6 A

I_2
2.4 A

120 V

48 V

R_1
5 Ω

R_2
20 Ω

48 V

R_3
6 Ω

72 V

I_3
12 A

Series-Parallel voltage and current.

LEARNING OUTCOMES

▶ Recognize the series and parallel parts of DC combination circuits.

▶ Understand and correctly apply Kirchhoff's voltage and current laws to series-parallel circuits.

▶ Determine equivalent resistance in DC series-parallel circuits.

▶ Solve a DC series-parallel circuit for unknown circuit values.

▶ Troubleshoot DC series-parallel resistive circuits.

▶ Understand the operating principles of a three-wire circuit.

The two previous chapters discussed circuit loads that were connected in either series or in parallel. A series-parallel circuit consists of a group of series and parallel loads in which the total current flows through at least one of the loads. The complex circuits contain series-parallel-connected components. The method used to solve series-parallel circuits is mostly combinations of those used for series circuits and parallel circuits.

13.1 Series-Parallel Circuit Connection

A series-parallel circuit consists of combinations of both series and parallel current paths. If current has only one path to follow through a component, that component is connected in series. If the total current has two or more paths to follow, these components are connected in parallel.

Figure 13-1 shows an example of a **series-parallel combination** of resistive loads. Resistor R_1 is connected in series with the parallel combination of R_2 and R_3. Total current flows from the negative side of the voltage source through R_1 and then divides to form two parallel paths through R_2 and R_3. These two branch currents then recombine and the total current flows back into the positive side of the voltage source.

The series-parallel circuit is also referred to as a combination circuit in which both series and parallel conditions exist. Most electrical systems are combination systems. These circuits represent an increased level of difficulty in solving for unknown quantities, as more steps are usually required to arrive at the desired solution. A key factor in analyzing series-parallel circuits is the ability to recognize and isolate the series and parallel sections.

13.2 Determining Current Flow

Kirchhoff's laws are an extension of Ohm's law. They can be considered as additional tools for solving values for electric circuits. Used along with Ohm's law, these laws allow you to analyze DC series-parallel circuit networks.

Kirchhoff's current law is used to describe the current characteristics of a parallel DC circuit. The law refers to the relationship between the total current flow in a parallel circuit and the current flow through each of its branches.

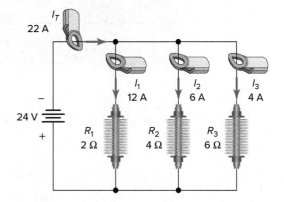

Figure 13-2 Kirchhoff's current law applied to a parallel circuit.

Recall that the sum of the branch currents is equal to the current flowing from the voltage source. This is referred to as Kirchhoff's current law and is illustrated in Figure 13-2.

Stated simply, Kirchhoff's current law reads, ***the current entering any point in a circuit is equal to the current leaving that same point.*** It becomes apparent that this statement is true when you analyze the currents in and out of different points in a parallel circuit. No matter which point is selected, the total current flow into the point is always equal to the current flowing out of the point. A connecting point or junction for two or more components is often referred to as a *node* and is illustrated in Figure 13-3.

Figure 13-4 illustrates how current flow can be used to establish series and parallel relationships in combination circuits. To determine which components are in series and which are in parallel, trace the flow of current through the circuit as follows:

- A series circuit has only one path for current flow, and a parallel circuit has two or more paths for current flow.
- Total current leaves the negative terminal of the power supply and all of this current has to travel through R_1, which is therefore a series-connected resistor.

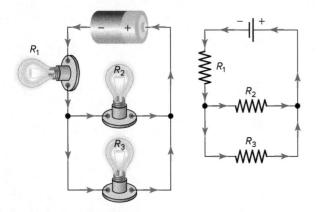

Figure 13-1 DC series-parallel circuit.

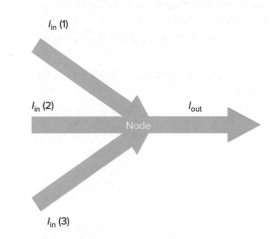

Figure 13-3 Tracing the currents in a parallel circuit.

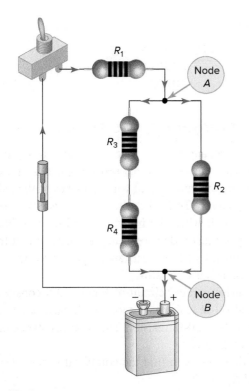

Figure 13-4 Current flow used to establish series and parallel relationships.

- The total current then splits at node A, and consequently, R_3 and R_4 with R_2 make up a parallel combination.
- The current that flows through R_3 will also flow through R_4; therefore, R_3 is in series with R_4.
- The two branch currents combine at node B to establish the total current, which is returned to the positive terminal of the power supply.
- In summary, it can be stated that R_3 and R_4 are in series with one another and both are in parallel with R_2, and this combination is in series with R_1.

In addition to Ohm's law, Kirchhoff's current law can be used to determine unknown currents in series-parallel circuit. In Figure 13-5, current in some of the conductors is given. The rest of the currents can be found using Kirchhoff's current law as follows:

- As 6 amperes of current returns to the one side of the power supply, the same amount must exit the other side. Therefore the total current (I_T) and the current through $R_1(I_1)$ are both 6 amperes.
- Since R_3 and R_4 are in series, the current entering R_3 is the same as the current leaving R_4 and is equal to 2 amperes.
- The current flow through R_2 is found by subtracting the 2-ampere current flow through R_3 and R_4 from the 6-ampere total current. Therefore the current through R_2 is 4 amperes.

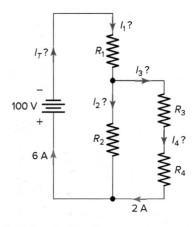

Figure 13-5 Using Kirchhoff's current law to determine unknown currents.

13.3 Determining Voltage

Kirchhoff's voltage law is used to describe the voltage characteristics of a series DC circuit. It states the relationship between the voltage drops and the applied voltage of a series circuit. Recall that the sum of all the voltage drops in a series circuit is equal to the applied voltage, as illustrated in Figure 13-6. This fact is referred to as Kirchhoff's voltage law.

The circuit of Figure 13-7 illustrates voltage relationships in a series-parallel circuit. Voltmeters are connected

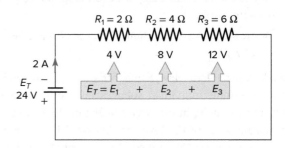

Figure 13-6 Kirchhoff's voltage law applied to a series circuit.

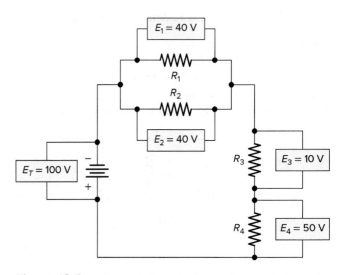

Figure 13-7 Voltage relationships in a series-parallel circuit.

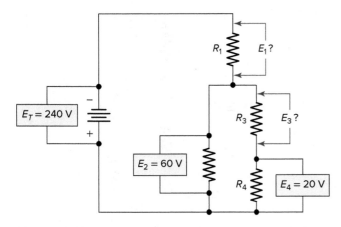

Figure 13-8 Using Kirchhoff's voltage law to determine unknown voltages.

to measure each of the load voltages and the readings are as indicated. The voltage relationships can be summarized as follows:

- The voltages across parallel branches have the same value.
- Voltages E_1 and E_2 have the same value (40 V) because R_1 and R_2 are connected in parallel with each other.
- The sum of voltages $E_1 + E_3 + E_4$ equals E_T because, by Kirchhoff's voltage law, the sum of the voltage drop around a single closed loop must equal the source voltage.
- For the same reason, the sum of voltages $E_2 + E_3 + E_4$ equals E_T.

In addition to Ohm's law, Kirchhoff's voltage law can be used to determine individual voltages in a series-parallel circuit. For example, in the circuit of Figure 13-8, voltage drops across some of the loads are known. The unknown voltage drops can be determined using Kirchhoff's voltage law as follows:

$E_T = E_1 + E_2$ (according to Kirchhoff's voltage law)

therefore

$$E_1 = E_T - E_2$$
$$= 240 \text{ V} - 60 \text{ V}$$
$$= 180 \text{ V}$$

$E_T = E_1 + E_3 + E_4$ (according to Kirchhoff's voltage law)

therefore

$$E_3 = E_T - E_1 - E_4$$
$$= 240 \text{ V} - 180 \text{ V} - 20 \text{ V}$$
$$= 40 \text{ V}$$

13.4 Determining Resistance

To the source, any resistive circuit is always a single resistor. Knowing the value of that resistance can be the key to determining the current and voltage values of the individual components. To find the total resistance of a series-parallel circuit when the individual resistance values are known, you must reduce or combine loads so that the resulting equivalent total resistance appears to be nothing more than a **single resistor.**

The series-parallel circuit of Figure 13-9 consists of a 40-Ω and 60-Ω resistor in series with each other and in parallel with a 20-Ω resistor. The total resistance can be found as follows.

The first step is to find the combined series resistance value of R_1 and R_2:

$$R_{1,2} = R_1 + R_2$$
$$= 40 \ \Omega + 60 \ \Omega$$
$$= 100 \ \Omega$$

The next step is to find the total equivalent circuit resistance consisting of $R_{1,2}$ connected in parallel with R_3:

$$R_T = \frac{R_{1,2} \times R_3}{R_{1,2} + R_3}$$
$$= \frac{100 \ \Omega \times 20 \ \Omega}{100 \ \Omega + 20 \ \Omega}$$
$$= \frac{2,000}{120}$$
$$= 16.7 \ \Omega$$

More complex series-parallel networks are solved for total resistance in a similar manner. General procedures to follow are:

- Reduce only one part at a time.
- After each circuit reduction, redraw the circuit and exchange the equivalent resistor for the original resistors.

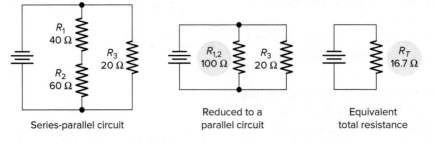

Series-parallel circuit

Reduced to a parallel circuit

Equivalent total resistance

Figure 13-9 Series-parallel circuit equivalent total resistance.

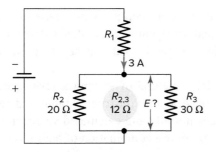

Figure 13-10 Determining voltage drop across a parallel branch.

- Be sure that all series resistors have been combined before a parallel portion is reduced.
- Combine parallel portions to a single resistor.
- Repeat combining equivalent resistors until the circuit is reduced to one equivalent total resistance.

Whenever the applied voltage and total current of any series-parallel circuit are known, the following Ohm's law equation can be used to determine the equivalent total resistance:

$$R_T = \frac{E_T}{I_T}$$

Whenever the equivalent resistance and total current of a parallel branch are known, Ohm's law can be used to determine the voltage drop across the parallel portion. For the parallel branch of the series-parallel circuit shown in Figure 13-10, the voltage can be calculated as follows:

E (parallel branch) = I (parallel branch) × R (parallel branch)
$$= 3\text{ A} \times 12\ \Omega$$
$$= 36\text{ V}$$

13.5 Determining Power

The total power supplied to a DC resistive circuit—whether series, parallel, or a combination of both—is equal to the sum of the power dissipated by the individual load resistors:

$$P_T = P_1 + P_2 + P_3 \cdots$$

Therefore, each circuit component contributes to the total power dissipated by a DC series-parallel circuit. By adding all the power dissipated by the individual load components, you can derive the total power used by the circuit. If two of the three quantities of each component are known (voltage, current, and/or resistance), power may be calculated for each using one of the following Ohm's law equations:

$$P = E \times I$$
$$P = I^2 \times R$$
$$P = \frac{E^2}{R}$$

1. Explain how Kirchhoff's current law is applied to determine current at a connecting point for two or more components.

2. Explain how Kirchhoff's voltage law is applied to determine voltage in a series-parallel circuit.

3. Determine, using Kirchhoff's laws, E_T, E_1, I_1, and I_3 for the circuit shown in Figure 13-11.

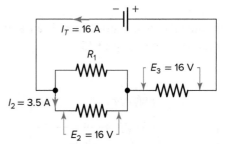

Figure 13-11 Circuit for review question 3.

4. Determine, using Kirchhoff's laws, I_1, E_2, I_4, and E_4 for the circuit shown in Figure 13-12.

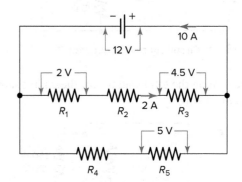

Figure 13-12 Circuit for review question 4.

5. Determine, using Kirchhoff's laws, E_1, E_4, E_7, I_7, and I_T for the circuit shown in Figure 13-13.

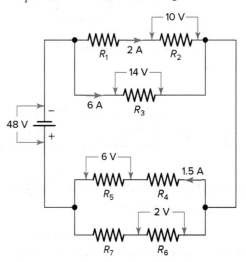

Figure 13-13 Circuit for review question 5.

Chapter 13 Solving the DC Series-Parallel Circuit 143

6. Determine the total equivalent resistance for the circuit shown in Figure 13-14.

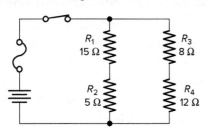

Figure 13-14 Circuit for review question 6.

7. Determine the total equivalent resistance for the circuit shown in Figure 13-15.

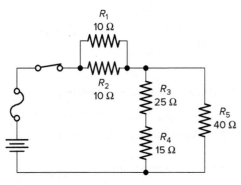

Figure 13-15 Circuit for review question 7.

8. Determine the total wattage for the circuit shown in Figure 13-16.

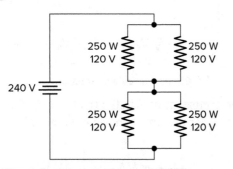

Figure 13-16 Circuit for review question 8.

PART 2 SOLVING THE DC SERIES-PARALLEL CIRCUIT

13.6 Solving for Current, Voltage, Resistance, and Power

When components are series-connected in some parts of a circuit and parallel-connected in others, you cannot apply the same set of rules to every part of that circuit.

Instead, you have to identify which parts of that circuit are series and which parts are parallel and then selectively apply series and parallel rules as necessary to determine parameters.

One helpful method of solving a series-parallel circuit is to once again construct a table and record all given values initially and calculated values in turn. Because the circuit is a combination of both series and parallel components, we cannot directly apply the rules for voltage, current, and resistance for the vertical columns, as was the case when the circuits were either series or parallel. The table will still help you manage the different values for series-parallel combination circuits, but you must be careful how and where you apply the different rules for series and parallel. Ohm's law, of course, still works just the same for determining values within a table row.

EXAMPLE 13-1

Problem: Both Ohm's law and Kirchhoff's laws are used in solving a series-parallel circuit. The general procedure followed is to find the equivalent resistance of each parallel group of components and then treat the result as a series circuit. Find all unknown values of E, I, R, and P for the series-parallel circuit of Figure 13-17.

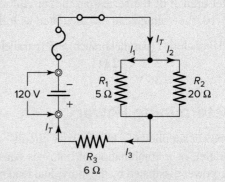

Figure 13-17 Circuit for example 13-1.

Solution:

Step 1. Draw a table that includes spaces for all values to be determined and record all known quantities as follows:

	Voltage	Current	Resistance	Power
R_1			5 Ω	
R_2			20 Ω	
R_3			6 Ω	
Total	120 V			

Step 2. Calculate R_T and enter its value in the table. Begin by calculating the equivalent resistance of R_1 and R_2 in parallel as follows:

$$R_{1,2} = \frac{R_1 \times R_2}{R_1 + R_2}$$
$$= \frac{5\,\Omega \times 20\,\Omega}{5\,\Omega + 20\,\Omega}$$
$$= \frac{100\,\Omega}{25\,\Omega}$$
$$= 4\,\Omega$$

Replacing parallel resistors R_1 and R_2 with one of an equivalent value produces the equivalent series circuit shown in Figure 13-18. The total resistance then becomes

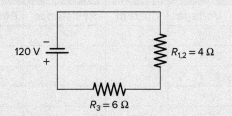

Figure 13-18 Equivalent series circuit.

	Voltage	Current	Resistance	Power
R_1			5 Ω	
R_2			20 Ω	
R_3			6 Ω	
Total	120 V		10 Ω	

$$R_T = R_{1,2} + R_3$$
$$= 4\,\Omega + 6\,\Omega$$
$$= 10\,\Omega$$

Step 3. Calculate I_T and I_3 and enter their values in the table. Total current is calculated by applying Ohm's law as follows:

	Voltage	Current	Resistance	Power
R_1			5 Ω	
R_2			20 Ω	
R_3		12 A	6 Ω	
Total	120 V	12 A	10 Ω	

$$I_T = \frac{E_T}{R_T}$$
$$= \frac{120\,\text{V}}{10\,\Omega}$$
$$= 12\,\text{A}$$

According to Kirchhoff's current law

$$I_3 = I_T$$
$$= 12\,\text{A}$$

Step 4. Calculate E_3, E_2, and E_1 and enter their values in the table. According to Ohm's law

$$E_3 = I_3 \times R_3$$
$$= 12\,\text{A} \times 6\,\Omega$$
$$= 72\,\text{V}$$

According to Kirchhoff's voltage law

	Voltage	Current	Resistance	Power
R_1	48 V		5 Ω	
R_2	48 V		20 Ω	
R_3	72 V	12 A	6 Ω	
Total	120 V	12 A	10 Ω	

$$E_2 = E_T - E_3$$
$$= 120\,\text{V} - 72\,\text{V}$$
$$= 48\,\text{V}$$

$$E_1 = E_T - E_3$$
$$= 120\,\text{V} - 72\,\text{V}$$
$$= 48\,\text{V}$$

Step 5. Calculate I_1 and I_2 and enter their values in the table. According to Ohm's law

	Voltage	Current	Resistance	Power
R_1	48 V	9.6 A	5 Ω	
R_2	48 V	2.4 A	20 Ω	
R_3	72 V	12 A	6 Ω	
Total	120 V	12 A	10 Ω	

$$I_1 = \frac{E_1}{R_1}$$
$$= \frac{48\,\text{V}}{5\,\Omega}$$
$$= 9.6\,\text{A}$$

$$I_2 = \frac{E_2}{R_2}$$
$$= \frac{48\,\text{V}}{20\,\Omega}$$
$$= 2.4\,\text{A}$$

Step 6. Calculate P_T, P_1, P_2, and P_3 as follows and enter their values in the table:

	Voltage	Current	Resistance	Power
R_1	48 V	9.6 A	5 Ω	461 W
R_2	48 V	2.4 A	20 Ω	115 W
R_3	72 V	12 A	6 Ω	864 W
Total	120 V	12 A	10 Ω	1,440 W

$$P_T = E_T \times I_T$$
$$= 120 \text{ V} \times 12 \text{ A}$$
$$= 1,440 \text{ W}$$

$$P_1 = E_1 \times I_1$$
$$= 48 \text{ V} \times 9.6 \text{ A}$$
$$= 461 \text{ W}$$

$$P_2 = E_2 \times I_2$$
$$= 48 \text{ V} \times 2.4 \text{ A}$$
$$= 115 \text{ W}$$

$$P_3 = E_3 \times I_3$$
$$= 72 \text{ V} \times 12 \text{ A}$$
$$= 864 \text{ W}$$

EXAMPLE 13-2

Problem: If certain key voltage and current measurements of a circuit are known, all unknown voltages and currents can be found by using only Kirchhoff's laws. Given the circuit in Figure 13-19 with voltages and currents as indicated, find all unknown values of E, I, R, and P.

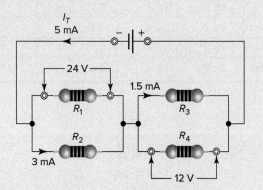

Figure 13-19 Circuit for example 13-2.

Solution:

Step 1. Draw a table that includes spaces for all values to be determined and record all known quantities as follows:

	Voltage	Current	Resistance	Power
R_1	24 V			
R_2		3 mA		
R_3		1.5 mA		
R_4	12 V			
Total		5 mA		

Step 2. Find all unknown voltage values using Kirchhoff's voltage law and enter their values in the table.

	Voltage	Current	Resistance	Power
R_1	24 V			
R_2	24 V	3 mA		
R_3	12 V	1.5 mA		
R_4	12 V			
Total	36 V	5 mA		

$$E_T = E_1 + E_4$$
$$= 24 \text{ V} + 12 \text{ V}$$
$$= 36 \text{ V}$$

$$E_2 = E_T - E_4$$
$$= 36 \text{ V} - 12 \text{ V}$$
$$= 24 \text{ V}$$

$$E_3 = E_T - E_1$$
$$= 36 \text{ V} - 24 \text{ V}$$
$$= 12 \text{ V}$$

Step 3. Find all unknown current values using Kirchhoff's current law and enter their values in the table.

	Voltage	Current	Resistance	Power
R_1	24 V	2 mA		
R_2	24 V	3 mA		
R_3	12 V	1.5 mA		
R_4	12 V	3.5 mA		
Total	36 V	5 mA		

$$I_1 = I_T - I_2$$
$$= 5 \text{ mA} - 3 \text{ mA}$$
$$= 2 \text{ mA}$$

$$I_4 = I_T - I_3$$
$$= 5 \text{ mA} - 1.5 \text{ mA}$$
$$= 3.5 \text{ mA}$$

Step 4. Use Ohm's law to determine all resistance values and enter their values in the table.

	Voltage	Current	Resistance	Power
R_1	24 V	2 mA	12 kΩ	
R_2	24 V	3 mA	8 kΩ	
R_3	12 V	1.5 mA	8 kΩ	
R_4	12 V	3.5 mA	3.43 kΩ	
Total	36 V	5 mA	7.2 kΩ	

$$R_T = \frac{E_T}{I_T}$$
$$= \frac{36\ \text{V}}{5\ \text{mA}}$$
$$= 7.2\ \text{k}\Omega$$

$$R_1 = \frac{E_1}{I_1}$$
$$= \frac{24\ \text{V}}{2\ \text{mA}}$$
$$= 12\ \text{k}\Omega$$

$$R_2 = \frac{E_2}{I_2}$$
$$= \frac{24\ \text{V}}{3\ \text{mA}}$$
$$= 8\ \text{k}\Omega$$

$$R_3 = \frac{E_3}{I_3}$$
$$= \frac{12\ \text{V}}{1.5\ \text{mA}}$$
$$= 8\ \text{k}\Omega$$

$$R_4 = \frac{E_4}{I_4}$$
$$= \frac{12\ \text{V}}{3.5\ \text{mA}}$$
$$= 3.43\ \text{k}\Omega$$

Step 5. Calculate P_T, P_1, P_2, P_3 and P_4 as follows and enter their values in the table:

	Voltage	Current	Resistance	Power
R_1	24 V	2 mA	12 kΩ	48 mW
R_2	24 V	3 mA	8 kΩ	72 mW
R_3	12 V	1.5 mA	8 kΩ	18 mW
R_4	12 V	3.5 mA	3.43 kΩ	42 mW
Total	36 V	5 mA	7.2 kΩ	180 mW

$$P_1 = E_1 \times I_1$$
$$= 24\ \text{V} \times 2\ \text{mA}$$
$$= 48\ \text{mW}$$

$$P_2 = E_2 \times I_2$$
$$= 24\ \text{V} \times 3\ \text{mA}$$
$$= 72\ \text{mW}$$

$$P_3 = E_3 \times I_3$$
$$= 12\ \text{V} \times 1.5\ \text{mA}$$
$$= 18\ \text{mW}$$

$$P_4 = E_4 \times I_4$$
$$= 12\ \text{V} \times 3.5\ \text{mA}$$
$$= 42\ \text{mW}$$

$$P_T = P_1 + P_2 + P_3 + P_4$$
$$= 48 + 72 + 18 + 42$$
$$= 180\ \text{mW}$$

EXAMPLE 13-3

Problem: The series-parallel circuit shown in Figure 13-20 consists of series components connected in parallel. Find all unknown values of *E*, *I*, *R*, and *P* for this circuit.

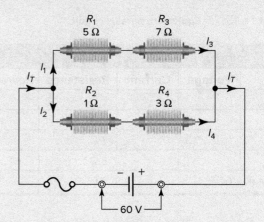

Figure 13-20 Circuit for example 13-3.

Solution:

Step 1. Draw a table that includes spaces for all values to be determined and record all known quantities as follows:

	Voltage	Current	Resistance	Power
R_1			5 Ω	
R_2			1 Ω	
R_3			7 Ω	
R_4			3 Ω	
Total	60 V			

Step 2. Calculate R_T and enter this value in the table. R_1 and R_3 in series become

$$R_{1,3} = R_1 + R_3$$
$$= 5\,\Omega + 7\,\Omega$$
$$= 12\,\Omega$$

R_2 and R_4 in series become

$$R_{2,4} = R_2 + R_4$$
$$= 1\,\Omega + 3\,\Omega$$
$$= 4\,\Omega$$

Replace series resistors with one of an equivalent value. The circuit is thus reduced to the equivalent parallel circuit of Figure 13-21. The total resistance then becomes

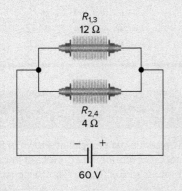

Figure 13-21 **Equivalent parallel circuit.**

	Voltage	Current	Resistance	Power
R_1			5 Ω	
R_2			1 Ω	
R_3			7 Ω	
R_4			3 Ω	
Total	60 V		3 Ω	

$$R_T = \frac{R_{1,3} \times R_{2,4}}{R_{1,3} + R_{2,4}}$$
$$= \frac{12\,\Omega \times 4\,\Omega}{12\,\Omega + 4\,\Omega}$$
$$= \frac{48\,\Omega}{16\,\Omega}$$
$$= 3\,\Omega$$

Step 3. Calculate all currents and enter their values into the table. According to Ohm's law

$$I_T = \frac{E_T}{R_T}$$
$$= \frac{60\text{ V}}{3\,\Omega}$$
$$= 20\text{ A}$$

The equivalent current flow, $I_{1,3}$ and $I_{2,4}$, can be calculated using Ohm's law and their equivalent series resistance values as follows:

	Voltage	Current	Resistance	Power
R_1		5 A	5 Ω	
R_2		15 A	1 Ω	
R_3		5 A	7 Ω	
R_4		15 A	3 Ω	
Total	60 V	20 A	3 Ω	

$$I_{1,3} = \frac{E_T}{R_{1,3}}$$
$$= \frac{60\text{ V}}{12\,\Omega}$$
$$= 5\text{ A}$$

$$I_{2,4} = \frac{E_T}{R_{2,4}}$$
$$= \frac{60\text{ V}}{4\,\Omega}$$
$$= 15\text{ A}$$

Step 4. Calculate all unknown voltages and enter their values into the table. According to Ohm's law

	Voltage	Current	Resistance	Power
R_1	25 V	5 A	5 Ω	
R_2	15 V	15 A	1 Ω	
R_3	35 V	5 A	7 Ω	
R_4	45 V	15 A	3 Ω	
Total	60 V	20 A	3 Ω	

$$E_1 = I_1 \times R_1$$
$$= 5\text{ A} \times 5\,\Omega$$
$$= 25\text{ V}$$

$$E_2 = I_2 \times R_2$$
$$= 15\text{ A} \times 1\,\Omega$$
$$= 15\text{ V}$$

$$E_3 = I_3 \times R_3$$
$$= 5\text{ A} \times 7\,\Omega$$
$$= 35\text{ V}$$

$$E_4 = I_4 \times R_4$$
$$= 15\text{ A} \times 3\,\Omega$$
$$= 45\text{ V}$$

Step 5. Calculate all power dissipation and enter their values into the table. According to Ohm's law

	Voltage	Current	Resistance	Power
R_1	25 V	5 A	5 Ω	125 W
R_2	15 V	15 A	1 Ω	225 W
R_3	35 V	5 A	7 Ω	175 W
R_4	45 V	15 A	3 Ω	675 W
Total	60 V	20 A	3 Ω	1,200 W

$$P_T = E_T \times I_T$$
$$= 60 \text{ V} \times 20 \text{ A}$$
$$= 1,200 \text{ W}$$

$$P_1 = E_1 \times I_1$$
$$= 25 \text{ V} \times 5 \text{ A}$$
$$= 125 \text{ W}$$

$$P_2 = E_2 \times I_2$$
$$= 15 \text{ V} \times 15 \text{ A}$$
$$= 225 \text{ W}$$

$$P_3 = E_3 \times I_3$$
$$= 35 \text{ V} \times 5 \text{ A}$$
$$= 175 \text{ W}$$

$$P_4 = E_4 \times I_4$$
$$= 45 \text{ V} \times 15 \text{ A}$$
$$= 675 \text{ W}$$

EXAMPLE 13-4

Problem: Some series-parallel circuits are constructed so that it is difficult to identify which components are in parallel and which are in series. The most difficult part of solving this type of circuit may be in identifying the series and parallel parts. In this case, it is best to redraw the circuit in a simpler schematic form before starting any calculations. An example of this method is illustrated in the following problem. Find all unknown values of *E*, *I*, and *R* for the series-parallel circuit shown in Figure 13-22.

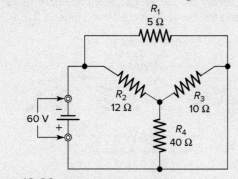

Figure 13-22 **Circuit for example 13-4.**

Solution:

Step 1. Simplify the circuit by redrawing it as shown in Figure 13-23.

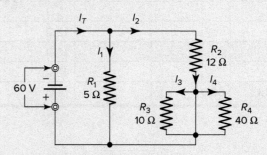

Figure 13-23 **Simplified circuit.**

Step 2. Draw a table that includes spaces for all values to be determined and record all known quantities as follows:

	Voltage	Current	Resistance
R_1			5 Ω
R_2			12 Ω
R_3			10 Ω
R_4			40 Ω
Total	60 V		

Step 3. Calculate R_T, as follows, and enter its value in the table. R_3 and R_4 in parallel become

$$R_{3,4} = \frac{R_3 \times R_4}{R_3 + R_4}$$
$$= \frac{10 \text{ Ω} \times 40 \text{ Ω}}{10 \text{ Ω} + 40 \text{ Ω}}$$
$$= \frac{400 \text{ Ω}}{50 \text{ Ω}}$$
$$= 8 \text{ Ω}$$

Replace parallel resistors R_3 and R_4 with one of equivalent size ($R_{3,4}$). This combination is then connected in series with R_2. The equivalent resistance of this branch then becomes

$$R_{2,3,4} = R_2 + R_{3,4}$$
$$= 12 \text{ Ω} + 8 \text{ Ω}$$
$$= 20 \text{ Ω}$$

The total resistance is then found by solving the equivalent parallel circuit, shown in Figure 13-24, as follows:

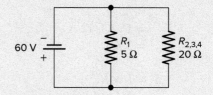

Figure 13-24 **Equivalent parallel circuit.**

	Voltage	Current	Resistance
R_1			5 Ω
R_2			12 Ω
R_3			10 Ω
R_4			40 Ω
Total	60 V		4 Ω

$$R_T = \frac{R_1 \times R_{2,3,4}}{R_1 + R_{2,3,4}}$$
$$= \frac{5\ \Omega \times 20\ \Omega}{5\ \Omega + 20\ \Omega}$$
$$= \frac{100\ \Omega}{25\ \Omega}$$
$$= 4\ \Omega$$

Step 4. Calculate all unknown currents and voltages and enter their values into the table.

	Voltage	Current	Resistance
R_1	60 V	12 A	5 Ω
R_2	36 V	3 A	12 Ω
R_3	24 V	2.4 A	10 Ω
R_4	24 V	0.6 A	40 Ω
Total	60 V	15 A	4 Ω

$$I_T = \frac{E_T}{R_T}$$
$$= \frac{60\ \text{V}}{4\ \Omega}$$
$$= 15\ \text{A}$$

$$I_1 = \frac{E_1}{R_1}$$
$$= \frac{60\ \text{V}}{5\ \Omega}$$
$$= 12\ \text{A}$$

$$I_2 = I_T - I_1$$
$$= 15\ \text{A} - 12\ \text{A}$$
$$= 3\ \text{A}$$

$$E_2 = I_2 \times R_2$$
$$= 3\ \text{A} \times 12\ \Omega$$
$$= 36\ \text{V}$$

$$E_3 = E_T - E_2$$
$$= 60\ \text{V} - 36\ \text{V}$$
$$= 24\ \text{V}$$

$$I_3 = \frac{E_3}{R_3}$$
$$= \frac{24\ \text{V}}{10\ \Omega}$$
$$= 2.4\ \text{A}$$

$$E_4 = E_3 = 24\ \text{V}$$

$$I_4 = \frac{E_4}{R_4}$$
$$= \frac{24\ \text{V}}{40\ \Omega}$$
$$= 0.6\ \text{A}$$

EXAMPLE 13-5

Conductors carry current from the power supply to the loads. A conductor should have as little resistance as possible for it to carry this current with minimal voltage drop and power loss. All conductors have resistance that in certain instances must be taken into consideration.

Problem: Determine the voltage across the loads, total power of the circuit, wasted power dissipated in the conductors, and the total power delivered to the loads, for the electrical distribution system of Figure 13-25.

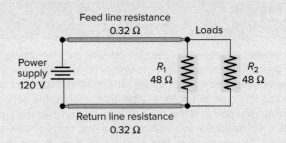

Figure 13-25 Circuit for example 13-5.

Step 1. Simplify the circuit, as shown in Figure 13-26, by representing the line wires as resistive loads.

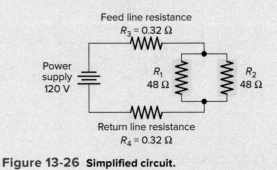

Figure 13-26 Simplified circuit.

Step 2. Draw a table that includes spaces for all values to be determined and record all known quantities as follows:

	Voltage	Current	Resistance	Power
R_1			48 Ω	
R_2			48 Ω	
R_3			0.32 Ω	
R_4			0.32 Ω	
Total	120 V			

Step 3. Calculate R_T, as follows, and enter its value in the table.

	Voltage	Current	Resistance	Power
R_1			48 Ω	
R_2			48 Ω	
R_3			0.32 Ω	
R_4			0.32 Ω	
Total	120 V		24.6 Ω	

$$R_{1,2} = \frac{R_N}{2}$$
$$= \frac{48 \ \Omega}{2}$$
$$= 24 \ \Omega$$

$$R_{3,4} = R_3 + R_4$$
$$= 0.32 \ \Omega + 0.32 \ \Omega$$
$$= 0.64 \ \Omega$$

$$R_T = R_{1,2} + R_{3,4}$$
$$= 24 \ \Omega + 0.64 \ \Omega$$
$$= 24.6 \ \Omega$$

Step 4. Calculate all unknown currents and voltages and enter their values into the table.

	Voltage	Current	Resistance	Power
R_1	117 V	2.44 A	48 Ω	
R_2	117 V	2.44 A	48 Ω	
R_3	1.56 V	4.88 A	0.32 Ω	
R_4	1.56 V	4.88 A	0.32 Ω	
Total	120 V	4.88 A	24.6 Ω	

$$I_T = \frac{E_T}{R_T}$$
$$= \frac{120 \ \text{V}}{24.6 \ \Omega}$$
$$= 4.88 \ \text{A}$$

$$I_T = I_3 = I_4 = 4.88 \ \text{A}$$

$$E_3 = I_3 \times R_3$$
$$= 4.88 \ \text{A} \times 0.32 \ \Omega$$
$$= 1.56 \ \text{V}$$

$$E_4 = I_4 \times R_4$$
$$= 4.88 \ \text{A} \times 0.32 \ \Omega$$
$$= 1.56 \ \text{V}$$

$$E_1 = E_T - (E_3 + E_4)$$
$$= 120 \ \text{V} - (1.56 \ \text{V} + 1.56 \ \text{V})$$
$$= 120 \ \text{V} - 3.12 \ \text{V}$$
$$= 117 \ \text{V}$$

$$E_2 = E_1 = 117 \ \text{V}$$

$$I_1 = \frac{E_1}{R_1}$$
$$= \frac{117 \ \text{V}}{48 \ \Omega}$$
$$= 2.44 \ \text{A}$$

$$I_2 = I_1 = 2.44 \ \text{A}$$

Step 5. Calculate all power dissipation and enter their values into the table.

	Voltage	Current	Resistance	Power
R_1	117 V	2.44 A	48 Ω	285 W
R_2	117 V	2.44 A	48 Ω	285 W
R_3	1.56 V	4.88 A	0.32 Ω	7.61 W
R_4	1.56 V	4.88 A	0.32 Ω	7.61 W
Total	120 V	4.88 A	24.6 Ω	586 W

$$P_T = E_T \times I_T$$
$$= 120 \ \text{V} \times 4.88 \ \text{A}$$
$$= 586 \ \text{W}$$

$$P_1 = E_1 \times I_1$$
$$= 117 \ \text{V} \times 2.44 \ \text{A}$$
$$= 285 \ \text{W}$$

$$P_2 = E_2 \times I_2$$
$$= 117 \ \text{V} \times 2.44 \ \text{A}$$
$$= 285 \ \text{W}$$

$$P_3 = E_3 \times I_3$$
$$= 1.56 \text{ V} \times 4.88 \text{ A}$$
$$= 7.61 \text{ W}$$

$$P_4 = E_4 \times I_4$$
$$= 1.56 \text{ V} \times 4.88 \text{ A}$$
$$= 7.61 \text{ W}$$

Step 6. As determined from the table:

- Voltage across the loads = 117 V
- Total power of the circuit = 586 W
- Wasted power dissipated in the conductors = 7.61 W + 7.61 W = 15.2 W
- Total power delivered to the loads = 285 W + 285 W = 570 W

13.7 Polarity in DC Series-Parallel Circuits

Polarity is important in all types of DC circuits. Failure to observe the correct polarity when connecting polarity-sensitive electrical and electronic devices may result in failure of operation and/or damage to the device itself.

As was the case in both the series and parallel circuits, polarity in a series-parallel circuit is associated with the direction of the current flow. Using the electron theory, as electrons pass through a circuit component, the end from which they enter is always negative (−) and the end through which they leave is always positive (+).

Figure 13-27 shows the correct connection for an LED (light-emitting diode) connected in a DC series-parallel circuit. Unlike incandescent light bulbs, which illuminate regardless of the electrical polarity, LEDs will only light with correct electrical polarity.

13.8 Troubleshooting a Series-Parallel Circuit

Steps to follow when troubleshooting series-parallel circuits are similar to those used in troubleshooting series circuits and parallel circuits:

- First, solve the circuit for all normal operating voltage, current, and resistance values. In some instances the

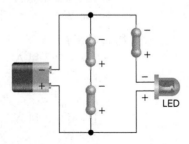

Figure 13-27 LED connected as part of a series-parallel circuit.

manufacturer may specify these normal operating parameters.

- Next, measure the actual circuit values, and compare these values to your calculated or specified values. You usually measure voltages first because these measurements are easy to make and do not normally require that you disturb the circuit. Often voltage measurements give you enough information to determine the faulty component.

- Generally, if the measured and normal operating values disagree by more than 10 percent, you can assume some part of the circuit is defective. The percentage of error allowable can vary. With very sensitive circuits, a smaller percentage difference can indicate a faulty circuit.

When troubleshooting a series-parallel circuit, it is important to remember the symptoms associated with open and shorted components in both series circuits and parallel circuits. Even though series-parallel circuits are more complex, the symptoms associated with each series and parallel portion are very similar. Table 13-1 summarizes the symptoms associated with open and shorted faults.

Table 13-1 Troubleshooting Series-Parallel Circuits		
Circuit Type	**Fault**	**Symptoms**
Series	Open	Circuit current drops to zero.
		The applied voltage is dropped across the open component.
		The voltage across each remaining component drops to zero.
	Short	Circuit current increases.
		The voltage across the shorted component drops to zero.
		The voltage across the remaining components increases.
Parallel	Open	Current through the open branch drops to zero.
		Total circuit current decreases.
		Other branches continue to operate normally.
	Short	The shorted branch causes the power supply protection device to open.
		The measured resistance across all branches is 0 Ω. (The faulty branch shorts out all parallel branches.)
		The voltage measure across any branch is zero.

EXAMPLE 13-6

Problem: An open fault has developed in load resistor R_1 of the circuit drawn in Figure 13-28. Complete a table that shows the conditions that would exist with R_1 faulted open.

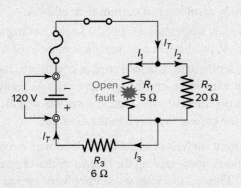

Figure 13-28 Circuit for example 13-6.

Solution: The malfunction of R_1 results in the formation of a series circuit made up of R_2 and R_3 in series.

	Voltage	Current	Resistance	Power
R_1	92.4 V	Zero	Infinite	Zero
R_2	92.4 V	4.62 A	20 Ω	427 W
R_3	27.7 V	4.62 A	6 Ω	128 W
Total	120 V	4.62 A	26 Ω	555 W

EXAMPLE 13-7

Problem: A shorted fault has developed in load resistor R_2 of the circuit drawn in Figure 13-29. Complete a table that shows the conditions that would exist with R_2 faulted shorted.

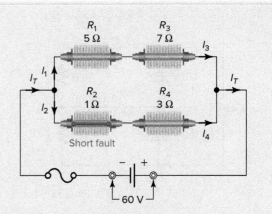

Figure 13-29 Circuit for example 13-7.

Solution: The malfunction of R_2 results in a series-parallel circuit made up of resistors R_1 and R_3 in series and that combination in parallel with R_4.

	Voltage	Current	Resistance	Power
R_1	25 V	5 A	5 Ω	125 W
R_2	Zero	20 A	Zero	Zero
R_3	35 V	5 A	7 Ω	175 W
R_4	60 V	20 A	3 Ω	1,200 W
Total	60 V	25 A	2.4 Ω	1,500 W

13.9 Three-Wire Circuits

A conventional two-wire circuit uses two wires to form a current path from the source to the load and back, as illustrated in Figure 13-30. The equivalent **three-wire circuit** requires only three wires to accomplish the same result. This saves one wire, reduces voltage drop, and provides both 120-V and 240-V circuits. The 120/240-AC version of the three-wire is standard for residential services and multi-wire branch circuits.

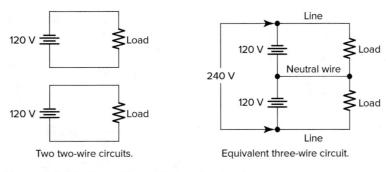

Two two-wire circuits.   Equivalent three-wire circuit.

Figure 13-30 Two-wire and three-wire circuit.

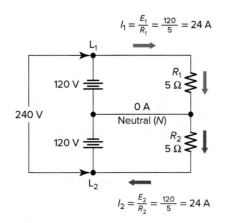

$$I_1 = \frac{E_1}{R_1} = \frac{120}{5} = 24 \text{ A}$$

Figure 13-31 Balanced three-wire circuit.

Three-wire circuits can be classified as being balanced or unbalanced. A **balanced three-wire circuit** is shown in Figure 13-31. In a balanced three-wire circuit, the loads are **equal,** so there is **no current** flow in the neutral wire. The operation of the circuit can be summarized as follows:

- The loads, R_1 and R_2, between the two 120-V circuits are equal.
- Each load has a voltage of 120 V and a resistance of 5 Ω, resulting in a current flow of 24 A through each ($I = E/R$).
- Under these balanced current conditions through line L_1 and line L_2, there will be no current flow in the neutral wire.

An **unbalanced three-wire circuit** is shown in Figure 13-32. In an unbalanced three-wire circuit, the loads are **not equal,** and the neutral wire returns the **unbalanced current** to the source. In order to achieve a balanced load condition, the loads need to be calculated and divided as equally as possible between the two 120-V circuits. It is rare, however, for residential services to have the same number of lights or devices turned on for the current to be the same on each of the live line wires. The operation of the unbalanced three-wire circuit can be summarized as follows:

- The loads, R_1 and R_2, between the two 120-V circuits are not equal.
- Load R_1 has a resistance of 5 Ω and a voltage of 120 V, resulting in a current flow of 24 A.
- Load R_2 has a resistance of 10 Ω and a voltage of 120 V, resulting in a current flow of only 12 A.
- Under these unbalanced current conditions, the neutral wire will carry the difference (12 A) between the two line currents. Note the amount and direction of the current indicated in the diagram.

A current analysis of unbalanced three-wire circuits can be helpful in understanding the characteristics of three-wire circuits. Figure 13-33 shows a three-wire circuit where 240 V and 120 V are being supplied simultaneously. The current analysis of this circuit can be summarized as follows:

- Current through R_1 equals: $I_1 = \frac{E_1}{R_1} = \frac{120}{10} = 12A$
- Current through R_2 equals: $I_2 = \frac{E_2}{R_2} = \frac{120}{12} = 10A$
- Current through R_3 equals: $I_3 = \frac{E_3}{R_3} = \frac{240}{48} = 5A$
- The amount of current flowing into any point in a circuit is equal to the amount leaving that point.
- At point A, 17 A flows in from line L_1, 12 A flows out to the 10-Ω, 120-V resistor load, and 5 A flows out to the 48-Ω, 240-V resistor load.
- At point B, 12 A flows in, 10 A flows out to the 12-Ω, 120-V resistor load, and 2 A is returned to the source via the neutral.
- Point C has 10 A and 5 A coming in and returns 15 A to the source via line L_2.

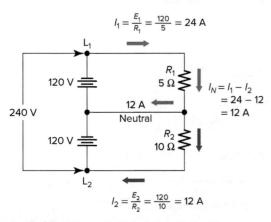

$$I_1 = \frac{E_1}{R_1} = \frac{120}{5} = 24 \text{ A}$$

$$I_N = I_1 - I_2$$
$$= 24 - 12$$
$$= 12 \text{ A}$$

$$I_2 = \frac{E_2}{R_2} = \frac{120}{10} = 12 \text{ A}$$

Figure 13-32 Unbalanced three-wire circuit.

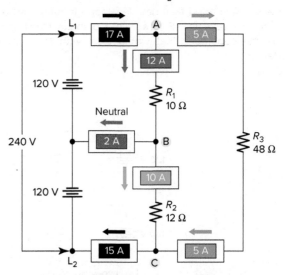

Figure 13-33 Current analysis of a 120/240-volt unbalanced three-wire circuit.

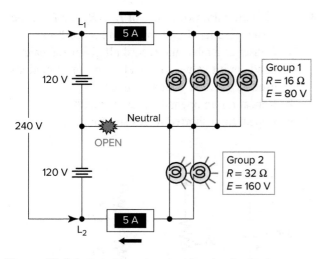

Figure 13-34 Unbalanced three-wire circuit with faulted open neutral.

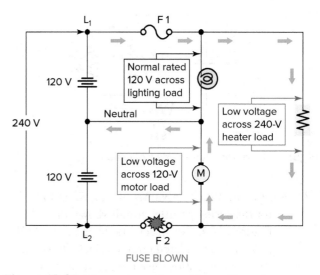

Figure 13-35 Three-wire system with one blown fuse.

If an unbalanced three-wire circuit is faulted so that the **neutral wire is opened** in any way (Figure 13-34), there is **danger** to the electrical equipment in the circuit. The operation of the unbalanced three-wire circuit with an open neutral can be summarized as follows:

- Calculations are based on each lamp having the same wattage rating at 120 volts.

- When the neutral wire is disconnected, there will no longer be a 120-V circuit.

- The two line sides of the system will now be in **series** with each other and have an applied voltage of 240 V.

- Since voltage is divided in a series circuit, the Group 2 side of the circuit, with fewer devices on (and, therefore, **higher resistance**), will receive a higher than normal voltage of 160 V ($E = I \times R = 5 \times 32 = 160$ V).

- The Group 1 side of the circuit, with more devices on (and therefore, **lower resistance**), will receive a lower than normal voltage of 80 V ($E = I \times R = 5 \times 16 = 80$ V).

- As a result, the Group 2 light bulbs will glow much more **brightly** with the increase in voltage and may burn out more quickly. The Group 1 light bulbs will be much **dimmer** than normal because of the decrease in voltage across them.

- For lamp loads no damage would occur as a result of the lower voltage, but the bulbs with the higher voltage may burn out. Motor loads, however, can be damaged when lower than rated voltages are applied to them. The danger of an unbalanced voltage condition is also one of the reasons for never fusing the neutral wire. A blown fuse on a neutral wire will result in an unbalanced voltage situation.

Another type of three-wire circuit problem can arise when **only one** of the two line wire fuses blows due to an overload or fault on the circuit. This scenario is illustrated in the three-wire circuit of Figure 13-35. If only fuse F2 opens, it is possible to have voltages that no longer match the requirements of the loads. The operation of this condition can be summarized as follows:

- With fuse F2 open, a single 120-V supply will be applied to the circuit.

- The value of the voltage across the light will be 120 V, so the lamp will operate under normal operating conditions.

- The 240-V heater and 120-V motor loads will now be connected in series with one another to the 120-V supply.

- This results in the single 120-V supply voltage being dropped across these loads in direct proportion to their resistance values.

- As a result, a lower than normal operating voltage will appear across both the heater and the motor.

Part 2 Review Questions

1. **a.** Find all unknown values of E, I, R, and P for the series-parallel circuit shown in Figure 13-36. Construct a table and record all given and calculated circuit values. Show all steps and equations used to arrive at your answers.
   b. Repeat for an opened R_3 resistor fault.
   c. Repeat for a shorted R_2 resistor fault.

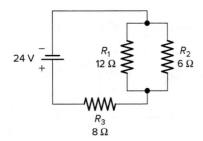

Figure 13-36 Circuit for review question 1.

2. **a.** Find all unknown values of E, I, R, and P for the series-parallel circuit shown in Figure 13-37. Construct a table and record all given and calculated circuit values. Show all steps and equations used to arrive at your answers.
   b. Repeat for an opened R_3 resistor fault.
   c. Repeat for a shorted R_1 resistor fault.

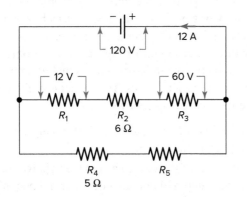

Figure 13-37 Circuit for review question 2.

3. **a.** Find all unknown values of E, I, R, and P for the series-parallel circuit shown in Figure 13-38. Construct a table and record all given and calculated circuit values. Show all steps and equations used to arrive at your answers.
   b. Repeat for an opened R_1 resistor fault.
   c. Repeat for a shorted R_4 resistor fault.

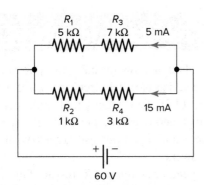

Figure 13-38 Circuit for review question 3.

4. **a.** Find all unknown values of E, I, R, and P for the series-parallel circuit shown in Figure 13-39. Construct a table and record all given and calculated circuit values. Show all steps and equations used to arrive at your answers.
   b. Repeat for an opened R_3 resistor fault.
   c. Repeat for a shorted R_1 resistor fault.

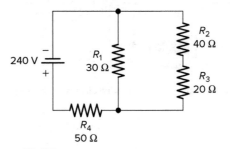

Figure 13-39 Circuit for review question 4.

5. **a.** Find all unknown values of E, I, R, and P for the series-parallel circuit shown in Figure 13-40. Construct a table and record all given and calculated circuit values. Show all steps and equations used to arrive at your answers.
   b. In what way, if any, are the voltage and current values of the other resistors affected by an open fault in resistor R_5. Why?

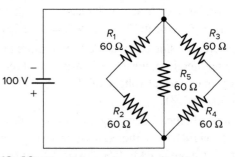

Figure 13-40 Circuit for review question 5.

6. **a.** Find all unknown values of E, I, R, and P for the series-parallel circuit shown in Figure 13-41. Construct a table and record all given and calculated circuit values. Show all steps and equations used to arrive at your answers.
   b. Repeat for an opened R_3 resistor fault.
   c. Repeat for a shorted R_5 resistor fault.

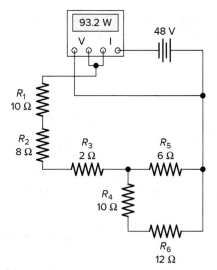

Figure 13-41 Circuit for review question 6.

7. Redraw the circuit shown in Figure 13-42 indicating the polarity of the voltage drop across each load resistor.

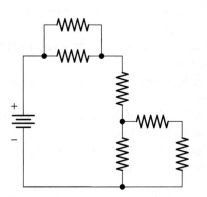

Figure 13-42 Circuit for review question 7.

8. Identify the series and parallel relationships for the circuit of Figure 13-43.

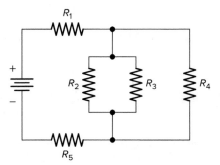

Figure 13-43 Circuit for review question 8.

9. Find the total equivalent resistance for the circuit shown in Figure 13-44.

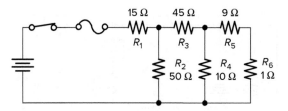

Figure 13-44 Circuit for review question 9.

10. If all resistors are 6 Ω in value, determine the current reading for each ammeter shown in Figure 13-45.

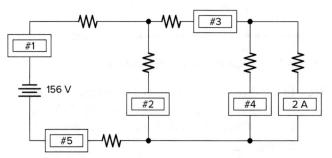

Figure 13-45 Circuit for review question 10.

11. Find all unknown values of *E, I,* and *R* for the circuit shown in Figure 13-46. Construct a table and record all given and calculated values.

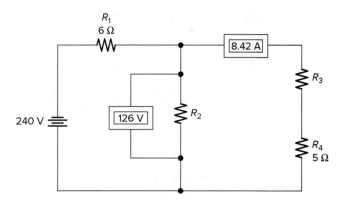

Figure 13-46 Circuit for review question 11.

12. What are advantages of using a three-wire circuit in place of 2 two-wire circuits?

13. Under what normal operating condition does no current flow in the neutral wire of a three-wire circuit?

14. For a three-wire system, what voltage is available between:
    a. The two line wires?
    b. A line wire and neutral?

15. Determine the current reading for each ammeter shown in the three-wire circuit of Figure 13-47.

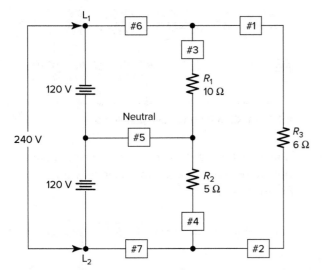

Figure 13-47 Circuit for review question 15.

16. In the three-wire circuit shown in Figure 13-48, the resistors represent the different loads in the circuit. Calculate the value of the current flow through
   a. Each of the loads.
   b. Line 1 and Line 2.
   c. The neutral conductor.

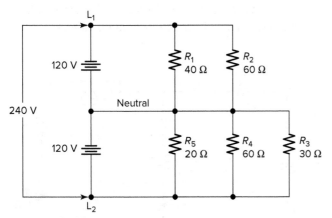

Figure 13-48 Circuit for review question 16.

17. For the three-wire circuit of Figure 13-49, each resistor load has a resistance of 120 Ω. Determine:
   a. The total equivalent resistance of Group 1 and Group 2.
   b. With the neutral **not faulted:**
      i. The voltage across the Group 1 loads.
      ii. The L_1 current flow.
      iii. The voltage across the Group 2 loads.
      iv. The L_2 current flow.
      v. The neutral current flow.
   c. With the neutral **faulted open:**
      i. The voltage across the Group 1 loads.
      ii. The L1 current flow.

iii. The voltage across the Group 2 loads.
iv. The L2 current flow.
v. The neutral current flow.

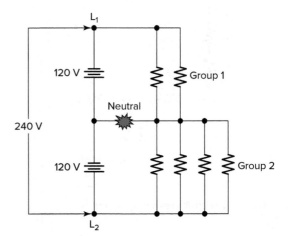

Figure 13-49 Circuit for review question 17.

18. Explain why the neutral wire of a three-wire system is never fused.

19. For the blown fuse three-wire circuit of Figure 13-50,
   a. Which, if any, of the loads would receive no voltage?
   b. Which, if any, of the loads would receive normal voltage?
   c. Which, if any, of the loads would receive higher than normal voltage?
   d. Which, if any, of the loads would receive lower than normal voltage?

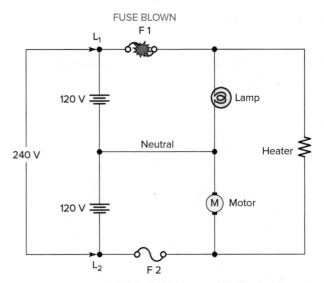

Figure 13-50 Circuit for review question 19.

CHAPTER FOURTEEN

Network Theorems

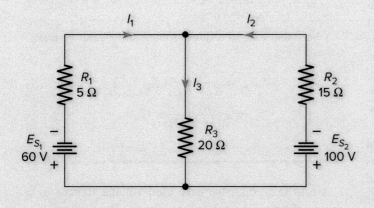

Network voltage and current.

There are several **theorems** that allow you to calculate circuit parameters (volts, amps, ohms, etc.) in complex DC circuit networks. A network is a combination of interconnected components such as resistors connected in a way to produce a particular end result. However, networks generally need more than the rules of series and parallel circuits for analysis. In this chapter, you learn how complex DC resistive networks can be reduced to a simpler form using the superposition, Thevenin, and Norton theorems.

PART 1 SUPERPOSITION THEOREM

14.1 Voltage and Current Sources

A **constant-voltage source** (also called an **ideal voltage source**) has zero internal resistance and supplies any amount of current with no change in its terminal voltage. The ideal voltage source will maintain the same voltage across its terminals no matter what current is drawn from it. Figure 14-1 shows the operation of an ideal voltage source. Note that the open- and closed-circuit terminal voltages are the same value regardless of the value of the load resistance. Actual voltage sources are not true constant-voltage sources; however, for practical applications many voltage sources are assumed to be constant-voltage sources.

In practice, every voltage source has a **series internal resistance (R_S),** which causes some loss of voltage at its terminals when a current flows. A voltage source should have the smallest possible internal resistance so that its output will remain constant regardless of whether a light load or heavy load is connected to its output terminals. When the internal resistance of a voltage source becomes a significant part of the total circuit resistance, you must take it into account when analyzing the circuit.

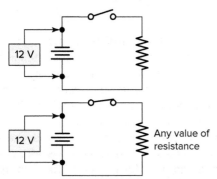

Figure 14-1 Constant-voltage source.

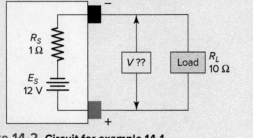

EXAMPLE 14-1

Problem: Assume a battery source has a no-load voltage of 12 volts and an internal resistance (R_S) of 1 Ω, as illustrated in Figure 14-2. Calculate the output voltage when the battery is connected to a load resistance (R_L) of 10 Ω.

Figure 14-2 Circuit for example 14-1.

Solution: The voltage divider rule is useful in determining the voltage drop across a resistance within a series circuit. The general formula is:

$$E_{R_X} = \frac{R_X}{R_T} \times E_S$$

where R_T = the total resistance of the series circuit
E_S = the value of the supply voltage applied to the circuit
R_X = the resistance across which you are calculating the voltage

From the voltage divider formula, the output voltage would be

$$E_{R_L} = \frac{R_L}{R_L + R_S} \times E_S$$
$$= \frac{10\ \Omega}{10\ \Omega + 1\ \Omega} \times 12\ \text{V}$$
$$= 10.9\ \text{V}$$

Up to this point, we have used voltage sources as a means of providing power for a circuit. The purpose of a DC voltage source is to provide an output voltage that remains relatively constant over a wide range of load resistance values. A **current source** is a source designed to provide an output current value that remains relatively constant over a wide range of load resistance values.

The circuit for a DC current source is shown in Figure 14-3. The arrow in the symbol indicates direction of the conventional current flow (+ to −) produced by the current source. Just as a voltage source has a voltage rating, the current source has a current rating (I_S), which in this case is 2 amperes. The rated current does not change regardless of the value of the resistance connected to it. Since the current remains constant, the terminal voltage changes as the load is changed.

An **ideal current source** has infinite internal resistance, which keeps circuit current constant when the load resistance changes. In practice, every current source has a parallel internal resistance (R_S), which is less than infinite, so its output current does vary slightly when load resistance changes. Figure 14-4 illustrates differences between ideal and practical current sources. Their operating characteristics can be summarized as follows:

- The primary difference between the ideal and practical sources is the effect of source resistance.

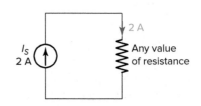

Figure 14-3 DC current source.

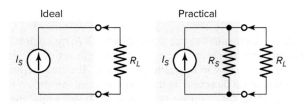

Figure 14-4 Ideal and practical current sources.

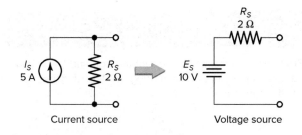

Figure 14-6 Current source to voltage source conversion.

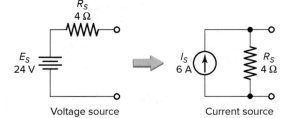

Figure 14-5 Voltage source to current source conversion.

- The ideal current source produces a constant load current, regardless of the value of the output load resistance.
- The load current from a practical current source varies inversely with the load resistance due to current division with the source resistance. Where the load resistance is much smaller than the source resistance, the entire source current is assumed to flow through the load.

For every voltage source, there exists an equivalent current source and vice versa. Source conversions let you replace a voltage source with a current source, or vice versa. In some cases, this can make a particular circuit easier to analyze. The method used to convert a voltage source to a current source is illustrated in Figure 14-5. The value of resistance R_S remains the same and the value of I_S is found as

$$I_S = \frac{E_S}{R_S}$$
$$= \frac{24 \text{ V}}{4 \text{ }\Omega}$$
$$= 6 \text{ A}$$

The method used to convert a current source to a voltage source is illustrated in Figure 14-6. The value of resistance R_S remains the same and the value of E_S is found as

$$E_S = I_S \times R_S$$
$$= 5 \text{ A} \times 2 \text{ }\Omega$$
$$= 10 \text{ V}$$

14.2 Superposition Theorem

All of the DC circuits covered to this point have had only one voltage source. There are circuits, however, that contain more than one voltage (or current) source. When analyzing these multisource circuits, the effect of each power source must be taken into account.

The **superposition theorem** is very useful in solving resistive networks that have more than one source of voltage or current. This technique treats each source as an independent source in the network and then combines the individual results. The **superposition theorem states that in a linear circuit with several sources, the current and voltage for any element in the circuit is the algebraic sum of the currents and voltages produced by each source acting independently.**

The following steps are used when applying the superposition theorem:

- Zero all voltage or current sources but one.
- In order to zero a voltage source, you replace it with a short circuit since the voltage across a short circuit is zero volts.
- In order to zero a current source, you replace it with an open circuit since the current through an open circuit is zero ampere.
- If the internal resistance is a significant part of the total circuit resistance, replace the source with a resistor equal to its internal resistance.
- Determine the current or voltage you need, along with its correct direction or polarity, by putting each source back into the circuit one at a time.
- Repeat for all the remaining sources in the circuit.
- To find a specific current or voltage, algebraically combine the currents or voltages due to the individual sources.
- If the currents act in the same direction or the voltages are of the same polarity, add them. If the reverse is true, subtract them with the direction of the resultant current or voltage being the same as the larger of the original quantity.

EXAMPLE 14-2

Problem: Calculate the current through and the voltage across R_1 and R_2 for the circuit of Figure 14-7, using the superposition theorem.

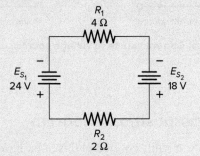

Figure 14-7 Circuit for example 14-2.

Solution:

Step 1. Find the current through R_1 due to source E_{S_1} by replacing E_{S_2} with a *short* circuit, as shown in Figure 14-8.

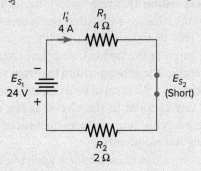

Figure 14-8 Circuit for step 1.

$$I_1' = \frac{E_{S_1}}{R_1 + R_2}$$

$$= \frac{24 \text{ V}}{6 \text{ }\Omega}$$

$$= 4 \text{ A}$$

Step 2. Find the current through R_1 due to source E_{S_2} by replacing E_{S_1} with a *short* circuit as shown in Figure 14-9.

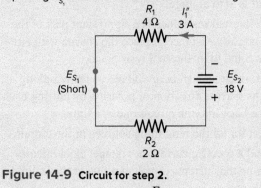

Figure 14-9 Circuit for step 2.

$$I_1'' = \frac{E_{S_2}}{R_T}$$

$$= \frac{18 \text{ V}}{6 \text{ }\Omega}$$

$$= 3 \text{ A}$$

Step 3. Calculate the net current through R_1 by subtracting the smaller current from the larger current since the two currents flow in opposite directions. The net current is in the same direction as the larger current, as shown in Figure 14-10. The current flow through R_2 will equal that through R_1 since the two are in series.

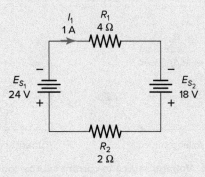

Figure 14-10 Circuit for step 3.

$$I_1 = I_1' - I_1''$$

$$= 4 \text{ A} - 3 \text{ A}$$

$$= 1 \text{ A}$$

Step 4. The voltage across R_1 and R_2 is then found using Ohm's law. The solved circuit is shown in Figure 14-11.

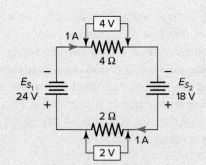

Figure 14-11 Solved circuit for example 14-2.

$$E_1 = I_1 \times R_1$$

$$= 1 \text{ A} \times 4 \text{ }\Omega$$

$$= 4 \text{ V}$$

$$E_2 = I_2 \times R_2$$

$$= 1 \text{ A} \times 2 \text{ }\Omega$$

$$= 2 \text{ V}$$

EXAMPLE 14-3

Problem: Calculate all the currents and the voltage drops for the circuit of Figure 14-12, using the superposition theorem.

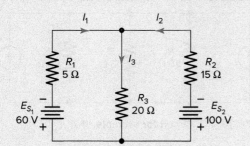

Figure 14-12 Circuit for example 14-3.

Solution:

Step 1. Find the currents I_1', I_2', and I_3' due to source E_{S_2} by replacing E_{S_1} with a **short** circuit as shown in Figure 14-13. Note that R_1 and R_3 appear connected in parallel with each other, and this combination is in series with R_2.

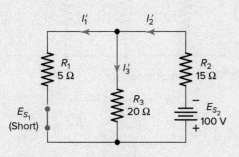

Figure 14-13 Circuit for step 1.

$$R_{1,3} = \frac{R_1 \times R_3}{R_1 + R_3}$$
$$= \frac{100\ \Omega}{25\ \Omega}$$
$$= 4\ \Omega$$

$$R_T = R_{1,3} + R_2$$
$$= 4\ \Omega + 15\ \Omega$$
$$= 19\ \Omega$$

$$I_T = \frac{E_{S_2}}{R_T}$$
$$= \frac{100\ \text{V}}{19\ \Omega}$$
$$= 5.26\ \text{A}$$

$$I_2' = I_T = 5.26\ \text{A}$$
$$E_1 = R_{1,3} \times I_2'$$
$$= 4\ \Omega \times 5.26\ \text{A}$$
$$= 21\ \text{V}$$

$$I_1' = \frac{E_1}{R_1}$$
$$= \frac{21\ \text{V}}{5\ \Omega}$$
$$= -4.2\ \text{A (negative)}$$

(Negative sign indicates current flow is in a direction opposite to that indicated in the original circuit.)

$$I_3' = I_2' - I_1'$$
$$= 5.26\ \text{A} - 4.2\ \text{A}$$
$$= 1.06\ \text{A}$$

Step 2. Find the currents I_1'', I_2'', and I_3'' due to source E_{S_1} by replacing E_{S_2} with a **short** circuit as shown in Figure 14-14. Note that R_2 and R_3 now appear connected in parallel with each other, and this combination is in series with R_1.

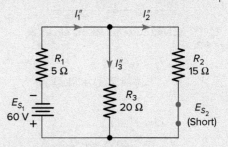

Figure 14-14 Circuit for step 2.

$$R_{2,3} = \frac{R_2 \times R_3}{R_2 + R_3}$$
$$= \frac{300\ \Omega}{35\ \Omega}$$
$$= 8.57\ \Omega$$

$$R_T = R_{2,3} + R_1$$
$$= 8.57\ \Omega + 5\ \Omega$$
$$= 13.6\ \Omega$$

$$I_T = \frac{E_{S_1}}{R_T}$$
$$= \frac{60\ \text{V}}{13.6\ \Omega}$$
$$= 4.41\ \text{A}$$

$$I_1'' = I_T = 4.41\ \text{A}$$
$$E_2 = R_{2,3} \times I_1''$$
$$= 8.57\ \Omega \times 4.41\ \text{A}$$
$$= 37.8\ \text{V}$$

$$I_2'' = \frac{E_2}{R_2}$$
$$= \frac{37.8\ \text{V}}{15\ \Omega}$$
$$= -2.52\ \text{A (negative)}$$

(Negative sign indicates current flow is in a direction opposite to that indicated in the original circuit.)

$$E_3 = E_2 = 37.8 \text{ V}$$
$$I_3'' = I_1'' - I_2''$$
$$= 4.41 \text{ A} - 2.52 \text{ A}$$
$$= 1.89 \text{ A}$$

Step 3. Calculate the net current through each resistor. The solved current values and directions are shown in Figure 14-15.

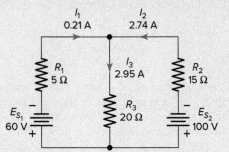

Figure 14-15 Solved currents for the circuit of example 14-3.

$$I_1 = I_1' + I_1''$$
$$= (-4.2 \text{ A}) + (4.41 \text{ A})$$
$$= 0.21 \text{ A}$$
$$I_2 = I_2' + I_2''$$
$$= 5.26 \text{ A} + (-2.52 \text{ A})$$
$$= 2.74 \text{ A}$$
$$I_3 = I_3' + I_3''$$
$$= 1.06 \text{ A} + 1.89 \text{ A}$$
$$= 2.95 \text{ A}$$

Step 4. The voltage across each of the resistors is then found using Ohm's law. The solved voltage values and polarities are shown in Figure 14-16.

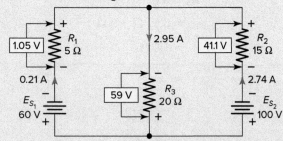

Figure 14-16 Solved voltages for the circuit of example 14-3.

$$E_1 = I_1 \times R_1$$
$$= 0.21 \text{ A} \times 5 \text{ }\Omega$$
$$= 1.05 \text{ V}$$
$$E_2 = I_2 \times R_2$$
$$= 2.74 \text{ A} \times 15 \text{ }\Omega$$
$$= 41.1 \text{ V}$$
$$E_3 = I_3 \times R_3$$
$$= 2.95 \text{ A} \times 20 \text{ }\Omega$$
$$= 59 \text{ V}$$

EXAMPLE 14-4

Problem: This example contains three resistors and two current sources. Calculate the value of current through resistor R_3, for the circuit of Figure 14-17, using the superposition theorem.

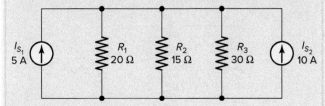

Figure 14-17 Circuit for example 14-4.

Solution:

Step 1. Find the current I_3' due to source I_{S_1} by replacing I_{S_2} with an **open** circuit as shown in Figure 14-18. Note that all three resistors appear connected in parallel with the current source I_{S_1}.

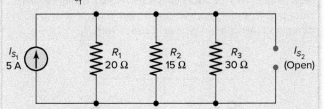

Figure 14-18 Circuit for step 1.

$$R_T = \cfrac{1}{\cfrac{1}{R_1} + \cfrac{1}{R_2} + \cfrac{1}{R_3}}$$
$$= \cfrac{1}{\cfrac{1}{20} + \cfrac{1}{15} + \cfrac{1}{30}}$$
$$= 6.67 \text{ }\Omega$$

$$E' = I_{S_1} \times R_T$$
$$= 5 \text{ A} \times 6.67 \text{ }\Omega$$
$$= 33.4 \text{ V}$$

$$I_3' = \frac{E'}{R_3}$$
$$= \frac{33.4 \text{ V}}{30 \text{ }\Omega}$$
$$= 1.11 \text{ A}$$

Step 2. Find the current I_3'' due to source I_{S_2} by replacing I_{S_1} with an **open** circuit as shown in Figure 14-19. Note that all three resistors still appear connected in parallel.

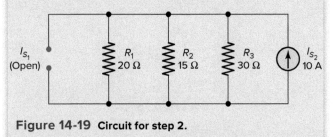

Figure 14-19 Circuit for step 2.

$$R_T = 6.67 \ \Omega$$
$$E'' = I_{S_2} \times R_T$$
$$= 10 \ \text{A} \times 6.67 \ \Omega$$
$$= 66.7 \ \text{V}$$
$$I''_3 = \frac{E''}{R_3}$$
$$= \frac{66.7 \ \text{V}}{30 \ \Omega}$$
$$= 2.22 \ \text{A}$$

Step 3. The net current is then found by superimposing the two currents for I_3. Because the current I'_3 acts in the same direction as I''_3 the net current is the sum of these two currents.

$$I_3 = I'_3 + I''_3$$
$$= 1.11 \ \text{A} + 2.22 \ \text{A}$$
$$= 3.33 \ \text{A}$$

EXAMPLE 14-5

Problem: This example contains one voltage source and one current source. Calculate the value of current through resistor R_2 for the circuit of Figure 14-20 using the superposition theorem.

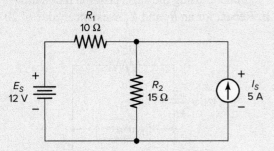

Figure 14-20 Circuit for example 14-5.

Solution:

Step 1. Find the value of the current I' through R_2 due to the voltage source E_S by replacing the current source I_S with an open circuit as shown in Figure 14-21. Note that resistors R_1 and R_2 appear connected in series with the voltage source E_S.

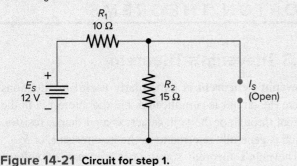

Figure 14-21 Circuit for step 1.

$$I' = \frac{E_S}{R_1 + R_2}$$
$$= \frac{12 \ \text{V}}{25 \ \Omega}$$
$$= 0.48 \ \text{A}$$

Step 2. Find the value of the current I'' through R_2 due to the current source I_S by replacing the voltage source E_S with a short circuit (Figure 14-22) and using the current divider formula. Note that resistors R_1 and R_2 now appear connected in parallel with the current source I_S. Current dividers are the inverse of voltage dividers. This is due to the fact that each branch current is **inversely** proportional to its resistance. The formulas for a current divider with two branch resistances are as follows:

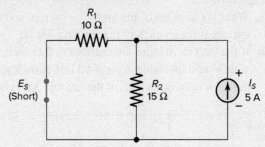
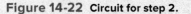

Figure 14-22 Circuit for step 2.

$$I_1 = \frac{R_2}{R_1 + R_2} \times I_T$$
$$I_2 = \frac{R_1}{R_1 + R_2} \times I_T$$
$$I' = \frac{R_1}{R_1 + R_2} \times I_S$$
$$= \frac{10 \ \Omega}{10 \ \Omega + 15 \ \Omega} \times 5 \ \text{A}$$
$$= 2 \ \text{A}$$

Step 3. Because the current flow I' flows in the same direction through R_2 as I'', the current flow through R_2 is the sum of these two currents.

$$I_2 = I' + I''$$
$$= 0.48 \ \text{A} + 2 \ \text{A}$$
$$= 2.48 \ \text{A}$$

The superposition theorem works only for circuits that are reducible to series or parallel combinations for each of the power sources one at a time. Also, this theorem cannot be applied in circuits where the resistance of a component changes with voltage or current. Hence, networks containing components like incandescent lamps could not be analyzed. Another prerequisite for using the superposition theorem is that all components must be **bilateral,** meaning that they behave the same with electrons flowing either direction through them. The resistor circuits we've been studying so far all meet these criteria.

1. In what way does the operation of an ideal voltage source differ from that of a real voltage source.

2. A constant-current source has a rating of 10 A. Explain what this implies.

3. Assume a voltage source has a no-load voltage of 100 V and an internal resistance of 0.5 Ω. Calculate the output voltage when a load of 20 Ω is connected across its output terminals.

4. For the constant-current source circuit shown in Figure 14-23:
   a. What is the value of the total line current and the current through each of the resistive loads?
   b. If the load resistances are changed to 2 Ω each, what would the value of the total line current and the current through each of the resistive loads be?

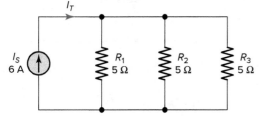

Figure 14-23 Circuit for review question 4.

5. State the superposition theorem.

6. Calculate the current through and the voltage across R_1 and R_2 for the circuit of Figure 14-24 using the superposition theorem.

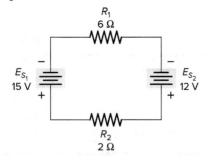

Figure 14-24 Circuit for review question 6.

7. Calculate the value of current I_1, I_2, and I_3 for the circuit of Figure 14-25 using the superposition theorem.

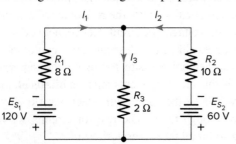

Figure 14-25 Circuit for review question 7.

8. Calculate the value of current through resistor R_2 for the circuit of Figure 14-26 using the superposition theorem.

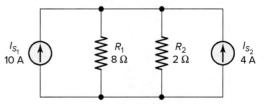

Figure 14-26 Circuit for review question 8.

9. Calculate the value of current through resistor R_1, for the circuit of Figure 14-27, using the superposition theorem.

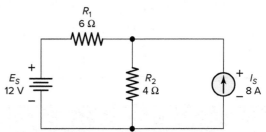

Figure 14-27 Circuit for review question 9.

10. For the circuit shown in Figure 14-28:
    a. Calculate the value of current flow through the ammeter using the superposition theorem.
    b. Repeat for an R_1 and R_2 resistance value of 20 Ω.

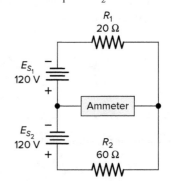

Figure 14-28 Circuit for review question 10.

PART 2   THEVENIN AND NORTON THEOREMS

14.3 Thevenin's Theorem

Thevenin's theorem is particularly useful in situations where the circuit is complicated, but the interest is in the current through or the voltage across a particular resistor, which is generally referred to as the load resistor, or R_L.

Thevenin's theorem states: *Any network of voltage sources and resistors can be replaced by a single equivalent voltage source (E_{TH}) in series with a single equivalent*

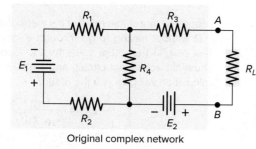

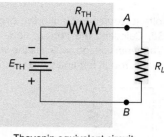

Original complex network

Thevenin equivalent circuit

Figure 14-29 Thevenin's theorem.

resistance ($\mathbf{R}_{TH}$) and the load resistor ($\mathbf{R}_L$). In other words, it is possible to simplify any linear circuit, no matter how complex, to an equivalent circuit with just a single voltage source in series with a resistance connected to a load, as shown Figure 14-29. Any load resistor connected between the terminals of a Thevenin equivalent circuit will have the same current through it and the same voltage across it as if it were connected to the terminals of the original circuit.

The following steps are used in converting a circuit to its Thevenin equivalent. These steps can be done either by calculating the values or by measuring the values if the circuit actually exists.

1. Select the resistor (R_L) you wish to know the current through, and remove it from the circuit.

2. Measure or calculate the voltage across the points in the circuit where the resistor was connected. This is the Thevenin voltage (E_{TH}).

3. Remove each voltage source and replace it with a short circuit. Be careful not to short across an active source in a working circuit.

4. Measure or calculate the resistance across the points in the circuit where the R_L resistor was connected. This is the Thevenin resistance (R_{TH}).

5. Create a series circuit with the Thevenin voltage (E_{TH}) as the source and the Thevenin resistance (R_{TH}) placed in series with the resistor (R_L) that was removed. The polarity of E_{TH} must be such as to produce the same voltage polarity across R_{TH} as in the original circuit. This is called the Thevenin equivalent circuit.

6. Finally, calculate the current through and the voltage across the load resistor R_L in the Thevenin circuit using Ohm's law.

Thevenin's theorem is especially useful in analyzing power systems and other circuits where one particular resistor in the circuit (called the **load resistor**) is subject to change, and recalculation of the circuit is necessary with each trial value of load resistance to determine voltage across it and current through it. The following are examples of how Thevenin's theorem is applied to different circuit configurations.

EXAMPLE 14-6

Problem: A voltage divider circuit is used to supply power to a load as shown in Figure 14-30. Determine the value of current through and the voltage across load resistor R_L for the circuit values shown using Thevenin's theorem. One advantage of using a Thevenin equivalent circuit is that the effect of different values of R_L can be calculated easily. To illustrate this, determine the value of load current and voltage if R_L were changed to 4 Ω.

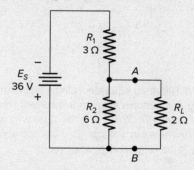

Figure 14-30 Circuit for example 14-6.

Solution:

Step 1. Disconnect resistor R_L, and short-circuit the voltage source, as shown in Figure 14-31. Looking back into terminals *A-B*, notice that R_1 and R_2 appear in parallel. This becomes the Thevenin resistance R_{TH}.

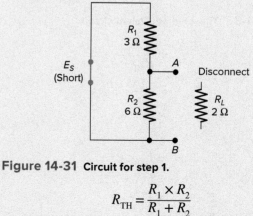

Figure 14-31 Circuit for step 1.

$$R_{TH} = \frac{R_1 \times R_2}{R_1 + R_2}$$

$$= \frac{3\,\Omega \times 6\,\Omega}{3\,\Omega + 6\,\Omega}$$

$$= 2\,\Omega$$

Step 2. Insert the voltage source back into the circuit as shown in Figure 14-32. Calculate the current flow in this circuit. Calculate the Thevenin voltage, which is the voltage across resistor R_2 as measured at terminals *A-B*.

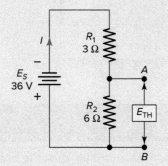

Figure 14-32 Circuit for step 2.

$$I = \frac{E_S}{R_1 + R_2}$$

$$= \frac{36 \text{ V}}{3 \text{ } \Omega + 6 \text{ } \Omega}$$

$$= 4 \text{ A}$$

$$E_{TH} = I \times R_2$$
$$= 4 \text{ A} \times 6 \text{ } \Omega$$
$$= 24 \text{ V}$$

Step 3. The Thevenin equivalent circuit is shown in Figure 14-33. Using this circuit, calculate the current through and the voltage across R_L. Note that the Thevenin resistance (R_{TH}) and the Thevenin voltage (E_{TH}) do not apply to any component in the original complex circuit.

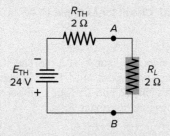

Figure 14-33 Thevenin equivalent circuit.

$$I_{R_L} = \frac{E_{TH}}{R_{TH} + R_L}$$

$$= \frac{24 \text{ V}}{4 \text{ } \Omega}$$

$$= 6 \text{ A}$$

$$E_L = I_{R_L} \times R_L$$
$$= 6 \text{ A} \times 2 \text{ } \Omega$$
$$= 12 \text{ V}$$

Step 4. Using the Thevenin equivalent circuit, with R_L changed to 4 Ω, calculate the value of the load current and voltage. Note that you can quickly determine what would

happen to that single resistor if it were of a value other than 2 Ω without having to go through a lot of analysis again. Just plug in that other value for the load resistor into the Thevenin equivalent circuit, and a little bit of series circuit calculation will give you the result.

$$I_{R_L} = \frac{E_{TH}}{R_{TH} + R_L}$$

$$= \frac{24 \text{ V}}{6 \text{ } \Omega}$$

$$= 4 \text{ A}$$

$$E_{R_L} = I_{R_L} \times R_L$$
$$= 4 \text{ A} \times 4 \text{ } \Omega$$
$$= 16 \text{ V}$$

EXAMPLE 14-7

Problem: Use Thevenin's theorem to determine the current through and the voltage across R_L of the dual-voltage circuit shown in Figure 14-34.

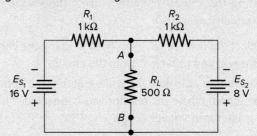
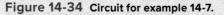

Figure 14-34 Circuit for example 14-7.

Solution:

Step 1. Remove R_L and short-circuit voltage source E_{S_2} as shown in Figure 14-35. Calculate the voltage drop at terminals *A-B* due to source E_{S_1} (this will be the same as the voltage drop across R_2). Note the polarity of the voltage at terminal *A* with respect to *B*.

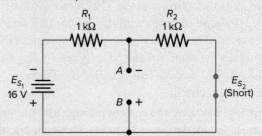

Figure 14-35 Circuit for step 1. Voltage at *A* is negative with respect to *B*.

$$E'_{A\text{-}B} = \frac{R_2}{R_1 + R_2} \times E_{S_1}$$

$$= \frac{1 \text{ k}\Omega}{2 \text{ k}\Omega} \times 16 \text{ V}$$

$$= 8 \text{ V}$$

Step 2. Short-circuit voltage source E_{S_1} as shown in Figure 14-36. Calculate the voltage drop at terminals *A-B* due to source E_{S_2} (this will be the same as the voltage drop across R_1). Note the polarity of the voltage at terminal *A* with respect to *B*.

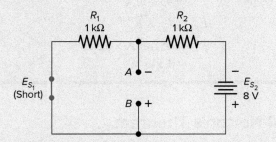

Figure 14-36 Circuit for step 2. Voltage at *A* is negative with respect to *B*.

$$E''_{A\text{-}B} = \frac{R_1}{R_1 + R_2} \times E_{S_2}$$
$$= \frac{1\text{ k}\Omega}{2\text{ k}\Omega} \times 8\text{ V}$$
$$= 4\text{ V}$$

Step 3. Calculate the net voltage at terminals *A-B* by adding the two voltages since they both have the same polarity. This is the Thevenin voltage.

$$E_{TH} = E'_{A\text{-}B} + E''_{A\text{-}B}$$
$$= 8\text{ V} + 4\text{ V}$$
$$= 12\text{ V}$$

Step 4. Calculate the Thevenin resistance by shorting both voltage sources, as shown in Figure 14-37. Looking back at the circuit from terminals *A-B*, the Thevenin equivalent resistance consists of R_1 and R_2 connected in parallel.

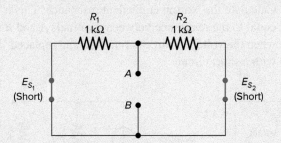

Figure 14-37 Circuit for step 4.

$$R_{TH} = \frac{R_1 \times R_2}{R_1 + R_2}$$
$$= \frac{1\text{ k}\Omega \times 1\text{ k}\Omega}{1\text{ k}\Omega + 1\text{ k}\Omega}$$
$$= 0.5\text{ k}\Omega$$
$$= 500\ \Omega$$

Step 5. The Thevenin equivalent circuit is shown in Figure 14-38. Using this circuit, calculate the current through and the voltage across R_L.

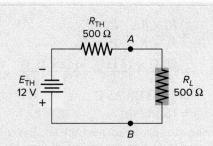
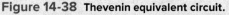

Figure 14-38 Thevenin equivalent circuit.

$$I_{R_L} = \frac{E_{TH}}{R_{TH} + R_L}$$
$$= \frac{12\text{ V}}{1{,}000\ \Omega}$$
$$= 0.012\text{ A}$$
$$= 12\text{ mA}$$
$$E_{R_L} = I_{R_L} \times R_L$$
$$= 0.012\text{ A} \times 500\ \Omega$$
$$= 6\text{ V}$$

EXAMPLE 14-8

Problem: A resistance bridge circuit can be difficult to analyze because it is not a straightforward series-parallel arrangement. Use Thevenin's theorem to determine the current through and the voltage across R_L of the resistance bridge circuit shown in Figure 14-39.

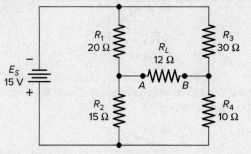

Figure 14-39 Circuit for example 14-8.

Solution:

Step 1. Remove resistor R_L, and short-circuit the voltage source, as shown in Figure 14-40, to calculate the Thevenin resistance. Looking back at the circuit from terminals *A-B*, the equivalent R_{TH} consists of R_1 and R_2 in parallel with each other and in series with the parallel combination of R_3 and R_4.

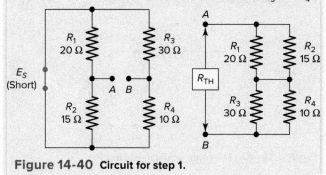

Figure 14-40 Circuit for step 1.

$$R_{TH} = \frac{R_1 \times R_2}{R_1 + R_2} + \frac{R_3 \times R_4}{R_3 + R_4}$$
$$= \frac{20\ \Omega \times 15\ \Omega}{20\ \Omega + 15\ \Omega} + \frac{30\ \Omega \times 10\ \Omega}{30\ \Omega + 10\ \Omega}$$
$$= 8.57\ \Omega + 7.5\ \Omega$$
$$= 16.1\ \Omega$$

Step 2. Insert the voltage source back into the circuit, as shown in Figure 14-41. Use the voltage divider rule to find the value and polarity of the voltage across R_1 and R_3.

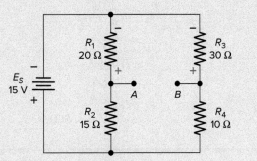

Figure 14-41 Circuit for step 2.

$$E_1 = \frac{R_1}{R_1 + R_2} \times E_S$$
$$= \frac{20\ \Omega}{20\ \Omega + 15\ \Omega} \times 15\ V$$
$$= 8.57\ V$$

$$E_3 = \frac{R_3}{R_3 + R_4} \times E_S$$
$$= \frac{30\ \Omega}{30\ \Omega + 10\ \Omega} \times 15\ V$$
$$= 11.25\ V$$

Step 3. The potential difference between terminals A and B is the difference in voltage between E_1 and E_3. This is also the Thevenin voltage (E_{TH}).

$$E_{TH} = E_{A-B}$$
$$= E_1 - E_3$$
$$= 8.57\ V - 11.25\ V$$
$$= -2.68\ V$$

(Negative sign indicates that point A is negative with respect to point B.)

Step 4. The Thevenin equivalent circuit is shown in Figure 14-42. Using this circuit, calculate the current through and the voltage across R_L.

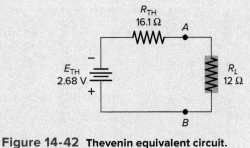

Figure 14-42 Thevenin equivalent circuit.

$$I_{R_L} = \frac{E_{TH}}{R_{TH} + R_L}$$
$$= \frac{2.68\ V}{16.1\ \Omega + 12\ \Omega}$$
$$= 95.4\ mA$$

$$E_{R_L} = I_{R_L} \times R_L$$
$$= 95.4\ mA \times 12\ \Omega$$
$$= 1.145\ V$$

14.4 Norton's Theorem

Norton's theorem is used for simplifying a network in terms of currents instead of voltages. This theorem can be used with either single-source or multiple-source circuits. In certain cases, analyzing the division of currents may be easier than voltage analysis. Norton's theorem simplifies a resistive network and represents it with a **Norton equivalent current source (I_N)** in parallel with an equivalent **Norton resistance (R_N)**, as shown in Figure 14-43. The basis of Norton's theorem is the use of a current source to supply a total load current that is divided among parallel branches.

As with similar theorems, certain steps have to be performed to arrive at an equivalent circuit. The following steps are used to convert a resistive network into its Norton equivalent:

1. Calculate the Norton equivalent current source. This is equal to the current that would flow between terminals A and B if the load resistor was removed and replaced with a short circuit.

2. Calculate the Norton equivalent resistance. This is equal to the resistance between terminals A and B when the voltage source is removed and replaced with a short circuit.

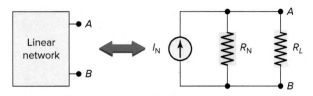

Figure 14-43 Norton equivalent circuit.

EXAMPLE 14-9

Problem: Derive the Norton equivalent circuit for the simple resistive network circuit shown in Figure 14-44. Calculate the load current and voltage for the 6-Ω load resistor R_L. Repeat the calculations for a load resistor of 3 Ω.

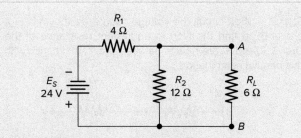

Figure 14-44 Circuit for example 14-9.

Solution:

Step 1. Short-circuit the load terminals *A-B* as shown in Figure 14-45. Calculate the resulting current flow. This is the value of the Norton equivalent current source I_N. Note that the short circuit across terminals *A-B* short-circuits both R_L and the parallel R_2. Then the only resistance in the circuit is R_1 in series with the voltage source.

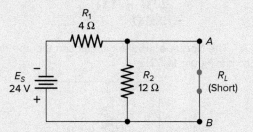

Figure 14-45 Circuit for step 1.

$$I_N = \frac{E_S}{R_1}$$
$$= \frac{24\text{ V}}{4\,\Omega}$$
$$= 6\text{ A}$$

Step 2. The Norton equivalent resistance (R_N) is equal to the resistance between terminals *A* and *B* with the load removed and the voltage source replaced with a short circuit, as shown in Figure 14-46. The resistance seen looking back into the circuit from terminals *A-B* is then R_1 in parallel with R_2.

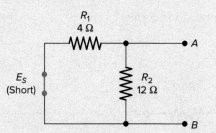

Figure 14-46 Circuit for step 2.

$$R_N = \frac{R_1 \times R_2}{R_1 + R_2}$$
$$= \frac{4\,\Omega \times 12\,\Omega}{4\,\Omega + 12\,\Omega}$$
$$= 3\,\Omega$$

Step 3. The resultant Norton equivalent circuit is shown in Figure 14-47. It consists of a 6-A current source (I_N) in parallel with a 3-Ω resistance (R_N).

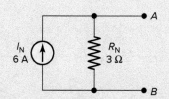

Figure 14-47 Norton equivalent circuit.

Step 4. To calculate I_{R_L} and E_{R_L}, connect the load resistor to the equivalent Norton circuit, as shown in Figure 14-48. The current source still delivers 6 A, but now that current is divided between the two branches of R_N and R_L. The load current can be calculated using the current divider rule, and the load voltage can be calculated using Ohm's law.

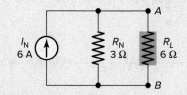

Figure 14-48 Circuit for step 4.

$$I_{R_L} = \frac{R_N}{R_N + R_L} \times I_N$$
$$= \frac{3\,\Omega}{3\,\Omega + 6\,\Omega} \times 6\text{ A}$$
$$= 2\text{ A}$$

$$E_{R_L} = I_{R_L} \times R_L$$
$$= 2\text{ A} \times 6\,\Omega$$
$$= 12\text{ V}$$

Step 5. When the value of R_L is changed, the values of I_N and R_N remain the same. Therefore the load current and voltage for a 3-Ω load can be determined as follows without recalculating the entire circuit.

$$I_{R_L} = \frac{R_N}{R_N + R_L} \times I_N$$
$$= \frac{3\,\Omega}{3\,\Omega + 3\,\Omega} \times 6\text{ A}$$
$$= 3\text{ A}$$

$$E_{R_L} = I_{R_L} \times R_L$$
$$= 3\text{ A} \times 3\,\Omega$$
$$= 9\text{ V}$$

EXAMPLE 14-10

Problem: Derive the Norton equivalent circuit for resistive network shown in Figure 14-49.

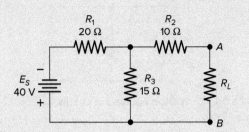

Figure 14-49 Circuit for example 14-10.

Solution:

Step 1. Short-circuit the load, as shown in Figure 14-50, to determine the Norton equivalent current. Note that the short circuit current I_N in this example is a branch current, not the main line current. Resistors R_2 and R_3 are now in parallel so the total resistance seen by the voltage source is

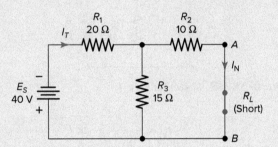

Figure 14-50 Circuit for step 1.

$$R_T = \frac{R_2 \times R_3}{R_2 + R_3} + R_1$$

$$= \frac{10\ \Omega \times 15\ \Omega}{10\ \Omega + 15\ \Omega} + 20\ \Omega$$

$$= 26\ \Omega$$

The total current is then

$$I_T = \frac{E_S}{R_T}$$

$$= \frac{40\ \text{V}}{26\ \Omega}$$

$$= 1.54\ \text{A}$$

The Norton equivalent current is then found by using the current divider rule:

$$I_N = \frac{R_3}{R_2 + R_3} \times I_T$$

$$= \frac{15\ \Omega}{10\ \Omega + 15\ \Omega} \times 1.54\ \text{A}$$

$$= 0.924\ \text{A}$$

$$= 924\ \text{mA}$$

Step 2. Short-circuit the voltage source, as shown in Figure 14-51 to find the Norton equivalent resistance of the circuit. R_1 and R_3 are now in parallel, and R_2 is in series with this parallel combination.

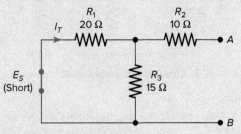

Figure 14-51 Circuit for step 2.

$$R_N = \frac{R_1 \times R_3}{R_1 + R_3} + R_2$$

$$= \frac{20\ \Omega \times 15\ \Omega}{20\ \Omega + 15\ \Omega} + 10\ \Omega$$

$$= 18.6\ \Omega$$

Step 3. The Norton equivalent circuit can then be drawn as shown in Figure 14-52.

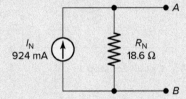

Figure 14-52 Norton equivalent circuit.

The Norton equivalent circuit may also be determined directly from the Thevenin equivalent circuit, and vice versa. Figure 14-53 shows the relationship between the two circuits. The following formulas can be used to convert from one equivalent circuit to the other:

From Thevenin to Norton

$$R_N = R_{TH}$$

$$I_N = \frac{E_{TH}}{R_{TH}}$$

From Norton to Thevenin

$$R_{TH} = R_N$$

$$E_{TH} = I_N \times R_N$$

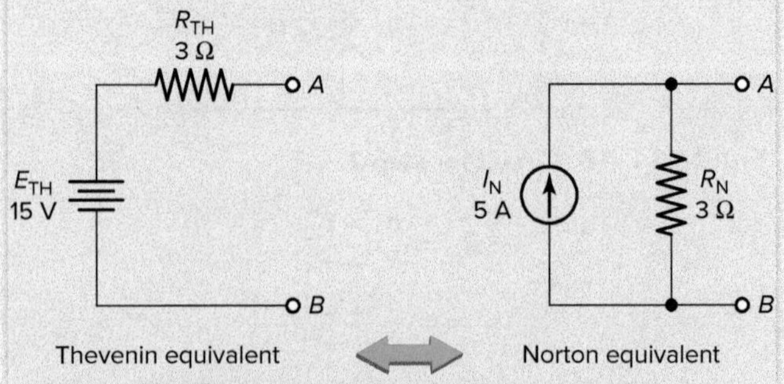

Thevenin equivalent   Norton equivalent

Figure 14-53 Norton–Thevenin conversions.

1. State Thevenin's theorem.

2. For the circuit of Figure 14-54:
   a. Determine the value of current through and the voltage across load resistor R_L using Thevenin's theorem.
   b. Repeat for a load resistor of 9 ohms.

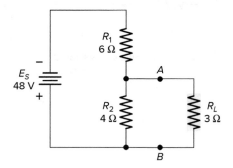

Figure 14-54 Circuit for review question 2.

3. Use Thevenin's theorem to determine the current through and the voltage across R_L of the dual-voltage circuit shown in Figure 14-55.

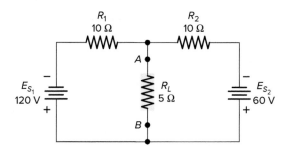

Figure 14-55 Circuit for review question 3.

4. Use Thevenin's theorem to determine the current through and the voltage across R_L of the resistance bridge circuit shown in Figure 14-56.

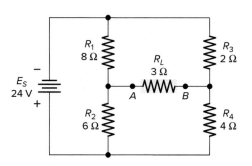

Figure 14-56 Circuit for review question 4.

5. For the circuit of Figure 14-57:
   a. Determine the value of current through and the voltage across load resistor R_L using Norton's theorem.
   b. Repeat for a load resistor of 8 ohms.

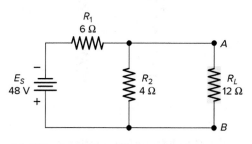

Figure 14-57 Circuit for review question 5.

6. Derive the Norton equivalent circuit for the resistive network shown in Figure 14-58.

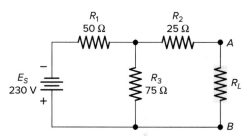

Figure 14-58 Circuit for review question 6.

7. Derive the Norton equivalent circuit for the Thevenin equivalent circuit shown in Figure 14-59.

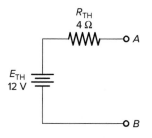

Figure 14-59 Circuit for review question 7.

8. Applying the superposition theorem to the circuit of Figure 14-60:
   a. What determines the value of the current flow through conductor C?
   b. Under what normal operating circuit condition would it be possible to have

current flowing through conductors A and B but none through C?

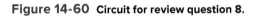

Figure 14-60 Circuit for review question 8.

9. Determine the Thevenin equivalent circuit for the resistive circuit of Figure 14-61.

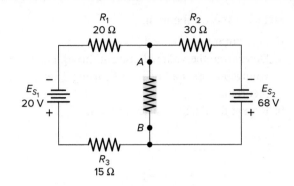

Figure 14-61 Circuit for review question 9.

Phase angle regulating transformers.
©emel82/123RF

Alternating Current (AC) Circuits

SECTION OUTLINE

CHAPTER FIFTEEN

Alternating Current Fundamentals

Standby generator.
©Stepan Popov/123RF

A generator is a machine that uses magnetism to convert mechanical energy into electric energy. Practical generators of electricity fall into two general groups: direct current (DC) and alternating current (AC). Direct current is produced by those generators that do not change the polarity of their terminals. Because the polarity of their terminals is constantly reversing, AC generators produce an alternating current flow. Most electric energy is distributed in the form of alternating current and voltage. In this chapter, we examine the ways in which AC voltages are generated and measured.

15.1 Differences between DC and AC

The supply of current for electrical devices may come from a direct current source or an alternating current source. The main difference between DC and AC is in how the current flows in each type of circuit. In a **direct current (DC)** circuit, electrons flow continuously in one direction from the source of power through a conductor to a load and back to the source of power. An example of direct current flow is illustrated in Figure 15-1. Voltage is supplied by the battery, the polarity of which is fixed. Based on electron theory, electrons flow from the negative to the positive terminals of the battery. If you reverse the two leads to the battery terminals, the current will flow in the opposite direction. Direct current can also be supplied by other sources, such as solar cells, fuel cells, rectifiers, and generators.

Direct current flow need not necessarily be constant, but it must travel in the same direction at all times. A DC waveform never crosses the zero reference line. The different types of direct current waveforms shown in Figure 15-2 are:

- **Pure** direct current has no variations in magnitude over a period of time. This type is associated with a battery.

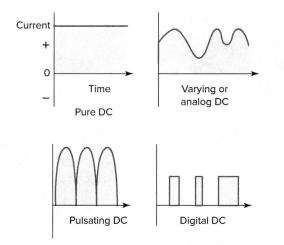

Figure 15-1 Direct current flow.

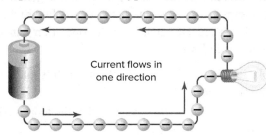

Figure 15-2 Types of direct current.

- **Varying or analog** direct currents have a value which changes when plotted against time. They have no definite pattern and are associated with certain types of electrical sensor signals.
- **Pulsating** direct current variations are uniform and repeat at regular time intervals. These are associated with electronic rectifier circuits and are used to convert AC to DC.
- **Digital** direct current is an electronic signal transmitted in discrete steps. These are associated with computers and used to represent information.

Direct current was the first source of electric power to be widely used. However, as more of the advantages of using alternating current became known, it gradually replaced DC as the primary source of electric power. There are, however, many present-day applications for DC. These include DC motors because of their variable speed and torque characteristics, as well as certain electric and electronic equipment, where only DC can perform the desired function.

An **alternating current (AC)** circuit is one in which the direction and amplitude of the current flow changes at regular intervals. The polarity of an AC voltage source changes at regular intervals, resulting in a reversal of the circuit current flow. The most popular graphic representation for AC is the sine wave, shown in Figure 15-3. A sine wave can represent AC current or voltage:

- The vertical axis represents the direction and magnitude of current or voltage.
- The horizontal axis represents time.
- When the waveform is above the time axis, current is flowing in one direction. This is sometimes referred to as the positive direction.
- When the waveform is below the time axis, current is flowing in the opposite direction. This is sometimes referred to as the negative direction.

The main advantage of AC over DC is that AC can be **transformed** and DC cannot. Overall, the transformation ability of AC makes it less expensive and more functional than DC. As an example, for purposes of transmission of electricity, a transformer can be used to step up the voltage

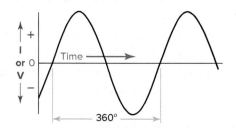

Figure 15-3 Sine wave graphic representation for AC.

for transmission and step down the voltage for use at the utilization point. The advantage of high-voltage transmission is that it takes less current for the same amount of power. This reduction in current allows smaller wire to be used for transmission purposes lowering the overall cost of the system.

When AC flows in a conductor, the alternating current flowing through the conductor causes certain voltages to be set up or to develop within the conductor. These voltages create small currents that, although they are independent from the current flowing through the conductor from one end to the other, have an effect on the power loss associated with the conductor. These currents are called **eddy currents.** The power losses associated with eddy currents increase the overall resistance of the conductor and the circuit.

In a DC circuit, the electrons travel through the whole conductor. However, in an AC circuit conductor, besides setting up eddy currents, the voltage that creates the eddy current also causes the current flow in the conductor to be repelled away from the center of the conductor toward the outside of the conductor. The current is forced to travel near the surface of the conductor. This effect, known as **skin effect,** creates the same consequence as reducing the cross-sectional area of the conductor because the electrons are forced to flow in a smaller area concentrated near the surface of the conductor. The skin effect also causes an increase in the conductor resistance. Both eddy currents and skin effect are directly related to the frequency of the circuit. Generally, the effects of eddy current and skin effect do not have a critical negative impact on circuits except at higher frequencies.

15.2 Alternating Current Generation

A generator is a machine which converts mechanical energy into electric energy. The most common method of obtaining AC is by use of an **AC generator** (also called an **alternator**). An AC generator generates a voltage by rotation of a loop of wire within a magnetic field, as illustrated in Figure 15-4. Its operation can be summarized as follows:

- The prime mover is the source of mechanical power for relative motion.
- Relative motion between the wire and the magnetic field causes a voltage to be induced between the ends of the wire.
- The generated voltage changes in magnitude as the loop cuts the magnetic lines of force at different angles.
- The generated voltage changes in polarity as the loop cuts the magnetic lines of force in two different directions.
- Slip rings are attached to the armature wire and rotate with it.

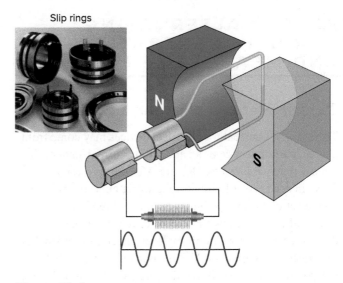

Figure 15-4 **Simplified AC generator.**
Photo: ©The Electric Materials Company

- Carbon brushes ride against the slip rings to conduct current from the armature to the resistor load.
- A complete rotation of 360 degrees results in the generation of one cycle of alternating current.
- This sequence is repeated as long as the generator is in operation.

Generating DC is basically the same as generating AC. The only difference is the manner in which the generated voltage is supplied to the output terminals. Figure 15-5 shows a simplified DC generator. Its operation can be summarized as follows:

- Voltage is induced into the coil when it cuts magnetic flux lines.
- The shape of the voltage generated in the loop is still that of an AC sine wave.

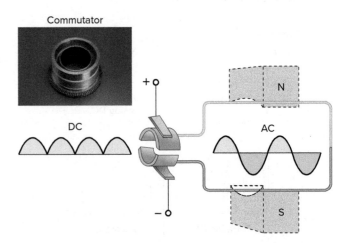

Figure 15-5 **Simplified DC generator.**
Photo: ©The Electric Materials Company

- The two slip rings of the AC generator are replaced by a single segmented ring called a **commutator.**

- The commutator acts like a mechanical switch which converts the generated AC voltage into DC voltage.

- As the armature starts to develop a negative alternation, the commutator switches the polarity of the output terminals via the brushes. This keeps all positive alternations on one terminal and all negative alternations on the other.

- The only essential difference between an AC and a DC generator is the use of slip rings on the one and a commutator on the other.

- Brushes move from one segment of the commutator to the next during the period of zero induced voltage when the coil is moving parallel to the magnetic flux or in what is called the neutral plane.

The amount of voltage induced in a conductor as it moves through a magnetic field depends on:

- The strength of the magnetic field: The stronger the field, the more voltage induced.

- The speed at which the conductor cuts through the flux: Increasing the conductor speed increases the amount of voltage induced.

- The angle at which the conductor cuts the flux: Maximum voltage is induced when the conductor cuts the flux at 90 degrees, and less voltage is induced when the angle is less than 90 degrees.

- The length of the conductor in the magnetic field: If the conductor is wound into a coil of several turns, its effective length increases and so the induced voltage will increase.

There are two basic types of alternators: the revolving-armature type and the revolving-field type. With the **revolving-armature** alternator, the armature is the movable part of the generator and is made up of a number of coils wound on an iron core. The AC voltage induced in the armature coils is connected to a set of slip rings from which the external circuit receives the voltage through a set of brushes. Electromagnets are used to produce a strong magnetic field for the generator and are referred to as field coils or windings. The revolving-armature alternator has very limited practical applications. This is because the brushes and slip rings used to remove power from the generator can handle only a relatively low level of voltage and kilovolt-ampere (kVA) capacity. The **revolving-field** alternator configuration, illustrated in Figure 15-6, permits higher kVA output capacity because the load is connected directly to the stator and not routed through brushes and slip rings.

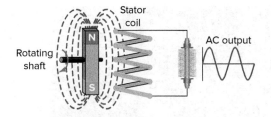

Figure 15-6 **Revolving-field alternator.**

The rotor or the rotating part of a revolving-field alternator is an electromagnet that provides the magnetic field needed to induce a voltage into the stator windings. This rotor may also have a squirrel-cage winding which dampens rotor speed from sudden changes in the load. Rotor windings require DC excitation to establish a magnetic field. The two types of rotor designs used are salient pole and cylindrical. Figure 15-7 shows a **salient pole** alternator rotor, the characteristics of which can be summarized as follows:

- The salient pole rotor is constructed such that the poles stick out from the surface of the rotor.

- Used for alternators with speeds less than 1,800 revolutions per minute (rpm).

- Typical prime movers include diesel engines and water turbines.

- Centrifugal force prevents the use of salient poles on high-speed machines.

- Salient pole rotors often consist of several separately wound pole pieces bolted to the frame of the rotor.

- Rotors are normally constructed of large diameter to allow for mounting of the pole pieces.

Figure 15-7 **Salient pole alternator rotor.**
©Hyundai Ideal Electric Co.

Figure 15-8 Cylindrical alternator rotor.
©IRD LLC

Figure 15-8 shows a **cylindrical** alternator rotor, the characteristics of which can be summarized as follows:

- The cylindrical poles of the rotor are constructed flush with the surface of the rotor.
- Used for high-speed machines.
- Typical prime movers include steam and natural gas turbines.
- The coils are located in the slots of the laminated rotor core and firmly embedded to withstand the tremendous centrifugal forces encountered at high speeds.
- Rotors are constructed of small diameter due to centrifugal force limitations and the fact that at high speed few poles are required.
- The long rotor increases the conductor length for increased voltage.

AC generators can be further classified as induction or synchronous generators. The **synchronous** generator is a rotating-field type consisting of a magnetic field on the rotor that rotates and a stationary stator containing multiple windings that supplies the generated power. The rotors magnetic field system (excitation) is created by using either permanent magnets mounted directly onto the rotor or energized electromagnetically by an external DC current flowing in the rotor field windings. **Induction** generators differ from synchronous in that they receive field excitation from the power system to which the generator is connected.

Single-phase generators are limited to applications that have small power demands. **Three-phase** alternating current generators of the rotating-field type produce most of the electric power in the world. The three-phase AC generator is effectively three single-phase AC generators combined into one machine, as illustrated in Figure 15-9. There are three equally spaced windings and three output voltages that are all 120 degrees out of phase with one another. The voltage and current rating of this type of alternator can be relatively high because the output of the alternator is taken directly from the three-phase stator windings through heavy, well-insulated cables to the external circuit without the use of slip rings.

The **prime mover** refers to the source of mechanical energy required to operate a generator. Two classes of generator prime movers are **high speed,** which includes steam and gas turbines and **low speed,** which include internal combustion engines and water turbines. Figure 15-10 shows a diesel engine–driven standby generator. Small AC generator windings are normally air-cooled by an attached fan that circulates air through the entire assembly. Most utility-scale generators use hydrogen gas to cool the generator windings because of its superior cooling characteristics.

Different methods of field excitation for the rotating-field alternator have been developed and used. Most large alternators use a **brushless exciter** system, in which a separate small armature is added to the shaft of the rotor. This rotates within the wound electromagnet stator field. The current generated is then rectified and used as the excitation current for the main alternator. The field excitation of the main alternator is controlled by the field current of the exciter alternator.

The output **frequency (number of voltage cycles per second)** of an alternator is determined by the number of stator poles and the speed of the rotation of the rotor:

$$\text{Frequency} = \frac{\text{poles} \times \text{speed (rpm)}}{120}$$

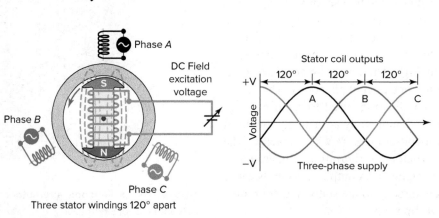

Figure 15-9 Three-phase alternating current generator.

Figure 15-10 **Diesel engine–driven standby generator.**
Courtesy of Baldor Electric Co.

EXAMPLE 15-1

Problem: What is the output frequency of an alternator with a rotor that has six poles and turns at 1,000 rpm?

Solution:

$$f = \frac{p \times S}{120}$$

$$= \frac{6 \times 1,000}{120}$$

$$= 50 \text{ Hz}$$

The standard alternator output frequency is typically **60 Hz** in North America and 50 Hz in Europe. The output frequency of an alternator must remain constant at a value that matches that of the electric grid. Generator frequency can vary due to load fluctuations. Frequency regulation is usually accomplished by adjusting the speed of the prime mover.

Most alternators are designed to operate at a constant speed so the amount of **output voltage** is controlled by increasing or decreasing the strength of the magnetic field. Under normal conditions, the field excitation is varied automatically. It responds to the load changes so as to maintain a constant AC line voltage to the system.

15.3 Alternating Current Measurements

A DC power source, such as a battery, outputs a constant voltage and current over time. By contrast, an AC source of electric power changes constantly in amplitude and regularly changes polarity. The **sine wave** is the most basic and widely used AC waveform. Common electrical terms and measurements associated with AC sine wave values are summarized as follows.

Cycle

A **cycle** is one complete wave of alternating voltage or current. During the generation of one cycle of output voltage there are two changes or **alternations** in the polarity of the voltage. These equal but opposite halves of a complete cycle are referred to as alternations. The terms *positive* and *negative* are used to distinguish one alternation from the other, as illustrated in Figure 15-11.

Period

The time required to complete one cycle is known as the **period** of a waveform (Figure 15-12). The period is usually measured in seconds (s) or smaller units of time such as milliseconds (ms) or microseconds (μs).

Frequency

The **frequency** is the number of complete cycles that occur in one second. Frequency, then, refers to how rapidly the current reverses or how often the voltage changes polarity. The base unit of frequency is the **hertz (Hz).** One hertz equals one cycle per second. Figure 15-13 illustrates 15 cycles being generated in ¼ second, which is equivalent to 60 Hz.

The frequency is the same as the number of rotations per second if the magnetic field is produced by only two

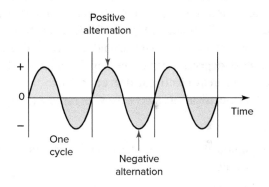

Figure 15-11 **Sine wave cycle.**

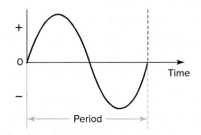

Figure 15-12 **Sine wave period.**

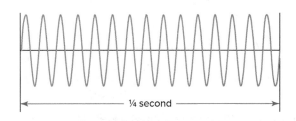
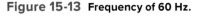

Figure 15-13 **Frequency of 60 Hz.**

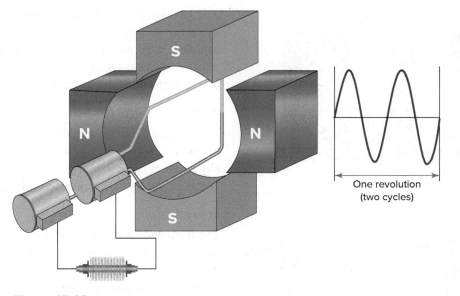

Figure 15-14 Four-pole generator.

poles. An increase in the number of poles causes a corresponding increase in the number of cycles completed in a revolution. There is always an even number of field poles. A two-pole generator completes one cycle per revolution (360°) and a four-pole generator completes two cycles per revolution, as illustrated in Figure 15-14. Hence there is a distinction between mechanical and electrical degrees. **Mechanical degrees** refer to the angular rotation of the armature, whereas **electrical degrees** refer to the movement of an electrical AC sine wave with respect to time.

Amplitude

The amplitude of an AC waveform can be specified in the following ways.

Peak Value The **maximum** value of voltage or current is the value of voltage or current from the zero reference level to the peak (Figure 15-15).

Also called the **peak value,** it occurs twice each cycle, once at the positive maximum value and once at the negative maximum value. All other amplitude values in the waveform are a function of this maximum or peak value.

Peak-to-Peak Value The **peak-to-peak** value of an AC waveform is the range from the positive peak to the negative peak (Figure 15-16). It is equal to double the peak value.

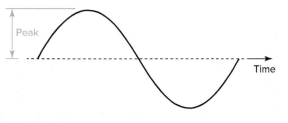

Figure 15-15 Peak value of an AC waveform.

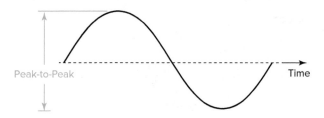

Figure 15-16 Peak-to-peak value of an AC waveform.

Both peak and peak-to-peak voltage values are used commonly when taking AC signal voltage measurements with an oscilloscope.

Instantaneous Value Each alternation of the sine wave is made up of a number of instantaneous values (Figure 15-17). The **instantaneous value** is the value at any one point on the sine wave. The voltage waveform produced as the armature of a basic two-pole AC generator rotates through 360 degrees is called a sine wave, because the instantaneous voltage or current is related to the **sine trigonometric function.** The instantaneous voltage and current at any point on the sine wave are equal to the peak value *times* the sine of the angle.

EXAMPLE 15-2

Problem: A sine wave voltage has a peak value of 100 V. What is the voltage after 60 degrees of rotation?

Solution:

$$E_{\text{instant}} = E_{\text{peak}} \times \text{sine of } 60°$$
$$= 100 \text{ V} \times 0.866$$
$$= 86.6 \text{ V}$$

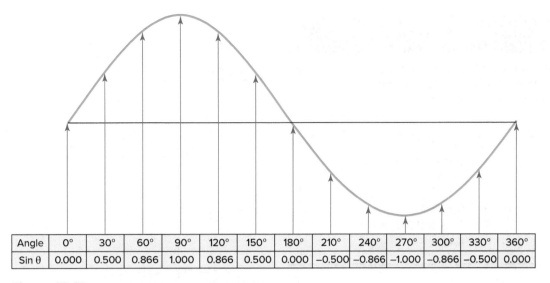

Angle	0°	30°	60°	90°	120°	150°	180°	210°	240°	270°	300°	330°	360°
Sin θ	0.000	0.500	0.866	1.000	0.866	0.500	0.000	−0.500	−0.866	−1.000	−0.866	−0.500	0.000

Figure 15-17 Instantaneous values of an AC waveform.

EXAMPLE 15-3

Problem: What is the instantaneous value of an AC current at 150 degrees if the maximum current is 40 A?

Solution:

$$I_{instant} = I_{peak} \times \text{sine of } 150°$$
$$= 40 \text{ A} \times 0.500$$
$$= 20 \text{ A}$$

Effective Value The **effective value** of AC is defined in terms of an **equivalent heating effect** when compared to DC. Figure 15-18 shows an example of this relationship, which can be summarized as follows:

- When 1 ampere of DC current flows through a resistor, a certain amount of energy is dissipated by the resistor in the form of heat.
- An AC current that will produce the same amount of heat in the resistor is considered to have an effective current value of 1 A.

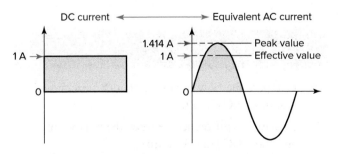

Figure 15-18 Equivalent effective value of an AC current.

- Obviously, the AC current must have a peak value that is higher than 1 A in order to be equivalent to a constant DC current of 1 A.
- In this case, the effective value of the AC current would be 1 A, while the peak value would be 1.414 A.

The effective value of a sine wave current or voltage is also called the **root-mean-square (rms)** value because of the way it is derived. It is equal to the square root of the average value of the squares of all the instantaneous values of voltage current during one-half cycle. By using this process, it can be proved that the rms or effective value of a pure or undistorted sine wave of current or voltage is always equal to **0.707** times its peak value. Instruments designed to measure AC voltage and current normally display the effective value of the measured voltage or current.

$$I_{rms} = I_{peak} \times 0.707$$
$$E_{rms} = E_{peak} \times 0.707$$

EXAMPLE 15-4

Problem: What is the rms value of an AC sine wave current with a peak value of 8 A?

Solution:

$$I_{rms} = I_{peak} \times 0.707$$
$$= 8 \text{ A} \times 0.707$$
$$= 5.66 \text{ A}$$

Since an alternating current is produced by an alternating voltage, the same rule applies in terms of its effective value. The effective value of a voltage sine wave is equal to 0.707 times its peak value.

EXAMPLE 15-5

Problem: A peak AC voltage of 160 V produces a peak current of 12 A through a resistive circuit. Determine the effective AC voltage and current.

Solution:

$$E_{eff} = E_{peak} \times 0.707$$
$$= 160\ V \times 0.707$$
$$= 113\ V$$

$$I_{eff} = I_{peak} \times 0.707$$
$$= 12\ A \times 0.707$$
$$= 8.48\ A$$

The equations can be transposed so that the peak value of a sine wave can be determined if its effective value is known. Accordingly, the peak value of a sine wave is equal to **1.414** times its effective value:

$$E_{peak} = E_{rms} \times 1.414$$
$$I_{peak} = I_{rms} \times 1.414$$

EXAMPLE 15-6

Problem: The voltage that is available at a convenience electric outlet is rated for an effective or rms value of 120 V. Find the peak and peak-to-peak value of this voltage.

Solution:

$$E_{peak} = E_{rms} \times 1.414$$
$$= 120\ V \times 1.414$$
$$= 170\ V$$

$$E_{peak-to-peak} = E_{peak} \times 2$$
$$= 170\ V \times 2$$
$$= 340\ V$$

The effective or rms value of a sine wave is the one most frequently used when referring to an AC voltage or current. It is common practice to assume that all AC voltage and current readings are effective values unless otherwise stated. Likewise, voltmeters and ammeters are normally calibrated to read effective (rms) values unless otherwise stated. It is common practice to write the symbols for effective voltage and current without the use of a subscript. To represent the effective values for current and voltage, we would simply write I and E, respectively, instead of I_{rms} and E_{rms}.

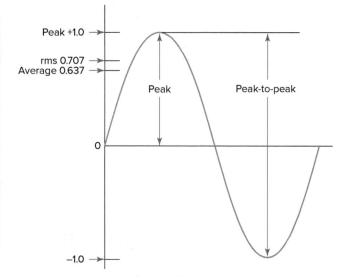

Figure 15-19 Amplitude values of a sine wave.

Average Value Figure 15-19 shows the four different ways that the amplitude of a sine wave can be specified. The **average value** of an AC sine wave is the average of all the instantaneous values during one-half cycle. It has been determined that the average value for one-half cycle of a pure sine wave is always equal to **0.637** times the peak or maximum value.

$$E_{average} = E_{peak} \times 0.637$$
$$I_{average} = I_{peak} \times 0.637$$

The bridge **rectifier** circuit, shown in Figure 15-20, converts an AC voltage to DC voltage using both half-cycles of the input AC sine wave voltage. It essentially converts the negative input alternations to positive alterations. The circuit has an AC input voltage of 120 volts rms with a peak value of 170 volts. Once the AC voltage has been rectified, the average DC output voltage as measure by the DC voltmeter would be

$$E_{average} = E_{peak} \times 0.637$$
$$= 170\ V \times 0.637$$
$$= 108\ VDC$$

Part 1 Review Questions

1. Compare the flow of current in a DC circuit with that in an AC circuit.

2. Explain the difference between the polarity of an AC and a DC voltage source.

3. Define the term *generator*.

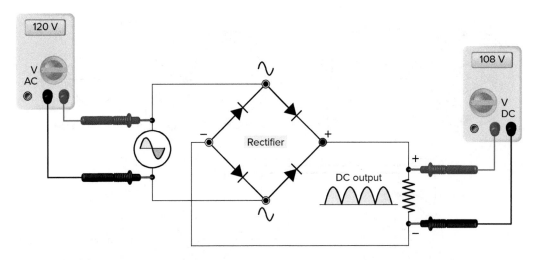

Figure 15-20 Bridge rectifier circuit.

4. Define the term *alternator*.

5. With reference to the generation of a sine wave voltage in the rotating coil of a two-pole generator:
   a. What causes the generated voltage to change in magnitude?
   b. What causes the generated voltage to change in polarity?

6. What is the essential difference in construction between an AC and a DC revolving-armature generator?

7. The revolving-field alternator is the type most widely used. Why?

8. Compare the construction of a salient pole and cylindrical alternator rotor.

9. What is the function of the prime mover in the generating process?

10. How are alternators with a small kVA rating usually cooled?

11. Utility-scale generators are often enclosed in a hydrogen atmosphere for cooling purposes. Why?

12. Determine the frequency of the voltage generated by a six-pole alternator which is rotated at a speed of 1,200 rpm.

13. Determine the rotation speed an eight-pole alternator would have to be operated at in order to generate an AC output voltage of 60 Hz.

14. In general, how is the output voltage produced by an alternator controlled?

15. Define each of the following as they apply to a sine wave:
    a. Cycle.
    b. Alternation.
    c. Period.
    d. Frequency.

16. What is the standard frequency of the AC voltage generated in North America and Europe?

17. A sine wave has an instantaneous voltage of 60 volts after 90 degrees of rotation. What is the instantaneous voltage reached by this waveform after 120 degrees of rotation?

18. What is the peak voltage of 240-V rms?

19. What is the average voltage of a sine wave if its peak-to-peak voltage is 300 V?

20. Determine the average value of a 120-V electric outlet.

21. An AC sine wave voltage is measured using an AC voltmeter and found to be 10 volts.
    a. What is the peak value of this voltage?
    b. What is the rms value?
    c. What is the peak-to-peak value?

22. Explain the main advantage of using AC over DC.

23. What effect do eddy currents and skin effect have on conductor resistance?

PART 2   AC SYSTEMS AND RESISTIVE CIRCUITS

15.4 Single- and Three-Phase Systems

An alternator can be designed to generate **single-phase** or **polyphase** AC voltages. Figure 15-21 illustrates the basic configurations used to generate single-phase, two-phase, and three-phase AC voltages. The stator coil or coils provide the output voltage and current, and the rotor is actually a rotating electromagnet, providing both the magnetic field and relative motion.

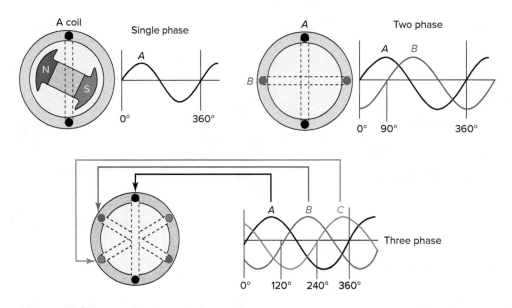

Figure 15-21 Generation of single-phase and polyphase voltages.

The **electric service** supplied to a residence normally consists of a single-phase (1∅) voltage supply with a center tap transformer, as shown in Figure 15-22. The center tap allows two different voltages (120/240 V) from the single-phase supply. This type of voltage supply is normally adequate for most lighting and power appliances in the home. It is also known as a **single-phase, three-wire supply system** and **split-phase** system. Its main advantage is that it saves conductor material over a single-ended, single-phase system. The *transformer* supplying a three-wire distribution system has a single-phase input (primary) winding. The output (secondary) winding is center-tapped and the center tap

connected to a grounded *neutral*. The secondary transformer voltage is 120 volts on either side of the center tap, giving 240 volts between the two live conductors.

Most commercial and industrial electrical installations require **three-phase** distribution systems. A **three-phase (3∅)** voltage supply is a combination of three, single-phase voltages. The single-phase voltage supplied to residential homes is, in fact, one of the phases taken from a three-phase distribution system. As load requirements increase, the use of single-phase power is no longer practical. Advantages of three-phase circuits include:

- Compared to an equivalent single-phase system, the three-phase system transmits 73 percent more power but uses only 50 percent more wire.

- The power delivered by a single-phase source is pulsating, whereas the power delivered by a three-phase system is relatively constant at all times. This means that even though the power in each phase is pulsating, the total power at any instant will be relatively constant. Therefore, the operating characteristics of three-phase machines will be superior to single-phase devices with similar ratings.

- A motor or transformer may be the same physical size, but the horsepower rating of three-phase motors and the kilovolt-ampere rating of three-phase transformers are 150 percent greater than for single-phase motors and transformers.

- A three-phase distribution system can be used to supply both three-phase and single-phase service.

A three-phase alternator contains three sets of coils positioned 120 degrees apart, and its output voltage consists of three voltage waves **120 electrical degrees** apart. The order in which these voltages succeed one another is called

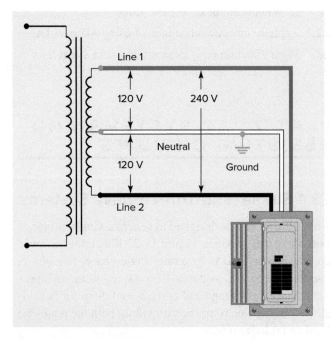

Figure 15-22 Single-phase, three-wire supply system.

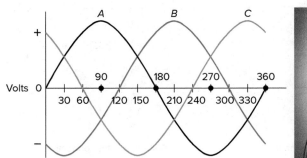

Phase rotation tester

Figure 15-23 **Phase rotation.**
Photo: ©Fluke Corporation

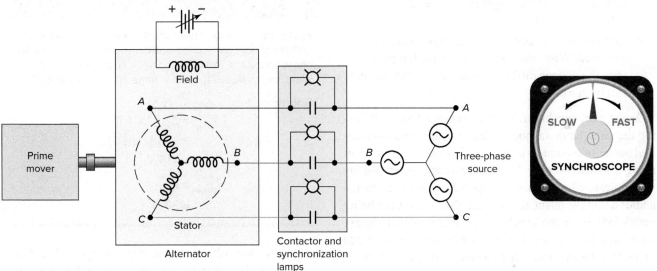

Figure 15-24 **Alternator synchronization.**

the **phase rotation or phase sequence.** With reference to the three-phase voltage waveform shown in Figure 15-23:

- Phase *A* starts to rise in a positive direction at zero electrical degrees.
- 120 electrical degrees later, phase *B* starts to increase in a positive direction.
- Phase *C* follows at another interval of 120 degrees.
- The phase sequence for this system is *A, B, C, A, B, C . . .*
- The phase sequence may be reversed by either reversing the direction of rotation of the alternator or interchanging the connections of any two of the three wires used for the transmission of the three-phase voltage.
- If two phases of the supply to a three-phase induction motor are interchanged, the motor will reverse its direction of rotation. A phase rotation tester can be used for measuring phase rotation where three phase supplies are used to feed motors, drives, and electrical systems.

Alternator synchronization is the process of connecting a three-phase alternator to an other alternator or to a

power grid. The conditions that must be meet when paralleling a three-phase alternator to another alternator or to a power grid system are:

- The phase sequence or rotation of the machine must be the same as that of the system.
- The alternator voltage must be equal to the system voltage.
- The alternator voltage must be in phase with the system voltage.
- The alternator frequency must be equal to the system frequency.

Figure 15-24 illustrates the method of synchronizing an alternator to a three-phase system using the three-dark lamp method. The operation of the synchronizing process is summarized as follows:

- Three lamps, one for each phase, are connected to detect when the direction of the phase rotation of the alternator is matched with that of the three-phase source.
- If all three lamps blink on and off in unison, the phase rotation of the two are matched.
- If they do not blink in unison, the two are not matched.

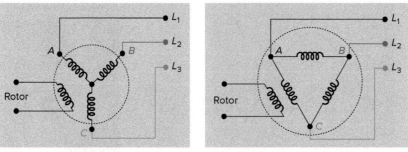

Wye alternator connection   Delta alternator connection

Figure 15-25 Wye and delta alternator connections.

- When the two are in phase and synchronized with each other and the voltages match, the three lamps will be dark or off.
- At this point, the contactor paralleling contacts can be closed. When the voltage across the lamps is zero, the voltage difference between the two systems will be zero.

In the past, synchronizing was performed manually using the three-lamp method. Today, synchronization is carried out more accurately at the exact instant of synchronism using a **synchroscope.** The synchroscope meter display will indicate whether there is a speed/frequency mismatch and/or a voltage mismatch between the two sources being paralleled. Since the synchroscope meter movement can travel in a 360-degree direction, when changes are made that effect the frequency and voltage output, the effects of those changes can be observed as they are made.

Utility generators are regularly being connected and disconnected from a large power grid in response to customer demand. Such a grid is said to be an infinite bus because it contains so many generators essentially connected in parallel that neither its voltage nor its frequency can be altered. Considerable disturbance in the electrical system and damage to the alternator windings can occur if proper synchronization procedures are not followed.

The three-phase alternator contains three sets of stator coils positioned 120 degrees apart. The output from these coils is three separate voltages with the same frequency and

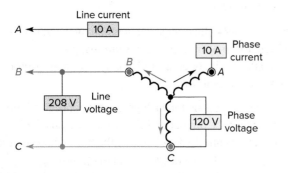

Figure 15-26 Voltages and currents in a wye-connected configuration.

EXAMPLE 15-7

Problem: Voltage and current measurements taken for a three-phase, wye-connected alternator output indicate a phase voltage of 240 volts and a line current of 30 amperes. What is the value of the line voltage and phase current?

Solution:

$$E_{line} = E_{phase} \times 1.73$$
$$= 240 \text{ V} \times 1.73$$
$$= 415 \text{ V}$$

$$I_{phase} = I_{line}$$
$$= 30 \text{ A}$$

---

magnitude, but 120 electrical degrees apart. The three windings of the alternator are connected so that only three or four wires, instead of six wires, are required for the transmission of the three-phase voltage. The three sets of stator coils of the three-phase alternator may be connected in **wye** (also known as star) or **delta,** as shown in Figure 15-25.

The **wye connection** is made by connecting one end of each of the three-phase alternator windings together. Figure 15-26 shows an example of the voltages and currents in a wye-connected configuration. The voltage across and the current through a single winding or the phase are known as the **phase voltage (E_{phase})** and **phase current (I_{phase}).** Similarly, the voltage between and the current through the line wires are known as the **line voltage (E_{line})** and **line current (I_{line}).**

In a wye connection, the line current and the phase current are the same because they act in series with each other: $I_{line} = I_{phase}$. However, the phase voltage is less than the line voltage by a factor of the square root of 3 ($\sqrt{3}$), which equals 1.73. This is because the voltage of each phase has a 120-degree delay or shift that has to be taken into account:

$$E_{phase} = \frac{E_{line}}{\sqrt{3}}$$
$$= \frac{E_{line}}{1.73}$$

or

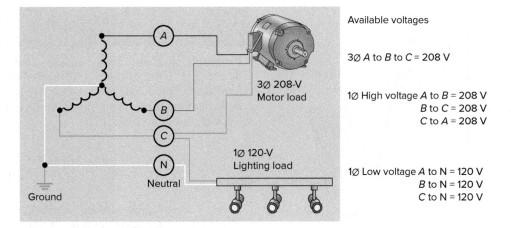

Available voltages

3∅ A to B to C = 208 V

1∅ High voltage A to B = 208 V
B to C = 208 V
C to A = 208 V

1∅ Low voltage A to N = 120 V
B to N = 120 V
C to N = 120 V

3∅ 208-V
Motor load

1∅ 120-V
Lighting load

Neutral

Ground

Figure 15-27 **Three-phase four-wire wye system.**

$$E_{line} = E_{phase} \times \sqrt{3}$$
$$= E_{phase} \times 1.73$$

The three-phase, four-wire wye system is very common and is the standard system supplied by many power utilities to commercial and industrial customers. Figure 15-27 shows a **four-wire, wye-connected** alternator capable of supplying power to a building. It can deliver both single-phase and three-phase power and both 208 V and 120 V without the use of a transformer.

The phase windings of a delta connection are all connected in series and resemble the Greek letter delta (Δ). Figure 15-28 shows an example of the voltages and currents in a **delta-connected configuration.** Since the phase windings form a closed loop, it may seem that a high current will continuously flow through the windings, even when no load is connected. Actually, because of the phase difference between the three generated voltages, negligible or no current flows in the windings under no-load conditions.

In a delta connection the line voltage and the phase voltage are the same because they act in parallel with each other: $E_{line} = E_{phase}$. The line current and phase current, however, are different. The line current of a delta connection is higher than the phase current by a

EXAMPLE 15-8

Problem: Voltage and current measurements taken for a three-phase, delta-connected alternator output indicate a line voltage of 480 volts and a line current of 100 amperes. What is the value of the voltage across each phase and the current through each phase?

Solution:

$$E_{phase} = E_{line}$$
$$= 480 \text{ V}$$
$$I_{phase} = \frac{I_{line}}{\sqrt{3}}$$
$$= \frac{100 \text{ A}}{1.73}$$
$$= 57.8 \text{ A}$$

factor of $\sqrt{3}$ or 1.73. This is because the current of each phase has a 120-degree delay or shift that has to be taken into account:

$$I_{line} = I_{phase} \times \sqrt{3}$$
$$= I_{phase} \times 1.732$$
or
$$I_{phase} = \frac{I_{line}}{\sqrt{3}}$$
$$= \frac{I_{line}}{1.73}$$

15.5 Resistive Circuits

When an alternating voltage is applied to a circuit, it causes an alternating current of the same frequency to flow through the circuit. For purely **resistive** AC circuits the voltage and current are **in phase** with each other, as illustrated in Figure 15-29. The voltage and current are

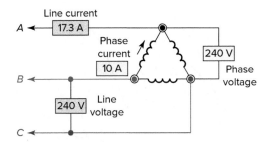

Line current

A

17.3 A

Phase current

10 A

240 V

Phase voltage

B

240 V

Line voltage

C

Figure 15-28 **Voltages and currents in a delta-connected configuration.**

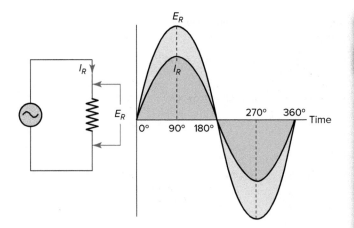

Figure 15-29 AC resistive circuit voltage and current.

considered to be in phase because the two waveforms pass through their zero values and increase in the same direction to their maximum values at the same time. Hence, the phase difference between waves that are in phase is zero.

For the most part, pure resistive circuits react much the same for AC or DC circuits. Resistive heating units and incandescent lighting are considered to be pure resistive loads. Resistive loads are characterized by the fact that they **produce heat** and the current and voltage are in phase with each other. Theoretically, no AC circuit can contain only resistance. Other properties can affect voltage and current to a much lesser degree, allowing the circuit to be treated as being purely resistive.

In general, all the laws and formulas that apply to DC circuits also apply to AC circuits. Furthermore, they apply **exactly the same way** to AC resistive circuits. This is true, because resistors are linear components and their characteristics do not depend on frequency. In a resistive DC circuit, both current and voltage are fixed, steady values. In an AC resistive circuit, the current alternates exactly in step with the voltage. The Ohm's law formula for an AC circuit can be expressed as

$$I_{rms} = \frac{E_{rms}}{R} \quad \text{or} \quad \text{simply} \quad I = \frac{E}{R}$$

Unless otherwise stated, all AC voltage and current values are given as effective or rms values. With this in mind, the formula for Ohm's law for an AC circuit can also be expressed as

$$I_{peak} = \frac{E_{peak}}{R} \quad \text{or} \quad I_{average} = \frac{E_{average}}{R}$$

When solving for quantities in AC-resistive circuits, the important thing to keep in mind is not to mix AC values. When you solve for effective values, all values you use in the formula must be effective values. Similarly, when you solve for peak or average values, all values you use must be peak or average values.

EXAMPLE 15-9

Problem: A series circuit consists of three resistors ($R_1 = 6\ \Omega$, $R_2 = 4\ \Omega$, and $R_3 = 5\ \Omega$) and an alternating voltage source of 120 volts, as shown in Figure 15-30. Determine the total resistance of the circuit and the effective value of the current flow.

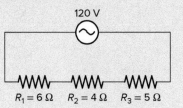

Figure 15-30 Circuit for example 15-9.

Solution:

$$R_T = R_1 + R_2 + R_3$$
$$= 6\ \Omega + 4\ \Omega + 5\ \Omega$$
$$= 15\ \Omega$$
$$I_T = \frac{E_T}{R_T}$$
$$= \frac{120\ \text{V}}{15\ \Omega}$$
$$= 8\ \text{A}$$

Recall that in a DC circuit the power is equal to the voltage *times* the current ($P = E \times I$). This is also true in an AC circuit when the current and voltage are in phase; that is, when the circuit is resistive. With alternating current, the values of current and voltage vary with time. At any instant the power is equal to the current at that instant multiplied by the voltage at that instant. Plotting all the instantaneous powers, as illustrated in Figure 15-31, produces a power waveform. Notice that the power waveform is always **positive** for in-phase currents and voltages. A negative current multiplied by a negative voltage yields a positive power. This means that the resistive load is converting electric energy into heat energy during the complete cycle. Then the power dissipated in a purely resistive load fed

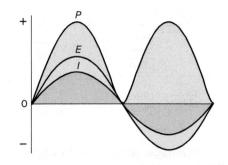

Figure 15-31 Power in an AC resistive circuit.

from an AC rms supply is the **same** as that for a resistor connected to a DC supply and is given as

$$P = IE$$
$$P = I^2R$$
$$P = \frac{E^2}{R}$$

where I and E are the effective or rms current and voltage.

EXAMPLE 15-10

Problem: Find the effective total power consumed by the single-phase resistive circuit of Figure 15-32.

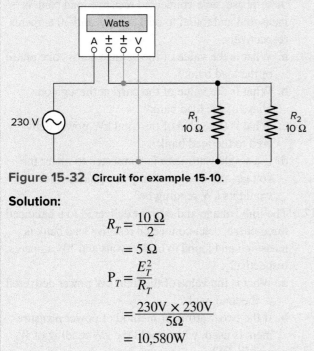

Figure 15-32 Circuit for example 15-10.

Solution:

$$R_T = \frac{10\ \Omega}{2}$$
$$= 5\ \Omega$$
$$P_T = \frac{E_T^2}{R_T}$$
$$= \frac{230V \times 230V}{5\Omega}$$
$$= 10,580W$$

For three-phase resistive circuits, voltages and currents are normally expressed as rms or effective values, as in single-phase analysis. The power in watts delivered to three-phase balanced resistive loads is calculated as follows:

$$P = \sqrt{3} \times E_{\text{line}} \times I_{\text{line}}$$
$$= 1.73 \times E_{\text{line}} \times I_{\text{line}}$$

EXAMPLE 15-11

Problem: Find the effective total power consumed by the balanced three-phase resistive heating load circuit of Figure 15-33.

Solution:

$$P = \sqrt{3} \times E_{\text{line}} \times I_{\text{line}}$$
$$= 1.73 \times 208\ \text{V} \times 20\ \text{A}$$
$$= 7,200\ \text{W}$$

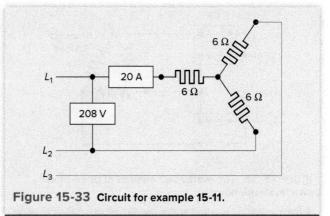

Figure 15-33 Circuit for example 15-11.

It is common practice to determine three-phase power in terms of the circuit's line voltage and line current because it is generally easier to measure the line voltage and current in a three-phase system. However, in a three-phase balanced resistive system, the total circuit power is also equal to the sum of the power of each of the three phases or three times the power of one phase:

$$P = 3 \times E_{\text{phase}} \times I_{\text{phase}}$$

Figure 15-34 shows the connection of a **single wattmeter** used to measure the three-phase power of a balanced three-phase resistive load bank. Note that the wattmeter is connected to meter the phase current and voltage. Accordingly, three times the measured value of the wattmeter is equal to the three-phase power, as long as the three phases are balanced.

Two wattmeters can be used to measure the three-phase power in systems which contain only three-phase conductors by metering the line current and voltage. Figure 15-35 shows the connection for the so-called **two-wattmeter** method of power measurement in three-phase systems. If the three-phase system is balanced and resistive, then the two wattmeters will have the same readings, and the total circuit power will be equal to the sum of the two wattmeter readings W_1 and W_2. It is important to observe the polarity marks ($\pm$) on the voltage and current coils of the wattmeters and make the connections exactly as shown.

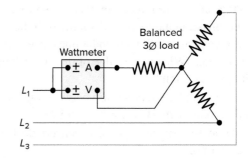

Figure 15-34 Single wattmeter used to measure the three-phase power.

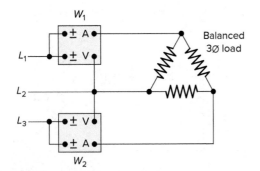

Figure 15-35 Two-wattmeter method of three-phase power measurement.

Part 2 Review Questions

1. What are the two single-phase voltages available from a standard residential service?
2. Compare the use of single-phase and three-phase systems with regards to the:
   a. The amount of wire required for an equivalent size load.
   b. The variation in the amount of power delivered to a load.
   c. The physical size of three-phase and single-phase equivalent-rated motors.
3. How does phase rotation affect the direction of rotation of a three-phase induction motor?
4. What conditions must be met when connecting a three-phase alternator to a power grid system?
5. How many degrees apart are the three stator coils of a three-phase alternator positioned?
6. Name the two basic types of three-phase alternator stator coil connections.
7. How do the line-and-phase voltages and currents for a wye-connected alternator compare in value?
8. How do the line-and-phase voltages and currents for a delta-connected alternator compare in value?
9. Voltage and current measurements taken for a three-phase, wye-connected alternator indicate a line voltage of 208 volts and a line current of 20 amperes. What is the value of the phase voltage and phase current?
10. Voltage and current measurements taken for a three-phase, delta-connected alternator indicate a phase voltage of 240 volts and a line current of 10 amperes. What is the value of the line voltage and phase current?
11. What is the phase relationship between the voltage and current in an AC resistive circuit?
12. What do all resistive-type loads convert electric energy into?
13. An AC voltage of 340 V peak-to-peak is connected to a 10-ohm heater. Determine:

    a. The peak value of the voltage.
    b. The effective value of the voltage.
    c. The wattage of the heater.
14. A series circuit consists of two 4-ohm resistors connected to a DC voltage source of 24 volts.
    a. Determine the total resistance and current flow.
    b. Repeat for the same circuit connected to a 24-volt AC source.
15. A 240-VAC furnace heating unit consists of four 20-Ω heating elements connected in parallel. Determine the kW rating of the unit.
16. The line voltage and current delivered to a balanced three-phase, wye-connected resistive load bank is measured and found to be 415 volts and 30 amperes, respectively.
    a. What is the value of the voltage across one phase of the load bank?
    b. What is the value of the current through one phase of the load bank?
    c. What is the value of the total kW power delivered to the load bank?
    d. If a single wattmeter is connected to meter the voltage and current of any one phase, what would its kW reading be?
17. The line voltage and current delivered to a balanced three-phase, delta-connected resistive load bank is measured and found to be 480 volts and 104 amperes, respectively.
    a. What is the value of the total kW power delivered to the load bank?
    b. If the two-wattmeter method of power measurement is used, what would the kW reading of W_1 and W_2 be?
18. Compare the voltages available from a four-wire 208 wye-connected system with that of a four-wire 480 wye-connected system.
19. For the AC circuit of Figure 15-36 determine all values of voltage, current, resistance, and power. Record your answers in table form.

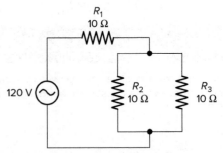

Figure 15-36 Circuit for review question 19.

20. Compare the method used to energize the magnetic field on the rotor of a synchronous AC generator with that of an induction type.

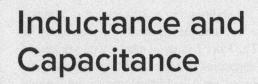
Inductance and Capacitance

Inductance coil on circuit board
©Evgeny Prokofyev/123RF

LEARNING OUTCOMES

▶ Understand the effects of inductance and capacitance in AC circuits.

▶ Explain the concept of inductive reactance and the factors that affect it.

▶ Calculate the inductive reactance of a coil.

▶ Explain the concept of capacitive reactance and the factors that affect it.

▶ Calculate the inductive and capacitive reactance.

▶ Define and calculate *time constants*.

▶ Compare phase shift in an inductor and capacitor.

Resistance, inductance, and capacitance are the three basic circuit properties that we use to control voltages and currents in AC electric circuits. Each behaves in a different way. Resistance opposes current, inductance opposes any change in current, while capacitance opposes any change in voltage. Also, resistance dissipates energy in the form of heat, while inductance and capacitance both store energy. In this chapter, we study the unique electrical properties of inductance and capacitance.

PART 1   INDUCTANCE

16.1 Inductance

Inductance is the ability of an electric circuit or component to oppose any **change (increase or decrease) in current** flow. The letter L is the symbol used to represent inductance. Components that are used to provide the inductance in a circuit are called inductors. An **inductor** is simply a coil of wire also known as a choke, reactor, or coil.

Inductance is the primary type of load found in alternating current (AC) circuits. There is always some inductance present in all AC circuits because of the continuously changing magnetic field created by the changing current through the conductor. When applied to a coil, this changing magnetic field cuts through the turns of the coil conductors inducing a voltage in the coil that opposes the change in current.

One way of classifying inductors is by the type of material used for the core of the inductor. The construction and symbol for air-core, iron-core, and ferrite-core inductor are shown in Figure 16-1. Basically, the air-core-type inductor consists of a coil wrapped around a form with nothing but air in the middle.

- Air-core inductors are often used in high-frequency communication circuits.
- Iron-core inductors are constructed with the coil wound around a laminated shell-type core. Nearly all inductors used at power frequencies (60 Hz, for example) are of the laminated iron-core type.
- Ferrite-core inductors consist of mixed metal oxides of iron and other elements which exhibit magnetic properties. They are extensively used in the cores of radio frequency transformers and inductors.

Figure 16-2 shows a three-phase line **reactor** used as an input filter for a variable-frequency motor drive. The reactor functions to adsorb power line disturbances capable of damaging the drive's sensitive electronic components.

An inductor opposes any change in current due to its ability to store and release energy from its magnetic field.

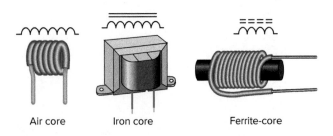

Air core   Iron core   Ferrite-core

Figure 16-1 Air-core, iron-core, and ferrite-core inductors.

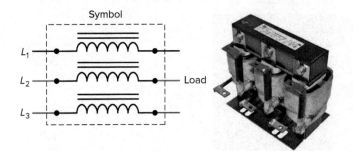

Figure 16-2 Three-phase line reactor.
Photo: ©Hammond Power Solutions Inc.

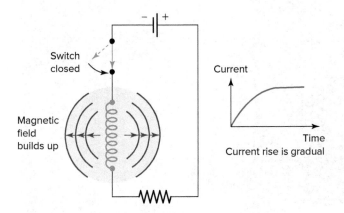

Figure 16-3 Applying DC power to a coil.

Figure 16-3 illustrates the effect of inductance in a DC circuit when power to a coil is first applied. The reaction to this change in current is summarized as follows:

- When direct current is initially applied to the coil, a magnetic field builds up around it.
- The expansion of the magnetic field cutting across the coil windings causes a counter voltage to be **induced** in the coil.
- This counter voltage tends to oppose the original applied voltage; therefore, current rise is gradual.
- Once the current reaches its Ohm's law value, the magnetic field remains constant and the inductive effect stops.

Figure 16-4 illustrates the effect of inductance in the DC circuit when power to the coil is removed. The reaction to this change in current is summarized as follows:

- When current to the coil is switched OFF, the inductance effect can once again be seen.
- As the current falls to a zero value, the collapsing magnetic field causes a high voltage to be induced in the coil.
- This results in a noticeable arc at the switch contacts.
- The arc produced at the contacts is caused by the high induced voltage attempting to maintain the current in the circuit.

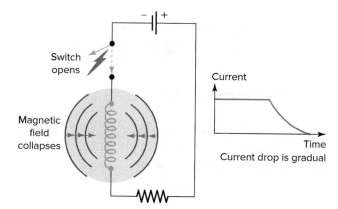

Figure 16-4 **Removing DC power from a coil.**

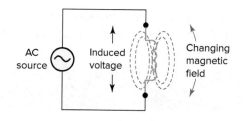

Figure 16-6 **Lenz's law.**

Figure 16-6 **Lenz's law.**

Faraday's law of induction explains the working principle of inductors. Simply put, Faraday's law states that whenever there is relative motion between conductor and a magnetic field, the flux linkage with a coil changes and this change in flux induce a voltage across a coil. The amount of voltage induced into a conductor depends on the number of turns of the coils, the strength of the magnetic field, and the speed of the relative motion.

Inductance is present only when the current changes. In a DC circuit, this occurs each time the circuit is turned ON or OFF. This inductive effect takes place in the coil itself and is called **self-inductance.** If a second coil is magnetically linked to the first, the inductive effect will also be felt in the second coil. **Mutual inductance** is the term used when the effect of induction is such that current change in one circuit produces an induced voltage in another circuit, as illustrated in Figure 16-5.

In AC circuits the magnetic field created by the varying current flow expands and collapses, resulting in a continuous inductive effect. This change in the magnetic field in turn produces an induced voltage within the coil, as illustrated in Figure 16-6. There is an important relationship between the direction of the current change and the in-

duced voltage. This relationship is summarized by **Lenz's law** and stated as follows:

The induced voltage always acts in a direction to oppose the current change that produced it.

The voltage induced in a conductor or coil by its own magnetic field is called a **counter electromotive force (cemf).** Another term used to describe this voltage is **back EMF** since it is always 180 degrees out of phase with the change in magnitude of the current and wants to push current back in the other direction. When the source voltage is increasing, the polarity of the cemf is such that it opposes the source voltage. When the source voltage is decreasing, the cemf aids the source voltage and tries to keep the current constant.

You can also view inductance in terms of energy conversion and storage. Current flow in an inductor generates a magnetic field, and this magnetic field stores energy. The amount of energy stored in the magnetic field is a function of the current and inductance. An **ideal inductor** (assuming no winding resistance) does not dissipate energy; it only stores it. When an AC voltage is applied to an inductor, energy is stored by the inductor during a portion of the cycle; then the stored energy is returned to the circuit during another portion of the cycle. No net energy is lost in an ideal inductor due to conversion to heat. This is in contrast to a pure resistive load in an AC circuit in which all the energy is converted to heat.

The inductance value of an inductor refers to its ability to generate a countervoltage that opposes a change in the current flow. The amount of inductance of a coil is measured by a unit called the **henry (H).** One henry (1 H) represents the inductance of a coil in which a current change of one ampere per second will produce a countervoltage of one volt. This can be represented by the formula:

$$E_{cemf} = L \times \frac{\Delta I}{\Delta T}$$

or

$$E_{cemf} = L \times A/s$$

where L = inductance in henrys
     ΔI = change in current
     ΔT = change in time

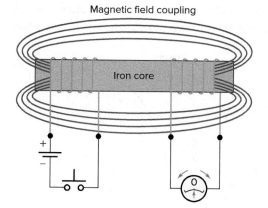

Figure 16-5 **Mutual inductance.**

Note that the greater the inductance value or the faster the rate of change of the current, the greater is the cemf induced in the circuit.

EXAMPLE 16-1

Problem: What voltage is induced across an inductor of 4 H when the current is changing at a rate of 10 amperes per second?

Solution:

$$E_{cemf} = L \times A/s$$
$$= 4\,H \times 10\,A/s$$
$$= 40\,V$$

Inductance of inductors varies in value from the henrys (H) down to microhenrys (μH). In addition to being rated for inductance, inductors are also rated for:

- **DC resistance,** which specifies the resistance of the wire in the winding of the inductor.
- **Current** rating, which indicates how much current the inductor can continuously carry without overheating.
- **Voltage** rating, which indicates how much voltage the insulation on the inductor winding can continuously withstand.
- **Tolerance,** which is specified as a percentage of the stated inductance.

An inductor's ability to store magnetic energy is measured by its inductance and depends entirely on the **physical construction** of both the core and the windings around the core. Factors that determine the amount of inductance created include:

- **Coil turns.** A greater number of turns of wire in the coil result in greater inductance; fewer turns of wire in the coil result in less inductance.
- **Coil length.** The longer the coil's length, the less inductance; the shorter the coil's length, the greater the inductance.
- **Core material.** The greater the magnetic permeability of the core which the coil is wrapped around, the greater the inductance; the less the permeability of the core, the less the inductance. Variable inductors are made by use of a sliding core that can be moved in and out of the coil as illustrated in Figure 16-7.
- **Core area.** The greater cross-section area of the core, the greater inductance.

Since the inductor's basic action is to develop a voltage that opposes a change in its current, it follows that current cannot change instantaneously in an inductor. A certain

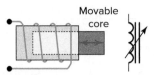

Figure 16-7 **Variable inductor.**

time is required for the current to make a change from one value to another. The rate at which the current changes is determined by the *L/R* time constant:

$$T = \frac{L}{R}$$

where T = the time constant in seconds
  L = the inductance in henrys
  R = the resistance in ohms (The resistance is that in series with L, being the coil resistance plus any external resistance.)

EXAMPLE 16-2

Problem: What is the *L/R* time constant of a 20-H coil having 100 Ω of series resistance?

Solution:

$$T = \frac{L}{R}$$
$$= \frac{20\,H}{100\,\Omega}$$
$$= 0.2\,s$$

The *L/R* time constant is a measure of how much time it takes the current to change by **63.2 percent** or approximately 63 percent. Figure 16-8 illustrates how time constants apply when the DC circuit to an inductor is first energized. The current will rise to approximately 63 percent of its full value in one time constant period after the switch is closed. This buildup of current follows an exponential curve and reaches maximum value after 5 time constant periods. Just as the circuit takes **5 time constants** to reach the maximum value when the switch is closed, the same circuit will also take 5 time constants to reach zero value once the switch is opened.

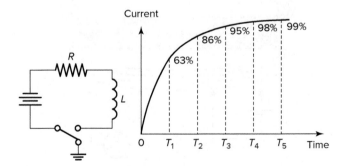

Figure 16-8 *L/R* **time constant.**

16.2 Inductive Reactance

In a DC circuit, the only changes in current occur when the circuit is closed to start current and when it is opened to stop current. However, in an AC circuit the current is continually changing each time the voltage alternates. Since inductance in a circuit opposes a change in current and AC current is continually changing, there is an opposition offered by the inductor at all times. This opposition to the AC current is called inductive reactance.

Inductive reactance is measured in **ohms (Ω)** and is represented by the symbol X_L. Current flow through a coil connected to a DC source is only limited by the wire resistance of the coil. Current flow through the same coil connected to an **AC source** is limited by both the wire **resistance** and the **inductive reactance** of the coil. This fact is illustrated by the circuits of Figure 16-9 and summarized as follows:

- In the DC circuit, the coil has low resistance and no inductive reactance (X_L).
- In the AC circuit, the coil has low resistance and high inductive reactance.
- As a result, compared to the AC circuit, the current flow and brightness of the lamp are greater for the DC circuit.

The inductive reactance of a coil is directly proportional to the inductance value of the coil. That is, ***the higher the inductance (L), the greater the inductive reactance (X_L).*** This fact is illustrated by the circuits of Figure 16-10 and summarized as follows:

- The 5-mH and 10-mH inductors are connected in series with a lamp to an AC source of the same voltage and frequency.
- The larger 10-mH inductor will produce more inductive reactance than the 5-mH inductor.

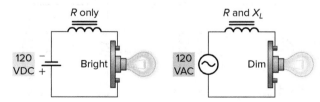

Figure 16-9 Effect of inductive reactance in DC and AC circuits.

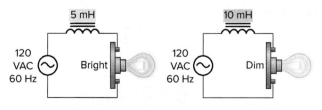

Figure 16-10 Effect of inductance on current flow.

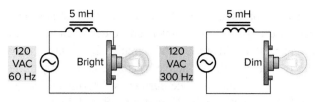

Figure 16-11 Effect of frequency on current flow.

- As a result compared to the 10-mH circuit, the current flow and the brightness of the lamp is greater for the 5-mH circuit.

The inductive reactance of a coil is also ***directly proportional to the frequency*** of the AC supply source. For any given coil, increasing the frequency of the voltage source increases the rate of change of current through the coil. This results in the magnetic field created by the current cutting across the coil windings at an increased rate. This, in turn, increases the countervoltage produced. This fact is illustrated by the circuits of Figure 16-11 and summarized as follows:

- The inductor and lamp are connected in series and operated from 60-Hz and 300-Hz AC sources.
- The higher frequency generates more inductive reactance with more ohms of opposition.
- As a result when the circuit is operated at 60-Hz, the current flow and brightness of the lamp are greater than when operated at 300 Hz.

The amount of inductive reactance in an AC circuit then depends on the amount of inductance and the frequency of the circuit current. An increase in the size of the inductor and/or the frequency will cause a higher opposition to current flow. To calculate inductive reactance for AC circuits apply the formula

$$X_L = 2\pi fL$$

where X_L = the inductive reactance in ohms
f = the frequency of the AC circuit in hertz
L = the inductance in henrys
$2\pi = 6.28$ (indicates 2π radians, 360°, or 1 cycle)

EXAMPLE 16-3

Problem: At a frequency of 60 Hz, what is the inductive reactance of a 5-henry inductor?

Solution:

$$\begin{aligned} X_L &= 2\pi fL \\ &= 2\pi(60\ \text{Hz})(5\ \text{H}) \\ &= 6.28(60)(5) \\ &= 1{,}884\ \Omega \end{aligned}$$

A **pure or ideal** inductor is an inductor that has **zero resistance.** It does not convert any electric energy into heat energy. In theoretical AC circuits that contain only pure inductance, the inductive reactance is the only thing that limits the current. The current is determined by the Ohm's law equation with X_L replacing R, as follows:

$$I = \frac{E}{X_L}$$

In general, the opposition offered to current flow by the DC resistance of AC coils such as relays and solenoids is very small in comparison to that of the inductive reactance. When the ratio of inductive reactance is 10 times or greater than that of the DC resistance, the wire resistance is considered negligible.

EXAMPLE 16-4

Problem: Calculate the current flowing through a 100-mH (0.1-H) coil connected to a 48-VAC, 60-Hz source. Assume the resistance of the wire used in the coil to be negligible.

Solution:

$$X_L = 2\pi f L$$
$$= 2\pi(60\ Hz)(0.1\ H)$$
$$= 6.28(60)(0.1)$$
$$= 37.7\ \Omega$$

$$I = \frac{E}{X_L}$$
$$= \frac{48\ V}{37.7\ \Omega}$$
$$= 1.27\ A$$

Finding the total inductance of a **series** circuit composed totally of inductors is the same as finding the total resistance of a series resistor circuit. You simply add all the individual inductances; this is assuming there is no magnetic interaction between the inductors:

$$L_T = L_1 + L_2 + L_3 \cdots$$

EXAMPLE 16-5

Problem: Determine the current flow for the series-connected inductor circuit of Figure 16-12.

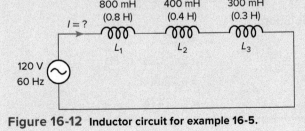

Figure 16-12 Inductor circuit for example 16-5.

Solution:

$$L_T = L_1 + L_2 + L_3$$
$$= 0.8\ H + 0.4\ H + 0.3\ H$$
$$= 1.5\ H$$

$$X_{L(total)} = 2\pi f L$$
$$= 2\pi(60\ Hz)(1.5\ H)$$
$$= 6.28(60)(1.5)$$
$$= 565\ \Omega$$

$$I = \frac{E}{X_{L(total)}}$$
$$= \frac{120\ V}{565\ \Omega}$$
$$= 0.212\ A$$
$$= 212\ mA$$

Inductors connected in **parallel** are also treated like resistors in parallel. You have your choice of three formulas for solving parallel inductance: the same-value formula, the product-over-sum formula, and the reciprocal formula.

$$L_T = \frac{L}{n} \qquad \text{(same-value formula)}$$

$$L_T = \frac{L_1 \times L_2}{L_1 + L_2} \qquad \text{(product-over-sum formula)}$$

$$L_T = \frac{1}{\frac{1}{L_1} + \frac{1}{L_2} + \frac{1}{L_3}} \qquad \text{(the reciprocal formula)}$$

EXAMPLE 16-6

Problem: Determine the current flow for the parallel-connected inductor circuit of Figure 16-13.

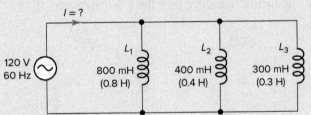

Figure 16-13 Inductor circuit for example 16-6.

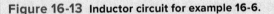

Solution:

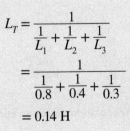

$$L_T = \frac{1}{\frac{1}{L_1} + \frac{1}{L_2} + \frac{1}{L_3}}$$
$$= \frac{1}{\frac{1}{0.8} + \frac{1}{0.4} + \frac{1}{0.3}}$$
$$= 0.14\ H$$

$$X_L = 2\pi fL$$
$$= 2\pi(60 \text{ Hz})(0.14 \text{ H})$$
$$= 6.28(60)(0.14)$$
$$= 52.8 \ \Omega$$

$$I_{(total)} = \frac{E}{X_{L(total)}}$$
$$= \frac{120 \text{ V}}{52.8 \ \Omega}$$
$$= 2.27 \text{ A}$$

16.3 Phase Shift in an Inductor

In a **purely resistive** circuit the voltage and the current are **in phase.** What this means is that voltage and current rise and fall at the same time, as illustrated in Figure 16-14. When the voltage is zero, the current is zero, and so on.

The **Ohm's law** triangle used for DC circuits can only be used for AC circuits if the load is **purely resistive.** Most AC circuits contain series or parallel combinations of resistance, capacitance, and inductance. This leads to the voltage and currents being **out of phase,** and load calculations become more complex.

In a **purely inductive** circuit, *voltage leads current by 90 degrees* and are said to be out of phase. The current cannot increase immediately with the applied voltage because it is opposed by the induced countervoltage in the inductor. As a result the current peaks occur a quarter cycle (90°) after the voltage peaks, as illustrated in Figure 16-15.

16.4 Inductive Reactive Power

In a purely resistive AC circuit, power in *watts is equal to the voltage times the current.* This is because the voltage and current are in phase, and as a result, the product of the instantaneous voltage and current will always be positive. The relationship between the voltage, current, and power of a purely resistive circuit is illustrated in Figure 16-16. All energy delivered by the source is consumed by the circuit and dissipated in the form of heat. This kind of positive power is often called **true power** and is measured in watts.

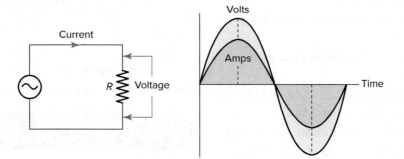

Figure 16-14 **Voltage and current for a purely resistive load.**

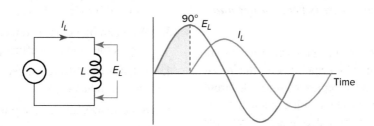

Figure 16-15 **Voltage and current for a purely inductive load.**

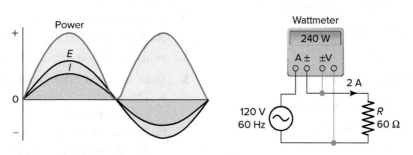

Figure 16-16 **Power in a purely resistive AC circuit.**

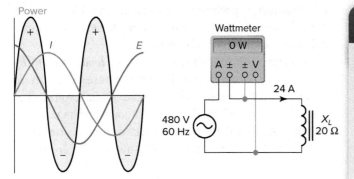

Figure 16-17 **Power in a purely inductive AC circuit.**

In a purely inductive AC circuit, we have seen that the current lags the voltage by 90 degrees. The power wave is plotted by multiplying voltage by current at every instant through 360 degrees, as shown in Figure 16-17. The circuit can be analyzed in steps as follows:

- When the current and voltage are both of the same polarity, the power is positive.
- When the current and voltage are of the opposite polarity, the power is negative.
- The resulting power waveform is therefore both positive and negative.
- Positive values of power indicate that energy is stored by the inductor.
- Negative values of power indicate that energy is returned from the inductor to the source.
- The average value of power calculates out to be *zero*.
- This means that all the energy the circuit receives from the source is returned to the source.
- *The power in watts of a purely inductive circuit is always zero even though there is both voltage and current present.*

Since the product of the voltage and current for a pure inductance is not mathematically equal to zero, this product for a pure inductance must represent some different kind of power. The power associated with an inductance is a kind of magnetic power, and is called inductive reactive power. **Inductive reactive power** is measured in **VARs,** which is the abbreviation for volt-ampere reactive. To calculate inductive reactance power apply the formula:

$$\text{VARs} = E_L \times I_L$$

or

$$\text{VARs} = I_L^2 \times X_L$$

where E_L = the voltage applied across the inductor

I_L = the current flow through the inductor

X_L = the inductive reactance

EXAMPLE 16-7

Problem: An inductor with negligible resistance draws a current of 2 A when connected to a 120-VAC, 60-Hz source. Determine:

a. The inductive reactance of the inductor.

b. The inductive reactive power.

c. The true power or wattage.

Solution:

a.
$$X_L = \frac{E_L}{I_L}$$
$$= \frac{120 \text{ V}}{2 \text{ A}}$$
$$= 60 \ \Omega$$

b.
$$\text{VARs} = E_L \times I_L$$
$$= 120 \text{ V} \times 2 \text{ A}$$
$$= 240 \text{ VARs}$$

or

$$\text{VARs} = I_L^2 \times X_L$$
$$= 2 \times 2 \times 60$$
$$= 240 \text{ VARs}$$

c. The true power or wattage would be zero.

Part 1 Review Questions

1. Define the term *inductance*.

2. Explain the process by which a countervoltage is induced into an inductor in a DC circuit.

3. When an inductor is connected into an AC circuit, why is the inductive effect present at all times?

4. What is the difference between self-inductance and mutual inductance?

5. State Lenz's law as it applies to a coil.

6. What is the phase relationship between the applied voltage and the counter electromotive force (cemf) of a coil?

7. Compare the energy conversion that takes place in a purely resistive load with that of a purely inductive load.

8. What is the base unit used to measure inductance?

9. State the relationship between:
   a. The inductance value of a coil and the amount of cemf it produces.
   b. The rate of change of current through a coil and the amount of cemf it produces.

10. What does the DC resistance value of an inductor refer to?

11. What effect (increase or decrease) would each of the following changes have on the inductance of a coil?
    a. Increasing the number of turns of wire.
    b. Removal of its iron core.
    c. Spacing the turns of wire farther apart.

12. For a coil that has an inductance of 5 H and a DC resistance of 10 Ω:
    a. Calculate the L/R time constant.
    b. When DC voltage is first applied to this coil, approximately how long will it take for the current to reach its maximum value?

13. Define the term *inductive reactance.*

14. What is the base unit used to measure inductive reactance?

15. A coil is connected in turn to a 10-volt DC circuit and a 10-volt AC circuit. What difference in current value, if any, should there be for the two circuits? Why?

16. State whether the inductive reactance of a coil increases or decreases with each of the following changes:
    a. Increase in the frequency of the AC supply source.
    b. Decrease in the inductance of the coil.

17. Calculate the inductive reactance of a 2.5-H inductor when operated at a frequency of 50 Hz.

18. A 6-H inductor is connected to a 12-VDC source. What is the value of its inductive reactance?

19. An AC voltage of 240 volts with a frequency of 60 Hz is applied to a 0.5-H inductor. Neglecting its small amount of wire resistance, how much current would flow through it?

20. Determine the total inductance of a 6-H and a 4-H inductor connected in:
    a. Series.
    b. Parallel.

21. Inductors of 1 H and 2 H are connected in series to a 440-V, 60-Hz power supply.
    a. Determine the total current flow for the circuit.
    b. Repeat for the two inductors connected in parallel to the power supply.

22. Compare the phase relationship between the voltage across and the current flow through a purely resistive load and a purely inductive load.

23. Explain why the power in watts of a purely inductive load is always zero.

24. Determine the current, wattage, and volt-amperes-reactive for a purely resistive load of 6 Ω connected to a 208-V, 60-Hz source.

25. Determine the current, wattage, and volt-amperes-reactive for a purely inductive reactive load of 4 Ω connected to a 230-V, 60-Hz source.

26. An AC motor starter coil, rated for 230 volts AC, is removed from the motor starter enclosure and bench-tested by applying rated voltage to the coil without its laminated steel core piece in place. Why can this result in overheating of the coil?

27. Assume you have two identical-looking coils, one rated for 120 VDC at 2 A and the other rated for 120 VAC at 2 A. Explain how an ohmmeter resistance measurement taken of the two coils could be used to distinguish the DC coil from the AC coil.

28. Why is inductance the primary type of load found in alternating current (AC) circuits?

29. State Faraday's law as it applies to inductors.

30. What three factors determines the amount of voltage induced into an inductor?

31. Compare how energy is stored or dissipated in a resistor compared to an ideal inductor.

PART 2   CAPACITANCE

16.5 Capacitance

Capacitance (*C*) is the ability of an electric circuit or component to store electric energy by means of an electrostatic field. The **capacitor,** shown in Figure 16-18, is an electrical device specially designed for this purpose. A capacitor has the ability to store electrons and release them at a later time. Basically, a capacitor is made up of **metal foil or plates** (conductors) placed near each other and separated by an insulating material called the **dielectric.** The metal sheets are usually wound up into a spiral and separated by the dielectric which can be any highly nonconducting material such as paper, mica, or ceramic.

The operation of a capacitor depends on the electrostatic field that is built up between the two oppositely charged

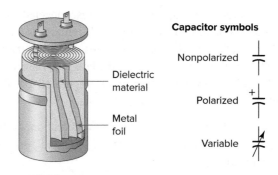

Figure 16-18 Capacitor.

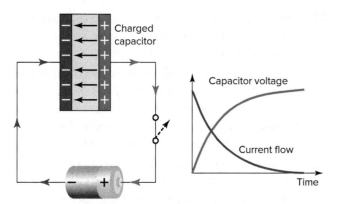

Figure 16-19 Capacitor charging circuit.

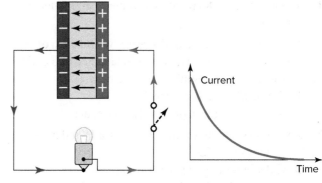

Direction of current flow is reversed

Figure 16-20 Capacitor discharging circuit.

parallel plates. When a capacitor has a potential difference or voltage between plates, it is said to be **charged.** A capacitor is charged by connecting its two plate leads to a DC voltage source, as illustrated in Figure 16-19. The circuit can be analyzed in steps as follows:

- The positive terminal of the voltage source attracts electrons from the plate connected to it.

- At the same time, the negative terminal of the voltage source repels an equal number of electrons into the other plate.

- This charging or flow of electrons continues until the voltage across the charged plates is equal to the applied voltage.

- Note that there is no flow of electrons through the dielectric.

- Electrons are simply removed from one plate and deposited on the other plate through the circuit that connects them.

- One plate (–) ends up with an excess of electrons, and the other with a lack of electrons (+).

- Because of the dielectric (insulation) between the two plates, electrons cannot flow internally from one plate to the other.

- However, there are electric lines of force between the plates, and this force is called the **electrostatic field.**

Once charged, the capacitor can be disconnected from the voltage source, and the energy will remain stored in the electrostatic field between its plates. To **discharge** a charged capacitor, the two charged plate leads are connected to a lamp load, as illustrated in Figure 16-20. The circuit can be analyzed in steps as follows:

- Electron flow is now in the opposite direction, and electrons travel from the negatively charged plate to the positively charged plate.

- Current flows only for the short duration of time it takes for the capacitor to discharge its charge.

- Current flow stops when both plates become neutralized, and the capacitor is said to be discharged.

Applications for capacitors include power factor correction of electrical systems, improving torque in motors, filters in AC circuits, and timing of control circuits. Article 460 of the NEC covers the installation of capacitors wired as part of electric circuits. An important safety aspect of this article, which pertains to all capacitors, is the discharging of the capacitor when equipment is deenergized. A capacitor will store a charge when power is shut off. If the charge is not removed from the capacitor, it can in some cases become a serious electrical hazard even though power has been disconnected. *Capacitors can store charges that can be fatal to anyone coming in contact with them.* The code requires that circuits be provided with a means of discharge of capacitors on removal of voltage from the line.

Capacitance exists between any two conductors separated by an insulating dielectric. The conductors do not have to be plates; they may be wires or conductors of any shape. The dielectric may be air or any other material that is an insulator. In any two-wire cable there will be capacitance between the two wires. Capacitance will also exist between circuit wiring and a metal chassis, between conductors on a printed circuit board, or between the leads of component parts. Capacitance resulting from these and other unwanted sources is referred to as **stray capacitance.**

16.6 Capacitor Ratings, Connections, and Types

The number of electrons that a capacitor can store for a given applied voltage is a measure of its **capacitance.** A large capacitance means that more charge can be stored. The base unit used to measure capacitance is the **farad (F).** A capacitor has a capacitance of one farad when it stores a charge of one coulomb when a voltage of one volt is applied to it. The farad is a very large unit of capacitance, so prefixes are generally used to show the smaller values. Three prefixes (multipliers) are used: μ (micro), n (nano), and p (pico):

- μ means 10^{-6} (millionth), so 1,000,000 μF = 1 F
- n means 10^{-9} (thousand-millionth), so 1,000 nF = 1 μF
- p means 10^{-12} (million-millionth), so 1,000 pF = 1 nF

Capacitors are limited in the amount of electric charge they can store. If the value of the applied voltage and capacitance of the capacitor are known, the amount of charge stored in the capacitor can be calculated. The equation used is

$$Q = CE$$

where Q = the charge in coulombs
C = the capacitance in farads
E = the voltage in volts

EXAMPLE 16-8

Problem: A 500-μF capacitor is charged by a 100-VDC source. Determine the amount of charge stored in the capacitor.

Solution:

$$Q = CE$$
$$= (500 \text{ μF})(10^{-6})(100 \text{ V})$$
$$= 0.05 \text{ C (coulombs)}$$

The factors that determine the capacitance value are illustrated in Figure 16-21 and summarized as follows:

- **Area of the plates.** The greater the plate area, the higher the capacitance value.
- **Type of dielectric.** The better the dielectric material, the higher the capacitance value. The dielectric constant (K) of a material measures its effectiveness when used as the dielectric of a capacitor. Air is assumed to have a dielectric constant of 1, and all others dielectrics are compared to this standard.
- **Spacing between plates.** The closer the plates, the higher the capacitance value.

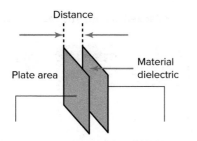

Figure 16-21 **Factors that determine the capacitance.**

Figure 16-22 **Capacitors are rated for both capacitance and voltage.**
©West Florida Components

Capacitors are rated for both capacitance and **voltage,** as shown in Figure 16-22. The voltage rating of a capacitor indicates the maximum voltage that can be safely applied to its plates and depends on the insulating strength of its dielectric. Voltages in excess of this value may break down the insulating dielectric material and permanently damage the capacitor. Always use a capacitor that has a voltage rating **higher** than the voltage of the circuit. A 50-V capacitor can be used on a 10-V or 25-V circuit, but a capacitor rated for only 25 V cannot be used on a 50-V circuit.

When capacitors are connected in **parallel,** the effective plate area is increased. Since capacitance is proportional to plate area, the capacitance will also increase. Capacitors are connected in parallel to obtain a greater total capacitance than is available in one unit. The total capacitance of parallel-connected capacitors is equal to the sum of all the individual capacitances connected in parallel:

$$C_T = C_1 + C_2 + C_3 \cdots$$

The largest voltage that can be applied safely to a group of capacitors in parallel can be determined easily. It is the voltage that can be applied safely to the capacitor having the **lowest** voltage rating.

EXAMPLE 16-9

Problem: Determine the total capacitance and maximum voltage rating for the parallel capacitance circuit of Figure 16-23.

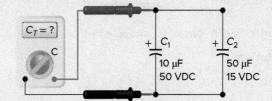

Figure 16-23 **Circuit for example 16-9.**

Solution:

$$C_T = C_1 + C_2$$
$$= 10 \text{ μF} + 50 \text{ μF}$$
$$= 60 \text{ μF}$$

Maximum voltage rating = 15 VDC

Capacitors are connected in **series** to enable the group to withstand a higher voltage than any individual capacitor is rated for. The voltage rating of a series-connected group of capacitors is equal to the **sum** of the individual capacitor voltage ratings. This increased voltage rating is accomplished at the expense of decreased total capacitance. The reason for this is that connecting capacitors in series effectively increases the distance between the plates, thereby reducing the total capacitance of the circuit. The formulas that can be used for calculating the total capacitance of capacitors in series is similar to that used to calculate the total resistance of resistors connected in parallel:

General formula for series capacitors

$$C_T = \frac{1}{\frac{1}{C_1} + \frac{1}{C_2} + \frac{1}{C_3} \cdots}$$

Formula for two capacitors in series

$$C_T = \frac{C_1 \times C_2}{C_1 + C_2}$$

Formula for n equal capacitors in series

$$C_T = \frac{C}{n}$$

EXAMPLE 16-10

Problem: Determine the total capacitance and maximum voltage rating for the series capacitance circuit of Figure 16-24.

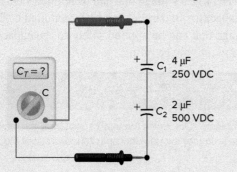

Figure 16-24 Circuit for example 16-10.

Solution:

$$C_T = \frac{C_1 \times C_2}{C_1 + C_2}$$

$$= \frac{4\ \mu F \times 2\ \mu F}{4\ \mu F + 2\ \mu F}$$

$$= 1.3\ \mu F$$

Maximum voltage rating $= 250\ V + 500\ V$

$$= 750\ V$$

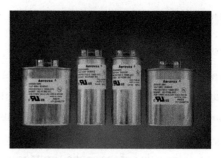

Figure 16-25 Oil-filled capacitors.
©Aerovox Corp

Capacitors are available in a wide variety of shapes, sizes, and ratings for different types of applications.

- **Oil-filled capacitors,** of the type shown in Figure 16-25, are used in AC circuits such as air-conditioners, motors, refrigerators, light appliances, and compressors to improve torque.

- **Polarized electrolytic** capacitors, of the type shown in Figure 16-26, are connected to DC circuits only. Electrolytic capacitors contain an electrolyte that makes it possible to produce high-capacity capacitors (1 μF and higher) that are relatively small in size. You can recognize a polarized capacitor by their polarity marking (negative or positive) symbols. The polarization of the capacitor must match the polarity of the voltage source. They must be installed with their positive terminal connected to the more positive voltage in the circuit. *If a polarized electrolytic capacitor is connected backward in a circuit, it will become extremely hot and can explode!*

- **Nonpolarized** capacitors, of the type shown in Figure 16-27, are small-value capacitors (up to 1 μF) that may be connected either way. It can be difficult to find the values of these small capacitors because there are many types of them and several different labeling systems. For example, a number code is often used on small capacitors where printing is difficult.

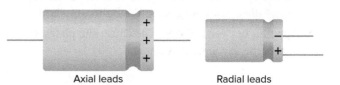

Axial leads   Radial leads

Figure 16-26 Polarized electrolytic capacitors.

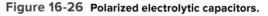

Figure 16-27 Nonpolarized capacitors.

16.7 *RC* Time Constant

When DC is applied directly to a capacitor, it charges almost instantly to the value of the source voltage. The time required to charge a capacitor can be controlled by connecting a resistor in series with the capacitor, as illustrated in Figure 16-28. With the resistance value set to zero, the capacitor charges almost instantaneously. With the resistance value set to 1 kΩ, the capacitor charges in a short time period. With the resistance value set to 10 kΩ, the capacitor charges in a longer time period with the same capacitor.

The charging rate of a resistor and capacitor connected in series is called the ***RC*** **time constant** and is the product of the resistance and the capacitance:

$$t = RC$$

where t = time constant in seconds
 R = resistance in MΩ
 C = capacitance in μF

Capacitors charge and discharge at an exponential rate. Each time constant represents the time, in seconds, it takes to charge a capacitor to 63.2 percent of its applied voltage value, as illustrated in Figure 16-29. It requires approximately five time constant periods for the capacitor to reach the applied voltage value. The discharge rate of a charged capacitor connected in series with a resistor is also predictable. The same *RC* time constant formula is used, and the only difference is that the capacitor loses 63.2 percent of its charge in each time constant. This characteristic makes a capacitor very useful in timing circuits.

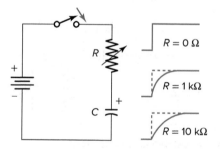

Figure 16-28 **Varying the charging rate of a capacitor.**

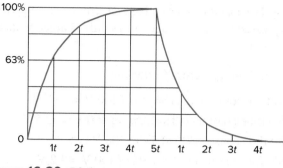

Figure 16-29 *RC* **time constants.**

Figure 16-30 illustrates the voltage waveforms that occur when a capacitor is charged from 0 to 100 volts and discharged from 100 to 0 volts. In this example we are using a resistance of 0.5 MΩ and a capacitance of 10 μF, which means the time constant, is equal to

$$t = RC$$
$$= 0.5 \text{ M}\Omega \times 10 \text{ μF}$$
$$= 5 \text{ s}$$

The waveforms are divided into 5 time constants. During the charge portion, each time constant will see the voltage across the capacitor experience a change equal to 63.2 percent of the amount left to reach the fully charged state as follows. The value of the voltage across the capacitor at the end of each time constant period is expressed as a percentage of the DC source voltage as follows:

Interval Voltage across Capacitor

1st time constant (5 s) 63.2% of 100 V = 63.2 V
2nd time constant (10 s) 86.5% of 100 V = 86.5 V
3rd time constant (15 s) 95% of 100 V = 95 V
4th time constant (20 s) 98.2% of 100 V = 98.2 V
5th time constant (25 s) 99.3% of 100 V = 99.3 V

Likewise, the capacitor discharges in a similar manner. The 10-μF capacitor, which charged to 100 V in 25 s (5 time constant periods), will discharge from 100 V to 0 in the same amount of time. The value of the voltage

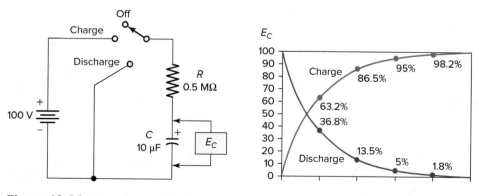

Figure 16-30 **Charging and discharging time constants.**

across the capacitor at the end of each time constant period is expressed as a percentage of the fully charged voltage as follows:

Interval Voltage across Capacitor

1st time constant (5 s) 36.8% of 100 V = 36.8 V
2nd time constant (10 s) 13.5% of 100 V = 13.5 V
3rd time constant (15 s) 5% of 100 V = 5 V
4th time constant (20 s) 1.8% of 100 V = 1.8 V
5th time constant (25 s) 0.7% of 100 V = 0.7 V

If a capacitor is not intentionally discharged either directly or through a load, it is not possible for a charged capacitor to remain charged for an indefinite period of time. Since no dielectric is a perfect insulator, electrons eventually move through the dielectric from the negative plate to the positive plate, causing the capacitor to discharge.

EXAMPLE 16-11

Problem: A 20-μF capacitor is connected in series with a 100-kΩ resistor and operated from a 12-VDC source, as shown in Figure 16-31.

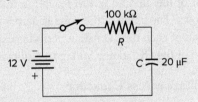

Figure 16-31 Circuit for example 16-11.

a. Calculate the *RC* time constant for the circuit.
b. When switch in initially closed, approximately how long would it take for the capacitor to fully charge to 12 volts?
c. If the fully charged capacitor is discharged through the 100-kΩ resistor, what would the value of the voltage across the capacitor be at the end of the second time constant period?

Solution:
a. $t = RC$
$$= \frac{100 \text{ k}\Omega \times 20 \text{ μF}}{1,000}$$
$$= 0.1 \text{ M}\Omega \times 20 \text{ μF}$$
$$= 2 \text{ s}$$

b. To completely charged:
$$\text{Time} = 5 \text{ time constant periods}$$
$$= 5 \times 2 \text{ s}$$
$$= 10 \text{ s}$$

c.
$$E_C = 13.5\% \text{ of } 12 \text{ V}$$
$$= 1.62 \text{ V}$$

16.8 Capacitive Reactance

Current flow to a capacitor only occurs during the time the capacitor is charging or discharging. In a DC circuit, capacitance prevents further current flow once the initial charging current has charged the capacitor. Capacitors are, therefore, said to **block DC,** as illustrated in Figure 16-32. When the switch is initially closed, the current flows to the capacitor to charge it. Once fully charged, the capacitor voltage will be equal to that of the source thus preventing any additional flow of current through the circuit. As a result, no voltage is developed across the load. In effect the capacitor in this application (once it is charged) provides DC blocking or isolation between the source and the load.

When a capacitor is connected to an AC power supply, the charges on the plates reverse with each change of the applied voltage polarity. The plates then are alternately charged and discharged. This results in a constant AC current flow, as illustrated in Figure 16-33. Again, electrons flow in and out of the plates through the external circuit. The circuit current does not pass through the dielectric but only appears to. This is why we sometimes say *capacitors pass AC and block DC.*

Similar to inductors, capacitors oppose current flow in an AC circuit. Different-size capacitors offer a different amount of opposition. The opposition to the flow of AC current offered by a capacitor is called capacitive reactance. **Capacitive reactance** is measured in **ohms** and is represented by the symbol X_C.

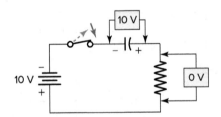

Figure 16-32 DC blocking capacitor.

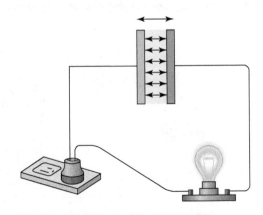

Figure 16-33 Capacitor passes AC current.

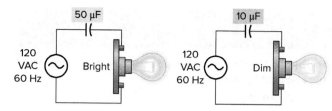

Figure 16-34 Effect of capacitance on current flow.

The *capacitive reactance of a capacitor is inversely proportional to the capacitance* of the capacitor. That is to say, the larger the capacitance (C) of the capacitor, the lower is its capacitive reactance (X_C) or opposition to AC current flow. This fact is illustrated by the circuits of Figure 16-34 and summarized as follows:

- The 10-µF and 50-µF capacitors are connected in series with a lamp to an AC source of the same voltage and frequency.
- The larger 50-µF capacitor can store more charge and discharge more current.
- As a result, compared to the 10-µF circuit, the current flow and the brightness of the lamp are greater for the 50-µF circuit.

The *capacitive reactance of a capacitor is inversely proportional to the frequency* of the AC supply source. For any given capacitor, increasing the frequency of the voltage source increases the rate at which the capacitor charges and discharges. This results in less capacitive reactance and a greater AC current flow. This fact is illustrated by the circuits of Figure 16-35 and summarized as follows:

- The capacitor and lamp are connected in series and operated from a 60-Hz and a 600-Hz AC source.
- When the frequency is increased from 60 Hz to 600 Hz, with the same capacitance in the circuit, the current flow and brightness of lamp increases.
- This is a direct result of a *decrease* in capacitive reactance due to the *increase* in frequency from 60 Hz to 600 Hz.

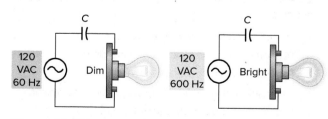

Figure 16-35 Effect of frequency on current flow.

The capacitive reactance of a capacitor can be calculated using the formula

$$X_C = \frac{1}{2\pi f C}$$

where X_C = capacitive reactance in ohms (Ω)
f = frequency in hertz (Hz)
C = capacitance in farads (F)

Unlike the resistor, which dissipates energy, ideal capacitors and inductors **store energy** rather than dissipating it. In AC circuits that contain only ideal capacitors, the capacitive reactance is the only thing that limits the current and will replace resistance in the Ohm's law equation as follows:

$$I = \frac{E}{R} \quad \text{(purely resistive circuit)}$$

$$I = \frac{E}{X_C} \quad \text{(purely capacitive circuit)}$$

EXAMPLE 16-12

Problem: A 100-µF capacitor is connected to a 120-V, 60-Hz source.

   a. Determine the capacitive reactance of the capacitor.

   b. What is the value of the current flow for this circuit?

Solution:

a. $X_C = \dfrac{1}{2\pi f C}$

$= \dfrac{1}{2 \times 3.14 \times 60 \times (100 \times 10^{-6})}$

$= 26.5\ \Omega$

b. $I = \dfrac{E}{X_C}$

$= \dfrac{120\ \text{V}}{26.5\ \Omega}$

$= 4.53\ \text{A}$

When capacitors are connected in series, the total capacitance is less than the smallest capacitance value because the effective distance between the plates increases. The formula for total **series capacitance** is similar to the formula for total resistance of **parallel resistors**:

$$C_T = \frac{C}{n} \quad \text{(same-value formula)}$$

$$C_T = \frac{C_1 \times C_2}{C_1 + C_2} \quad \text{(product-over-sum formula)}$$

$$C_T = \frac{1}{\dfrac{1}{C_1} + \dfrac{1}{C_2} + \dfrac{1}{C_3}} \quad \text{(reciprocal formula)}$$

EXAMPLE 16-13

Problem: Determine the current flow for the series-connected capacitor circuit of Figure 16-36.

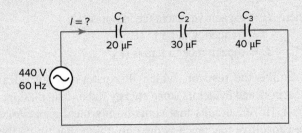

Figure 16-36 Circuit for example 16-13.

Solution:

$$C_T = \cfrac{1}{\cfrac{1}{C_1} + \cfrac{1}{C_2} + \cfrac{1}{C_3}}$$

$$= \cfrac{1}{\cfrac{1}{20} + \cfrac{1}{30} + \cfrac{1}{40}}$$

$$= 9.23 \ \mu F$$

$$X_{C(total)} = \frac{1}{2\pi f C}$$

$$= \frac{1}{2 \times 3.14 \times 60 \times (9.23 \times 10^{-6})}$$

$$= 288 \ \Omega$$

$$I = \frac{E}{X_{C(total)}}$$

$$= \frac{440 \ V}{288 \ \Omega}$$

$$= 1.53 \ A$$

The voltage drop across each capacitor in a series connection depends on its capacitive reactance and current value:

$$E_C = I_C \times X_C$$

Since the current is the same at any point in a series circuit, the largest-value capacitor in series (lowest X_C) will have the smallest voltage drop, and the smallest capacitance value (highest X_C) will have the largest voltage drop. The voltage drop across any individual capacitor in series can be determined directly using the formula

$$E_X = \frac{C_T}{C_X} \times E_S$$

where C_X = any capacitor in series, such as $C_1, C_2, C_3 \cdots$
E_X = the voltage across capacitor C_X
C_T = the total capacitance of the series circuit
E_S = the voltage of the source

EXAMPLE 16-14

Problem: Find the voltage drop across each capacitor for the circuit of Figure 16-37.

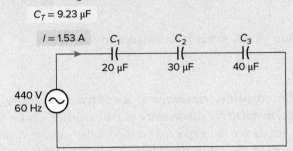
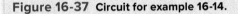

Figure 16-37 Circuit for example 16-14.

Solution:

$$E_{C_1} = \frac{C_T}{C_1} \times E_S$$

$$= \frac{9.23}{20} \times 440$$

$$= 203 \ V$$

$$E_{C_2} = \frac{C_T}{C_2} \times E_S$$

$$= \frac{9.23}{30} \times 440$$

$$= 135 \ V$$

$$E_{C_3} = \frac{C_T}{C_3} \times E_S$$

$$= \frac{9.23}{40} \times 440$$

$$= 102 \ V$$

When capacitors are connected in parallel, the total capacitance is the sum of the individual capacitances because the effective plate area increases. The formula for total **parallel capacitance** is similar to the formula for total resistance of **series resistors:**

$$C_T = C_1 + C_2 + C_3 \cdots \qquad \text{(for capacitors in parallel)}$$

EXAMPLE 16-15

Problem: For the parallel capacitor circuit of Figure 16-38, determine:

   a. The total capacitance of the circuit.

   b. The total capacitive reactance of the circuit.

   c. The total current flow of the circuit.

   d. The value of the voltage drop across each capacitor.

   e. The value of the current passed by C_3.

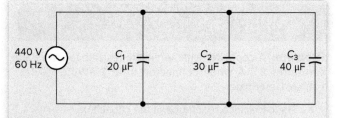

Figure 16-38 Circuit for example 16-15.

Solution:

a. $C_T = C_1 + C_2 + C_3$
$= 20\ \mu F + 30\ \mu F + 40\ \mu F$
$= 90\ \mu F$

b. $X_{C(total)} = \dfrac{1}{2\pi fC}$

$= \dfrac{1}{2 \times 3.14 \times 60 \times (90 \times 10^{-6})}$

$= 29.5\ \Omega$

c. $I = \dfrac{E}{X_{C(total)}}$

$= \dfrac{440\ V}{29.5\ \Omega}$

$= 14.9\ A$

d. The voltage drop across each component in any parallel circuit is equal to the voltage of the source. Therefore the voltage drop across each capacitor is 440 V.

e. $X_{C_3} = \dfrac{1}{2\pi fC_3}$

$= \dfrac{1}{2 \times 3.14 \times 60 \times (40 \times 10^{-6})}$

$= 66.3\ \Omega$

f. $I_{C_3} = \dfrac{E}{X_{C_3}}$

$= \dfrac{440\ V}{66.3\ \Omega}$

$= 6.64\ A$

16.9 Phase Shift in a Capacitor

Like inductors, capacitors cause the voltage and current to be out of phase. However, in the case of capacitors, the current leads the voltage. This can be explained by the fact that when a DC voltage is first applied to a capacitor, the current flow is maximum and then tapers off as the voltage across the capacitor increases. In other words, current is leading voltage.

Figure 16-39 illustrates the phase relationship between current flow in the external circuit and the voltage across the capacitor when an AC source is applied. The *current leads the voltage in the ideal capacitor by exactly 90 degrees.* This is so because the maximum value of current

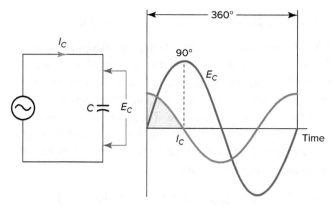

Figure 16-39 Current leads the voltage in a capacitor.

corresponds to the time when the capacitor is fully discharged (has zero volts across it). When the capacitor is fully charged, current flow stops or is at zero. Since an AC source is used, the voltage applied to the capacitor is constantly causing the capacitor to charge in one direction and then discharge in the other.

The phase shift in a capacitor is exactly opposite to the phase shift in an inductor. *Current leads voltage by 90 degrees in a capacitor and lags the voltage by 90 degrees in an inductor.* In a series circuit, the capacitor and inductor **voltages** will be 180 degrees out of phase with each other, while in a parallel circuit the capacitor and inductor **currents** will be 180 degrees out of phase with each other. In the case of the series circuit, the inductor and capacitor voltages act to directly oppose each other, as illustrated in Figure 16-40. In the case of the parallel circuit, the inductor and capacitor currents act to directly oppose each other, as illustrated in Figure 16-41.

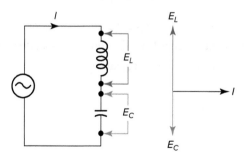

Figure 16-40 Voltage phase relationships in a series *LC* circuit.

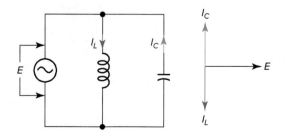

Figure 16-41 Current phase relationships in a parallel *LC* circuit.

16.10 Capacitive Reactive Power

Like the inductor, *a capacitor does not dissipate energy; it only stores energy temporarily.* Due to the 90-degree phase shift, the voltage and current have the same polarity for half the time during one cycle and have opposite polarities the other half of the time. The power wave is plotted by multiplying voltage by current at every instant through 360 degrees, as shown in Figure 16-42. The circuit can be analyzed in steps as follows:

- When the voltage and current have the same polarity, energy is stored in the electrostatic field of the capacitor.
- When the voltage and current have opposite polarities, the stored energy is discharged back into the circuit.
- If the values of current *times* voltage for one full cycle are added, the sum equals zero, just as it is with pure inductive circuits.
- *Therefore, there is no true power, or watts, produced in a purely capacitive circuit even though there is both voltage and current present.*

The power associated with a capacitor is **reactive power** and is measured in **VARs,** just as for an inductor. To calculate the capacitive reactive power apply the formula

$$VARs = E_C \times I_C$$

or

$$VARs = I_C^2 \times X_C$$

where E_C = the voltage applied across the capacitor
I_C = the current flow through the capacitor
X_C = the capacitive reactance

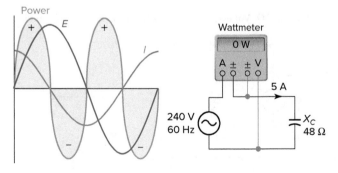

Figure 16-42 Power in a purely capacitive AC circuit.

EXAMPLE 16-16

Problem: A capacitor with negligible resistance draws a current of 2.5 A when connected to a 120-VAC, 60-Hz source. Determine:

  a. The capacitive reactance of the capacitor.
  b. The capacitive reactive power.
  c. The true power or wattage.

Solution:

a.
$$X_C = \frac{E_C}{I_C}$$
$$= \frac{120 \text{ V}}{2.5 \text{ A}}$$
$$= 48 \ \Omega$$

b.
$$VARs = E_C \times I_C$$
$$= 120 \text{ V} \times 2.5 \text{ A}$$
$$= 300 \text{ VARs}$$

or

$$VARs = I_C^2 \times X_C$$
$$= 2.5 \times 2.5 \times 48$$
$$= 300 \text{ VARs}$$

c. The true power or wattage would be zero.

For a pure inductance, the current lags the voltage by 90 degrees, and the reactive power in VARs is considered to be **positive.** However, for a pure capacitance, the current leads the voltage by 90 degrees. Therefore, the reactive power in VARs for a pure capacitance will be opposite to that for a pure inductance. Hence, the reactive power for a pure capacitance is considered to be **negative.** An inductance is often said to **consume VARs,** while a capacitance is said to **produce VARs.**

16.11 Troubleshooting Inductors and Capacitors

Inductors almost always fail due to **open-circuit or short-circuit** conditions. One method of determining if an inductor is open or shorted is to measure its DC resistance. Resistance values depend on the wire size (diameter) and number of turns (length) of the wire that makes up the inductor. Some coils with fine wire and a large number of turns will have hundreds of ohms of resistance; large coils with large wire and a small number of turns will have tens of ohms of resistance. *If no resistance is measured at all, the coil is open.*

The test procedure for capacitors is significantly different from the procedure used to test inductors. A very important consideration when troubleshooting capacitors is the **dielectric**

leakage current. Since a perfect insulator does not exist, a certain amount of current trickles through the dielectric even in the best of capacitors. Normally leakage current is so small that it is hardly measurable and has no effect on the operation of the circuit. However, as a capacitor ages, its dielectric resistance may decrease, resulting in high values of leakage current. High leakage current can alter the normal operation of a circuit because the faulty capacitor is now providing a DC current path instead of having infinite resistance.

A quick way of determining if a capacitor is open or shorted is to measure its DC resistance. If the capacitor is shorted or there is leakage in the dielectric, the ohmmeter will read near zero ohms. If checking by measuring capacitance, as illustrated in Figure 16-43, the capacitance value will change as the dielectric deteriorates.

One of the important code requirements for capacitor installations is that the stored charge of a capacitor must be drained by a discharge circuit either permanently connected to the capacitor or automatically connected when the line voltage of the capacitor is removed. Figure 16-44 shows one method of safely discharging a capacitor out of circuit. This technique uses a low-resistance, high-wattage resistor with a short enough time constant that will drop the voltage to a safe value in a few seconds.

Figure 16-43 **Measuring capacitance.**
©Fluke Corporation

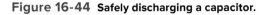

Figure 16-44 **Safely discharging a capacitor.**

1. Define *capacitance*.

2. Compare the manner in which energy is stored in an inductor and a capacitor.

3. When is the only time that current will flow in a capacitor circuit that is connected to a DC source? Why?

4. List four common practical applications for capacitors.

5. What important safety aspect of working with circuits that contains capacitors is covered in the National Electrical Code?

6. What is the capacitance that exists between wires of a two-conductor cable called?

7. What is the base unit used to measure capacitance?

8. List three factors that determine the capacitance of a capacitor.

9. What factor determines the voltage rating of a capacitor?

10. How are two 50-µF capacitors connected to provide a total capacitance of 100 µF?

11. Determine the total capacitance and maximum voltage rating for two 220-µF, 300-V capacitors connected in series.

12. With regards to electrolytic capacitors:
    a. What is their main advantage over other types?
    b. When connecting these polarized capacitors into DC circuits, what polarity rule must be followed?
    c. What can happen if you get the polarity wrong?

13. For a 25-kΩ resistor connected in series with a 1,000-µF capacitor and operated from a 12-VDC source:
    a. Calculate the *RC* time constant.
    b. When voltage is first applied to this circuit, approximately how long will it take for the voltage across the capacitor to reach 12 volts?
    c. If the fully charged capacitor is discharged through the 25-kΩ resistor, what would the value of the voltage across the capacitor be after the first 25 seconds of discharge?

14. Explain how a capacitor affects the flow of current in a DC circuit.

15. Explain how a capacitor affects the flow of current in an AC circuit.

16. Define *capacitive reactance*.

17. What is the base unit used to measure capacitive reactance?

18. In what way is capacitive reactance affected by capacitance?

19. In what way is capacitive reactance affected by frequency?

20. An AC voltage of 240 volts at a frequency of 60 Hz is applied to a 50-μF capacitor. Determine the amount of AC current flow in the circuit.

21. Why is the negative effect of stray capacitance more pronounced in very high frequency circuits?

22. A 60-μF and a 90-μF capacitor are connected in series to a 120-V, 60-Hz supply.
    a. Determine the total current flow for the circuit.
    b. Repeat for the two capacitors connected in parallel.

23. A 10-μF (C_1) and a 15-μF (C_2) capacitor are connected in series to a 230-V, 60-Hz source. Determine the value of the voltage drop across each capacitor.

24. A 20-μF (C_1) and a 40-μF (C_2) capacitor are connected in parallel to a 480-V, 60-Hz source.
    a. What is the inductive reactance and current flow through C_1?
    b. What is the inductive reactance and current flow through C_2?

25. What is the phase relationship between the current through and the voltage across an ideal capacitor?

26. What is the voltage phase relationship between an inductor and capacitor connected in series to an AC source?

27. What is the current phase relationship between an inductor and capacitor connected in parallel to an AC source?

28. Explain why there is no true power, or watts, produced in a purely capacitive circuit even though there is both voltage and current present.

29. What is the type of power associated with a capacitor called and in what unit is it measured?

30. A power factor correction capacitor bank, with negligible resistance, draws a current of 625 amperes when connected to its 480-VAC, 60-Hz source. Calculate the capacitive reactive power of the capacitor bank.

31. Specifications for a preassembled capacitor power factor correction unit states that this equipment comes with a built-in automatic discharge device. Explain what this means?

32. Compare how energy is stored or dissipated in a resistor with an ideal capacitor.

CHAPTER SEVENTEEN

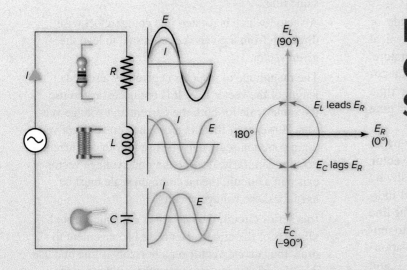

Resistive, Inductive, Capacitive (*RLC*) Series Circuits

Voltage and current relationship in a series *RLC* circuit.

LEARNING OUTCOMES

▶ Plot and interpret vector quantities.

▶ Solve for unknown quantities in a series *RL* circuit.

▶ Solve for unknown quantities in a series *RC* circuit.

▶ Solve for unknown quantities in a series *LC* circuit.

▶ Solve for unknown quantities in a series *RLC* circuit.

▶ Understand the relationship between resistance, reactance, and impedance.

▶ Understand the relationship between true power, reactance power, apparent power, and power factor.

Resistance, inductance, and capacitance are the three basic circuit properties that we use to control voltages and currents in AC electric circuits. In previous chapters you were shown how to analyze AC circuits that contain resistance, inductance or capacitance only. Most AC circuits contain two or more of these basic circuit properties. In this chapter, you will learn how to analyze and solve AC circuits that contain series combinations of resistance and inductance (*RL*); resistance and capacitance (*RC*); inductance and capacitance (*LC*); and resistance, capacitance, and inductance (*RLC*). This chapter also discusses impedance (*Z*), power factor (PF), and resonance as applied to series *RLC* circuits.

PART 1 VECTORS

17.1 Vector Diagrams

Any quantity that is defined only by its magnitude or amount is called a **scalar** quantity in mathematics. Typical examples of scalar quantities are time, speed, temperature, and volume. A scalar quantity has no directional component, only magnitude. For example, the units for time (minutes, days, hours, etc.) represent an amount of time only and tell nothing of direction.

Any quantity that needs to be defined not only by its magnitude but also by its direction is called a **vector** quantity. Vector quantities are represented graphically by an arrowhead at the end of a line. The arrowhead indicates the **direction** of the vector and the length of the line its **magnitude.** In electrical work, vectors are used to represent values of current and voltage in angular relationship to each other, as illustrated in Figure 17-1 and summarized as follows:

- From 0 degrees vectors are assumed to be rotating in a counterclockwise direction.

- The angle of the vector represents the phase shift in degrees between the waveforms in question.
- When two waveforms are in phase, they have the same direction and so their vectors are drawn on the same line.
- A vector, which is rotated in a counterclockwise direction from a given vector, is said to lead the given vector.
- The magnitude of a vector is given by a scaled length of the vector line. It is not necessary to use the same scale for both the current and voltage vectors. However, if there is more than one current vector, a common scale must be used for these current vectors. Similarly, if more than one voltage vector exists in a circuit, then a common scale must be used for these voltage vectors.
- In a series circuit, the current is constant through all parts of the circuit. Hence, it is convenient to draw the current vector on a horizontal line and use it as the reference vector for other vectors in the same diagram.
- In a parallel circuit, the voltage is the same across parallel branches. Therefore, it is convenient to draw the voltage vector on a horizontal line and use it as the reference vector for other vectors in the same diagram.

17.2 Combining Vectors

Vectors can be combined to obtain a single **resultant.** They can be added, subtracted, multiplied, and divided. The addition of vectors is different from the addition of scalar quantities. When adding vectors, both the **magnitude and direction** of the vector must be taken into account, whereas in scalar addition only the magnitude of the numbers is used.

When **in-phase** AC voltage sources are connected in series, their voltages **add** like DC battery voltages, as illustrated in Figure 17-2. Note the positive (+) and negative (−) polarity marks next to the leads of the two AC sources. Even though AC doesn't have polarity in the same sense that DC does, these marks are essential in order to relate multiple AC voltages at different phase angles to each other. The net output is a voltage of 50 volts at 0 degrees.

If directly **opposing** AC voltage sources (180 degrees out of phase) are connected in series, their voltages **subtract** similar to DC batteries connected in an opposing fashion, as illustrated in Figure 17-3. Determining whether or not these voltage sources are opposing each other requires an examination of their polarity markings and their phase angles. The net output is a voltage of 10 volts at 180 degrees.

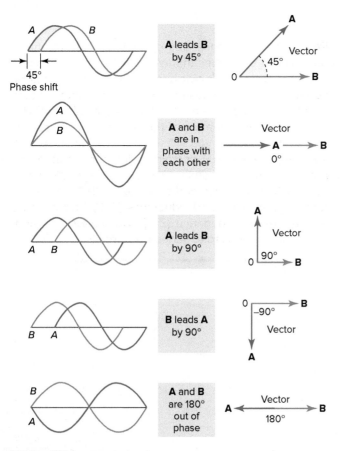

Figure 17-1 Vector diagrams.

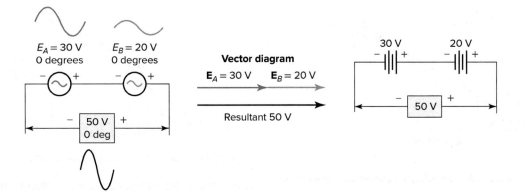

Figure 17-2 **Adding in-phase voltages.**

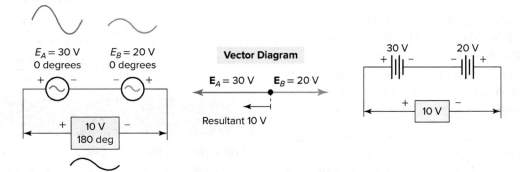

Figure 17-3 **Directly opposing AC voltages.**

EXAMPLE 17-1

Problem: A series circuit contains three voltage drops ($E_1 =$ 60 V; $E_2 = 80$ V; and $E_3 = 30$ V), and E_1 is in phase with E_2 and 180 degrees out of phase with E_3. What is the resultant vector sum of these three voltages?

Solution:

$$E_{resultant} = E_1 + E_2 - E_3$$
$$= 60 \text{ V} + 80 \text{ V} - 30 \text{ V}$$
$$= 110 \text{ V}$$

Vectors that have directions other than in phase or 180 degrees out of phase with each other can also be added. One manner of addition is the **triangular** method, shown in Figure 17-4, and which is applied as follows:

- The two AC voltages, E_A and E_B, have amplitudes of 20 V and 16 V respectively.
- Voltage E_A lags voltage E_B by 30 degrees.
- The vector sum may be obtained by drawing each of the vectors to length with respect to a convenient scale.

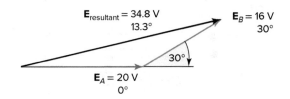

Figure 17-4 **Vector addition using the triangular method.**

- The first vector is started from the zero or the origin point and the second vector is then drawn starting at the end of the first vector.
- The resultant voltage is the line that closes the figure.
- The resultant voltage is 34.8 volts at 13.3 degrees.

In the **parallelogram** method for vector addition, the vectors are moved to a common origin and a parallelogram is constructed to determine the resultant sum. A parallelogram is a four-sided figure whose opposite sides are parallel to each other. Figure 17-5 illustrates how the parallelogram method is used to find the resultant and is applied as follows:

- The two 120-volt AC voltage sources (E_A and E_B) are 60 degrees out of phase with each other.
- Draw the first vector (E_A) at 0 degrees.

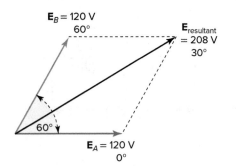

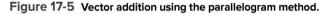

Figure 17-5 Vector addition using the parallelogram method.

- Draw the second vector (E_B), with its tail at the tail of the first vector.
- Make a parallelogram by drawing two additional sides, each passing through the head of one of the vectors and parallel to the other vector.
- The resultant sum is drawn along the diagonal from the common tail to the intersection of the two lines.
- The resultant sum is 208 volts at 30 degrees.

When two vectors have a phase difference of exactly 90 degrees (a right angle), the sum of the two vectors can be most easily found using the **Pythagorean theorem.** If you know the lengths of any two sides of a right-angle triangle (Figure 17-6), you can find the length of the third side by using one of the following Pythagorean theorem equations:

$$C = \sqrt{A^2 + B^2}$$
$$A = \sqrt{C^2 - B^2}$$
$$B = \sqrt{C^2 - A^2}$$

Generally speaking, trigonometry is a branch of mathematics that involves the relationship between the angles and sides of a triangle. The relationship between an angle and side ratios in a right triangle is one of the most important trigonometry applications. In order to define these relationships, we must give them a certain reference, which is the **angle theta (θ),** shown in Figure 17-7. There are a total of six relationships. The three most popular are

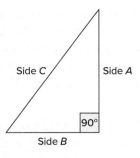

Figure 17-6 Right-angle triangle.

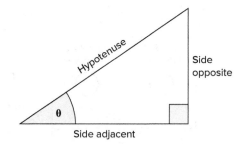

Figure 17-7 Basic trigonometry functions.

expressed as functions, called **sine, cosine,** and **tangent.** These functions are expressed as follows:

$$\text{sine } \theta = \frac{\text{side opposite}}{\text{hypotenuse}}$$

$$\text{cosine } \theta = \frac{\text{side adjacent}}{\text{hypotenuse}}$$

$$\text{tangent } \theta = \frac{\text{side opposite}}{\text{side adjacent}}$$

EXAMPLE 17-2

Problem: For the vector diagram shown in Figure 17-8:

a. Apply the Pythagorean theorem to find the resultant voltage sum of the two voltages E_1 and E_2.

b. Use a scientific calculator to determine the angle theta (θ) between E_1 and the resultant voltage sum.

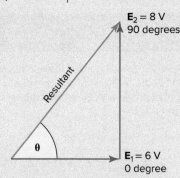

Figure 17-8 Vector diagram for example 17-2.

Solution:

a.
$$E_{\text{resultant}} = \sqrt{E_1^2 + E_2^2}$$
$$= \sqrt{6^2 + 8^2}$$
$$= \sqrt{100}$$
$$= 10 \text{ V}$$

b.
$$\text{Sine } \theta = \frac{\text{side opposite}}{\text{hypotenuse}}$$
$$= \frac{8}{10}$$
$$= 0.8$$
$$\text{Angle} = \text{sine } 0.8$$
$$= 53.13°$$

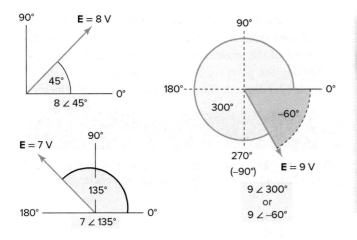

Figure 17-9 Vectors represented in polar notation.

A **phasor** is a line whose direction represents the phase angle in electrical degrees and whose length represents the magnitude of the electrical quantity. From the definition of a phasor, you can see that phasors and vectors are very similar. In electricity the term *phasor* is used to emphasize the fact that the position of the line changes with time as the sine wave rotates through its cycle from 0 to 360 degrees.

A **complex number** is a number made up of a real part and an imaginary part. Alternating current is often represented by a complex number. There are two basic forms of complex number notation: polar and rectangular. Using **polar notation,** the value of a vector is expressed as a magnitude followed by an angle. Figure 17-9 shows examples of vectors expressed in polar notation. Standard orientation for vector angles in AC circuit calculations defines 0 degrees as being to the right (horizontal), making 90 degrees straight up, 180 degrees to the left, and 270 degrees straight down. Note that vectors **angled down** can have angles represented in polar form as positive numbers in excess of 180 degrees, or negative numbers less than 180 degrees.

Complex numbers in polar notation are used for carrying out multiplication and division operations. When multiplying one polar value by another, the result equals the product of the magnitudes followed by the sum of the angles. Similarly when dividing one polar value by another, the result equals the quotient of the magnitudes followed by the difference between the angles.

EXAMPLE 17-3

Problem: Determine the product of $8 \angle 35°$ and $12 \angle 45°$.

Solution:

$$(8 \angle 35°)(12 \angle 45°) = (8 \times 12) \angle (35° + 45°)$$
$$= 96 \angle 80°$$

EXAMPLE 17-4

Problem: Determine the quotient of the following complex number fraction:

$$\frac{50 \angle 60°}{10 \angle 25°}$$

Solution:

$$\frac{50 \angle 60°}{10 \angle 25°} = \frac{50}{10} \angle (60° - 25°)$$
$$= 5 \angle 35°$$

Using **rectangular notation,** the value of a vector is expressed by its respective horizontal and vertical components. Basically, the angled vector is taken to be the hypotenuse of a right triangle, described by the lengths of the adjacent and opposite sides. These two-dimensional figures (horizontal and vertical) are symbolized by two numerical figures. In order to distinguish the horizontal and vertical dimensions from each other, the vertical is prefixed with a lowercase i (in pure mathematics) and j (in electric circuits).

The so-called **j operator** does not represent a real number, but rather is a mathematical operator used to distinguish the vector's vertical component from its horizontal component. As a complete complex number, the horizontal and vertical quantities are written as a sum. For example, 3 + j4 is a complex number including 3 units on the real axis added to 4 units 90 degrees out of phase on the j axis. Figure 17-10 shows a series of points on a complex

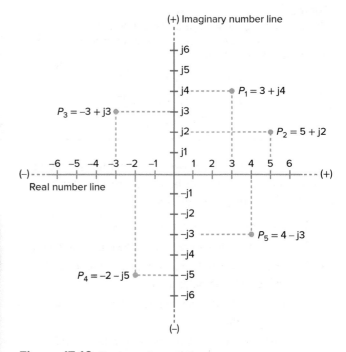

Figure 17-10 Rectangular notation.

number plane. Using the real and imaginary numbers given we can express each point in rectangular notation.

Complex numbers in rectangular notation are used for carrying out addition and subtraction operations. The sum of two complex numbers expressed in rectangular notation is equal to the sum of their real and imaginary parts. Similarly, their difference is equal to the difference between the separate real and imaginary parts.

EXAMPLE 17-5

Problem: Calculate the sum of two vectors having the following values:

$$E_1 = 4 + j6$$
$$E_2 = -7 + j4$$

Solution:
Since we are dealing with complex numbers in rectangular notation, the values can be added directly as follows:

E_1:	$4 + j6$
E_2:	$-7 + j4$
$(E_1 + E_2)$:	$-3 + j10$

EXAMPLE 17-6

Problem: Two vectors have the following values: $E_1 = 2 - j4$ and $E_2 = -2 + j5$. Calculate the value of $(E_1 - E_2)$.

Solution:
Since we are dealing with complex numbers in rectangular notation, the value of E_2 can be subtracted directly from the value of E_1 as follows:

E_1:	$2 - j4$
E_2:	$-(-2 + j5)$
$(E_1 - E_2)$:	$4 - j9$

Conversion between the two notational forms involves basic trigonometry and can be related graphically in the form of a right triangle as follows:

- The hypotenuse represents the vector itself.
- The hypotenuse length represents the polar magnitude.
- The angle between the horizontal side and magnitude represents the polar angle.
- The horizontal side (x) represents the rectangular *real* component.
- The vertical side (j) represents the rectangular *imaginary* component.

To convert from polar-to-rectangular notation, you find the real component by multiplying the polar magnitude by the cosine of the angle, and the imaginary component by multiplying the polar magnitude by the sine of the angle. Example 17-7 illustrates how this conversion can be achieved.

EXAMPLE 17-7

Problem: The vector, shown in Figure 17-11, has a value of $E_1 = 10 \angle 45°$. Convert this value to rectangular form.

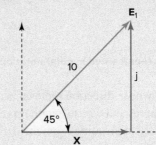

Figure 17-11 Vector diagram for example 17-7.

Solution:
Real component (x)   = (10) cosine 45°
                     = (10) (0.707)
                     = 7.07

Imaginary component (j) = (10) sine 45°
                        = (10) (0.707)
                        = 7.07

Therefore, $10 \angle 45°$ = 7.07 + j7.07

To convert from rectangular to polar, find the polar magnitude through the use of the Pythagorean theorem and the angle by taking the arctangent of the imaginary component divided by the real component. The arctan is the inverse of the tangent function. Example 17-8 illustrates how this conversion can be achieved.

EXAMPLE 17-8

Problem: Convert the rectangular number 4 + j3 to polar form.

Solution:
$$\text{Magnitude} = \sqrt{4^2 + 3^2}$$
$$= \sqrt{25}$$
$$= 5$$

Angle = arctan (3/4)
      = 36.9°

Therefore, rectangular number $4 + j3$ = polar number $5 \angle 36.9°$

1. Compare scalar and vector quantities.

2. What factors determine the length and angle of a vector representing a voltage waveform?

3. In which direction are vectors assumed to rotate?

4. In a series circuit, what circuit quantity is normally used as the reference vector? Why?

5. In a parallel circuit, what circuit quantity is normally used as the reference vector? Why?

6. Draw a vector diagram (to approximate scale) that could be used to represent each of the following electrical circuit conditions:
   a. A voltage of 10 V, in phase with a voltage of 15 V.
   b. A current of 12 A, 180 degrees out of phase with a current of 6 A.
   c. A current of 4 A, leading a current of 12 A by 45 degrees.
   d. A voltage of 120 V, lagging a voltage of 240 V by 90 degrees.

7. Find the vector sum of a current of 3 A and a current of 4 A, the two currents being in phase.

8. A voltage E_1 of 120 volts leads a voltage E_2 of 100 volts by 180 degrees. Find their vector sum.

9. A current I_2 of 8 A leads a current I_1 of 6 A by 90 degrees, as shown in Figure 17-12. Determine the magnitude and angle of the resultant current.

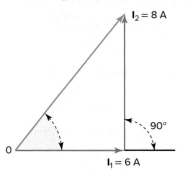

Figure 17-12 Vector diagram for review question 9.

10. Determine the resultant voltage and angle for the vector diagram of Figure 17-13.

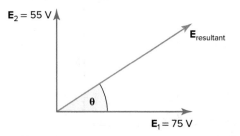

Figure 17-13 Vector diagram for review question 10.

11. In what form is the value of a vector expressed in polar notation?

12. In what form is the value of a vector expressed in rectangular notation?

PART 2   SERIES *RL* CIRCUIT

17.3 Series *RL* Circuits

In a purely resistive AC circuit, any inductive effects are considered negligible. Similarly, in a purely inductive AC circuit, any resistive effects are considered extremely small, and as a result they are omitted from any calculations. In many AC circuits, however, the load is actually a ***combination of both resistance and inductance.*** That is, the circuit can no longer be treated as either purely resistive or as purely inductive.

The combination of a resistor and inductor connected in series to an AC source is called a **series *RL* circuit.** Figure 17-14 shows a resistor and a pure or ideal inductor connected in series with an AC voltage source. The current flow in the circuit causes voltage drops to be produced across the inductor and the resistor. These voltages are proportional to the current in the circuit and the individual resistance and inductive reactance values. As in any series circuit the current will be the same value throughout the circuit. The resistor voltage (E_R) and the inductor voltage (E_L) expressed in terms of Ohm's law are

$$E_R = I \times R$$

$$E_L = I \times X_L$$

The total opposition to current flow in any AC circuit is called **impedance.** In a series *RL* circuit, this total opposition is due to a combination of both resistance (R) and inductive reactance (X_L). The symbol for impedance is **Z,** and like resistance and reactance, it too is measured in **ohms.**

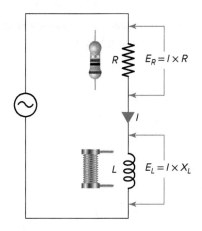

Figure 17-14 Series *RL* circuit.

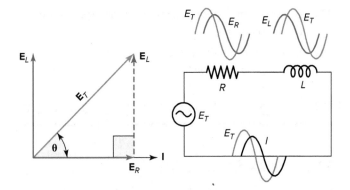

Figure 17-15 Series *RL* circuit vector diagram.

From Ohm's law, the impedance of a circuit will be equal to the total supply voltage (E_T) divided by the circuit current:

$$Z = \frac{E_T}{I}$$

It was previously shown that the current flowing through a pure resistance was in phase with the voltage across the resistance and that the current through a pure inductance lagged the voltage across the inductance by 90 degrees. For this reason, in the series *RL* circuit the two voltage drops will not be directly additive but will be a **vector sum.**

The relationship between the current and voltages in a series *RL* circuit is shown in the vector diagram of Figure 17-15 and can be summarized as follows:

- The *reference vector is labeled I* and represents the current in the circuit, which is common to all circuit elements.
- Since the voltage across the resistor is in phase with the current flowing through it, the voltage vector E_R, it is drawn superimposed on the current vector.
- The inductor voltage E_L leads the current by 90 degrees and is drawn leading the current vector by 90 degrees.
- The total supply voltage (E_T) is the vector sum of the resistor and inductor voltages:

$$E_T = \sqrt{E_R^2 + E_L^2}$$

- The phase shift between the applied voltage and current is between 0 and 90 degrees.

- As the frequency increases, the inductive reactance (X_L) increases, which causes the phase angle, or shift between the applied voltage and current, to increase.

Due to the phase shift created by the inductor, the impedance of a series *RL* circuit cannot be found by simply adding the resistance and inductive reactance values. The total impedance of a series *RL* circuit, similar to its total voltage, is the vector sum of the resistance and inductive reactance.

The **impedance triangle** for a series *RL* circuit is shown in Figure 17-16. Note that the impedance triangle is geometrically similar to the circuit vector diagram and will have the **same phase angle** theta (θ). The reason for this is that the voltage drops for the resistor and the inductor are a result of the current flow in the circuit and their respective opposition. Equations used to solve the impedance triangle include:

$$Z = \sqrt{R^2 + X_L^2}$$

$$\theta = \tan^{-1}\left(\frac{X_L}{R}\right)$$

EXAMPLE 17-9

Problem: An AC series *RL* circuit is made up of a resistor that has a resistance value of 150 Ω and an inductor that has an inductive reactance value of 100 Ω. Calculate the impedance and the phase angle theta (θ) of the circuit.

Solution:

$$Z = \sqrt{R^2 + X_L^2}$$
$$= \sqrt{150^2 + 100^2}$$
$$= \sqrt{32,500}$$
$$= 180 \ \Omega$$

$$\theta = \tan^{-1}\left(\frac{X_L}{R}\right)$$
$$= \tan^{-1}\left(\frac{100}{150}\right)$$
$$= \tan^{-1}(0.667)$$
$$= 33.7°$$
or

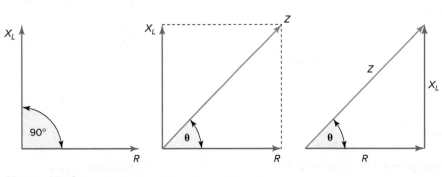

Figure 17-16 Series *RL* circuit Impedance triangle.

$$Z = R + jX_L$$
$$= 150 + j100$$
$$= 180 \ \Omega \ \angle 33.7°$$

Once the impedance of a circuit is found it is possible to find the current by using Ohm's law and substituting Z for R as follows:

$$I = \frac{E_T}{Z}$$

Since the current is the same throughout the series circuit, the individual voltage drops across the inductor and resistor can be calculated by applying Ohm's law as follows:

$$E_R = I \times R$$
$$E_L = I \times X_L$$

EXAMPLE 17-10

Problem: For the series *RL* circuit shown in Figure 17-17:

a. Calculate the value of the current flow.

b. Calculate the value of the voltage drop across the resistor.

c. Calculate the value of the voltage drop across the inductor.

d. Calculate the circuit phase angle based on the voltage drops across the resistor and inductor.

e. Express all voltages in polar notation.

f. Use a calculator to convert all voltages to rectangular notation.

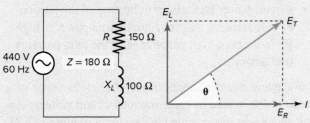

Figure 17-17 *RL* series circuit for example 17-10.

Solution:

a.
$$I = \frac{E_T}{Z}$$
$$= \frac{440 \text{ V}}{180 \ \Omega}$$
$$= 2.44 \text{ A}$$

b.
$$E_R = I \times R$$
$$= 2.44 \text{ A} \times 150 \ \Omega$$
$$= 366 \text{ V}$$

c.
$$E_L = I \times X_L$$
$$= 2.44 \text{ A} \times 100 \ \Omega$$
$$= 244 \text{ V}$$

d.
$$\theta = \tan^{-1} \frac{E_L}{E_R}$$
$$= \tan^{-1} \frac{244 \text{ V}}{366 \text{ V}}$$
$$= \tan^{-1} (0.667)$$
$$= 33.7°$$

e.
$$E_T = 440 \text{ V} \ \angle 33.7°$$
$$E_R = 366 \text{ V} \ \angle 0°$$
$$E_L = 244 \text{ V} \ \angle 90°$$

f.
$$E_T = 366 + j244 \text{ V}$$
$$E_R = 366 + j0 \text{ V}$$
$$E_L = 0 + j244 \text{ V}$$

The various power components associated with the series *RL* circuit are shown in Figure 17-18 and can be identified as follows:

- **True power** is measured in watts (W) and is the power drawn by the resistive component of the circuit. For a pure resistor the voltage and current are in phase, and power dissipated as heat is calculated by multiplying voltage by current ($W = E_R \times I_R$).

- **Reactive** power is measured in volt-amperes reactive (VARs). Reactive power is the power continually stored and discharged by the magnetic field of the inductive load. For purely inductive loads, the voltage and current are 90 degrees out of phase, and true power in watts is zero. The inductive reactive power is calculated by multiplying the inductor voltage by its current ($VARs = E_L \times I_L$).

- **Apparent power** is measured in volt-amperes (VA) and is the combination of the reactive and true power. For a series *RL* circuit the phase shift between the applied voltage and current is between 0 and 90 degrees. The apparent power or volt-amps is calculated by multiplying the applied voltage by the current flow ($VA = E_T \times I_T$).

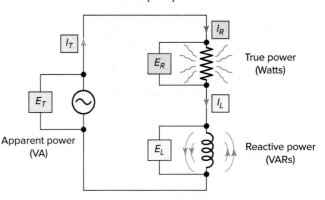

Figure 17-18 Power components associated with the *RL* series circuit.

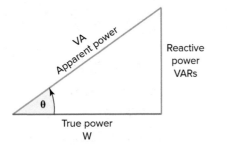

Figure 17-19 Series *RL* circuit power triangle.

The **power triangle** of Figure 17-19 shows the relationship between the various power components of a series *RL* circuit. In this triangle:

- The length of the hypotenuse of a right-angle triangle represents the apparent power.
- The angle theta (θ) is used to represent the phase difference.
- The side adjacent to theta (θ) represents the true power.
- The side opposite theta (θ) represents the reactive power.
- The power triangle is geometrically similar to the impedance triangle and the series *RL* circuit vector diagram.

EXAMPLE 17-11

Problem: For the series *RL* circuit shown in Figure 17-20, determine:

a. True power.
b. Inductive reactive power.
c. Apparent power.

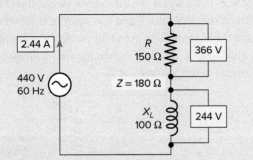

Figure 17-20 *RL* series circuit for example 17-11.

Solution:

a. True power $= E_R \times I_R$
$$= 366 \text{ V} \times 2.44 \text{ A}$$
$$= 893 \text{ W}$$

b. Inductive reactive power $= E_L \times I_L$
$$= 244 \text{ V} \times 2.44 \text{ A}$$
$$= 595 \text{ VARs}$$

c. Apparent power $= E_T \times I_T$
$$= 440 \text{ V} \times 2.44 \text{ A}$$
$$= 1{,}074 \text{ VA}$$

The **power factor (PF)** for any AC circuit is the ratio of the true power (also called real power) to the apparent power:

$$\text{PF} = \frac{\text{watts (W)}}{\text{volt-amperes (VA)}} = \frac{\text{true power}}{\text{apparent power}} = \cos \angle\theta$$

Power factor is a measure of how **effectively** equipment converts electric current to useful power output, such as heat, light, or mechanical motion. The power factor for a *RL* circuit is the ratio of the actual power dissipation to apparent power and can be summarized as follows:

- The power factor ranges from 0 to 1 and is sometimes expressed as a percentage.
- A 0 percent PF indicates a purely reactive load, while 100 percent PF indicates a purely resistive load.
- For circuits containing both resistance and inductive reactance, the power factor is said to be **lagging** (current lags) in some value between 0 and 1.
- The greater the power factor, the more resistive the circuit; the lower power factor, the more reactive the circuit.
- Circuit power factor is an indication of the portion of volt-amperes that are actually true power; a high PF indicates a high percentage of the total power is true power.

For many practical applications, the power factor of a circuit is determined by metering total circuit voltage, current, and power, as illustrated in the circuit of Figure 17-21.

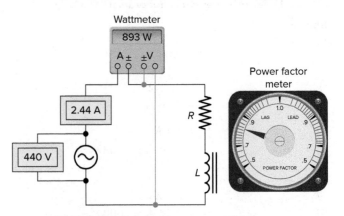

Figure 17-21 Determining circuit power factor.

The power factor can then be determined by dividing the reading of the wattmeter by the product of the voltmeter and ammeter readings as follows:

$$PF = \frac{\text{watts (W)}}{\text{volt-amperes (VA)}}$$

$$= \frac{893 \text{ W}}{2.44 \text{ A} \times 440 \text{ V}}$$

$$= 0.832$$

$$= 83.2\% \text{ (lagging)}$$

The power factor is **not** an angular measure but a numerical ratio with a value between 0 and 1. As the phase angle between the source voltage and current increases, the power factor decreases, indicating an increasingly reactive circuit. Any of the following equations can be used to calculate the power factor of a series *RL* circuit:

$$PF = \cos \angle\theta$$

$$PF = \frac{E_R}{E_T}$$

$$PF = \frac{R}{Z}$$

$$PF = \frac{W}{VA}$$

EXAMPLE 17-12

Problem: For the series *RL* circuit shown in Figure 17-22, determine:

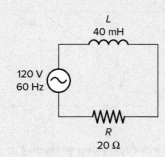

Figure 17-22 Circuit for example 17-12.

a. Inductive reactance (X_L).
b. Impedance (Z).
c. Current (I).
d. Voltage drop across the resistor (E_R) and inductor (E_L).
e. The angle theta (θ) and power factor (PF) for the circuit.
f. True power (W), reactive power (VARs), apparent power (VA).

Solution:

Step 1. Make a table and record all known values.

	E	I	L	R/X$_L$/Z	W/VA/VARs	∠θ	PF
R			N/A	20 Ω		0°	N/A
L			40 mH			90°	
Total	120 V		N/A				

Step 2. Calculate X_L and enter the value in the table.

$$X_L = 2\pi f L$$

$$= 2 \times 3.14 \times 60 \times 0.04$$

$$= 15.1 \ \Omega$$

	E	I	L	R/X$_L$/Z	W/VA/VARs	∠θ	PF
R			N/A	20 Ω		0°	N/A
L			40 mH	15.1 Ω		90°	
Total	120 V		N/A				

Step 3. Calculate Z and enter the value in the table.

$$Z = \sqrt{R^2 + X_L^2}$$

$$= \sqrt{20^2 + 15.1^2}$$

$$= \sqrt{400 + 228}$$

$$= 25.1 \ \Omega$$

	E	I	L	R/X$_L$/Z	W/VA/VARs	∠θ	PF
R			N/A	20 Ω		0°	N/A
L			40 mH	15.1 Ω		90°	
Total	120 V		N/A	25.1 Ω			

Step 4. Calculate I_T, I_R, and I_L and enter the values in the table.

$$I_T = \frac{E_T}{Z}$$

$$= \frac{120}{25.1}$$

$$= 4.78 \text{ A}$$

$$I_T = I_R = I_L = 4.78 \text{ A}$$

	E	I	L	R/X$_L$/Z	W/VA/VARs	∠θ	PF
R		4.78 A	N/A	20 Ω		0°	N/A
L		4.78 A	40 mH	15.1 Ω		90°	
Total	120 V	4.78 A	N/A	25.1 Ω			

Step 5. Calculate E_R and E_L and enter the values in the table.

$$E_R = I \times R$$
$$= 4.78 \times 20$$
$$= 95.6 \text{ V}$$

$$E_L = I \times X_L$$
$$= 4.78 \times 15.1$$
$$= 72.2 \text{ V}$$

	E	I	L	R/X_L/Z	W/VA/ VARs	∠θ	PF
R	95.6 V	4.78 A	N/A	20 Ω		0°	N/A
L	72.2 V	4.78 A	40 mH	15.1 Ω		90°	
Total	120 V	4.78 A	N/A	25.1 Ω			

Step 6. Calculate the angle θ and PF for the circuit and enter the values in the table.

$$\text{Cosine } \theta = \frac{R}{Z}$$
$$= \frac{20}{25.1}$$
$$= 0.797$$
$$\text{Angle } \theta = 37.1°$$
$$\text{Power factor} = \cos \theta$$
$$= 0.797 \text{ or } 79.7\% \text{ lagging}$$

	E	I	L	R/X_L/Z	W/VA/ VARs	∠θ	PF
R	95.6 V	4.78 A	N/A	20 Ω		0°	N/A
L	72.2 V	4.78 A	40 mH	15.1 Ω		90°	
Total	120 V	4.78 A	N/A	25.1 Ω		37.1°	79.7%

Step 7. Calculate the W, VARs, and VA for the circuit and enter the values in the table.

$$W = E_R \times I_R$$
$$= 95.6 \times 4.78$$
$$= 457 \text{ watts}$$

$$\text{VARs} = E_L \times I_L$$
$$= 72.2 \times 4.78$$
$$= 345 \text{ VARs}$$

$$\text{VA} = E_T \times I_T$$
$$= 120 \times 4.78$$
$$= 574 \text{ VA}$$

	E	I	L	R/X_L/Z	W/VA/ VARs	∠θ	PF
R	95.6 V	4.78 A	N/A	20 Ω	457 W	0°	N/A
L	72.2 V	4.78 A	40 mH	15.1 Ω	345 VARs	90°	
Total	120 V	4.78 A	N/A	25.1 Ω	574 VA	37.1°	79.7%

A real inductor has resistance due to the wire. It is impossible to have a pure inductance because all coils, relays, or solenoids will have a certain amount of resistance, no matter how small, associated with the coils turns of wire being used. This being the case, we can consider our simple coil as being a resistance in series with a pure inductance.

Part 2 Review Questions

1. Define the term *impedance* as it applies to AC circuits.
2. What symbol is use to represent impedance?
3. A circuit consists of a resistance of 20 Ω and an inductive reactance of 40 Ω connected in series and supplied from a 240-volt, 60-Hz source. Determine:
   a. The circuit impedance.
   b. Amount of current flow.
   c. The phase angle theta (θ) of the circuit.
4. For the series *RL* circuit vector (phasor) diagram shown in Figure 17-23, determine the value of the voltage drop across the inductor.

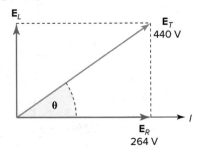

Figure 17-23 Vector for review question 4.

5. The known quantities in a given series *RL* circuit are as follows: Resistance equals 8 Ω, inductive reactance equals 39 Ω, current equals 3 A, and the applied voltage is 120 volts, 60 Hz. Determine the following unknown quantities:
   a. Impedance.
   b. Voltage across the resistor.
   c. Voltage across the inductor.
   d. Angle by which the applied voltage leads the current.

6. A wattmeter connected to a 240-volt, 60-Hz series *RL* circuit indicates a reading of 691 watts. A clamp-on ammeter used to measure current flow indicates a current of 4.8 A. Determine the:
   a. True power.
   b. Apparent power.
   c. Reactive power.
   d. Circuit power factor.

7. For the series *RL* circuit shown in Figure 17-24, determine:
   a. Apparent power.
   b. True power.
   c. Reactive power.
   d. Circuit power factor.

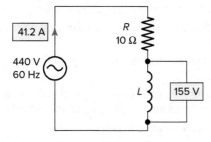

Figure 17-24 **Circuit for review question 7.**

8. Complete a table for all given and unknown quantities for the series *RL* circuit shown in Figure 17-25.

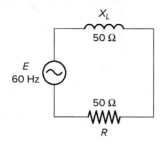

Figure 17-25 **Circuit for review question 8.**

9. The frequency to an *RL* series circuit is decreased. What effect will this have on the phase angle between the applied voltage and current? Why?

PART 3 SERIES *RC* CIRCUIT

17.4 Series *RC* Circuits

Recall that current and voltage are in phase for purely resistive AC circuits, while current leads voltage by 90 degrees in purely capacitive circuits. Therefore, when resistance and capacitance are combined, the overall difference in angle between circuit voltage and current is an angular difference between 0 and 90 degrees.

The combination of a resistor and capacitor connected in series to an AC source is called a **series *RC* circuit.**

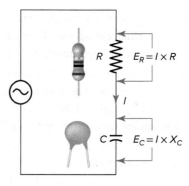

Figure 17-26 **Series *RC* circuit.**

Figure 17-26 shows a resistor and pure or ideal capacitor connected in series with an AC voltage source. The current flow in the circuit causes voltage drops to be produced across the capacitor and the resistor. These voltages are proportional to the current in the circuit and the individual resistance and capacitor values. The resistor voltage (E_R) and the capacitor voltage (E_C) expressed in terms of Ohm's law are

$$E_R = I \times R$$
$$E_C = I \times X_C$$

Series *RC* circuits are similar to series *RL* circuits. The formulas are basically the same with capacitance values substituted for inductance values. In a series *RC* circuit, the total opposition or impedance is due to a combination of both resistance (R) and capacitive reactance (X_C). The formula for the impedance of a series *RC* circuit based on Ohm's law remains as

$$Z = \frac{E_T}{I}$$

The relationship between the current and voltages in a series *RC* circuit is illustrated in the vector diagram of Figure 17-27 and can be summarized as follows:

- The applied source voltage and current will be out of phase by some amount between 0 and 90 degrees.
- The size of the phase angle will be determined by the ratio between resistance and capacitance.
- The series *RC* circuit vector is similar to the *RL* circuit in that it uses the common current element as the reference vector.

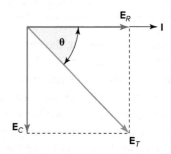

Figure 17-27 **Series *RC* circuit vector diagram.**

- As the frequency increases, the capacitive reactance (X_C) decreases, which causes the phase angle, or shift between the applied voltage and current, to decrease.
- For a capacitor, the current I leads the voltage E by 90 degrees; therefore, the only change made is that the capacitor voltage $\mathbf{E}_C$ *lags the current* $\mathbf{I}$ *by 90 degrees* and is drawn lagging the current vector by 90 degrees.
- The total supply voltage (E_T) is the vector sum of the resistor and capacitor voltages:

$$E_T = \sqrt{E_R^2 + E_C^2}$$

The resistance (R) and capacitive reactance (X_C) are 90 degrees out of phase with each other, and this forms the impedance triangle shown in Figure 17-28. Once again, the impedance triangle is geometrically similar to the circuit vector diagram and will have the same phase angle theta (θ).

When the resistance and capacitive reactance of a series RC circuit are known, the impedance is found using the equation:

$$Z = \sqrt{R^2 + X_C^2}$$

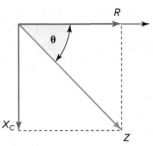

Figure 17-28 Series *RC* circuit impedance triangle.

EXAMPLE 17-13

Problem: An AC series *RC* circuit is made up of a resistor that has a resistance value of 20 Ω and a capacitor that has a capacitive reactance value of 30 Ω. Calculate the impedance and the phase angle theta (θ) of the circuit.

Solution:

$$Z = \sqrt{R^2 + X_C^2}$$
$$= \sqrt{20^2 + 30^2}$$
$$= \sqrt{400 + 900}$$
$$= \sqrt{1{,}300}$$
$$= 36\ \Omega$$

$$\theta = \tan^{-1}\frac{X_C}{R} \qquad Z = R + jX_C$$
$$= \tan^{-1}\frac{30}{20} \quad \text{or} \quad = 20 + j30$$
$$= \tan^{-1}(1.5) \qquad = 36\ \Omega \angle -56.31°$$
$$= 56.31°$$

Therefore, the circuit can be said to have a total impedance of 36 Ω ∠−56.31° (relative to the circuit current).

Once the impedance is known, the current and voltage drops can be determined as outlined in example 17-14. Note that the equations used are basically the same as that used for the series *RL* circuit with the capacitive component values used in place of the inductive values. One difference is that the current will be in phase with the resistance voltage drop, E_R, but will **lead** the capacitance voltage drop, E_C, by 90 degrees.

EXAMPLE 17-14

Problem: For the series *RC* circuit shown in Figure 17-29:

a. Calculate the value of the current flow.
b. Calculate the value of the voltage drop across the resistor.
c. Calculate the value of the voltage drop across the capacitor.
d. Calculate the circuit phase angle based on the voltage drops across the resistor and capacitor.
e. Express all voltages in polar notation.
f. Use a calculator to convert all voltages to rectangular notation.

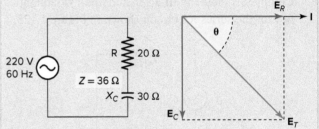

Figure 17-29 Circuit for example 17-14.

Solution:

a.
$$I = \frac{E_T}{Z}$$
$$= \frac{220\ \text{V}}{36\ \Omega}$$
$$= 6.11\ \text{A}$$

b.
$$E_R = I \times R$$
$$= 6.11\ \text{A} \times 20\ \Omega$$
$$= 122\ \text{V}$$

c.
$$E_C = I \times X_C$$
$$= 6.11\ \text{A} \times 30\ \Omega$$
$$= 183\ \text{V}$$

d.
$$\theta = \tan^{-1}\frac{E_C}{E_R}$$
$$= \tan^{-1}\frac{183}{122}$$
$$= \tan^{-1}(1.5)$$
$$= 56.31°$$

The calculated circuit phase angle is the same as that determined in the previous example using the ratio of X_C/R.

e.
$$E_T = 220 \text{ V} \angle -56.31°$$
$$E_R = 122 \text{ V} \angle 0°$$
$$E_C = 183 \text{ V} \angle -90°$$

f.
$$E_T = 122 - j183 \text{ V}$$
$$E_R = 122 + j0 \text{ V}$$
$$E_C = 0 - j183 \text{ V}$$

Like *RL* circuits, *RC* circuits have significant values of apparent power, reactive power, and true power. True power (watts) is found in the resistive part of the circuit. Capacitive reactive power (VARs) is found in the capacitive part of the circuit. The total or apparent power (VA) will contain both a true power component and a reactive power component. The power equations for a series *RC* circuit are similar to those for series *RL* circuits and are calculated as shown in example 17-15.

EXAMPLE 17-15

Problem: For the series *RC* circuit shown in Figure 17-30, determine:

a. True power.
b. Capacitive reactive power.
c. Apparent power.

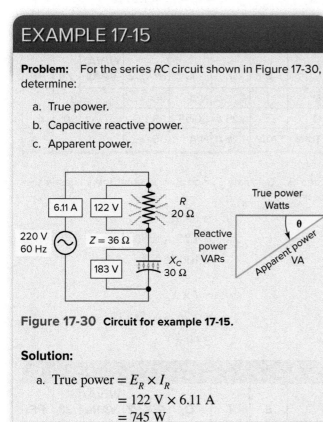

Figure 17-30 Circuit for example 17-15.

Solution:

a. True power $= E_R \times I_R$
$$= 122 \text{ V} \times 6.11 \text{ A}$$
$$= 745 \text{ W}$$

b. Capacitive reactive power $= E_C \times I_C$
$$= 183 \text{ V} \times 6.11 \text{ A}$$
$$= 1{,}118 \text{ VARs}$$

c. Apparent power $= E_T \times I_T$
$$= 220 \text{ V} \times 6.11 \text{ A}$$
$$= 1{,}344 \text{ VA}$$

Recall the power factor of any AC circuit is equal to the ratio of the true power to apparent power:

$$\text{PF} = \frac{\text{watts (W)}}{\text{volt-amperes (VA)}} = \frac{\text{true power}}{\text{apparent power}} = \cos \angle\theta$$

In a series *RL* circuit it was shown that the current lags the applied voltage and the power factor and, in this case, is described as **lagging.** For a series *RC* circuit, the current leads the applied voltage and the power factor is described as **leading.**

EXAMPLE 17-16

Problem: Determine the power factor for the series *RC* circuit shown in Figure 17-31.

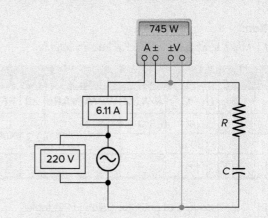

Figure 17-31 Circuit for example 17-16.

Solution:
$$\text{Apparent power} = E_T \times I_T$$
$$= 220 \text{ V} \times 6.11 \text{ A}$$
$$= 1{,}344 \text{ VA}$$

$$\text{PF} = \frac{\text{watts (W)}}{\text{volt-amperes (VA)}}$$
$$= \frac{745}{1{,}344}$$
$$= 0.554 \text{ or } 55.4\% \text{ (leading)}$$

EXAMPLE 17-17

Problem: For the series RC circuit shown in Figure 17-32, determine:

a. Capacitive reactance (X_C).
b. Impedance (Z).
c. Current (I).
d. Voltage drop across the resistor (E_R) and capacitor (E_C).
e. The angle theta (θ) and power factor (PF) for the circuit.
f. True power (W), reactive power (VARs), apparent power (VA).

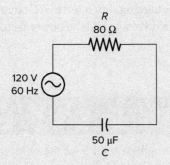

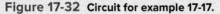

Figure 17-32 Circuit for example 17-17.

Solution:

Step 1. Make a table and record all known values.

	E	I	C	R/X_C/Z	W/VA/VARs	∠θ	PF
R			N/A	80 Ω		0°	N/A
C			50 µF			90°	
Total	120 V		N/A				

Step 2. Calculate X_C and enter the value in the table.

$$X_C = \frac{1}{2\pi fC}$$

$$= \frac{1,000,000}{2 \times 3.14 \times 60 \times 50}$$

$$= 53.1 \ \Omega$$

	E	I	C	R/X_C/Z	W/VA/VARs	∠θ	PF
R			N/A	80 Ω		0°	N/A
C			50 µF	53.1 Ω		90°	
Total	120 V		N/A				

Step 3. Calculate Z and enter the value in the table.

$$Z = \sqrt{R^2 + X_C^2}$$

$$= \sqrt{80^2 + 53.1^2}$$

$$= 96 \ \Omega$$

	E	I	C	R/X_C/Z	W/VA/VARs	∠θ	PF
R			N/A	80 Ω		0°	N/A
C			50 µF	53.1 Ω		90°	
Total	120 V		N/A	96 Ω			

Step 4. Calculate I_T, I_R, and I_C and enter the values in the table.

$$I_T = \frac{E_T}{Z}$$

$$= \frac{120 \text{ V}}{96 \ \Omega}$$

$$= 1.25 \text{ A}$$

$$I_T = I_R = I_C = 1.25 \text{ A}$$

	E	I	C	R/X_C/Z	W/VA/VARs	∠θ	PF
R		1.25 A	N/A	80 Ω		0°	N/A
C		1.25 A	50 µF	53.1 Ω		90°	
Total	120 V	1.25 A	N/A	96 Ω			

Step 5. Calculate E_R and E_C and enter the values in the table.

$$E_R = I \times R$$

$$= 1.25 \text{ A} \times 80 \ \Omega$$

$$= 100 \text{ V}$$

$$E_C = I \times X_C$$

$$= 1.25 \text{ A} \times 53.1 \ \Omega$$

$$= 66.4 \text{ V}$$

	E	I	C	R/X_C/Z	W/VA/VARs	∠θ	PF
R	100 V	1.25 A	N/A	80 Ω		0°	N/A
C	66.4 V	1.25 A	50 µF	53.1 Ω		90°	
Total	120 V	1.25 A	N/A	96 Ω			

Step 6. Calculate the angle θ and PF for the circuit and enter the values in the table.

$$\text{Inverse cosine } \angle = \frac{R}{Z}$$

$$= \frac{80\ \Omega}{96\ \Omega}$$

$$= 0.833$$

$$\text{Angle } \theta = 33.6°$$

$$\text{Power factor} = \cos\theta$$

$$= 0.833 \text{ or } 83.3\% \text{ leading}$$

	E	I	C	R/X$_C$/Z	W/VA/VARs	∠θ	PF
R	100 V	1.25 A	N/A	80 Ω		0°	N/A
C	66.4 V	1.25 A	50 µF	53.1 Ω		90°	
Total	120 V	1.25 A	N/A	96 Ω		33.6°	83.3%

Step 7. Calculate the W, VARs, and VA for the circuit and enter the values in the table.

$$W = E_R \times I_R$$

$$= 100\ V \times 1.25\ A$$

$$= 125 \text{ watts}$$

$$\text{VARs} = E_C \times I_C$$

$$= 66.4\ V \times 1.25\ A$$

$$= 83 \text{ VARs}$$

$$VA = E_T \times I_T$$

$$= 120\ V \times 1.25\ A$$

$$= 150 \text{ VA}$$

	E	I	C	R/X$_C$/Z	W/VA/VARs	∠θ	PF
R	100 V	1.25 A	N/A	80 Ω	125 W	0°	N/A
C	66.4 V	1.25 A	50 µF	53.1 Ω	83 VARs	90°	
Total	120 V	1.25 A	N/A	96 Ω	150 VA	33.6°	83.3%

It is impossible to have a pure AC capacitance because all capacitors will have a certain amount of internal resistance across their plates, giving rise to a leakage current. When the capacitor resistance is to be taken into consideration, then we need to represent the total impedance of the capacitor as a resistance in series with a capacitance.

Part 3 Review Questions

1. For the series *RC* circuit shown in Figure 17-33:
   a. Calculate the value of the applied voltage.
   b. Determine the angle displacement theta (θ) by which the current leads the applied voltage.

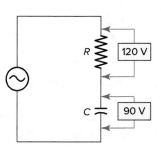

Figure 17-33 Circuit for review question 1.

2. Determine the value of the current flow for the series *RC* circuit of Figure 17-34. Show all steps required to arrive at the answer.

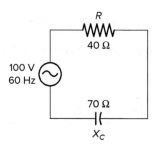

Figure 17-34 Circuit for review question 2.

3. The known quantities in a given series *RC* circuit are as follows: Resistance equals 30 Ω, capacitive reactance equals 40 Ω, and the applied voltage is 200 volts, 60 Hz. Determine the following unknown quantities:
   a. Impedance.
   b. Voltage across the resistor.
   c. Voltage across the capacitor.
   d. Angle by which the current leads the applied voltage.
   e. Circuit power factor.

4. For the circuit of Figure 17-35, determine:
   a. Wattmeter reading.
   b. E_R.
   c. E_C.
   d. Apparent power.
   e. Reactive power.
   f. Power factor.

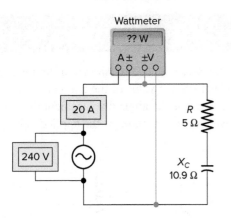

Figure 17-35 Circuit for review question 4.

5. Complete a table for all given and unknown quantities for the series *RC* circuit shown in Figure 17-36.

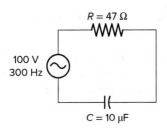

Figure 17-36 Circuit for review question 5.

6. Why is it impossible to have pure AC capacitance?

PART 4   SERIES *RLC* CIRCUIT

17.5 Series *LC* Circuits

A series *LC* circuit consists of an inductance and a capacitance connected in series, as shown in Figure 17-37. The characteristics of an *LC* series circuit can be summarized as follows:

• It is assumed that there is no resistance in the circuit, only pure inductance and capacitance. As such it is an ideal circuit which only really exists in theory.

• As in all series circuits the current has the same value at all points. This means the current in the inductor is the same as, and therefore in phase with, the current in the capacitor.

• The voltage across the inductor leads the current by 90 degrees, and voltage across capacitor lags the current by 90 degrees.

• Since the current through both is the same, the voltage across the inductor leads the voltage across the capacitor by 180 degrees.

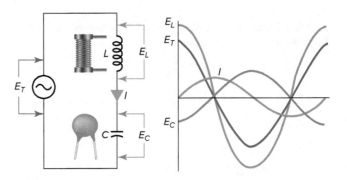

Figure 17-37 Series *LC* circuit.

• The voltages dropped across the inductor and the capacitor depend on the circuit current and the values of X_L and X_C:

$$E_L = I \times X_L \quad \text{and} \quad E_C = I \times X_C$$

The circuit vector diagram for a series *LC* circuit is shown in Figure 17-38 and is constructed as follows:

• The vector diagram is drawn starting with a horizontal line representing the current vector **I,** which is the common quantity.

• The voltage vector, E_L, is placed 90 degrees ahead of *I* since the voltage leads the current by exactly 90 degrees in an inductor.

• The voltage vector, E_C, is placed 90 degrees behind that of *I* since the voltage lags the current by exactly 90 degrees in a capacitor.

• The vector addition of E_L and E_C gives a resultant that represents the applied voltage E_T.

• Since the E_L and E_C voltages are 180 degrees out-of-phase, the total applied voltage E_T is the difference

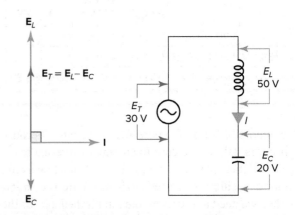

Figure 17-38 Series *LC* circuit vector diagram.

between these two voltage values and is in phase with the larger of the two voltages, which in this case is E_L.

- **The larger of voltages $\mathbf{E_L}$ and $\mathbf{E_C}$, or their respective reactance values, determines if the circuit is inductive or capacitive.**

- Any time X_L is greater than X_C, the voltage across the inductor will be greater than that across the capacitor, and the circuit will be inductive in nature. The reverse is true whenever X_C is greater than X_L.

- In this example, the circuit would be inductive.

- Note that either or both of the voltages, E_L or E_C may be greater than the total applied voltage in an AC series circuit consisting of only L and C.

The series LC circuit voltage vector and reactance vector are similar to each other, except for the units by which they are measured. Both X_L and X_C are 180 degrees out of phase with each other; therefore, the value of one subtracts from the other, leaving the circuit either inductive or capacitive, depending on which reactance is larger. The equivalent **total reactance** *(represented by the symbol **X**)* of a series LC circuit is equal to the difference between the values of inductive and capacitive reactance:

$$X_L = 2\pi f L$$

$$X_C = \frac{1}{2\pi f C}$$

$$X = X_L - X_C$$

where
X_L = inductive reactance in ohms
X_C = capacitive reactance in ohms
X = equivalent total reactance in ohms

The total opposition to the current flow in a series LC circuit is equal to the equivalent total reactance (X). Therefore, the impedance (Z) of the circuit will be the same value as the total equivalent reactance. According to Ohm's law, the following formulas then apply:

$$X_L = \frac{E_L}{I}$$

$$X_C = \frac{E_C}{I}$$

$$I = \frac{E_T}{X}$$

$$Z = X = \frac{E_T}{I}$$

EXAMPLE 17-18

Problem: For the series LC circuit of Figure 17-39, determine:

a. Equivalent total reactance (X).
b. Circuit current flow (I).
c. Voltage drop across the inductor (E_L).
d. Voltage drop across the capacitor (E_C).

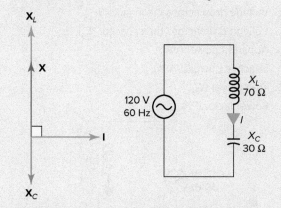

Figure 17-39 Circuit for example 17-18.

Solution:

a.
$$X = X_L - X_C$$
$$= 70\ \Omega - 30\ \Omega$$
$$= 40\ \Omega$$

b.
$$I = \frac{E_T}{X}$$
$$= \frac{120\ \text{V}}{40\ \Omega}$$
$$= 3\ \text{A}$$

c.
$$E_L = I \times X_L$$
$$= 3\ \text{A} \times 70\ \Omega$$
$$= 210\ \text{V}$$

d.
$$E_C = I \times X_C$$
$$= 3\ \text{A} \times 30\ \Omega$$
$$= 90\ \text{V}$$

The total applied voltage E_T is always in phase with either E_L or E_C (whichever is greater), and therefore will always be 90 degrees out of phase with the current. There is a constant interchange of power, or energy, between the source and the circuit, but *no power consumption.* This means the circuit will react as if it contained only inductance or capacitance. Therefore, for a series LC circuit:

- The applied voltage and current will always be 90 degrees out of phase.
- The true power or watts will always equal zero.
- The power factor will always equal zero.

EXAMPLE 17-19

Problem: For the series *LC* circuit of Figure 17-40, determine:

a. Inductive reactance (X_L).
b. Capacitive reactance (X_C).
c. Equivalent total reactance (X).
d. Circuit current (I).
e. Voltage drop across the inductor (E_L).
f. Voltage drop across the capacitor (E_C).
g. Apparent power (VA).
h. Reactive power (VARs).
i. True power (W).
j. Power factor (PF).

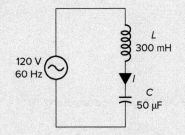

Figure 17-40 Circuit for example 17-19.

Solution:

a.
$$X_L = 2\pi f L$$
$$= 2 \times 3.14 \times 60 \times 0.3$$
$$= 113\ \Omega$$

b.
$$X_C = \frac{1}{2\pi f C}$$
$$= \frac{1,000,000}{2 \times 3.14 \times 60 \times 50}$$
$$= 53.1\ \Omega$$

c.
$$X = X_L - X_C$$
$$= 113\ \Omega - 53.1\ \Omega$$
$$= 59.9\ \Omega$$

d.
$$I = \frac{E_T}{X}$$
$$= \frac{120\text{ V}}{59.9\ \Omega}$$
$$= 2\text{ A}$$

e.
$$E_L = I \times X_L$$
$$= 2\text{ A} \times 113\ \Omega$$
$$= 226\text{ V}$$

f.
$$E_C = I \times X_C$$
$$= 2\text{ A} \times 53.1\ \Omega$$
$$= 106\text{ V}$$

g.
$$\text{VA} = E_T \times I$$
$$= 120\text{ V} \times 2\text{ A}$$
$$= 240\text{ VA}$$

h.
$$\text{VARs} = \text{VA}$$
$$= 240\text{ VARs}$$

i.
$$W = 0 \text{ (pure reactive load)}$$

j.
$$PF = \frac{W}{VA}$$
$$= \frac{0}{240}$$
$$= 0$$

17.6 Series *RLC* Circuits

A **series *RLC*** circuit contains elements of resistance, inductance, and capacitance connected in series with an AC source, as shown in Figure 17-41. The characteristics of the *RLC* series circuit can be summarized as follows:

- The current is the same through all components, but the voltage drops across the elements are out of phase with each other.
- The voltage dropped across the resistance is in phase with the current.
- The voltage dropped across the inductor leads the current by 90 degrees.
- The voltage dropped across the capacitor lags the current by 90 degrees.
- The voltages dropped across the resistor, inductor, and capacitor depends on the circuit current and the values of R, X_L, and X_C:

$$E_R = I \times R$$
$$E_L = I \times X_L$$
$$E_C = I \times X_C$$

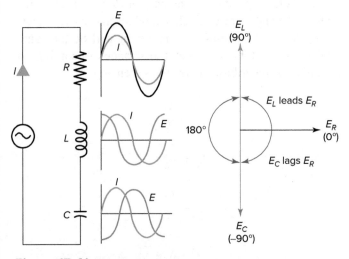

Figure 17-41 Series *RLC* circuit.

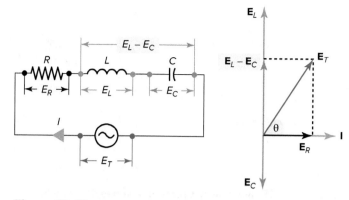

Figure 17-42 Voltage vector diagram for a series *RLC* circuit.

The three voltages of a series *RLC* circuit are combined, as shown in the circuit voltage vector diagram of Figure 17-42 and constructed as follows:

- A horizontal reference line representing the common current element is drawn first.
- The voltage across the resistor is in phase with the current and, therefore, is placed directly on the current line.
- The voltage across the inductor leads the current by 90 degrees and so is drawn upward at a 90-degree angle from the current.

- The voltage across the capacitor lags the current by 90 degrees and so is drawn downward at a 90-degree angle from the current.
- To combine the voltages, the two reactive voltage values, which are 180 degrees out of phase with each other, are subtracted.
- The total applied voltage (E_T) is the vector sum of the voltage across the resistor (E_R) and the difference in voltage between E_L and E_C. This voltage is computed using the following formula:

$$E_T = \sqrt{E_R^2 + (E_L - E_C)^2}$$

The **circuit's phase angle** theta (θ) is always the angle that separates the circuit's current and the applied voltage source, as summarized in Table 17-1.

- A series *RLC* circuit will be **inductive** and have a **positive phase angle** when the inductive reactance and resulting voltage across the inductor is greater than the capacitive reactance and the resulting voltage across the capacitor.
- A series *RLC* circuit will be **capacitive** and have a **negative phase angle** when the capacitive reactance and resulting voltage across the capacitor is greater than the inductive reactance and the resulting voltage across the inductor.

Table 17-1 Series *RLC* Circuit Phase Angle

Circuit Elements	Impedance Z	Phase Angle ϕ
R	$Z = R$	$0°$ I → V_R
L	$Z = X_L$	V_L $+90°$ I
C	$Z = X_C$	I $-90°$ V_C
R L	$Z = \sqrt{R^2 + X_L^2}$	Positive, between $0°$ and $90°$
R C	$Z = \sqrt{R^2 + X_C^2}$	Negative, between $-90°$ and $0°$
R L C	$Z = \sqrt{R^2 + (X_L - X_C)^2}$	Negative if $X_C > X_L$ Positive if $X_C < X_L$

EXAMPLE 17-20

Problem: For the series *RLC* circuit shown in Figure 17-43:

a. Determine the value of the applied voltage E_T.

b. Draw a voltage vector diagram for the circuit.

c. Is the circuit inductive or capacitive? Why?

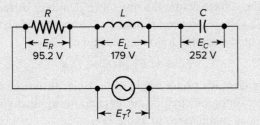

Figure 17-43 Circuit to example 17-20.

Solution:

a. $E_T = \sqrt{E_R^2 + (E_C - E_L)^2}$

$= \sqrt{95.2^2 + (252 - 179)^2}$

$= \sqrt{95.2^2 + 73^2}$

$= \sqrt{14{,}392}$

$= 120 \text{ V}$

or

$E_T = 95.2 + j0 + 0 - j252 + 0 + j179$

$= 95.2 - j73$

$= 120 \text{ V} \angle{-37.48°}$

b. The voltage vector diagram is shown in Figure 17-44.

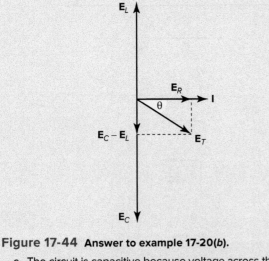

Figure 17-44 Answer to example 17-20(b).

c. The circuit is capacitive because voltage across the capacitor is greater than the voltage across the inductor.

When $\mathbf{X_L}$ *is greater than* $\mathbf{X_C}$*, the net reactance is inductive and the circuit acts essentially as a* **RL** *series circuit.* This means that the impedance, which is the vector sum of the net reactance and resistance, will have an angle between 0 and 90 degrees. Similarly, *when* $\mathbf{X_C}$ *is*

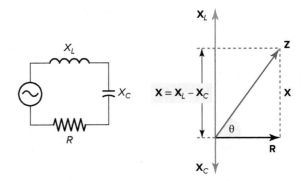

Figure 17-45 Impedance vector for a series *RLC* circuit.

greater than $\mathbf{X_L}$*, the net reactance is capacitive and the circuit acts as a* **RC** *series circuit.* The impedance, therefore, has an angle somewhere between 0 and 90 degrees. In both cases, the value of the impedance angle depends on the relative values of the net reactance (X) and the resistance (R). The angle can be found by the equation

$$\theta = \tan^{-1} \frac{X}{R}$$

The impedance vector for a typical series *RLC* circuit, inductive in nature, is shown in Figure 17-45 and can be summarized as follows:

- The total impedance (Z) is equal to the vector sum of the circuit's reactance and resistance.

- Since the inductive reactance (X_L) and the capacitive reactance (X_C) are 180 degrees out of phase, the total net reactance is found first by subtracting the smaller of the two reactive values from the larger.

- The smaller reactance value is canceled out, and the larger value is reduced by the amount of the smaller value.

- The vector sum of the total reactive value (X) and resistance (R) is equal to the impedance and can be computed as follows:

$$Z = \sqrt{R^2 + (X_L - X_C)^2}$$

or

$$Z = \sqrt{R^2 + X^2}$$

where $X = X_L - X_C$

If the impedance and applied voltage of a series *RLC* circuit are known, the current can be found using the Ohm's law equation:

$$I = \frac{E_T}{Z}$$

Similarly, once the current is known, the various voltage drops can be found using the Ohm's law equations:

$$E_R = I \times R$$
$$E_L = I \times X_L$$
$$E_C = I \times X_C$$

EXAMPLE 17-21

Problem: For the series *RLC* circuit shown in Figure 17-46, determine:

a. Impedance (*Z*).

b. Current (*I*).

c. Voltage drop across the inductor (*E_L*), capacitor (*E_C*), and resistor (*E_R*).

d. The phase angle (θ) of the circuit.

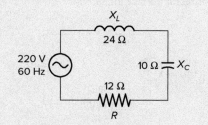

Figure 17-46 Circuit for example 17-21.

Solution:

a.
$$X = X_L - X_C$$
$$= 24\ \Omega - 10\ \Omega$$
$$= 14\ \Omega$$

$$Z = \sqrt{R^2 + X^2}$$
$$= \sqrt{12^2 \times 14^2}$$
$$= \sqrt{340}$$
$$= 18.4\ \Omega$$

b.
$$I = \frac{E_T}{Z}$$
$$= \frac{220\ V}{18.4\ \Omega}$$
$$= 12\ A$$

c.
$$E_L = I \times X_L$$
$$= 12\ A \times 24\ \Omega$$
$$= 288\ V$$

$$E_C = I \times X_C$$
$$= 12\ A \times 10\ \Omega$$
$$= 120\ V$$

$$E_R = I \times R$$
$$= 12\ A \times 12\ \Omega$$
$$= 144\ V$$

d.
$$\theta = \tan^{-1} \frac{X}{R}$$
$$= \tan^{-1} \frac{14}{12}$$
$$= 49.4°$$

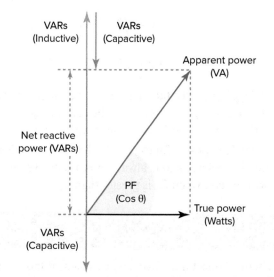

Figure 17-47 Power in a series *RLC* circuit.

Power in a series *RLC* circuit is illustrated in Figure 17-47 and summarized as follows:

- **True power** is dissipated by the resistive component only and may be calculated using any one of the following equations:

$$W = I^2 \times R$$
$$W = \frac{E_R^2}{R}$$
$$W = E_R \times I$$
$$W = E_T \times I_T \times PF\ (\cos \theta)$$

- **Reactive power** is produced by the circuit's inductive and capacitive components. Since these two reactive components are 180 degrees out of phase, the net or total volt-amps reactive (VARs) is equal to the difference between the two and may be calculated using any one of the following equations:

$$VARs = \text{inductive VARs} - \text{capacitive VARs}$$
$$VARs = I^2 X_L - I^2 X_C$$
$$VARs = (I \times E_L) - (I \times E_C)$$

- When the supply voltage and current are out of phase because of reactance, the product of the applied voltage and current is called the **apparent power** and may be calculated as follows:

$$VA = E_T \times I$$

- The circuit's **power factor (PF)** is equal to the cosine of the angle that separates the circuit's current and applied voltage. The PF of a series *RLC* circuit can be found using any of the following equations:

$$PF = \cos \theta$$
$$PF = \frac{R}{Z}$$

$$PF = \frac{E_R}{E_T}$$

$$PF = \frac{W}{VA}$$

- A **lagging power factor** means the current lags the applied voltage and is always the case in a series *RLC* circuit when X_L is greater than X_C.
- A **leading power factor** means the current leads the applied voltage and is always the case in a series *RLC* circuit when X_C is greater than X_L.

EXAMPLE 17-22

Problem: From the measurements taken of the series *RLC* circuit shown in Figure 17-48, determine:

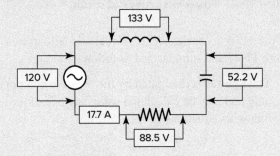

Figure 17-48 Circuit for example 17-22.

a. True power (W).
b. Reactive power (VARs).
c. Apparent power (VA).
d. Power factor (PF).
e. If the power factor is leading or lagging, explain why.

Solution:

a. $W = I \times E_R$
    $= 17.7 \text{ A} \times 88.5 \text{ V}$
    $= 1,566 \text{ watts}$

b. $VARs = (I \times E_L) - (I \times E_C)$
    $= (17.7 \text{ A} \times 133 \text{ V}) - (17.7 \text{ A} \times 52.2 \text{ V})$
    $= 2,354 - 924$
    $= 1,430 \text{ VARs}$

c. $VA = E_T \times I$
    $= 120 \text{ V} \times 17.7 \text{ V}$
    $= 2,124 \text{ volt-amps}$

d. $PF = \dfrac{W}{VA}$
    $= \dfrac{1,566}{2,124}$
    $= 0.737 \quad \text{or} \quad 73.7\%$

e. The circuit is inductive in nature because the voltage across the inductor is greater than that across the capacitor. Therefore, the power factor is said to be lagging.

EXAMPLE 17-23

Problem: For the series *RLC* circuit shown in Figure 17-49, determine:

a. Impedance (*Z*).
b. Current (*I*).
c. The voltage drop across the resistor (E_R), across the inductor (E_L) and the capacitor (E_C).
d. True power (W), Reactive power (VARs), Apparent power (VA).
e. Power factor (PF) for the circuit.

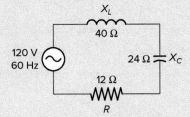

Figure 17-49 Circuit for example 17-23.

Solution:

Step 1. Make a table and record all known values.

	E	**I**	**R/X_L C_C/Z**	**W/VA/VARs**	**PF**
R			12 Ω		
L			40 Ω		N/A
C			24 Ω		
Total	120 V				

Step 2. Calculate *Z* and enter the value in the table.

$$Z = \sqrt{R^2 + (X_L - X_C)^2}$$
$$= \sqrt{12^2 + (40 - 24)^2}$$
$$= \sqrt{400}$$
$$= 20 \ \Omega$$

	E	**I**	**R/X_L C_C/Z**	**W/VA/VARs**	**PF**
R			12 Ω		
L			40 Ω		N/A
C			24 Ω		
Total	120 V	20 Ω			

Step 3. Calculate the *I* and enter the value in the table.

$$I = \frac{E_T}{Z}$$
$$= \frac{120 \text{ V}}{20 \ \Omega}$$
$$= 6 \text{ A}$$

	E	I	R/X$_L$C$_C$/Z	W/VA/VARs	PF
R		6 A	12 Ω		
L		6 A	40 Ω		N/A
C		6 A	24 Ω		
Total	120 V	6 A	20 Ω		

Step 4. Calculate the voltage drop across the resistor (E_R), across the inductor (E_L), and the capacitor (E_C) and enter the values in the table.

$$E_R = I \times R$$
$$= 6\ A \times 12\ \Omega$$
$$= 72\ V$$

$$E_L = I \times X_L$$
$$= 6\ A \times 40\ \Omega$$
$$= 240\ V$$

$$E_C = I \times X_C$$
$$= 6\ A \times 24\ \Omega$$
$$= 144\ V$$

	E	I	R/X$_L$/X$_C$/Z	W/VA/VARs	PF
R	72 V	6 A	12 Ω		
L	240 V	6 A	40 Ω		N/A
C	144 V	6 A	24 Ω		
Total	120 V	6 A	20 Ω		

Step 5. Calculate the true power (W), reactive power (VARs), and apparent power (VA) and enter the values in the table.

$$W = I \times E_R$$
$$= 6\ A \times 72\ V$$
$$= 432\ watts$$

$$VARs = (I \times E_L) - (I \times E_C)$$
$$= (6 \times 240) - (6 \times 144)$$
$$= 1{,}440 - 864$$
$$= 576\ VARs$$

$$VA = E_T \times I$$
$$= 120\ V \times 6\ A$$
$$= 720\ volt\text{-}amps$$

	E	I	R/X$_L$/X$_C$/Z	W/VA/VARs	PF
R	72 V	6 A	12 Ω	432 W	
L	240 V	6 A	40 Ω	1,440 VARs	N/A
C	144 V	6 A	24 Ω	864 VARs	
Total	120 V	6 A	20 Ω	720 VA	

Step 6. Calculate the circuit power factor (PF) and enter the value in the table.

$$PF = \frac{W}{VA}$$
$$= \frac{432}{720}$$
$$= 0.6 \quad \text{or} \quad 60\%$$

	E	I	R/X$_L$/X$_C$/Z	W/VA/VARs	PF
R	72 V	6 A	12 Ω	432 W	
L	240 V	6 A	40 Ω	1,440 VARs	N/A
C	144 V	6 A	24 Ω	864 VARs	
Total	120 V	6 A	20 Ω	720 VA	60%

Part 4 Review Questions

1. For the series *LC* circuit of Figure 17-50, determine:
   a. Equivalent total reactance (*X*).
   b. Circuit current flow (*I*).
   c. Voltage drop across the inductor (E_L).
   d. Voltage drop across the capacitor (E_C).

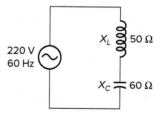

Figure 17-50 Circuit for review question 1.

2. For the series *LC* circuit of Figure 17-51, determine:
   a. Voltage drop across the inductor.
   b. Current flow.
   c. Impedance.
   d. Apparent power.
   e. Net reactive power.
   f. True power.

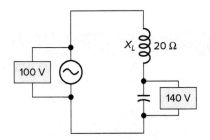

Figure 17-51 Circuit for review question 2.

g. Capacitive reactance.
h. Phase angle of the circuit.
i. Power factor of the circuit.

3. For the series *RLC* circuit of Figure 17-52, determine:
   a. The value of the applied voltage.
   b. The phase angle between the applied voltage and circuit current.

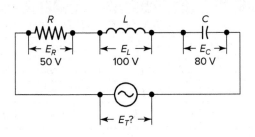

Figure 17-52 Circuit for review question 3.

4. For the series *RLC* circuit of Figure 17-53, determine:
   a. Impedance.
   b. Current.
   c. Voltage drop across the inductor, capacitor, and resistor.

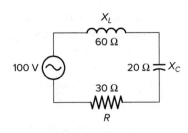

Figure 17-53 Circuit for review question 4.

5. For the series *RLC* circuit of Figure 17-54, determine:
   a. Current.
   b. Apparent power.
   c. Inductive reactive power.
   d. Capacitive reactive power.
   e. Net reactive power.
   f. Power factor.

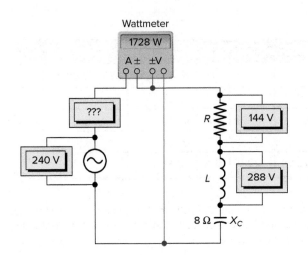

Figure 17-54 Circuit for review question 5.

6. Complete a table for all given and unknown quantities for the series *RLC* circuit shown in Figure 17-55.

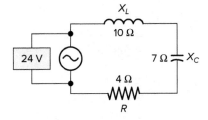

Figure 17-55 Circuit for review question 6.

PART 5 SERIES RESONANT CIRCUIT

17.7 Series Resonant Circuits

Circuits in which the ***inductive reactance equals the capacitive reactance*** $(X_L = X_C)$ are called **resonant circuits.** They can be series or parallel circuits and either *RLC* or *LC* circuits. The impedance vector for a typical **series *RLC* resonant circuit** is shown in Figure 17-56 and is

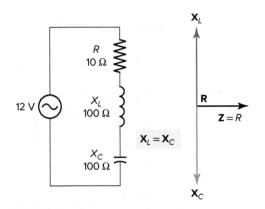

Figure 17-56 Impedance vector for a series *RLC* resonant circuit.

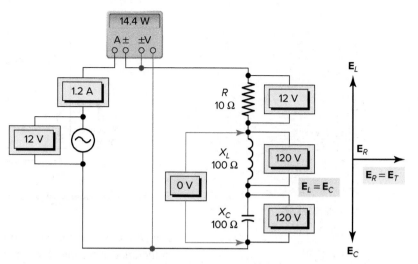

Figure 17-57 Voltage vector for the series *RLC* resonant circuit.

summarized as follows:

- X_L and X_C are 180 degrees out of phase.
- X_L and X_C are equal in value (100 Ω), resulting in a net reactance of zero ohm.
- The only opposition to current is then R (10 Ω).
- Z is equal to R and is at its **minimum** value, allowing the greatest amount of current to flow.

The voltage vector for the series *RLC* resonant circuit is shown in Figure 17-57 and is summarized as follows:

- The current as determined by Ohm's law is

$$I = \frac{E_T}{R}$$
$$= \frac{12\text{ V}}{10\text{ }\Omega}$$
$$= 1.2\text{ A}$$

- The same amount of current flows through all components.
- The voltage drop across the inductor is

$$E_L = I \times X_L$$
$$= 1.2\text{ A} \times 100\text{ }\Omega$$
$$= 120\text{ V}$$

- The voltage drop across the capacitor is

$$E_C = I \times X_C$$
$$= 1.2\text{ A} \times 100\text{ }\Omega$$
$$= 120\text{ V}$$

- Voltages across X_L and X_C are equal (120 V) and 180 degrees out of phase with each other so that each cancels the other.

- With the effects of both the inductor and capacitor canceled out, the only current-limiting component will be the resistor, and the total applied supply voltage appears across the resistor.
- Therefore, the **phase angle** between the circuit current and the supply voltage will be **zero** and the power factor will be 1, or 100 percent.
- The circuit can be considered to be **purely resistive** in nature with inductive reactive VARs of the inductor being canceled out by the inductive capacitive VARs of the capacitor. The true power is

True power (watts) = $E_R \times I$
$$= 12\text{ V} \times 1.2\text{ A}$$
$$= 14.4\text{ W}$$

EXAMPLE 17-24

Problem: For the resonant series *RLC* circuit shown in Figure 17-58, determine:

a. Impedance (*Z*).
b. Current (*I*).
c. Voltage drop across the resistor (E_R), inductor (E_L) and capacitor (E_C).
d. Apparent, true, and net reactive power.
e. Power factor.

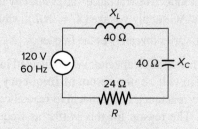

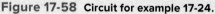

Figure 17-58 Circuit for example 17-24.

Solution:

a.
$$Z = R$$
$$= 24 \, \Omega$$

b.
$$I = \frac{E_T}{R}$$
$$= \frac{120 \text{ V}}{24 \, \Omega}$$
$$= 5 \text{ A}$$

c.
$$E_R = I \times R$$
$$= 5 \text{ A} \times 24 \, \Omega$$
$$= 120 \text{ V}$$

$$E_L = I \times X_L$$
$$= 5 \text{ A} \times 40 \, \Omega$$
$$= 200 \text{ V}$$

$$E_C = I \times X_C$$
$$= 5 \text{ A} \times 40 \, \Omega$$
$$= 200 \text{ V}$$

d.
$$\text{Apparent power} = E_T \times I$$
$$= 120 \text{ V} \times 5 \text{ A}$$
$$= 600 \text{ VA}$$

$$\text{True power} = E_R \times I$$
$$= 120 \text{ V} \times 5 \text{ A}$$
$$= 600 \text{ watts}$$

$$\text{Net reactive power} = (I \times E_L) - (I \times E_C)$$
$$= (5 \text{ A} \times 200 \text{ V}) - (5 \text{ A} \times 200 \text{ V})$$
$$= 1{,}000 - 1{,}000$$
$$= 0 \text{ VARs}$$

e.
$$\text{Power factor} = \frac{\text{watts}}{\text{VA}}$$
$$= \frac{600 \text{ W}}{600 \text{ VA}}$$
$$= 1 \qquad \text{or} \qquad 100\%$$

Recall that inductive reactance varies ***directly*** as the frequency of the AC supply voltage ($X_L = 2\pi f L$), while capacitive reactance varies ***inversely*** as the frequency $\left(X_C = \frac{1}{2\pi f C} \right)$. When an inductor and capacitor are connected in series in a circuit, there will be **one resonant frequency** at which the inductive reactance and capacitive reactance will become equal. The reason for this is that as frequency increases, inductive reactance increases and capacitive reactance decreases. The following formula is used to

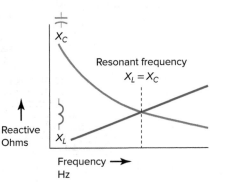

Figure 17-59 How X_L and X_C vary with change in frequency.

determine the resonant frequency when the values of inductance and capacitance are known:

$$f_R = \frac{1}{2\pi \sqrt{LC}}$$

where f_R = resonant frequency in hertz
L = inductance in henrys
C = capacitance in farads

As an example, suppose that a fixed AC voltage of variable frequency is applied to a series *RLC* circuit. As the frequency of the applied voltage is increased, the inductive reactance X_L **increases** but the capacitive reactance X_C **decreases,** as illustrated in Figure 17-59. You can see from this graph that at the resonant frequency $X_L = X_C$.

EXAMPLE 17-25

Problem: Calculate the resonant frequency of a *RLC* series circuit containing a 750-mH inductor and a 47-µF capacitor.

Solution:

$$f_R = \frac{1}{2\pi \sqrt{LC}}$$
$$= \frac{1}{2 \times 3.14 \sqrt{0.75 \times 0.000047}}$$
$$= \frac{1}{6.28 \times 0.00594}$$
$$= 26.8 \text{ Hz}$$

In certain applications a series resonant circuit is used to achieve an increase in voltage at the resonant frequency. As an example, in the series resonant circuit of Figure 17-60, the voltage across X_L and X_C is much ***higher than the applied total voltage.*** This seemingly impossible condition is caused by the interaction between the capacitor and inductor. The voltage across the inductor and capacitor is

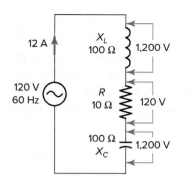

Figure 17-60 High voltage across reactive elements.

1,200 volts, while the applied voltage is only 120 volts. In some control applications, the voltage across either X_L or X_C is used as a signal voltage to perform some function.

A typical **frequency response curve** for a series *RLC* circuit is shown in Figure 17-61 and summarized as follows:

- The frequency is varied, and the values of current at the different frequencies plotted on the graph.
- *The magnitude of the current is a function of the frequency.*
- The response curve starts near zero, reaches maximum value at the resonance frequency, and then drops to near zero as the frequency becomes infinite.
- There is a small range of frequencies, called the resonant band or **bandpass,** on either side of resonance where the current is almost the same as it is at resonance.
- The circuit can be used to isolate or **filter** out certain frequencies.

In a series *RLC* circuit at resonance, the two reactances, X_L and X_C are equal and canceling. In addition, the two voltages representing V_L and V_C are also opposite and

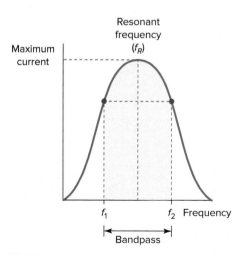

Figure 17-61 Frequency response curve for a series *RLC* circuit.

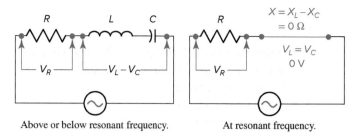

Figure 17-62 Series *RLC* circuit characteristics.

equal in value, thereby canceling each other out. Figure 17-62 compares the circuit condition that exists above or below resonance with that when the circuit is at resonance and is summarized as follows:

- When operating above its resonant frequency, a series *RLC* circuit has the dominate characteristics of a series R_L circuit.
- When operating below its resonant frequency, a series *RLC* circuit has the dominate characteristics of a series R_C circuit.
- When operating at its resonant frequency:
  - Reactance (*X*) is zero as $X_L = X_C$.
  - Impedance is minimum and current is maximum as $Z = R$.
  - The voltage measured across the two series reactive components *L* and *C* is zero.
  - All the supply voltage is dropped across the resistor.
  - The phase angle between the supply voltage and current is zero and the power factor is 1.

Part 5 Review Questions

1. In a series resonant *RLC* circuit how does the value of X_L compare with that of X_C?
2. Maximum current will flow when a series *RLC* circuit is at resonance. Why?
3. What are the circuit voltage conditions that always exist across the inductor, capacitor, and resistor in a resonant *RLC* circuit?
4. In a series resonant *RLC* circuit the apparent power (VA) is the same value as the true power (watts). Why?
5. State the phase relationship of each of the following for a series resonant *RLC* circuit:
   a. X_L and X_C.
   b. E_L and *I*.
   c. E_C and *I*.
   d. E_L and E_C.

e. *R* and *Z*.

f. E_T and E_R.

6. Determine each of the following for the *RLC* circuit shown in Figure 17-63, when at resonance:

a. The resonant frequency.

b. The circuit impedance.

c. The circuit current.

d. The voltage drops across the resistor, inductor and capacitor.

e. Net reactive power.

f. True power.

g. Power factor.

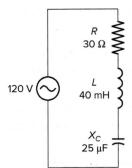

Figure 17-63 Circuit for review question 6.

7. Answer the following with reference to the circuit shown in Figure 17-64.

a. What is the resonant frequency of the circuit? Why?

b. What should the reading on the voltmeter be? Why?

c. Assume the frequency of the applied voltage is increased to 400 Hz. Would the circuit become inductive or capacitive? Why?

d. Assume the frequency of the applied voltage is decreased to 60 Hz. Would the circuit become inductive or capacitive? Why?

e. If the frequency of the applied voltage is increased above 318 Hz, what happens to the value of the impedance and current? Why?

f. If the frequency of the applied voltage is decreased below 318 Hz, what happens to the value of the impedance and current? Why?

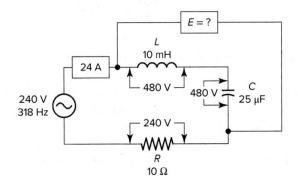

Figure 17-64 Circuit for review question 7.

8. A series *RLC* circuit is connected to a variable frequency AC power supply with an ammeter connected to measure current flow and a voltmeter connected to measure the voltage drop across the resistor. Outline how you would proceed to set the circuit to resonance.

9. What should you be aware of when measuring voltages in a series resonant circuit?

10. Give two practical applications for series resonant circuits.

11. A 24-Ω resistor, an inductor with a reactance of 120 Ω, and a capacitor with a reactance of 120 Ω are in series across a 60-V source. The circuit is at resonance. Determine the voltage across the inductor.

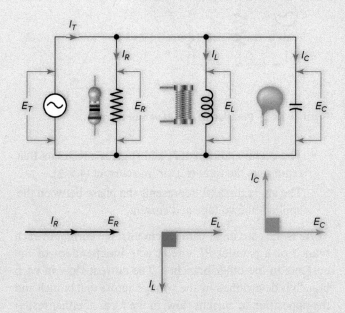

CHAPTER EIGHTEEN

Resistive, Inductive, Capacitive (*RLC*) Parallel Circuits

Voltage and current relationship in a parallel *RLC* circuit.

LEARNING OUTCOMES

▶ Solve for unknown quantities in a parallel *RL* circuit.

▶ Solve for unknown quantities in a parallel *RC* circuit.

▶ Solve for unknown quantities in a parallel *LC* circuit.

▶ Solve for unknown quantities in a parallel *RLC* circuit.

▶ Understand the relationship between resistance, reactance, and impedance in a parallel *RLC* circuit.

▶ Understand the relationship between true power, reactive power, apparent power, and power factor in a parallel *RLC* circuit.

▶ Apply the principles of power factor correction.

When an alternating current circuit has components of resistance, inductance, and capacitance connected in parallel, the characteristics associated with parallel circuits are applied. For parallel AC circuits the voltage drop across each of the components is in phase with each other and equal in value to that of the applied voltage. The currents, however, are not and will vary in phase and magnitude in accordance to the rules of parallel AC circuits. In this chapter, you will learn how to analyze and solve AC circuits that contain parallel combinations of resistance, inductance, and capacitance. This chapter also covers impedance (*Z*), power factor (PF), power factor correction, and resonance as applied to parallel *RLC* circuits.

243

PART 1 PARALLEL *RL* CIRCUIT

18.1 Parallel *RL* Circuits

The combination of a resistor and inductor connected in parallel to an AC source, as illustrated in Figure 18-1, is called a **parallel *RL* circuit.** In a parallel DC circuit, the voltage across each of the parallel branches is equal. This is also true of the AC parallel circuit. The voltages across each parallel branch are:

- The same value.
- Equal in value to the total applied voltage E_T.
- All in phase with each other.

Therefore, for a *RL* parallel circuit

$$E_T = E_R = E_L$$

In parallel DC circuits, the simple arithmetic sum of the individual branch currents equals the total current. The same is true in an AC parallel circuit if only pure resistors or only pure inductors are connected in parallel. However, when a resistor and inductor are connected in parallel, the two currents will be **out of phase** with each other. In this case, the total current is equal to the **vector sum** rather than the arithmetic sum of the currents.

Recall that the voltage and current through a resistor are in phase, but through a pure inductor the current lags the voltage by exactly 90 degrees. This is still the case when the two are connected in parallel. The relationship between the voltage and currents in a parallel *RL* circuit is illustrated in the vector diagram of Figure 18-2 and summarized as follows:

- The *reference vector is labeled E* and represents the voltage in the circuit, which is common to all elements.
- Since the current through the resistor is in phase with the voltage across it, I_R (2 A) is shown superimposed on the voltage vector.
- The inductor current I_L (4 A) lags the voltage by 90 degrees and is positioned in a downward direction lagging the voltage vector by 90 degrees.

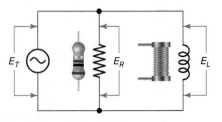

Figure 18-1 Parallel *RL* circuit.

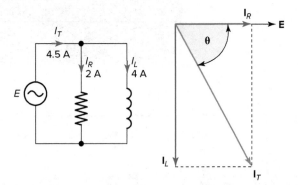

Figure 18-2 Parallel *RL* circuit vector diagram.

- The vector addition of I_R and I_L gives a resultant that represents the total (I_T), or line current (4.5 A).
- The angle theta (θ) represents the phase between the applied line voltage and current.

As is the case in all parallel circuits, the current in each branch of a parallel *RL* circuit acts **independent** of the currents in the other branches. The current flow in each branch is determined by the voltage across that branch and the opposition to current flow, in the form of either resistance or inductive reactance, contained in the branch. Ohm's law can then be used to find the individual branch currents as follows:

$$I_R = \frac{E}{R}$$

$$I_L = \frac{E}{X_L}$$

The resistive branch current has the same phase as the applied voltage, but the ***inductive branch current lags the applied voltage by 90 degrees.*** As a result, the total line current (I_T) consists of I_R and I_L 90 degrees out of phase with each other. The current flow through the resistor and the inductor form the legs of a right triangle, and the total current is the hypotenuse. Therefore, the Pythagorean theorem can be applied to add these currents together by using the equation:

$$I_T = \sqrt{I_R^2 + I_L^2}$$

In all parallel *RL* circuits, the phase angle theta (θ) by which the total current lags the voltage is somewhere between 0 and 90 degrees. The size of the angle is determined by whether there is ***more inductive current or resistive current.*** If there is more inductive current, the phase angle will be closer to 90 degrees. It will be closer to 0 degrees if there is more resistive current. From the circuit vector diagram you can see that the value of the phase angle can be calculated from the equation:

$$\theta = \tan^{-1} \frac{I_L}{I_R}$$

EXAMPLE 18-1

Problem: For the parallel *RL* circuit shown in Figure 18-3, determine:

a. Current flow through the resistor.
b. Current flow through the inductor.
c. The total line current.
d. The phase angle between the voltage and total current flow.
e. Express all currents in polar notation.
f. Use a calculator to convert all currents to rectangular notation.

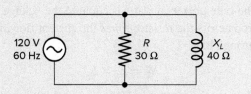

Figure 18-3 Circuit for example 18-1.

Solution:

a.
$$I_R = \frac{E}{R}$$
$$= \frac{120 \text{ V}}{30 \text{ }\Omega}$$
$$= 4 \text{ A}$$

b.
$$I_L = \frac{E}{X_L}$$
$$= \frac{120 \text{ V}}{40 \text{ }\Omega}$$
$$= 3 \text{ A}$$

c.
$$I_T = \sqrt{I_R^2 + I_L^2}$$
$$= \sqrt{4^2 + 3^2}$$
$$= \sqrt{25}$$
$$= 5 \text{ A}$$

d.
$$\theta = \tan^{-1} \frac{I_L}{I_R}$$
$$= \tan^{-1} \frac{3}{4}$$
$$= \tan^{-1} (0.75)$$
$$= 36.9°$$

e.
$$I_T = 5 \angle -36.9°$$
$$I_R = 4 \angle 0°$$
$$I_L = 3 \angle -90°$$

f.
$$I_T = 4 - j3$$
$$I_R = 4 + j0$$
$$I_L = 0 - j3$$

The **impedance** (*Z*) of a parallel *RL* circuit is the total opposition to the flow of current. It includes the opposition (*R*) offered by the resistive branch and the inductive reactance (*X_L*) offered by the inductive branch. The impedance of a parallel *RL* circuit is calculated similarly to a parallel resistive circuit. However, since X_L and *R* are vector quantities, they must be added vectorially. As a result, the equation for the impedance of a parallel *RL* circuit consisting of a single resistor and inductor is:

$$Z = \frac{RX_L}{\sqrt{R^2 + X_L^2}}$$

where the quantity in the denominator is the vector sum of the resistance and inductive reactance. If there is more than one resistive or inductive branch, *R* and X_L must equal the total resistance or reactance of theses parallel branches.

When the total current (*I_T*) and the applied voltage are known, the impedance is ***more easily calculated*** using the Ohm's law as follows:

$$Z = \frac{E}{I_T}$$

The impedance of a parallel *RL* circuit is ***always less*** than the resistance or inductive reactance of any one branch. This is because each branch creates a separate path for current flow, thus reducing the overall or total circuit opposition to the current flow. The branch that has the greater amount of current flow (or lesser amount of opposition) has the most effect on the phase angle. This is the ***opposite*** of a series *RL* circuit. In a parallel *RL* circuit, if X_L is larger than *R*, the resistive branch current is greater than the inductive branch current so the phase angle between the applied voltage and total current is closer to 0 degrees (***more resistive in nature***).

EXAMPLE 18-2

Problem: For the parallel *RL* circuit shown in Figure 18-4, determine:

a. Impedance (*Z*) based on the given *R* and X_L values.
b. Current flow through the resistor and inductor.
c. The total line current.
d. Impedance (*Z*) based on the total current (*I_T*) and the applied voltage values.

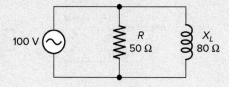

Figure 18-4 Circuit for example 18-2.

Solution:

a.
$$Z = \frac{RX_L}{\sqrt{R^2 + X_L^2}}$$
$$= \frac{50 \times 80}{\sqrt{50^2 + 80^2}}$$
$$= \frac{4{,}000}{94.4}$$
$$= 42.4\ \Omega$$

b.
$$I_R = \frac{E}{R}$$
$$= \frac{100\ \text{V}}{50\ \Omega}$$
$$= 2\ \text{A}$$

$$I_L = \frac{E}{X_L}$$
$$= \frac{100\ \text{V}}{80\ \Omega}$$
$$= 1.25\ \text{A}$$

c.
$$I_T = \sqrt{I_R^2 + I_L^2}$$
$$= \sqrt{2^2 + 1.25^2}$$
$$= \sqrt{5.56}$$
$$= 2.36\ \text{A}$$

d.
$$Z = \frac{E}{I_T}$$
$$= \frac{100\ \text{V}}{2.36\ \text{A}}$$
$$= 42.4\ \Omega$$

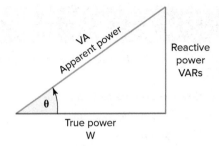

Figure 18-6 Power triangle for a *RL* parallel circuit.

as it is for the *RL* series circuit. The following is a summary of these formulas:

- The **true power** in watts is equal to the voltage drop across the resistor *times* the current flowing through it:

$$W = E_R \times I_R$$

- The **reactive power** in VARs is equal to the voltage drop across the inductor *times* the current flowing through it:

$$\text{VARs} = E_L \times I_L$$

- The **apparent power** in VA is equal to the applied voltage *times* the total current:

$$\text{VA} = E_T \times I_T$$

Figure 18-6 shows the power triangle for a *RL* parallel circuit. Apply the Pythagorean theorem, and the various power components can be determined using the following equations:

$$\text{VA} = \sqrt{W^2 + \text{VARs}^2}$$
$$W = \sqrt{\text{VA}^2 - \text{VARs}^2}$$
$$\text{VARs} = \sqrt{\text{VA}^2 - W^2}$$
$$\text{Power factor} = \frac{\text{true power (watts)}}{\text{apparent power (VA)}}$$

Power factor (PF) in a *RL* parallel circuit is the ratio of true power to the apparent power just as it is in the series *RL* circuit. There are, however, some differences in the

In the parallel *RL* circuit, the VA (apparent power) includes both the **watts** (true power) and the VARs (reactive power), as shown in Figure 18-5. The true power (W) is that power dissipated by the resistive branch, and the reactive power (VARs) is the power that is returned to the source by the inductive branch. The relationship of VA, W, and VARs is the *same* for the *RL* parallel circuit

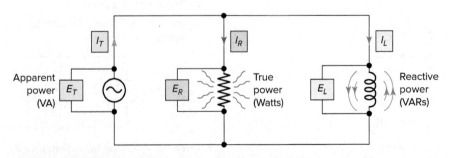

Figure 18-5 Power components of a *RL* parallel circuit.

other formulas used to calculate power factor in the series and parallel *RL* circuits. In a series *RL* circuit, the power factor could be found by dividing the voltage drop across the resistor by the total applied voltage. In a parallel circuit the voltage is the same but the currents are different, and power factor can be calculated using the formula

$$PF = \frac{I_R}{I_T}$$

Another power factor formula that is different involves resistance and impedance. In the parallel *RL* circuit, the impedance will be *less* than the resistance. Therefore, when PF is computed using resistance and impedance, the formula used is

$$PF = \frac{Z}{R}$$

EXAMPLE 18-3

Problem: For the parallel *RL* circuit shown in Figure 18-7, determine:

a. Current flow through the resistor.
b. True power in watts.
c. Current flow through the inductor.
d. Reactive power in VARs.
e. Inductance of the inductor.
f. Total current flow.
g. Circuit impedance.
h. Apparent power in VA.
i. Power factor.
j. The circuit phase angle θ.

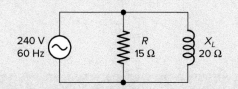

Figure 18-7 **Circuit for example 18-3.**

Solution:

Step 1. Make a table and record all known values.

	E	I	L	R/X$_L$/Z	W/VA/ VARs	∠ θ	PF
R	240 V		N/A	15 Ω		0°	N/A
L	240 V			20 Ω		90°	
Total	240 V		N/A				

Step 2. Calculate the current through the resistor and enter the value in the table.

$$I_R = \frac{E}{R}$$
$$= \frac{240 \text{ V}}{15 \text{ Ω}}$$
$$= 16 \text{ A}$$

	E	I	L	R/X$_L$/Z	W/VA/ VARs	∠ θ	PF
R	240 V	16 A	N/A	15 Ω		0°	N/A
L	240 V			20 Ω		90°	
Total	240 V		N/A				

Step 3. Calculate the true power and enter the value in the table.

$$W = E_R \times I_R$$
$$= 240 \text{ V} \times 16 \text{ A}$$
$$= 3,840 \text{ watts}$$

	E	I	L	R/X$_L$/Z	W/VA/ VARs	∠ θ	PF
R	240 V	16 A	N/A	15 Ω	3,840 W	0°	N/A
L	240 V			20 Ω		90°	
Total	240 V		N/A				

Step 4. Calculate the current through the inductor and enter the value in the table.

$$I_L = \frac{E}{X_L}$$
$$= \frac{240 \text{ V}}{20 \text{ Ω}}$$
$$= 12 \text{ A}$$

	E	I	L	R/X$_L$/Z	W/VA/ VARs	∠ θ	PF
R	240 V	16 A	N/A	15 Ω	3,840 W	0°	N/A
L	240 V	12 A		20 Ω		90°	
Total	240 V		N/A				

Step 5. Calculate the reactive power and enter the value in the table.

$$VARs = E_L \times I_L$$
$$= 240 \text{ V} \times 12 \text{ A}$$
$$= 2{,}880 \text{ VARs}$$

	E	I	L	$R/X_L/Z$	W/VA/ VARs	$\angle \theta$	PF
R	240 V	16 A	N/A	15 Ω	3,840 W	0°	N/A
L	240 V	12 A		20 Ω	2,880 VARs	90°	
Total	240 V		N/A				

Step 6. Calculate the inductance of the inductor and enter the value in the table.

$$X_L = 2\pi f L$$
$$L = \frac{X_L}{2\pi f}$$
$$= \frac{20}{377}$$
$$= 0.053 \text{ H } (53 \text{ mH})$$

	E	I	L	$R/X_L/Z$	W/VA/ VARs	$\angle \theta$	PF
R	240 V	16 A	N/A	15 Ω	3,840 W	0°	N/A
L	240 V	12 A	53 mH	20 Ω	2,880 VARs	90°	
Total	240 V		N/A				

Step 7. Calculate the total current and enter the value in the table.

$$I_T = \sqrt{I_R^2 + I_L^2}$$
$$= \sqrt{16^2 + 12^2}$$
$$= 20 \text{ A}$$

	E	I	L	$R/X_L/Z$	W/VA/ VARs	$\angle \theta$	PF
R	240 V	16 A	N/A	15 Ω	3,840 W	0°	N/A
L	240 V	12 A	53 mH	20 Ω	2,880 VARs	90°	
Total	240 V	20 A	N/A				

Step 8. Calculate the impedance and enter the value in the table.

$$Z = \frac{E}{I_T}$$
$$= \frac{240 \text{ V}}{20 \text{ A}}$$
$$= 12 \ \Omega$$

	E	I	L	$R/X_L/Z$	W/VA/ VARs	$\angle \theta$	PF
R	240 V	16 A	N/A	15 Ω	3,840 W	0°	N/A
L	240 V	12 A	53 mH	20 Ω	2,880 VARs	90°	
Total	240 V	20 A	N/A	12 Ω			

Step 9. Calculate the apparent power and enter the value in the table.

$$VA = E_T \times I_T$$
$$= 240 \text{ V} \times 20 \text{ A}$$
$$= 4{,}800 \text{ VA}$$

	E	I	L	$R/X_L/Z$	W/VA/ VARs	$\angle \theta$	PF
R	240 V	16 A	N/A	15 Ω	3,840 W	0°	N/A
L	240 V	12 A	53 mH	20 Ω	2,880 VARs	90°	
Total	240 V	20 A	N/A	12 Ω	4,800 VA		

Step 10. Calculate the power factor and enter the value in the table.

$$PF = \frac{I_R}{I_T}$$
$$= \frac{16}{20}$$
$$= 0.8 \quad \text{or} \quad 80\%$$

	E	I	L	$R/X_L/Z$	W/VA/ VARs	$\angle \theta$	PF
R	240 V	16 A	N/A	15 Ω	3840 W	0°	N/A
L	240 V	12 A	53 mH	20 Ω	2880 VARs	90°	
Total	240 V	20 A	N/A	12 Ω	4800 VA		80%

Step 11. Calculate the circuit phase angle θ and enter the value in the table.

$$\theta = \tan^{-1} \frac{I_L}{I_R}$$
$$= \tan^{-1} \frac{12}{16}$$
$$= \tan^{-1} (0.75)$$
$$= 36.9°$$

	E	I	L	R/X$_L$/Z	W/VA/VARs	∠ θ	PF
R	240 V	16 A	N/A	15 Ω	3,840 W	0°	
L	240 V	12 A	53 mH	20 Ω	2,880 VARs	90°	N/A
Total	240 V	20 A	N/A	12 Ω	4,800 VA	36.9°	80%

Part 1 Review Questions

1. List three characteristics of the voltage across each branch of a parallel *RL* circuit.

2. In a parallel *RL* circuit the total current is equal to the vector sum rather than the arithmetic sum. Why?

3. What is used as the reference vector in the vector diagram of a parallel *RL* circuit?

4. Assume the resistive element of a parallel *RL* circuit is increased. Will this cause the phase angle of the circuit to increase or decrease? Why?

5. In a parallel *RL* circuit the impedance or total opposition is always less than that of the individual resistance or inductive reactance. Why?

6. Define the terms *apparent power, reactive power,* and *true power* as they apply to the parallel *RL* circuit.

7. Current measurements of a parallel *RL* circuit indicate a current flow of 2 amperes through the resistive branch and 4 amperes through the inductive branch. Determine:
   a. The value of the total current flow.
   b. The phase angle between the voltage and total current.

8. For the parallel *RL* circuit shown in Figure 18-8, determine:
   a. Apparent power.
   b. True power.
   c. Reactive power.
   d. Circuit power factor.

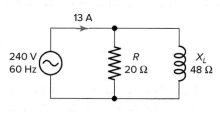

Figure 18-8 Circuit for review question 8.

9. Complete a table for all given and unknown quantities for the parallel *RL* circuit shown in Figure 18-9.

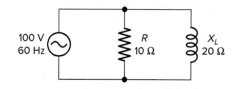

Figure 18-9 Circuit for review question 9.

PART 2 PARALLEL *RC* CIRCUIT

18.2 Parallel *RC* Circuits

The combination of a resistor and capacitor connected in parallel to an AC source, as illustrated in Figure 18-10, is called a **parallel *RC* circuit.** The conditions that exist in *RC* parallel circuits and the methods used for solving them are quite similar to those used for *RL* parallel circuits. The voltage is the same value across each parallel branch and provides the basis for expressing any phase differences. The principle *difference is one of phase relationship.* In a pure capacitor the current leads the voltage by 90 degrees, while in a pure inductor the current lags the voltage by 90 degrees.

The relationship between the voltage and currents in a parallel *RC* circuit is illustrated in the vector diagram of Figure 18-11 and summarized as follows:

- The *reference vector is labeled E* and represents the voltage in the circuit, which is common to all elements.
- Since the current through the resistor is in phase with the voltage across it, I_R (8 A) is shown superimposed on the voltage vector.
- The capacitor current I_C (12 A) leads the voltage by 90 degrees and is positioned in an upward direction, leading the voltage vector by 90 degrees.
- The vector addition of I_R and I_C gives a resultant that represents the total (I_T) or line current (14.4 A).

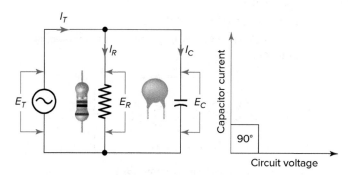

Figure 18-10 Parallel *RC* circuit.

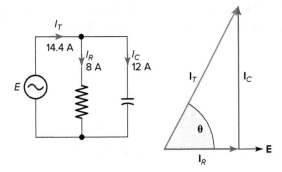

Figure 18-11 Parallel *RC* circuit vector diagram.

$$I_T = \sqrt{I_R^2 + I_C^2}$$
$$= \sqrt{8^2 + 12^2}$$
$$= \sqrt{208}$$
$$= 14.4 \text{ A}$$

- The angle theta (θ) represents the phase between the applied line voltage and current.

In a parallel *RC* circuit, the *line current leads the applied voltage* by some phase angle less than 90 degrees but greater than 0 degrees. The exact angle depends on whether the capacitive current or resistive current is greater. If there is *more capacitive current, the angle will be closer to 90 degrees,* while if the resistive current is greater, the angle is closer to 0 degrees. The value of the phase angle can be calculated from the values of the two branch currents using the following equation:

$$\theta = \tan^{-1} \frac{I_C}{I_R}$$

EXAMPLE 18-4

Problem: For the parallel *RC* circuit shown in Figure 18-12, determine:

a. Current flow through the resistor.

b. Current flow through the capacitor.

c. Total line current.

d. The phase angle between the voltage and total current flow.

e. Express all currents in polar notation.

f. Use a calculator to convert all currents to rectangular notation.

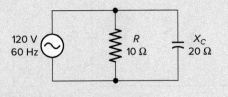

Figure 18-12 Circuit for example 18-4.

Solution:

a.
$$I_R = \frac{E}{R}$$
$$= \frac{120 \text{ V}}{10 \text{ }\Omega}$$
$$= 12 \text{ A}$$

b.
$$I_C = \frac{E}{X_C}$$
$$= \frac{120 \text{ V}}{20 \text{ }\Omega}$$
$$= 6 \text{ A}$$

c.
$$I_T = \sqrt{I_R^2 + I_C^2}$$
$$= \sqrt{12^2 + 6^2}$$
$$= \sqrt{180}$$
$$= 13.4 \text{ A}$$

d.
$$\theta = \tan^{-1} \frac{I_C}{I_R}$$
$$= \tan^{-1} \frac{6}{12}$$
$$= \tan^{-1} (0.5)$$
$$= 26.6°$$

e.
$$I_T = 13.4 \angle 26.6°$$
$$I_R = 12 \angle 0°$$
$$I_C = 6 \angle 90°$$

f.
$$I_T = 12 + j6$$
$$I_R = 12 + j0$$
$$I_C = 0 + j6$$

The **impedance (Z)** of a parallel *RC* circuit is similar to that of a parallel *RL* circuit and is summarized as follows:

- Impedance can be calculated directly from the resistance and capacitive reactance values using the equation

$$Z = \frac{RX_C}{\sqrt{R^2 + X_C^2}}$$

- Impedance can be calculated using the Ohm's law equation

$$Z = \frac{E}{I_T}$$

- The impedance of a parallel *RC* circuit is always less than the resistance or capacitive reactance of the individual branches.

EXAMPLE 18-5

Problem: For the parallel *RC* circuit shown in Figure 18-13 determine the:

a. Current flow through the resistor (I_R).

b. Current flow through the capacitor (I_C).

c. The total line current (I_T).

d. Impedance (Z).

e. Phase angle between the voltage and total current flow.

f. If the circuit is more resistive or capacitive.

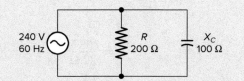

Figure 18-13 Circuit for example 18-5.

Solution:

a.
$$I_R = \frac{E}{R}$$
$$= \frac{240 \text{ V}}{200 \text{ }\Omega}$$
$$= 1.2 \text{ A}$$

b.
$$I_C = \frac{E}{X_C}$$
$$= \frac{240 \text{ V}}{100 \text{ }\Omega}$$
$$= 2.4 \text{ A}$$

c.
$$I_T = \sqrt{I_R^2 + I_C^2}$$
$$= \sqrt{1.2^2 + 2.4^2}$$
$$= \sqrt{7.20}$$
$$= 2.68 \text{ A}$$

d.
$$Z = \frac{E}{I_T}$$
$$= \frac{240 \text{ V}}{2.68 \text{ A}}$$
$$= 89.6 \text{ }\Omega$$

e.
$$\theta = \tan^{-1}\frac{I_C}{I_R}$$
$$= \tan^{-1}\frac{2.4}{1.2}$$
$$= \tan^{-1}(2)$$
$$= 63.4°$$

f. The circuit is more capacitive in nature because the capacitive current is greater than the resistive current.

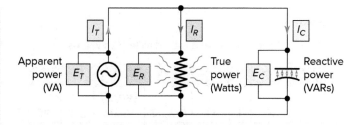

Figure 18-14 Power components of a *RC* parallel circuit.

The **power components** for a parallel *RC* circuit are illustrated in Figure 18-14. The formulas that apply are the same as that of a parallel *RL* circuit:

$$VA = \sqrt{W^2 + VARs^2}$$
$$W = \sqrt{VA^2 - VARs^2}$$
$$VARs = \sqrt{VA^2 - W^2}$$
$$\text{Power factor} = \frac{\text{true power (watts)}}{\text{apparent power (VA)}}$$

The **power factor** of a parallel *RC* circuit is ***always leading.*** Any time the branch resistance increases, less current flows through it and the circuit becomes more capacitive, resulting in a lower power factor. The reverse is true if the resistance decreases. With current or resistance and impedance values, the power factor can be determined as follows:

$$\text{Power factor} = \frac{I_R}{I_T}$$
$$\text{Power factor} = \frac{Z}{R}$$

EXAMPLE 18-6

Problem: For the parallel *RC* circuit shown in Figure 18-15, determine the:

a. Capacitive reactance of the capacitor (X_C).

b. Current flow through the capacitor (I_C).

c. Reactive power of the capacitor (VARs).

d. Current flow through the resistor (I_R).

e. True power (W).

f. Total line current flow (I_T).

g. Circuit impedance (Z).

h. Apparent power (VA).

i. Power factor (PF).

j. Circuit phase angle θ.

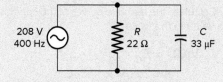

Figure 18-15 Circuit for example 18-6.

Solution:

Step 1. Make a table and record all known values.

	E	I	C	R/X_C/Z	W/VA/VARs	∠ θ	PF
R	208 V		N/A	22 Ω		0°	N/A
C	208 V		33 μF			90°	
Total	208 V		N/A				

Step 2. Calculate the capacitive reactance of the capacitor and enter the value in the table.

$$X_C = \frac{1}{2\pi fC}$$
$$= \frac{1,000,000}{2 \times 3.14 \times 400 \times 33}$$
$$= 12.1\ \Omega$$

	E	I	C	R/X_C/Z	W/VA/VARs	∠ θ	PF
R	208 V		N/A	22 Ω		0°	N/A
C	208 V		33 μF	12.1 Ω		90°	
Total	208 V		N/A				

Step 3. Calculate the current flow of the capacitor and enter the value in the table.

$$I_C = \frac{E_C}{X_C}$$
$$= \frac{208\ \text{V}}{12.1\ \Omega}$$
$$= 17.2\ \text{A}$$

	E	I	C	R/X_C/Z	W/VA/VARs	∠ θ	PF
R	208 V		N/A	22 Ω		0°	N/A
C	208 V	17.2 A	33 μF	12.1 Ω		90°	
Total	208 V		N/A				

Step 4. Calculate the reactive power of the capacitor and enter the value in the table.

$$\text{VARs} = E_C \times I_C$$
$$= 208\ \text{V} \times 17.2\ \text{A}$$
$$= 3,578\ \text{VARs}$$

	E	I	C	R/X_C/Z	W/VA/VARs	∠ θ	PF
R	208 V		N/A	22 Ω		0°	N/A
C	208 V	17.2 A	33 μF	12.1 Ω	3,578 VARs	90°	
Total	208 V		N/A				

Step 5. Calculate the current flow through the resistor and enter the value in the table.

$$I_R = \frac{E_R}{R}$$
$$= \frac{208\ \text{V}}{22\ \Omega}$$
$$= 9.45\ \text{A}$$

	E	I	C	R/X_C/Z	W/VA/VARs	∠ θ	PF
R	208 V	9.45 A	N/A	22 Ω		0°	N/A
C	208 V	17.2 A	33 μF	12.1 Ω	3,578 VARs	90°	
Total	208 V		N/A				

Step 6. Calculate the true power and enter the value in the table.

$$W = E_R \times I_R$$
$$= 208\ \text{V} \times 9.45\ \text{A}$$
$$= 1,966\ \text{W}$$

	E	I	C	R/X_C/Z	W/VA/VARs	∠ θ	PF
R	208 V	9.45 A	N/A	22 Ω	1,966 W	0°	N/A
C	208 V	17.2 A	33 μF	12.1 Ω	3,578 VARs	90°	
Total	208 V		N/A				

Step 7. Calculate the total line current and enter the value in the table.

$$I_T = \sqrt{I_R^2 + I_C^2}$$
$$= \sqrt{9.45^2 + 17.2^2}$$
$$= \sqrt{385}$$
$$= 19.6\ \text{A}$$

	E	I	C	R/X$_C$/Z	W/VA/VARs	∠θ	PF
R	208 V	9.45 A	N/A	22 Ω	1,966 W	0°	N/A
C	208 V	17.2 A	33 µF	12.1 Ω	3,578 VARs	90°	N/A
Total	208 V	19.6 A	N/A				

Step 8. Calculate the impedance and enter the value in the table.

$$Z = \frac{E_T}{I_T}$$
$$= \frac{208 \text{ V}}{19.6 \text{ Ω}}$$
$$= 10.6 \text{ Ω}$$

	E	I	C	R/X$_C$/Z	W/VA/VARs	∠θ	PF
R	208 V	9.45 A	N/A	22 Ω	1,966 W	0°	N/A
C	208 V	17.2 A	33 µF	12.1 Ω	3,578 VARs	90°	N/A
Total	208 V	19.6 A	N/A	10.6 Ω			

Step 9. Calculate the apparent power and enter the value in the table.

$$VA = E_T \times I_T$$
$$= 208 \text{ V} \times 19.6 \text{ A}$$
$$= 4,077 \text{ VA}$$

	E	I	C	R/X$_C$/Z	W/VA/VARs	∠θ	PF
R	208 V	9.45 A	N/A	22 Ω	1,966 W	0°	N/A
C	208 V	17.2 A	33 µF	12.1 Ω	3,578 VARs	90°	N/A
Total	208 V	19.6 A	N/A	10.6 Ω	4,077 VA		

Step 10. Calculate the power factor and enter the value in the table.

$$\text{Power factor} = \frac{\text{true power (watts)}}{\text{apparent power (VA)}}$$
$$= \frac{1,968 \text{ W}}{3,973 \text{ VA}}$$
$$= 49.5\% \quad \text{(leading)}$$

	E	I	C	R/X$_C$/Z	W/VA/VARs	∠θ	PF
R	208 V	9.45 A	N/A	22 Ω	1,966 W	0°	N/A
C	208 V	17.2 A	33 µF	12.1 Ω	3,578 VARs	90°	N/A
Total	208 V	19.6 A	N/A	10.6 Ω	4,077 VA		49.5%

Step 11. Calculate the circuit phase angle θ and enter the value in the table.

$$\theta = \tan^{-1} \frac{I_C}{I_R}$$
$$= \tan^{-1} \frac{17.2}{9.45}$$
$$= \tan^{-1} (1.82)$$
$$= 61.2°$$

	E	I	C	R/X$_C$/Z	W/VA/VARs	∠θ	PF
R	208 V	9.45 A	N/A	22 Ω	1,966 W	0°	N/A
C	208 V	17.2 A	33 µF	12.1	3,578 VARs	90°	N/A
Total	208 V	19.6 A	N/A	10.6	4,077 VA	61.2°	49.5%

Part 2 Review Questions

1. What is the main difference between a parallel *RL* and *RC* circuit?

2. Assume the resistance of the resistive component of a parallel *RC* circuit is increased. What effect, if any, will this have on the phase angle of the circuit?

3. A parallel *RC* circuit is connected to a 100-volt, 60-Hz source. The current flow through the resistor is measured and found to be 10 amps. The current flow through the capacitor is measured and found to be 10 amps. Determine:
   a. Line current (I_T).
   b. Impedance (Z).
   c. True power (W).
   d. Reactive power (VARs).
   e. Apparent power (VA).
   f. PF percentage.

4. For the circuit shown in Figure 18-16, determine:
   a. The amount of current flow through the resistor.
   b. The capacitive reactance of the capacitor.
   c. The amount of current flow through the capacitor.

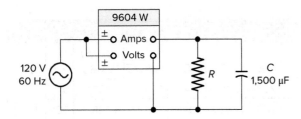

Figure 18-16 Circuit for review question 4.

d. The line current.

e. Apparent power.

f. PF percentage.

5. Complete a table for all given and unknown quantities for the parallel *RC* circuit shown in Figure 18-17.

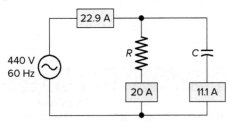

Figure 18-17 Circuit for review question 5.

PART 3 PARALLEL *RLC* CIRCUIT

18.3 Parallel *LC* Circuits

A **parallel *LC* circuit** consists of an inductance and a capacitance connected in parallel, as shown in Figure 18-18. The characteristics of an *LC* parallel circuit can be summarized as follows:

- It is assumed that there is no resistance in the circuit, only pure inductance and capacitance.
- The voltages across the branches of a parallel *LC* circuit are the same as the applied voltage, as they are in all parallel circuits.
- The voltage across the inductor is the same as, and therefore in phase with, the voltage across the capacitor.

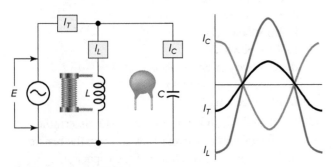

Figure 18-18 Parallel *LC* circuit.

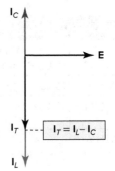

Figure 18-19 Current vector diagram for a parallel *LC* circuit.

- The capacitor current leads the voltage by 90 degrees, while the inductor current lags the voltage by 90 degrees. Therefore, there is 180-degree difference in phase between the capacitor and inductor current waveforms.

Figure 18-19 shows the current vector diagram for a parallel *LC* circuit, the construction of which is summarized as follows:

- ***The voltage, which is common to all circuit elements, is used as the reference vector.***
- The ***capacitor current (I_C) leads*** the voltage by 90 degrees and is drawn leading the voltage vector by 90 degrees.
- The ***inductor current (I_L) lags*** the voltage by 90 degrees and is drawn lagging the voltage vector by 90 degrees.
- Currents I_C and I_L are 180 degrees out of phase so the line current is equal to their vector sum.
- The vector addition is done by subtracting the smaller current, I_L or I_C, from the larger.
- The total line current has the phase characteristic of the ***larger branch current.***
- Therefore, if the inductive branch current is the larger, the line current is inductive and lags the applied voltage by 90 degrees. Similarly, if the capacitive branch current is the larger, the line current is capacitive and leads the applied voltage by 90 degrees.

The line current for a parallel *LC* circuit is always less than one of the branch currents and sometimes less than both. This characteristic is illustrated in the circuit of Figure 18-20 and summarized as follows:

- The ***total line current is only 6 amps,*** while the current through the inductive branch is ***10 amps.***
- This is different from all other previously discussed parallel circuits in which the line current was always greater than any one of the branch currents.
- The reason is due to the fact that the branch currents are 180 degrees out of phase.

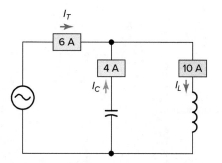

Figure 18-20 Total line current of a parallel *LC* circuit.

- As a result of the phase difference, some ***cancellation takes place*** between the two currents when they combine to produce the line current.

EXAMPLE 18-7

Problem: For the parallel *LC* circuit shown in Figure 18-21, determine the:

  a. Value of the total line current.

  b. Phase angle.

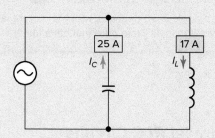

Figure 18-21 Circuit for example 18-7.

Solution:

a.
$$I_T = I_C - I_L$$
$$= 25 \text{ A} - 17 \text{ A}$$
$$= 8 \text{ A}$$

b. Since the capacitor current is greater than that of the inductor, the line current leads the applied voltage by 90 degrees. The value of the current written in polar notation would be

$$I_T = 8 \text{ A} \angle 90°$$

The magnitude of the branch currents depends on the reactance in the respective branches and can be found from the Ohm's law equations

$$I_L = \frac{E}{X_L}$$

$$I_C = \frac{E}{X_C}$$

The impedance of a parallel *LC* circuit can be calculated directly from the inductive and capacitive reactance values using the equations

$$Z = \frac{X_L \times X_C}{X_L - X_C} \quad \text{(for } X_L \text{ larger than } X_C\text{)}$$

$$Z = \frac{X_L \times X_C}{X_C - X_L} \quad \text{(for } X_C \text{ larger than } X_L\text{)}$$

The impedance can also be calculated using the Ohm's law equation

$$Z = \frac{E}{I_T}$$

EXAMPLE 18-8

Problem: For the parallel *LC* circuit shown in Figure 18-22, determine the:

  a. Impedance (*Z*) using the given reactance values.

  b. Current flow through the capacitor (*I*$_C$).

  c. Current flow through the inductor (*I*$_L$).

  d. Line current (*I*$_T$).

  e. Impedance (*Z*) using the applied voltage and calculated current.

  f. Phase angle between the applied voltage and the line current.

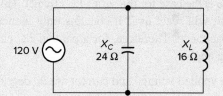

Figure 18-22 Circuit for example 18-8.

Solution:

a.
$$Z = \frac{X_L \times X_C}{X_C - X_L}$$
$$= \frac{16 \times 24}{24 - 16}$$
$$= \frac{384}{8}$$
$$= 48 \ \Omega$$

b.
$$I_C = \frac{E}{X_C}$$
$$= \frac{120 \text{ V}}{24 \ \Omega}$$
$$= 5 \text{ A}$$

c.

$$I_L = \frac{E}{X_L}$$

$$= \frac{120 \text{ V}}{16 \text{ }\Omega}$$

$$= 7.5 \text{ A}$$

d.

$$I_T = I_L - I_C$$

$$= 7.5 \text{ A} - 5 \text{ A}$$

$$= 2.5 \text{ A}$$

e.

$$Z = \frac{E}{I_T}$$

$$= \frac{120 \text{ V}}{2.5 \text{ A}}$$

$$= 48 \text{ }\Omega$$

f. Since the inductor current is greater than that of the capacitor, we know the line current lags the applied voltage by 90 degrees. The value of the current written in polar notation would be

$$I_T = 2.5 \text{ A} \angle -90°.$$

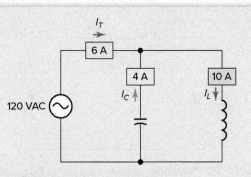

Figure 18-23 Circuit for example 18-9.

Solution:

a. Apparent power $= E \times I_T$

$$= 120 \text{ V} \times 6 \text{ A}$$

$$= 720 \text{ VA}$$

b. Reactive power = apparent power

$$= 720 \text{ VARs}$$

c. True power = zero (pure reactive load)

d. Power factor = zero (pure reactive load)

e. Phase angle is 90 degrees, with the current lagging the voltage (acts like a pure inductive load).

For any parallel *LC* circuit, the applied voltage and current are always 90 degrees out of phase with each other (the same is the case for the series *LC* circuit). This means the circuit will react as if it contains only inductance or only capacitance. Therefore, for a parallel *LC* circuit the following apply:

- The applied voltage and current are 90 degrees out of phase.
- The true power or watts is equal to zero.
- The apparent power (VA) is equal to the reactive power (VARs).
- The power factor is equal to zero.

EXAMPLE 18-9

Problem: For the parallel *LC* circuit shown in Figure 18-23, determine the:

a. Apparent power (VA).

b. Reactive power (VARs).

c. True power (W).

d. Power factor (PF).

e. Circuit phase angle θ.

18.4 Parallel *RLC* Circuits

A **parallel *RLC*** circuit contains elements of pure resistance, inductance, and capacitance connected in parallel, as shown in Figure 18-24. The characteristics of the *RLC* parallel circuit can be summarized as follows:

- The voltages across each component are all equal and in phase with one another: $E_T = E_R = E_L = E_C$.
- Their currents, however, are all out of phase with each other.
- The **resistive current** (I_R) is in phase with the voltage (E_R).
- The **inductive current** (I_L) lags the voltage (E_L) by 90 degrees.
- The **capacitive current** (I_C) leads the voltage (E_C) by 90 degrees.

The three currents of a parallel *RLC* circuit are combined, as shown in the vector diagram of Figure 18-25, and constructed as follows:

- A horizontal reference line representing the common voltage element is drawn first.

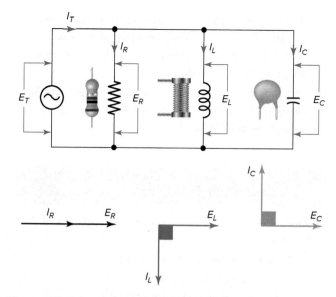

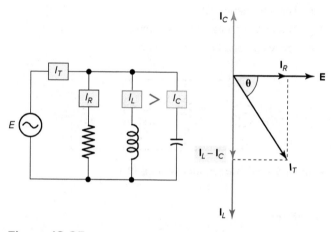

Figure 18-24 Parallel *RLC* circuit.

Figure 18-25 Current vector diagram for a parallel *RLC* circuit.

- The current through the resistor is in phase with the voltage and, therefore, is placed directly on the voltage line.
- The current through the capacitor leads the voltage by 90 degrees and so is drawn upward at a 90-degree angle from the voltage.
- The current through the inductor lags the voltage by 90 degrees and so is drawn downward at a 90-degree angle from the voltage.
- To combine the currents, the two reactive current values, which are 180 degrees out of phase with each other, are subtracted.
- The total line current (I_T) is the vector sum of the current flow through the resistor (I_R) and the difference in current between I_L and I_C.

- In the vector diagram illustrated, I_L is greater than I_C, so the complete circuit behaves as a resistor and inductor in parallel.
- The total line current lags the applied voltage by the phase angle theta (θ).
- Whenever I_C is greater than I_L, the circuit performs as a resistor and capacitor in parallel.

The **vector addition** of I_R, I_L, and I_C can be completed as follows:

- First, the two reactive currents are added, using the methods followed for parallel *LC* circuits:

$$I_X = I_L - I_C \quad \text{(if } I_L \text{ is larger than } I_C\text{)}$$

$$I_X = I_C - I_L \quad \text{(if } I_C \text{ is larger than } I_L\text{)}$$

- To find the total line current, the quantity I_X is then added to I_R, using the Pythagorean theorem:

$$I_T = \sqrt{I_R^2 + I_X^2}$$

- These equations can be combined to give the equation for total line current in terms of branch currents:

$$I_T = \sqrt{I_R^2 + (I_L - I_C)^2} \quad \text{(if } I_L \text{ is larger than } I_C\text{)}$$

$$I_T = \sqrt{I_R^2 + (I_C - I_L)^2} \quad \text{(if } I_C \text{ is larger than } I_L\text{)}$$

EXAMPLE 18-10

Problem:

a. Calculate the total line current for the parallel *RLC* circuit shown in Figure 18-26.

b. Draw a current vector diagram for the circuit (Figure 18-27). Is the circuit capacitive or inductive? Why?

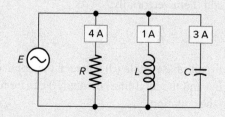

Figure 18-26 Circuit for example 18-10.

Solution:

a.

$$I_T = \sqrt{I_R^2 + (I_C - I_L)^2}$$

$$= \sqrt{4^2 + (3 - 1)^2}$$

$$= \sqrt{16 + 4}$$

$$= \sqrt{20}$$

$$= 4.47 \text{ A}$$

b. The circuit is capacitive in nature because the capacitor current is greater than the inductor current.

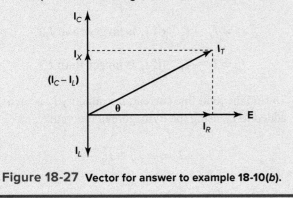

Figure 18-27 Vector for answer to example 18-10(*b*).

The **impedance** and **circuit phase angle** can be found as follows:

- The impedance, *Z*, is found using either Ohm's law or the product-sum method for a parallel circuit.
- Although the product-sum method can be used to determine the impedance from resistance and reactance values, it is more involved for a parallel *RLC* circuit. The most practical method used to find the impedance of a parallel *RLC* circuit is to apply Ohm's law for alternating current to the total circuit:

$$Z = \frac{E_T}{I_T}$$

- To find the total current (I_T), we must first calculate the individual branch currents (I_R, I_L, and I_C) and then add them vectorially:

$$I_R = \frac{E}{R} \qquad I_L = \frac{E}{X_L} \qquad I_C = \frac{E}{X_C}$$

- The circuit phase angle (θ) between the applied voltage (E_T) and the total line current (I_T) can be found using the equation

$$\theta = \tan^{-1} \frac{I_X}{I_R}$$

- When I_L is greater than I_C, the total current in a parallel *RLC* circuit has a **negative phase angle,** which indicates the circuit has the characteristics of a parallel *RL* circuit.
- When I_C is greater than I_L, the **phase angle will be positive,** indicating the circuit has the characteristics of a parallel *RC* circuit.

EXAMPLE 18-11

Problem: For the parallel *RLC* circuit shown in Figure 18-28, determine:

   a. Current flow through the resistor (I_R).
   b. Current flow through the inductor (I_L).
   c. Current flow through the capacitor (I_C).
   d. Net reactive current (I_X).
   e. Total line current (I_T).
   f. The impedance (*Z*).
   g. The phase angle (θ) of the circuit.
   h. The value of I_T in polar notation.

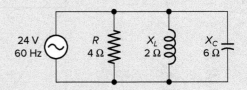

Figure 18-28 Circuit for example 18-11.

Solution:

a.

$$I_R = \frac{E}{R}$$

$$= \frac{24 \text{ V}}{4 \text{ }\Omega}$$

$$= 6 \text{ A}$$

$$= 6 \text{ A} \angle 0°$$

b.

$$I_L = \frac{E}{X_L}$$

$$= \frac{24 \text{ V}}{2 \text{ }\Omega}$$

$$= 12 \text{ A}$$

$$= 12 \text{ A} \angle -90°$$

c.

$$I_C = \frac{E}{X_C}$$

$$= \frac{24 \text{ V}}{6 \text{ }\Omega}$$

$$= 4 \text{ A}$$

$$= 4 \text{ A} \angle 90°$$

d.
$$I_X = I_L - I_C$$
$$= 12\ A - 4\ A$$
$$= 8\ A$$
$$= 8\ A \angle -90°$$

e.
$$I_T = \sqrt{I_R^2 + I_X^2}$$
$$= \sqrt{6^2 + 8^2}$$
$$= \sqrt{36 + 64}$$
$$= \sqrt{100}$$
$$= 10\ A$$

f.
$$Z = \frac{E_T}{I_T}$$
$$= \frac{24\ V}{10\ A}$$
$$= 2.4\ \Omega$$

g.
$$\theta = \tan^{-1}\frac{I_X}{I_R}$$
$$= \tan^{-1}\frac{8}{6}$$
$$= 53.1°$$

h.
$$I_T = 10\ A \angle -53.1°$$

Power in a parallel *RLC* circuit can be summarized as follows:

- The true power (watts) of any AC circuit is consumed only by the resistive component. Formulas used to find **true power** in a parallel *RLC* circuit are similar to those previously used for the series *RLC* circuit and are as follows:

$$W = I_R^2 \times R$$
$$W = \frac{E_R^2}{R}$$
$$W = E_R \times I_R$$
$$W = E_T \times I_T \times PF\ (\cos \theta)$$

- The **reactive component (VARs)** in a parallel *RLC* circuit will be produced by the circuit's inductive and capacitive reactances. Since these two reactances are 180 degrees out of phase, the net or total volt-amps reactive (VARs) is equal to the *difference* between the two. When the inductive VARs are greater than the capacitive VARs, it indicates the net circuit effect is inductive and the VARs can be found as follows:

VARs = (inductive VARs) − (capacitive VARs)
$$\textbf{VARs} = (I_L^2\,X_L) - (I_C^2\,X_C)$$
$$\textbf{VARs} = (E_L \times I_L) - (E_C \times I_C)$$

When the reverse is true, the net circuit effect is capacitive, and the inductive VARs are subtracted from the larger capacitive VARs.

- The circuit's **power factor (PF)** is always equal to the cosine of the angle theta (θ) that separates the circuit's applied voltage and total line current. The PF of a parallel *RLC* circuit can be found using any of the following equations:

$$PF = \cos(\theta) \qquad PF = \frac{Z}{R} \qquad PF = \frac{I_R}{I_T} \qquad PF = \frac{W}{VA}$$

A **lagging power factor** means the current lags the applied voltage and is always the case in a parallel *RLC* circuit when I_L is greater than I_C (or X_L is less than X_C). Similarly, a **leading power factor** means the current leads the applied voltage and is always the case in a parallel *RLC* circuit when I_C is greater than I_L (or X_C is less than X_L).

EXAMPLE 18-12

Problem: From the measurements taken of the parallel *RLC* circuit shown in Figure 18-29, determine:

a. True power (W).

b. Reactive power (VARs).

c. Apparent power (VA).

d. Power factor (PF).

e. If the power factor is leading or lagging.

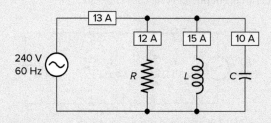

Figure 18-29 Circuit for example 18-12.

Solution:

a.
$$W = E_R \times I_R$$
$$= 240\ V \times 12\ A$$
$$= 2{,}880\ watts$$

b.
$$VARs\ (inductor) = E_L \times I_L$$
$$= 240\ V \times 15\ A$$
$$= 3{,}600\ VARs$$

$$VARs\ (capacitor) = E_C \times I_C$$
$$= 240\ V \times 10\ A$$
$$= 2{,}400\ VARs$$

$$\text{VARs (net)} = \text{(inductive VARs)} - \text{(capacitive VARs)}$$
$$= 3{,}600 - 2{,}400$$
$$= 1{,}200 \text{ VARs}$$

c.
$$\text{Volt-amps} = E_T \times I_T$$
$$= 240 \text{ V} \times 13 \text{ A}$$
$$= 3{,}120 \text{ VA}$$

d.
$$\text{PF} = \frac{\text{W}}{\text{VA}}$$
$$= \frac{2{,}880}{3{,}120}$$
$$= 0.923 \quad \text{or} \quad 92.3\%$$

e. The power factor is lagging because the inductive current is greater than the capacitive current.

EXAMPLE 18-13

Problem: For the parallel *RLC* circuit shown in Figure 18-30, determine:

a. Branch currents (I_R, I_L, and I_C).
b. Total line current (I_T).
c. Power factor (PF).
d. Impedance (*Z*).
e. True power (W).
f. Net reactive power (VARs).
g. Apparent power (VA).

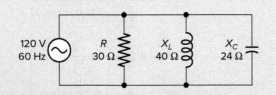

Figure 18-30 Circuit for example 18-13.

Solution:
Step 1. Make a table and record all known values.

	E	I	R/X$_L$/X$_C$/Z	W/VA/VARs	PF
R	120 V		30 Ω		
L	120 V		40 Ω		N/A
C	120 V		24 Ω		
Total	120 V				

Step 2. Calculate I_R, I_L, and I_C and enter the values in the table.

$$I_R = \frac{E}{R} \qquad I_L = \frac{E}{X_L} \qquad I_C = \frac{E}{X_C}$$
$$= \frac{120 \text{ V}}{30 \text{ Ω}} \qquad = \frac{120 \text{ V}}{40 \text{ Ω}} \qquad = \frac{120 \text{ V}}{24 \text{ Ω}}$$
$$= 4 \text{ A} \qquad\quad = 3 \text{ A} \qquad\quad = 5 \text{ A}$$

	E	I	R/X$_L$/X$_C$/Z	W/VA/VARs	PF
R	120 V	4 A	30 Ω		
L	120 V	3 A	40 Ω		N/A
C	120 V	5 A	24 Ω		
Total	120 V				

Step 3. Calculate the total line current (I_T) and enter the value in the table.

$$I_T = \sqrt{I_R^2 + (I_C - I_L)^2}$$
$$= \sqrt{4^2 + (5 - 3)^2}$$
$$= \sqrt{16 + 4}$$
$$= \sqrt{20}$$
$$= 4.47 \text{ A}$$

	E	I	R/X$_L$/X$_C$/Z	W/VA/VARs	PF
R	120 V	4 A	30 Ω		
L	120 V	3 A	40 Ω		N/A
C	120 V	5 A	24 Ω		
Total	120 V	4.47 A			

Step 4. Calculate the power factor (PF) and enter the value in the table.

	E	I	R/X$_L$/X$_C$/Z	W/VA/VARs	PF
R	120 V	4 A	30 Ω		
L	120 V	3 A	40 Ω		N/A
C	120 V	5 A	24 Ω		
Total	120 V	4.47 A			89.5%

$$PF = \frac{I_R}{I_T}$$
$$= \frac{4}{4.47}$$
$$= 0.895 \quad \text{or} \quad 89.5\% \quad \text{(leading)}$$

Step 5. Calculate the impedance (Z) and enter the value in the table.

$$Z = \frac{E_T}{I_T}$$
$$= \frac{120 \text{ V}}{4.47 \text{ A}}$$
$$= 26.8 \ \Omega$$

	E	I	$R/X_L/X_C/Z$	W/VA/ VARs	PF
R	120 V	4 A	30 Ω		
L	120 V	3 A	40 Ω		N/A
C	120 V	5 A	24 Ω		
Total	120 V	4.47 A	26.8 Ω		89.5%

Step 6. Calculate the true power (W) and enter the value in the table.

$$W = I_R^2 \times R$$
$$= 4^2 \times 30$$
$$= 480 \text{ W}$$

	E	I	$R/X_L/X_C/Z$	W/VA/ VARs	PF
R	120 V	4 A	30 Ω	480 W	
L	120 V	3 A	40 Ω		N/A
C	120 V	5 A	24 Ω		
Total	120 V	4.47 A	26.8 Ω		89.5%

Step 7. Calculate the reactive power (VARs) and enter the values in the table.

$$\text{VARs (inductor)} = E_L \times I_L$$
$$= 120 \text{ V} \times 3 \text{ A}$$
$$= 360 \text{ VARs}$$

$$\text{VARs (capacitor)} = E_C \times I_C$$
$$= 120 \text{ V} \times 5 \text{ A}$$
$$= 600 \text{ VARs}$$

$$\text{VARs (net)} = \text{(capacitive VARs)} - \text{(inductive VARs)}$$

$$= 600 - 360$$
$$= 240 \text{ VARs}$$

	E	I	$R/X_L/X_C/Z$	W/VA/ VARs	PF
R	120 V	4 A	30 Ω	480 W	
L	120 V	3 A	40 Ω	360 VARs	N/A
C	120 V	5 A	24 Ω	600 VARs	
Total	120 V	4.47 A	26.8 Ω		89.5%

Step 8. Calculate the apparent power (VA) and enter the value in the table.

$$\text{Volt-amps} = E_T \times I_T$$
$$= 120 \text{ V} \times 4.47 \text{ A}$$
$$= 536 \text{ VA}$$

	E	I	$R/X_L/X_C/Z$	W/VA/ VARs	PF
R	120 V	4 A	30 Ω	480 W	
L	120 V	3 A	40 Ω	360 VARs	N/A
C	120 V	5 A	24 Ω	600 VARs	
Total	120 V	4.47 A	26.8 Ω	536 VA	89.5%

When the R, X_L, and X_C values of a parallel RLC circuit are known, you can determine the impedance directly from these values. You must first find the net reactance (X) of the inductive and capacitive branches. Then, using X, you can find the impedance (Z) the same as you would in a parallel R_L or R_C circuit.

$$X = \frac{X_L \times X_C}{X_L + X_C}$$
$$Z = \frac{X \times R}{\sqrt{X^2 + R^2}}$$

Note:

- X_L is a positive quantity, and X_C is negative. Therefore, both X and Z will also be either negative (capacitive) or positive (inductive).
- Whenever Z is inductive, the line current will lag the applied voltage.
- Whenever Z is capacitive, the line current will lag the applied voltage.

- The angle of lead or lag depends on the relative values of X and R and can be found by the equations: $\tan\theta = \dfrac{R}{X}$ or $\cos\theta = \dfrac{Z}{R}$.
- Again, if the line current and the applied voltage are known, the impedance can also be found using the equation: $Z = \dfrac{E_{LINE}}{I_{LINE}}$.

EXAMPLE 18-14

Problem: For a parallel *RLC* circuit consisting of a 15-Ω resistor, an inductor with 8-Ω inductive reactance, and a capacitor with 12-Ω capacitive reactance, determine:

a. The reactance (X).

b. The impedance (Z).

c. The phase angle (θ).

Solution:

a.
$$X = \frac{X_L \times X_C}{X_L + X_C}$$
$$= \frac{8 \times (-12)}{8 + (-12)}$$
$$= 24$$

b.
$$Z = \frac{X \times R}{\sqrt{X^2 + R^2}}$$
$$= \frac{24 \times 15}{\sqrt{(24)^2 + (15)^2}}$$
$$= \frac{360}{28.3}$$
$$= 12.7\text{-}\Omega \; inductive$$

c.
$$\theta = \tan^{-1}\frac{R}{X}$$
$$= \tan^{-1}\frac{15}{24}$$
$$= 32°$$

EXAMPLE 18-15

Problem: For a parallel *RLC* circuit consisting of a 20-Ω resistor, an inductor with 10-Ω inductive reactance, and a capacitor with 5-Ω capacitive reactance, determine:

a. The reactance (X).

b. The impedance (Z).

c. The phase angle (θ).

Solution:

a.
$$X = \frac{X_L \times X_C}{X_L + X_C}$$
$$= \frac{10 \times (-5)}{10 + (-5)}$$
$$= -10$$

b.
$$Z = \frac{X \times R}{\sqrt{X^2 + R^2}}$$
$$= \frac{10 \times 20}{\sqrt{(10)^2 + (20)^2}}$$
$$= \frac{200}{22.4}$$
$$= 8.9\text{-}\Omega \; capacitive$$

c.
$$\theta = \tan^{-1}\frac{R}{X}$$
$$= \tan^{-1}\frac{20}{10} = 2$$
$$= 63.4°$$

Part 3 Review Questions

1. For the parallel *LC* circuit shown in Figure 18-31, determine:
   a. The value of the line current.
   b. The phase angle between the applied voltage and the line current.

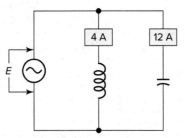

Figure 18-31 Circuit for review question 1.

2. For the parallel *LC* circuit shown in Figure 18-32, determine:
   a. The impedance (Z) using the given reactance values.
   b. Current flow through the capacitor (I_C).
   c. Current flow through the inductor (I_L).
   d. Line current (I_T).
   e. The impedance (Z) using the applied voltage and calculated current.
   f. Phase angle between the applied voltage and the line current.

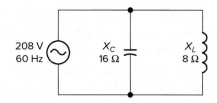

Figure 18-32 Circuit for review question 2.

3. For the parallel *LC* circuit shown in Figure 18-33, determine:
   a. Apparent power (VA).
   b. Reactive power (VARs).
   c. True power (W).
   d. Power factor (PF).
   e. The circuit phase angle θ.

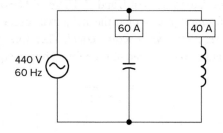

Figure 18-33 Circuit for review question 3.

4. a. Calculate the total line current for the parallel *RLC* circuit shown in Figure 18-34.
   b. Is the circuit capacitive or inductive in nature? Why?

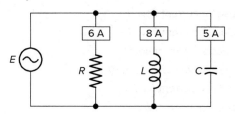

Figure 18-34 Circuit for review question 4.

5. For the parallel *RLC* circuit shown in Figure 18-35, determine:
   a. Current flow through the resistor (I_R).
   b. Current flow through the inductor (I_L).
   c. Current flow through the capacitor (I_C).
   d. Net reactive current (I_X).
   e. Total line current (I_T).
   f. The impedance (Z).
   g. The phase angle (θ) of the circuit.
   h. The value of I_T in polar notation.

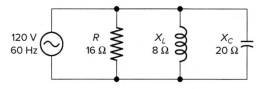

Figure 18-35 Circuit for review question 5.

6. From the measurements taken of the parallel *RLC* circuit shown in Figure 18-36, determine:
   a. Ammeter reading I_X.
   b. Ammeter reading I_T.
   c. True power (W).
   d. Reactive power (VARs).
   e. Apparent power (VA).
   f. Power factor (PF).
   g. Is the power factor leading or lagging? Why?

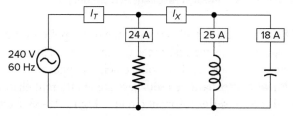

Figure 18-36 Circuit for review question 6.

7. Complete a table for all given and unknown quantities for the parallel *RLC* circuit shown in Figure 18-37.

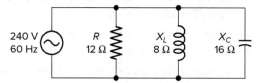

Figure 18-37 Circuit for review question 7.

8. For a parallel *RLC* circuit consisting of a 12-Ω resistor, an inductor with 15-Ω inductive reactance, and a capacitor with 6-Ω capacitive reactance, determine:
   a. The reactance (X).
   b. The impedance (Z).
   c. The phase angle (θ).

9. For a parallel *RLC* circuit consisting of a 20-Ω resistor, an inductor with 16-Ω inductive reactance, and a capacitor with 24-Ω capacitive reactance, determine:
   a. The reactance (X).
   b. The impedance (Z).
   c. The phase angle (θ).

PART 4 PARALLEL RESONANCE AND POWER FACTOR CORRECTION

18.5 Parallel Resonant Circuits

Any circuit in which the inductive reactance equals the capacitive reactance ($X_L = X_C$) is known as a **resonant** circuit. Recall that in a series *RLC* circuit the current is at its maximum value when resonance is reached; the effects

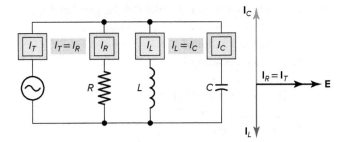

Figure 18-38 Parallel resonant *RLC* circuit.

of X_L and X_C are canceled out and the total applied voltage is applied across the resistance of the circuit. This results in the current and applied voltage being in phase.

Parallel resonant circuits have some different characteristics from series resonant circuits. Figure 18-38 shows a parallel resonant *RLC* circuit and accompanying current vector diagram, the characteristics of which are summarized as follows:

- The leading capacitive current is equal in value to the lagging inductive current.
- Capacitive and inductive currents are 180 degrees out of phase with each other and cancel each other.
- The AC source supplies only the in-phase current required by the resistive load of the circuit.
- Total line current and the resistive current have the same value.
- The source current and voltage are in phase, angle theta (θ) is zero, and the power factor (cos θ) is 1. Thus, the *RLC* parallel resonant circuit is purely resistive in nature.
- Total current to the circuit will be **minimum** as the only current drawn from the power source is that which flows through the resistor.

The combined inductive and capacitive branch of a parallel resonant *RLC* circuit is often called a **tank circuit.** The circuit of Figure 18-39 illustrates the concept of a resonant parallel tank circuit and is summarized as follows:

- The circuit stores energy in the magnetic field of the coil and in the electrostatic field of the capacitor.

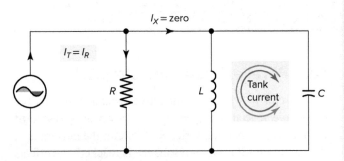

Figure 18-39 Resonant parallel tank circuit.

- The inductive and capacitive branches have circulating currents flowing through them *while using no current* from the power supply.
- The inductor and the capacitor are basically transferring energy back and forth between themselves.
- Stored energy is transferred back and forth on alternate quarter-cycles.
- Current goes first one way and then the other when the inductor deenergizes and the capacitor charges, and vice versa.

For any given values of inductance and capacitance, the frequency at which parallel resonance takes place is identical to the frequency at which series resonance would take place for the same values of *L* and *C*. There will be only **one resonant frequency** at which the inductive reactance and capacitive reactance will become equal. Therefore, the parallel resonant frequency can also be found by the equation

$$f_R = \frac{1}{2\pi\sqrt{LC}}$$

EXAMPLE 18-16

Problem: For the ideal parallel resonant *RLC* circuit shown in Figure 18-40, determine:

a. Resistor current flow (I_R).
b. Inductor current flow (I_L).
c. Capacitor current flow (I_C).
d. Net reactive current flow (I_X).
e. Total line current flow (I_T).
f. Apparent power (VA).
g. True power (W).
h. Net reactive power (VARs).
i. Power factor (PF).

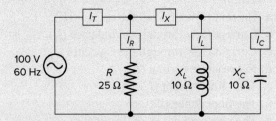

Figure 18-40 Circuit for example 18-16.

Solution:

a.
$$I_R = \frac{E}{R}$$
$$= \frac{100\ V}{25\ \Omega}$$
$$= 4\ A$$

b.

$$I_L = \frac{E}{X_L}$$

$$= \frac{100 \text{ V}}{10 \text{ }\Omega}$$

$$= 10 \text{ A}$$

c.

$$I_C = \frac{E}{X_C}$$

$$= \frac{100 \text{ V}}{10 \text{ }\Omega}$$

$$= 10 \text{ A}$$

d.

$$I_X = I_L - I_C$$

$$= 10 \text{ A} - 10 \text{ A}$$

$$= 0$$

e.

$$I_T = I_R$$

$$= 4 \text{ A}$$

f.

$$\text{Apparent power} = E \times I_T$$

$$= 100 \text{ V} \times 4 \text{ A}$$

$$= 400 \text{ VA}$$

g.

$$\text{True power} = E \times I_R$$

$$= 100 \text{ V} \times 4 \text{ A}$$

$$= 400 \text{ watts}$$

h.

$$\text{Net reactive power} = E \times I_X$$

$$= 100 \times 0$$

$$= 0 \text{ VARs}$$

i.

$$\text{Power factor} = \frac{\text{watts}}{\text{VA}}$$

$$= \frac{400}{400}$$

$$= 1$$

Table 18-1 contains a summary of some of the characteristics associated with series and parallel *RLC* circuits.

18.6 Power Factor Correction

A common practical application for the *RLC* parallel circuit is **power factor correction,** which is the process of changing the power factor to approach 1 or 100 percent. Recall that power factor is the ratio between the **true power** in watts and the **apparent total** power in volt-amps of an electrical load or system. Power factor is a measure of how *efficiently the line current* of a load or system is being converted into useful work output. The ideal power factor is unity (1 or 100%). Anything less than 1 means that extra power is required to achieve the actual task at hand.

The power triangle of Figure 18-41 illustrates the phase angle (θ) between the true power and apparent power. The cosine of the phase angle is known as the power factor (PF), and its value is inversely proportional to the amount

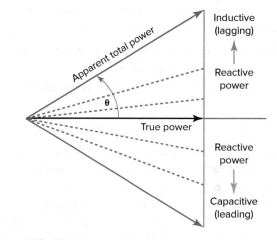

Figure 18-41 Power factor triangle.

Table 18-1 Characteristics Associated with Series and Parallel *RLC* Circuits		
Circuit Factor	***RLC* Series Circuits**	***RLC* Parallel Circuits**
Common element	Current.	Voltage.
$X_L > X_C$	Circuit considered inductive and I_T will lag E_T.	Circuit considered capacitive and I_T will lead E_T.
$X_C > X_L$	Circuit considered capacitive and I_T will lead E_T.	Circuit considered inductive and I_T will lag E_T.
Frequency increases	X_L will increase. X_C will decrease.	X_L will increase. X_C will decrease.
Frequency decreases	X_L will decrease. X_C will increase.	X_L will decrease. X_C will increase.
Voltage and current	Voltage drop across individual components can be greater than the applied source voltage.	Current flow through any single branch can be greater than the line current.
Impedance at resonance	Z is minimum.	Z is maximum.
Line current at resonance	I_T is maximum.	I_T is minimum.

of reactive power. The smaller the angle θ, the *less* reactive power present and the *greater* your power factor is.

Apparent power (VA) is equal to or greater than the true power (watts), depending on the power factor (PF). Therefore, when sizing circuits or equipment, it is important that you size the circuit equipment according to the apparent power (volt-amperes) and not the true power (watts). Formulas that can be used for determining apparent power are as follows:

Apparent power (VA) = volts × ampere (single phase)

Apparent power (VA) = volts × ampere × 1.73 (three phase)

$$\text{Apparent power} = \frac{\text{true power}}{\text{power factor}}$$

$$\text{Apparent power} = \sqrt{\text{true power}^2 + \text{reactive power}^2}$$

True power (watts) is the energy consumed by the resistive part of an AC circuit. The true power of a circuit that contains inductive and/or capacitive reactance in addition to resistance is measured with a **wattmeter** and can be calculated by use of one of the following formulas:

True power (W) = volts × ampere × power factor (single phase)

True power (W) = volts × ampere × power factor × 1.73 (three phase)

$$\text{True power} = \sqrt{\text{apparent power}^2 - \text{reactive power}^2}$$

Reactive power (VARs) is power supplied to a reactive load. Almost all AC circuits include reactive power in the form of inductive reactance and/or capacitive reactance. *Inductive reactance is by far the most common,* since all motors, transformers, solenoids, and coils have inductive reactance. The formulas for determining reactive power are

Reactive power (VARs) = volts × ampere × sin θ (single phase)

Reactive power (VARs) = volts × ampere × sin θ × 1.73 (three phase)

$$\text{Reactive power} = \sqrt{\text{apparent power}^2 - \text{true power}^2}$$

Power factor (PF) is the ratio of true power (watts) to apparent power (VA) and expressed as a percentage that does not exceed 100 percent. Power factor is a form of measurement of how far the voltage and current are **out of phase** with each other. Unity power factor (1 or 100%) can only occur if the AC circuit supplies resistive loads or when capacitive reactance (X_C) is equal to inductive reactance (X_L). When the power factor is less than 100 percent, the circuit is less efficient and has a higher operating cost because not all current is performing useful work.

The formulas for *determining power factor* are

$$\text{Power factor} = \frac{\text{true power}}{\text{apparent power}}$$

$$\text{Power factor} = \frac{\text{watts}}{\text{VA}}$$

The formulas for *determining current* are

$$I = \frac{\text{watts}}{\text{volts} \times \text{power factor}} \quad \text{(single phase)}$$

$$I = \frac{\text{watts}}{\text{volts} \times \text{power factor} \times 1.73} \quad \text{(three phase)}$$

EXAMPLE 18-17

Problem: What is the apparent power of a 15-ampere motor load operating from a 120-volt, single-phase source?

Solution:

$$\begin{aligned} \text{Apparent power} &= E \times I \\ &= 120 \text{ V} \times 15 \text{ A} \\ &= 1{,}800 \text{ VA} \end{aligned}$$

EXAMPLE 18-18

Problem: What is the apparent power of a 250-watt gas discharge lighting fixture that has a power factor of 80 percent?

Solution:

$$\begin{aligned} \text{Apparent power} &= \frac{\text{true power}}{\text{power factor}} \\ &= \frac{250 \text{ W}}{0.8} \\ &= 313 \text{ VA} \end{aligned}$$

EXAMPLE 18-19

Problem: Determine the power factor of the following electric circuits:

a. A three-phase motor operating at 55.95 kW and 74.6 kVA.

b. Single-phase inductive load operating at 750 kVA and 400 kVARs.

Solution:

a. $$\begin{aligned} \text{Power factor} &= \frac{\text{true power}}{\text{apparent power}} \\ &= \frac{55.95 \text{ kW}}{74.6 \text{ kVA}} \\ &= 0.75 \quad \text{or} \quad 75\% \end{aligned}$$

b. $$\begin{aligned} \text{True power} &= \sqrt{\text{apparent power}^2 - \text{reactive power}^2} \\ &= \sqrt{750 \text{ kVA}^2 - 400 \text{ kVARs}^2} \\ &= 634 \text{ kW} \end{aligned}$$

$$\text{Power factor} = \frac{\text{true power}}{\text{apparent factor}}$$

$$= \frac{634 \text{ kW}}{750 \text{ kVA}}$$

$$= 0.845 \quad \text{or} \quad 84.5\%$$

EXAMPLE 18-20

Problem: What is the current flow for a three-phase, 15-kW, 230-volt load that has a power factor of 90 percent?

Solution:

$$I = \frac{\text{watts}}{\text{volts} \times \text{power factor} \times 1.73}$$

$$= \frac{15,000 \text{ W}}{230 \text{ V} \times 0.9 \times 1.73}$$

$$= 41.9 \text{ A}$$

As the power factor of a system drops, the system becomes less efficient. As an example, a drop in power factor from 1.0 to 0.7 requires approximately 43 percent more current; and a power factor of 0.5 requires 100 percent (twice as much more current) to handle the same load. Any time the power factor drops to *less than 85 percent,* the circuit or system is considered to have a poor power factor.

A poor power factor can be the result of either a **phase difference** between the voltage and current, or it can be due to a high harmonic content or **distorted/discontinuous** current waveform. Phase difference is commonly the result of magnetic inductive loads, such as motors, while distorted current waveforms are caused by electronic loads, such as computer power supplies and electronic adjustable-speed motor drives.

Most inductive loads use a coil winding to produce an electromagnetic field, allowing the device to function. Motor loads, for example, require two kinds of power to operate:

- True power (watts): To produce the motive force
- Reactive power (VARs): To energize the magnetic field

Power factor correction is usually achieved by adding a **capacitive load** to offset the inductive load present in the power system. Figure 18-42 illustrates how capacitive power factor correction is applied to an induction motor, the operation of which is summarized as follows:

- The capacitor acts to reduce the inductive component of the line current, thereby reducing the current in the supply line.
- In this example, the capacitor has been sized to obtain a power factor of 1, or 100 percent.

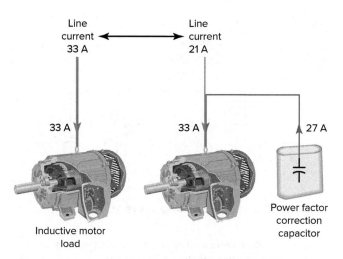

Figure 18-42 Capacitor power factor correction.

- Power factor correction is performed by connecting the capacitor in parallel with the inductive motor load.
- After the correction is made, the current flow to the motor remains unchanged, but the **line current** delivered by the source is *reduced* from 33 to 21 amps.
- This is accomplished with no effect on the operation of the motor itself.
- *Power factor correction does not decrease the power consumed by the motor.*

Some of the benefits of improving your power factor are as follows:

- Because the utility company must invest in oversized equipment to serve low power factor loads, a charge is commonly assessed on a facility's electric bill to recover the equipment costs and lost energy caused by the low power factor.
- The electrical system's branch *capacity will increase.* This can free up the electrical system's capacity, which can be used for additional loads.
- Uncorrected power factor will *cause power losses* in the distribution system. You may experience voltage drops as power losses increase. Excessive voltage drops can cause overheating and premature failure of motors and other inductive equipment.

EXAMPLE 18-21

Problem: For the motor power factor correction circuit shown in Figure 18-43, determine:

a. Apparent power, power factor, reactive power, without the capacitor connected.

b. Apparent power and reactive power, with capacitor connected for a corrected power factor of 90 percent.

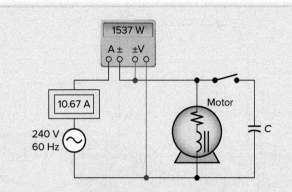

Figure 18-43 Circuit for example 18-21

c. Capacitive VARs with capacitor connected for a corrected power factor of 90 percent.

d. Capacitive reactance of the capacitor required to raise the power factor to 90 percent.

e. Size of capacitor connected in parallel with the motor required to raise the power factor to 90 percent.

f. Amount of reduction in the line current with the power factor corrected to 90 percent.

Solution:

a.
$$\text{Apparent power} = E_T \times I_T$$
$$= 240\text{ V} \times 10.67\text{ A}$$
$$= 2{,}561\text{ VA}$$

$$\text{PF} = \frac{\text{watts}}{\text{VA}}$$
$$= \frac{1{,}537}{2{,}561}$$
$$= 0.60 \qquad \text{or} \qquad 60\%$$

$$\text{Reactive power} = \sqrt{\text{apparent power}^2 - \text{true power}^2}$$
$$= \sqrt{2{,}561^2 - 1{,}537^2}$$
$$= 2{,}048\text{ VARs}$$

b.
$$\text{VA (at 90\% PF)} = \frac{\text{W}}{\text{PF}}$$
$$= \frac{1{,}537}{0.90}$$
$$= 1{,}708\text{ VA}$$

Reactive power (at 90% PF)
$$= \sqrt{\text{apparent power}^2 - \text{true power}^2}$$
$$= \sqrt{1{,}708^2 - 1{,}537^2}$$
$$= 745\text{ VARs}$$

c.
$$\text{VARs (cap)} = \text{VARs (original)} - \text{VARs (corrected)}$$
$$= 2{,}048 - 745$$
$$= 1{,}303\text{ VARs}$$

d.
$$X_C = \frac{E^2}{\text{VARs}}$$
$$= \frac{240^2}{1{,}303}$$
$$= 44.2\ \Omega$$

e.
$$C = \frac{1}{2\pi f X_C}$$
$$= \frac{1}{2 \times 3.14 \times 60 \times 44.2}$$
$$= 60.04\ \mu\text{F}$$

f.
$$I_T \text{ (at 90\% PF)} = \frac{\text{VA}}{E_T}$$
$$= \frac{1{,}708}{240}$$
$$= 7.117\text{ A}$$

$$\text{Reduction in line current} = I_T \text{ (original)} - I_T \text{ (corrected)}$$
$$= 10.67\text{ A} - 7.117\text{ A}$$
$$= 3.553\text{ A}$$

Capacitors used for power factor correction are generally rated in VARs or kVARs instead of farads (F), as it is far more convenient to simply match this with the amount of corrective VARs required. These capacitors can be found installed at motor starters, switchboards, and distribution panels, as shown in Figure 18-44.

A capacitive (leading) power factor can occur if an inductive (lagging) power factor has been ***overcompensated*** by putting in too much capacitance. Overcorrecting the power factor may cause high transient voltages, currents,

Figure 18-44 Power factor correction capacitors installed within a panel.

Courtesy of Power Survey International Inc.

and torques that can increase safety hazards to personnel and possibly damage motor-driven equipment. For these reasons, when using power factor correction capacitors, a leading power factor should be avoided.

Part 4 Review Questions

1. For the circuit shown in Figure 18-45 determine the:
   a. Line current (I_T).
   b. Impedance (Z).
   c. Reactive power of the inductor (VARs).
   d. Reactive power of the capacitor (VARs).
   e. Net reactive power (VARs).
   f. True power (watts).
   g. Apparent power (VA).
   h. Power factor (PF).

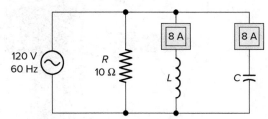

Figure 18-45 Circuit for review question 1.

2. For the power factor correction circuit shown in Figure 18-46:
   a. Determine the value of the wattmeter reading, line current reading, and power factor.
   b. Assume the switch is *opened* so that the capacitor is no longer in the circuit. Determine the new value of the wattmeter reading, line current reading, and power factor.

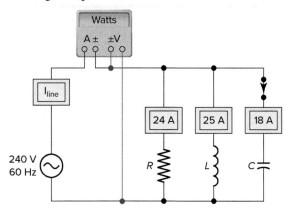

Figure 18-46 Circuit for review question 2.

3. Power factor correction is to be applied to the motor circuit shown in Figure 18-47. Determine the following:
   a. The circuit power factor (capacitor not in the circuit).
   b. Assume the switch is closed, connecting the capacitor in parallel with the motor. If the line current drops to 8 amps, what is the corrected power factor?

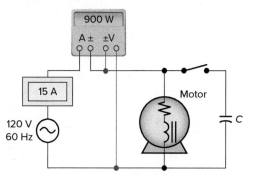

Figure 18-47 Circuit for review question 3.

4. A 24-kW load with a power factor of 92 percent is connected to a 230-volt, three-phase source. Determine the value of the line current flow.

5. What kVA-size transformer is required for a 30-kW load that has a power factor of 85 percent?

CHAPTER NINETEEN

Transformers

Pole mounted transformer installation.
©Montree Hanlue/123RF

LEARNING OUTCOMES

▶ Understand the operating characteristics of transformers.

▶ Calculate transformer turns, voltage, and current ratios.

▶ Be familiar with routine transformer installation and maintenance procedures.

▶ Connect transformers in common circuit configurations.

▶ Correctly interpret and apply transformer nameplate data.

Transformers transfer electric energy from one electric circuit to another by means of electromagnetic mutual induction. Their main purpose is to convert AC power at one voltage level to AC power of the same frequency at another voltage level. This chapter covers the operation and installation of single-phase and three-phase transformers.

PART 1 TRANSFORMER BASICS

19.1 Transformer Operation

Electrical transformers are used to transform voltages from one level to another. The operation of a transformer involves the transfer of energy from one AC circuit to another. This transfer of energy may involve an *increase or decrease in voltage,* but the frequency will be the same in both circuits. The electrical transformer, shown in Figure 19-1, is basically a multiple-winding inductor.

When the transformation takes place with an increase in voltage, it is called a **step-up transformer.** When voltage is decreased, it is called a **step-down transformer.** Transformers are static devices (having no moving parts) and as such accomplish the changing AC voltage levels with very little loss of power. Without transformers the widespread distribution of electric power would be impractical. High-voltage power transformers, shown in Figure 19-2, make it possible to generate power at a convenient voltage, step it up to a very high voltage for long-distance transmission, and then step it down for safe distribution.

Figure 19-1 Electrical transformer.
©Photonic 20/Alamy Stock Photo

Figure 19-2 High-voltage power transformers.
©Dannicolae/iStock/Getty Images

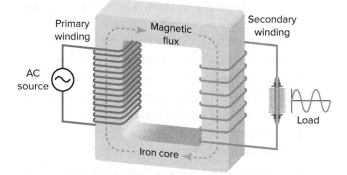

Figure 19-3 Mutual induction in a transformer.

The basic transformer, shown in Figure 19-3, consists of two coils wound around an iron core and linked together by magnetic flux. Its operation is summarized as follows:

- The operation is based on **mutual induction,** which occurs when the magnetic field surrounding one conductor cuts across another conductor, inducing a voltage in it.
- The winding that receives the power from the supply is called the **primary** winding.
- The winding that delivers power to the load is called the **secondary** winding.
- When an AC voltage is applied to the primary coil, the resultant current flow sets up a magnetic field that is *constantly changing.*
- As this field expands and collapses, it causes an AC voltage to be **induced** in the secondary winding.
- The AC frequency of the primary induces the *same frequency* in the secondary.

The **no-load** primary current provided to the transformer is called **magnetizing** current or **exciting** current. This current is used to produce the flux in the transformer core and ranges in value from 2 to 5 percent of the rated full-load primary current. The voltages induced in the transformer due to the magnetizing current are illustrated in Figure 19-4 and summarized as follows:

- The exciting current sets up an alternating flux that links the turns and induces a voltage in both the primary and secondary windings.

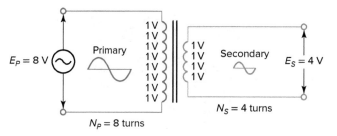

Figure 19-4 Transformer induced voltages.

- The voltage of **self-induction** induced in the primary is a countervoltage, opposite in polarity and almost equal in magnitude to the applied voltage. This limits the exciting current to a relatively low value.

- Primary exciting current lags the applied voltage by approximately 90 degrees because the coil is mainly inductive and has very little resistance.

- The voltage induced in the secondary winding is a result of **mutual inductance.**

- Since a typical power transformer has a flux linkage of almost 100 percent, the same voltage will be induced in each turn of the secondary coil.

- The total induced voltage, therefore, will be directly **proportional** to the **number of turns** on the primary and secondary windings.

Eddy currents are caused by the alternating current that induces a voltage in the core of the transformer itself. Because the iron core is a conductor, it produces a current from the induced voltage. Eddy currents represent **wasted power** dissipated as heat in the core. It is possible to reduce the eddy currents by laminating the core. Laminating means constructing it with many thin steel sheets that are electrically insulated from each other, as illustrated in Figure 19-5, rather than making it a single solid piece. The laminations confine the currents to a single laminate at a time, where the paths are long, making for higher resistance that results in less current flow and therefore less power loss.

The most common configurations of iron and copper in a transformer are the shell type and core type, illustrated in Figure 19-6. Most small power transformers are **shell type,** in which the iron surrounds the copper with low-voltage winding placed closest to the core and the high-voltage winding wound over the top of it. The **core-type** transformer configuration is commonly used with larger high-voltage transformers. With this configuration the copper surrounds the iron, and both primary and secondary windings are placed on each leg.

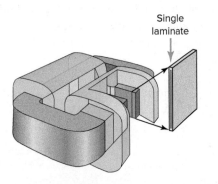

Figure 19-5 Reducing eddy currents with steel laminations.

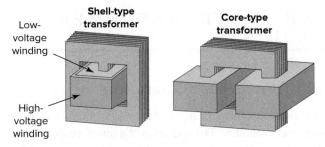

Figure 19-6 Configurations of iron and copper in a transformer.

19.2 Relationship of Voltage, Current, and Power

A transformer does the job of increasing or decreasing the voltage without any significant loss in power. The transformer *power output equals transformer power input* minus the internal losses and is the product of voltage and current. Transformers are rated in **kVA** (rather than kW) because this power rating is independent of power factor and always at least as large as the true power.

The total voltage induced into the secondary winding of a transformer is determined mainly by the **ratio** of the number of turns in the primary to the number of turns in the secondary and by the amount of voltage applied to the primary. The **turns ratio,** also called the **transformation ratio,** is expressed by the equation

$$\text{Turns ratio} = \frac{N_P}{N_S}$$

where N_P = *the number of turns in the primary winding*
N_S = *the number of turns in the secondary winding*

The voltage of the windings in a transformer is directly proportional to the turns ratio and is expressed by the equations

$$\frac{N_P}{N_S} = \frac{E_P}{E_S}$$

$$E_S = E_P \times \frac{N_S}{N_P}$$

$$E_P = E_S \times \frac{N_P}{N_S}$$

where E_P = *the primary voltage*
E_S = *the secondary voltage*

For example, if the number of secondary winding turns is twice the number of primary turns, the secondary voltage will be twice the primary voltage. Similarly, if the number of primary winding turns is twice the number of secondary turns, the secondary voltage will be half the primary voltage.

Transformers are classified as step up or step down in relation to their effect on voltage. A **step-up transformer** is one in which the secondary winding output voltage is *greater* than the primary winding input voltage. This type of transformer has more turns in the secondary coil than in the primary coil.

EXAMPLE 19-1

Problem: For the step-up transformer shown in Figure 19-7, determine:

a. The turns ratio.

b. The secondary output voltage.

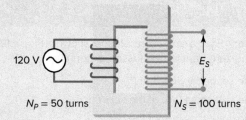

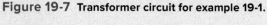

$N_P = 50$ turns $N_S = 100$ turns

Figure 19-7 Transformer circuit for example 19-1.

Solution:

a.   Turns ratio $= \dfrac{N_P}{N_S} = \dfrac{50}{100} = \dfrac{1}{2}$ or $1:2$

which indicates that there is one turn in the primary winding for every two turns in the secondary winding.

b.
$$E_S = E_P \times \frac{N_S}{N_P}$$
$$= 120 \text{ V} \times \frac{100}{50}$$
$$= 240 \text{ V}$$

A **step-down transformer** is one in which the secondary winding output voltage is less than the primary coil input voltage. This type of transformer has fewer turns in the secondary winding than in the primary winding. Again, the turns ratio of the primary winding to that of the secondary winding determines the input-to-output voltage ratio of the transformer.

EXAMPLE 19-2

Problem: For the step-down transformer shown in Figure 19-8, determine:

a. The turns ratio.

b. The secondary output voltage.

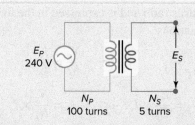

Figure 19-8 Transformer circuit for example 19-2.

Solution:

a.   Turns ratio $= \dfrac{N_P}{N_S} = \dfrac{100}{5} = \dfrac{20}{1}$ or $20:1$

which indicates that there are 20 turns in the primary winding for every one turn in the secondary winding.

b.
$$E_S = E_P \times \frac{N_S}{N_P}$$
$$= 240 \text{ V} \times \frac{5}{100}$$
$$= 12 \text{ V}$$

A transformer *automatically adjusts its input current* to meet the requirements of its output load current. With **no load** connected to the secondary, no current flows in the secondary winding and only magnetizing current flows in the primary winding. When a *load is connected* across the secondary the following chain of events occurs:

- The induced secondary coil voltage causes a load current to flow through the load and through the secondary coil.
- The load current flowing through the secondary coil sets up a magnetic flux in the core that opposes the flux produced by the magnetizing current in the primary coil.
- The primary's counter electromotive force (cemf) is therefore reduced, so the primary current increases.
- The amount of increase in primary current is equivalent to that required to strengthen the primary winding's magnetic field and overcome the effects of the opposing secondary winding's magnetic flux.

From the viewpoint of the primary, a load connected across the secondary winding of a transformer appears to have resistance that is not necessarily equal to the actual resistance of the load. The actual load is essentially **reflected** into the primary and determined by the turns ratio. This reflected load is what the source effectively sees, and it determines the amount of primary current. The current in the two windings is in inverse proportion to the voltage and turns ratio and is expressed in the form of the following equations:

$$\frac{E_P}{E_S} = \frac{I_S}{I_P} \quad \text{or} \quad \frac{N_P}{N_S} = \frac{I_S}{I_P}$$

$$I_P = \frac{E_S}{E_P} \times I_S \quad \text{or} \quad I_P = \frac{N_S}{N_P} \times I_S$$

$$I_S = \frac{E_P}{E_S} \times I_P \quad \text{or} \quad I_S = \frac{N_P}{N_S} \times I_P$$

EXAMPLE 19-3

Problem: The step-down soldering gun transformer shown in Figure 19-9 has a turns ratio of 200:1 and a secondary heating current of 400 amperes. Determine:

a. The value of the primary current.

b. The value of the secondary voltage when the primary is operated from a 120-volt AC source.

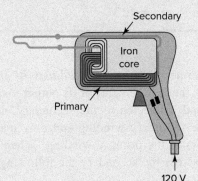

Figure 19-9 Soldering gun circuit for example 19-3.

Solution:

a.
$$I_P = \frac{N_S}{N_P} \times I_S$$
$$= \frac{1}{200} \times 400 \text{ A}$$
$$= 2 \text{ A}$$

b.
$$E_S = E_P \times \frac{N_S}{N_P}$$
$$= 120 \text{ V} \times \frac{1}{200}$$
$$= 0.6 \text{ V}$$

The power rating of a transformer is equal to the product of the voltage times the current. However, the result is not usually expressed in watts because not all loads are purely resistive. Instead, the ***power-handling capacity of a transformer is rated in volt-amps (VA) or kilovolt-amps (kVA).*** Only resistance consumes power measured in watts, but in effect, the temperature rise of a transformer is directly related to the apparent power or volt-amps that

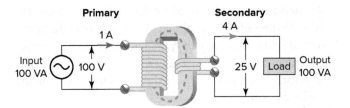

Figure 19-10 Input VA equals output VA.

flow through its windings. With zero losses assumed (an ideal transformer), the power in the secondary equals the power in the primary (Figure 19-10):

$$E_P \times I_P = E_S \times I_S$$

EXAMPLE 19-4

Problem: For the step-up transformer shown in Figure 19-11, assuming zero transformer losses, determine:

a. The secondary current (I_S).

b. The volt-amps (VA) of the secondary.

c. The volt-amps (VA) of the primary.

d. The primary current (I_P).

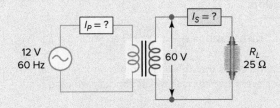

Figure 19-11 Transformer circuit for example 19-4.

Solution:

a.
$$I_S = \frac{E_S}{R_L}$$
$$= \frac{60 \text{ V}}{25 \text{ }\Omega}$$
$$= 2.4 \text{ A}$$

b. VA secondary $= E_S \times I_S$
$$= 60 \text{ V} \times 2.4 \text{ A}$$
$$= 144 \text{ VA}$$

c. VA secondary $=$ VA primary
$$= 144 \text{ VA}$$

d.
$$I_P = \frac{\text{VA}}{E_P}$$
$$= \frac{144 \text{ VA}}{12 \text{ V}}$$
$$= 12 \text{ A}$$

EXAMPLE 19-5

Problem: A transformer has a primary rated voltage of 120 volts and a secondary rated voltage of 24 volts. Determine the amount of voltage induced in the secondary if only 100 volts is applied to the primary.

Solution:

$$E_S = E_P \times \frac{E_S}{E_P}$$
$$= 100 \times \frac{24}{120}$$
$$= 20 \text{ V}$$

EXAMPLE 19-6

Problem: A single-phase 240-volt load is drawing 43 amps and is connected to the secondary of a 480-V primary to a 240-V secondary transformer. Determine how much current is flowing in the primary winding.

Solution:

$$I_P = I_S \times \frac{E_S}{E_P}$$
$$= 43 \times \frac{240}{480}$$
$$= 21.5 \text{ A}$$

Some transformers use **winding taps** to adjust the transformer voltage to the correct input or output voltage or to permit selecting various voltages for different applications. Figure 19-12 shows a secondary winding with three terminal points for the load connection. The operation of the **tap changer** is summarized as follows:

- In normal operation the switch connects terminal B to the load. With this connection assume the turns ratio is 1:10.
- If required, the switch can connect tap A to the load. This changes the turns ratio because fewer secondary turns are being accessed by the same number of

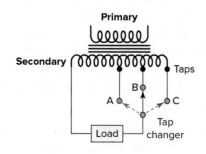

Figure 19-12 Transformer tap changer.

primary turns. The net effect will be for the secondary voltage to decrease.

- If required, the switch can connect tap C to the load. This changes the turns ratio because more secondary turns are being accessed by the same number of primary turns. The net effect will be for the secondary voltage to increase.

Isolation transformers have a turns ratio of 1:1, resulting in equal primary and secondary voltages. Their main function is to electrically isolate a piece of electrical equipment from the power distribution system. Although any transformer with a separate primary and secondary winding is an isolation transformer to some extent, the term usually refers to a special-purpose transformer built just for that use.

An **autotransformer** is a special transformer in which the primary and secondary circuits share a common winding. Figure 19-13 shows an autotransformer connected to step up and step down the voltage using a single-tapped inductor coil. Unlike two winding conventional transformers, autotransformers have their primary and secondary windings connected to each other electrically. Autotransformers offer the benefits of smaller size, lower weight, and lower cost. A disadvantage of the autotransformer is that the secondary is *not electrically isolated* from the primary. This is an important safety consideration when deciding to use an autotransformer in a given application.

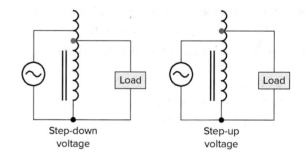

Figure 19-13 Autotransformer.

Part 1 Review Questions

1. Explain how electric energy is transferred from the primary to secondary windings of a transformer.
2. What distinguishes the primary winding from the secondary winding?
3. What is the no-load primary current of a transformer called?
4. What limits the value of the no-load primary current?
5. How are transformers designed to minimize eddy currents?

6. In which type of transformer core configuration is the high-voltage winding wound over the top of the low-voltage winding?

7. In an ideal transformer, what is the relationship between the:
   a. Turns ratio and the voltage ratio?
   b. Voltage ratio and the current ratio?
   c. Primary and secondary power?

8. What is the difference between a step-up and step-down transformer?

9. The turns ratio of a transformer is specified as being 1:5. Would this indicate that it is a step-up or a step-down transformer. Why?

10. Calculate the turns ratio of a transformer which is rated for a primary voltage of 4,800 volts and a secondary voltage of 240 volts.

11. A transformer is being designed to decrease the voltage from 120 volts to 12 volts. If the primary winding requires 800 turns of wire, how many turns would be required on the secondary winding?

12. Assuming a transformer has a turns ratio of 1:2 and the secondary voltage is 960 volts, how much voltage is applied to the primary?

13. What causes a transformer's primary current to increase when a load is connected to the secondary circuit?

14. Why is the power rating of a transformer normally given in volt-amps rather than watts?

15. For the ideal transformer shown in Figure 19-14, determine the:
   a. Turns ratio.
   b. Primary current flow.
   c. kVA of the primary.
   d. kVA of the secondary.

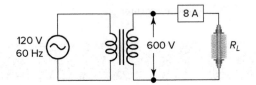

Figure 19-14 **Transformer circuit for review question 15.**

16. A single-phase 60-kVA transformer has a primary voltage rating of 2,400 volts and a secondary rating of 240 volts.
   a. What is the rated primary current?
   b. What is the rated secondary current?
   c. What is the turns ratio?
   d. When a resistive load is connected to the secondary, the primary current is measured and found

to be 16.5 amperes. Assuming ideal transformer conditions apply, what should the value of the secondary current flow be?

17. For which applications are winding taps used?

18. What is the main function of an isolation transformer?

19. What safety consideration limits the use of autotransformers?

20. A transformer has a primary rated voltage of 120 volts and a secondary rated voltage of 12 volts. Determine the amount of voltage induced in the secondary if only 80 volts is applied to the primary.

21. A single-phase 120-volt load is drawing 12 amps and connected to the secondary of a 480-V primary to 120-V secondary transformer. Determine how much current is flowing in the primary winding.

PART 2 SINGLE-PHASE TRANSFORMER

19.3 Transformer Losses

Since a transformer has no rotating parts, it has no mechanical losses. This contributes to its high operating efficiency of over 90 percent. Transformer electrical losses appear in the form of waste heat and are summarized as follows:

- **Copper loss:** This loss is caused by the resistance of the copper wire in the primary and secondary windings. As current flows through the windings, some power is dissipated in the form of heat.

- **Eddy current losses:** Eddy currents are caused by the alternating current inducing a current in the core of the transformer and are kept to a minimum by the use of laminated cores.

- **Hysteresis loss:** This loss originates from the core laminations resisting being magnetized and demagnetized by changes in polarity of the applied current.

- **Flux linkage:** This loss is the result of the leakage of the electromagnetic flux lines between the primary and secondary winding.

- **Saturation:** A saturation loss may occur if the transformer is loaded beyond its rated capacity. This happens when the core reaches its saturation point and an increase in current doesn't produce any additional flux lines.

An ideal transformer would have no losses and would therefore be 100 percent efficient. Transformer **efficiency** is a function of a transformer's copper and core losses and has nothing to do with power factor. These losses are all measured in watts. The efficiency of a transformer is found by dividing the output by the input and is expressed in the form of an equation as

$$\% \text{ efficiency} = \frac{\text{output watts}}{\text{input watts}} \times 100$$

In a standard power transformer the full-load efficiency is generally from **96 to 99 percent.** The no-load efficiency of a transformer is lower than its full-load efficiency. Therefore, sizing power transformers to meet their expected loading greatly influences transformer efficiency. Oversized transformers can contribute to inefficiency, but when transformers are appropriately matched to their loads, efficiency increases.

Voltage regulation is the measure of how well a power transformer can maintain constant secondary voltage given a constant primary voltage and wide variance in load current. Voltage regulation in transformers is the difference between the no-load voltage and the full-load voltage and is expressed in the form of an equation as

$\%$ voltage regulation

$$= \frac{\text{no-load voltage } (E_S) - \text{full-load voltage } (E_S)}{\text{full-load voltage } (E_S)} \times 100$$

EXAMPLE 19-7

Problem: Determine the percentage voltage regulation for a single-phase transformer that has a secondary output voltage of 100 volts at no load and 95 volts at full load.

Solution:

$\%$ voltage regulation

$$= \frac{\text{no-load voltage } (E_S) - \text{full-load voltage } (E_S)}{\text{full-load voltage } (E_S)} \times 100$$

$$= \frac{100 \text{ V} - 95 \text{ V}}{95 \text{ V}} \times 100$$

$$= \frac{5}{95} \times 100$$

$$= 5.26\%$$

It is important that a transformer be operated on an AC circuit with a **frequency** for which it is designed. When a lower frequency than what the transformer is designed for is used, the reactance of the primary winding will decrease, resulting in an increase in the exciting current and higher losses. Transformers below 2 kVA are usually

designed and rated to be used at 50 or 60 Hz. A 60-Hz-only transformer design is physically smaller than a 50-Hz design and should not be used on a 50-Hz circuit.

19.4 Transformer Ratings

When a transformer is to be used in an application, the following capabilities of the primary and secondary **windings** must be taken into consideration:

- When **nominal values** of voltage, current, and power are specified, they represent the middle point of the respective maximum and minimum rated values.
- The **maximum voltage** that can safely be applied to any winding is determined by the type and thickness of the insulation used.
- The **maximum current** that can be carried by a transformer winding is determined by the diameter of the wire used for the winding.
- **Transformer power** is rated in volt-amperes (VA) or kilovolt-amperes (kVA). The primary and secondary full-load currents usually are not given but can be calculated from the rated VA or kVA as follows:

Single phase: Full-load current $= \dfrac{\text{VA}}{\text{voltage}}$ or

$$\frac{\text{kVA} \times 1{,}000}{\text{voltage}}$$

Three phase: Full-load current $= \dfrac{\text{kVA} \times 1{,}000}{1.73 \times \text{voltage}}$

EXAMPLE 19-8

Problem: For the single-phase transformer shown in Figure 19-15, determine:

a. Full-load current rating of the primary.
b. Full-load current rating of the secondary.

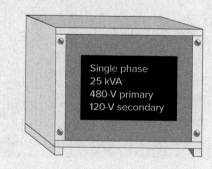

Single phase
25 kVA
480-V primary
120-V secondary

Figure 19-15 Transformer for example 19-6.

Solution:

a.   Primary full-load current $= \dfrac{\text{kVA} \times 1{,}000}{\text{voltage}}$

$= \dfrac{25 \times 1{,}000}{480}$

$= 52 \text{ A}$

b.   Secondary full-load current $= \dfrac{\text{kVA} \times 1{,}000}{\text{voltage}}$

$= \dfrac{25 \times 1{,}000}{120}$

$= 208 \text{ A}$

Excessive **temperature** is the main cause of transformer failure. Adequate cooling must be provided to prevent deterioration of the insulating materials inside a transformer and ensure its long life expectancy. Transformers are cooled using air, water, oil, or natural and forced convection. **Dry-type** transformers depend on the circulation of air over or through their enclosure. **Liquid-filled** transformers have the transformer's coils and core submerged in an approved insulating liquid such as mineral oil or synthetic fluid for cooling purposes. Forced-air cooling (Figure 19-16) involves a fan mounted near the radiator fins. This type of setup allows the transformer oil circulating in the fins to stay cooler and provide better performance within the transformer.

A transformer's rated **temperature rise** (in degrees Celsius) is the average temperature of the transformer's windings over an ambient temperature of 40°C. As an example, a transformer rated for 150°C temperature rise will operate at an average winding temperature of 190°C (150 + 40) when at full-rated load in a 40°C ambient environment.

Figure 19-16 Transformer forced-air cooling.
©KoKimk/iStock/Getty Images

The **impedance rating (Z)** of a transformer is a measure of its current-limiting characteristics and is generally expressed as a percentage. This is the percentage of normal-rated primary voltage that must be applied to the transformer to cause full-load-rated current to flow in the short-circuited secondary. For instance, if a 480-V/120-V transformer has an impedance of 5 percent, this means that 5 percent of 480 V, or 24 V, applied to its primary will cause rated load current flow in its secondary.

The impedance rating is used for determining the **interrupting capacity** of a circuit breaker or fuse employed to protect the primary of a transformer. The transformer short-circuit secondary current available can be determined using the following formula:

$$I_{\text{short circuit}} = \dfrac{I_{\text{full load}}}{\%Z}$$

EXAMPLE 19-9

Problem: A single-phase transformer rated for 25 kVA and an output voltage of 240 volts has a rated impedance of 2.2 percent. Calculate:

a.   The secondary full-load current.

b.   The secondary short-circuit current.

Solution:

a.   Secondary full-load current $= \dfrac{\text{kVA} \times 1{,}000}{\text{voltage}}$

$= \dfrac{25 \times 1{,}000}{240}$

$= 104 \text{ A}$

b.   Secondary short-circuit current $= \dfrac{I_{\text{full load}}}{\%Z}$

$= \dfrac{104}{2.2\%}$

$= 4{,}727 \text{ A}$

All transformers produce an **audible hum** when energized. The hum is produced by vibrations in the core laminations and other components as the magnetic fields vary with the applied voltage. The volume of the sound is determined by the transformer design. Test procedures have been established so that transformer manufacturers can publish the sound level ratings of their transformers. Sound levels are rated in decibels (dB), and the higher the decibel rating, the louder the sound.

Transformers are designed to operate at the **standard supply voltage.** When the supply voltage is constantly too low or too high (usually more than ±5 percent), the transformer fails to operate at maximum efficiency. When selecting a transformer to meet the requirements of a specific installation, it is important that:

- The nominal supply voltage matches the voltage rating of the transformer's primary winding.
- The secondary voltage of the transformer matches the voltage requirements of the load.
- The VA or kVA rating of the transformer is equal to, or greater than, that required by the load to which it will supply power. If it is *too much* greater, it will have poor efficiency.

19.5 Single-Phase Transformer Connections

Transformer polarity refers to the instantaneous voltage polarity obtained from the primary winding in relation to the secondary winding. An understanding of transformer polarity markings is essential when making single-phase and three-phase transformer connections.

On power transformers, the high-voltage winding leads are marked H1 and H2 and the low-voltage winding leads are marked X1 and X2, as illustrated in Figure 19-17. By convention, H1 and X1 have the same polarity, which means that when H1 is instantaneously positive, X1 is also instantaneously positive. These markings are used in establishing the proper terminal connections when single-phase transformers are connected in parallel, series, and three-phase configurations.

In practice, the four terminals on a single-phase transformer are mounted in a standard way so the

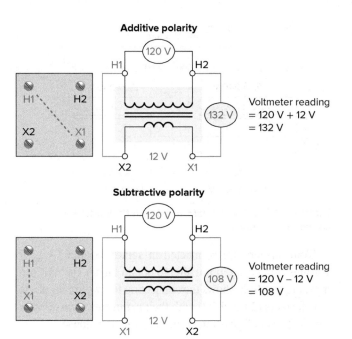

Figure 19-18 Additive and subtractive polarity test circuits.

transformer has either additive or subtractive polarity as follows:

- The location of the H and X terminals indicate whether the transformer was designed to be either additive or subtractive.
- A transformer is said to have **additive polarity** when terminal H1 is **diagonally opposite** terminal X1.
- A transformer is said to have **subtractive polarity** when terminal H1 is **adjacent** to terminal X1.
- Figure 19-18 illustrates additive and subtractive transformer terminal markings along with a test circuit that can be used to verify markings.

Another form of polarity marking is through the use of dots. **Dot notation** is used with schematic diagrams to express which terminals are positive at the same instant in time. Figure 19-19 illustrates how dot notation can be used to identify the polarity of a transformer's H1 and X1 leads.

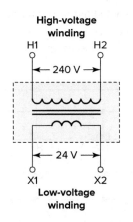

Figure 19-17 Transformer polarity markings.

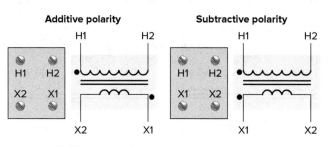

Figure 19-19 Dot notation polarity marking.

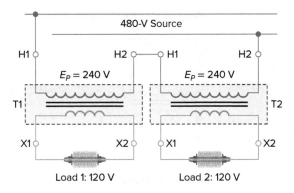

Figure 19-20 Two transformers with their primaries connected in series.

Transformers are connected in **series** to obtain higher **voltage** ratings and in **parallel** to obtain higher **current** ratings. Figure 19-20 shows the connection for two single-phase transformers (T1 and T2) with their *primaries connected in series.* The configuration is summarized as follows:

- The given parameters of the circuit are a source voltage of 480 volts with two load requirements of 120 volts each.
- The two transformer primary windings are rated at 240 volts and their secondaries at 120 volts.
- In order to drop 240 volts across each primary winding, with 480 volts being supplied by the source, the two primary windings are connected in series.
- H2 of transformer T1 is connected to H1 of transformer T2 to provide the series connection with the correct polarity.
- Each of the two loads requires a voltage of 120 volts, which is obtained directly from the X1 and X2 secondary leads of transformers T1 and T2 and connected to the feeders that supply the loads.
- *If connected improperly, it is possible to burn out the transformers when they are energized.*

A transformer with more than one primary or secondary winding is called a **dual-voltage transformer.** Figure 19-21 shows the connection for a dual-voltage transformer used to

step 240 or 480 volts down to 120 volts. The two different connections are summarized as follows:

- The primary connections on the transformer are identified as H1, H2, H3, and H4.
- The transformer winding between H1 and H2 and the one between H3 and H4 are rated for 240 V each.
- The low-voltage secondary connections on the transformer, X1 and X2, can have 120 V from either a 480- or 240-V line.
- If the transformer is to be used to step 480 V down to 120 V, the primary windings are connected in series by a jumper wire or metal link.
- When the transformer is to be used to step 240 V down to 120 V, the two primary windings must be connected in parallel with each other.

Two transformers of equal rating connected in parallel will handle twice the kVA rating of either one. Transformers connected in parallel must have the *same* voltage, turns ratio, and impedance ratings. Figure 19-22 shows the connection diagram for *two transformers connected in parallel.* The connection procedure is summarized as follows:

- Only terminals having the same polarity are connected together.
- Regardless if the transformers being connected are additive or subtractive, polarity will be correct if all "H" and "X" terminals with the *same numbers* are connected to the same line.
- An error in polarity produces a short circuit as soon as the transformers are excited.
- The resistance of the wire connections between the two transformers should be balanced.

Figure 19-23 shows the connection diagram for a dual-voltage distribution transformer used to supply electric power to a residence. The connection is summarized as follow:

- The two secondary windings are rated at 120 volts each.
- The secondary windings are connected in series, so the total voltage between the lines is 240 volts, while the voltage between the lines and the middle conductor is 120 volts.

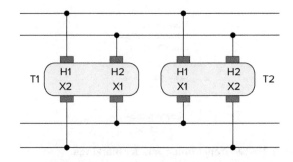

Figure 19-22 Two transformers connected in parallel.

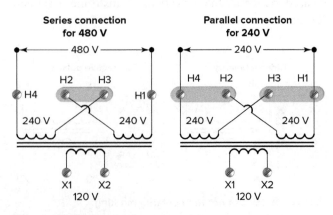

Figure 19-21 Dual-voltage transformer connection.

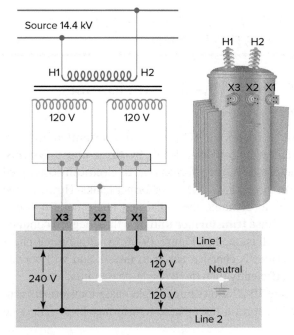

Figure 19-23 Dual-voltage distribution transformer.

- The middle conductor, called neutral, is always connected to ground.
- At first, when power requirements of homes were relatively small, only two-wire, 120-volt systems were used. With increased power demand, the three-wire 240/120-volt system became the standard.
- With this transformer configuration, the addition of the third wire provides a 100 percent increase in capacity with only a 50 percent increase in the cost of the conductors.

Buck-boost transformers are single-phase transformers designed to reduce (buck) or raise (boost) line voltage. The most common example is boosting 208 volts to accommodate 230-volt loads, as shown in Figure 19-24. The connection is summarized as follow:

- For this application, a transformer with the voltage ratings of 120/240:12/24 volts (turns ratio of 10:1) is used.
- The two primary coils are connected in series across the 208-volt source, and as such there will be a voltage drop of 104 volts across each of them.

- Due to the 10:1 turns ratio of the transformer, 10.4 volts will be induced across each secondary winding or 20.8 volts total across the two windings in series.
- The two secondary coils are in series with the source and load.
- The 208-volt source when added to the secondary boost voltage of 20.8 volts will provide a voltage of 228.8 volts for the load.

Part 2 Review Questions

1. Identify the type of transformer loss associated with each of the following:
   a. Magnetizing and demagnetizing of the transformer core.
   b. Induced current flow in the transformer core.
   c. Current flow through the windings.
2. If the output of the secondary of a transformer is 1,320 watts and the input of the primary is 1,800 watts, what is the percentage efficiency of the transformer?
3. Compare the no-load and full-load efficiency of a transformer.
4. If the no-load voltage of the secondary of a transformer is 480 volts and the full-load voltage is 465 volts, what is the percentage voltage regulation of the transformer?
5. What does the rated nominal current value of a transformer specify?
6. Why is a transformer rated for 60 Hz likely to overheat if used on a 50-Hz circuit?
7. What is the primary and secondary line current for a three-phase 37.5-kVA transformer, rated 480 volts for the primary and 208 volts for the secondary?
8. The following data are given on the nameplate of a transformer:
   - 35 kVA.
   - 60 Hz.
   - Phase single.
   - HV 480 V.
   - LV 240 V.
   - Impedance 2.6%.
   - Temperature rise 80°C.

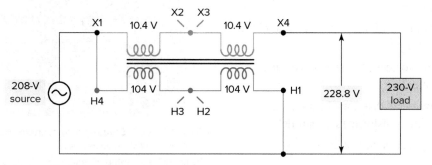

Figure 19-24 Buck-boost transformer connected to boost voltage.

Assuming the transformer is connected to step-down voltage, determine the following from the information given:

   a. What would be the markings for the primary terminals?

   b. What would be the markings for the secondary terminals?

   c. At what frequency is it rated?

   d. What is the value of the turns ratio?

   e. What is the maximum allowable current that it can deliver to a load?

   f. How much current is available at the output terminals under a short-circuit condition?

   g. What should the winding temperature be when the transformer is operated at rated full load in a 40°C ambient environment?

  9. List four ways in which a transformer may be cooled.

 10. What causes the sound produced by a transformer?

 11. Consider a 10-kVA transformer with a rated primary of 480 volts and a rated secondary of 24 volts. Which winding of the transformer has a larger conductor size? Why?

 12. A polarity test is being made on the transformer shown in Figure 19-25.

   a. What type of polarity is indicated?

   b. What is the value of the voltage across the secondary winding?

   c. Redraw the diagram with the unmarked leads of the transformer correctly labeled.

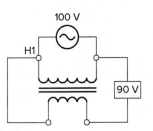

Figure 19-25 Transformer circuit for review question 12.

 13. How is a transformer's winding connected to obtain higher-voltage ratings?

 14. How is a transformer's winding connected to obtain higher-current ratings?

 15. What type of transformer is classified as a dual-voltage transformer?

 16. List three requirements that must be adhered to when connecting transformers in parallel.

 17. What type of application uses a buck-boost transformer?

PART 3 THREE-PHASE TRANSFORMER

19.6 Three-Phase Transformer Connections

All large blocks of AC power are transmitted and distributed at high-voltage levels by **three-phase systems.** Single-phase (1Ø) transformers can be electrically connected to form three-phase (3Ø) transformer banks. However, the norm is to combine the individual phase coils into one **three-phase transformer unit** on a three-legged common core, as illustrated in Figure 19-26. A single three-phase transformer is cheaper, easier to install, and will operate more efficiently than three single-phase units.

Figure 19-27 shows three single-phase transformers connected to form a three-phase transformer bank. The total

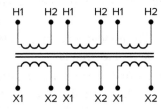

Figure 19-26 Common core of a three-phase transformer unit.
Photo: ©SNC Manufacturing Co., Inc.

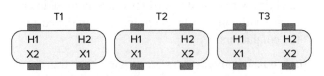

Figure 19-27 Single-phase transformers used to form a three-phase transformer bank.
Photo: ©Dora Gatlin, Gatlin Technical Service, Inc.

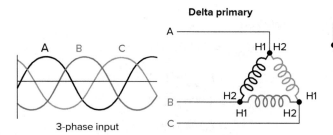
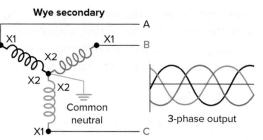

Figure 19-28 Delta-wye transformer connection.

kVA capacity of the three-phase bank is equal to the *sum* of the transformer capacity of each transformer. With this arrangement a malfunctioning transformer in the bank can be easily replaced rather than having to replace the entire unit, as would be the case if a common-core, three-phase transformer was used.

The bank of single-phase transformers or the common-core, three-phase transformer can be connected in **delta** or **wye** configurations. Figure 19-28 illustrates a delta-wye transformer connection. The polarity markings are fixed on any transformer, and the connections are made in accordance with them. The connections are summarized as follows:

- The transformer has six windings: three primary and three secondary.
- Each leg of the three-legged iron core has a respective primary and secondary winding.
- The *primary delta connection:*
  - Has the three primary windings connected in series to develop a closed circuit.

- Each primary winding end H2 is connected to the beginning H1 of another primary winding.
- At each point where windings are connected together, one of the three-phase lines is also connected so that each primary winding is connected directly across the line.

- The *secondary wye connection:*
  - Has one end of each secondary winding connected to a **common or neutral** point.
  - The winding ends with the polarity X2 are connected together at the common point.
  - The X1 terminal ends connect to the output line wires.
  - Wye connections provide two output voltages due to the common point or neutral connection.

A typical rating would be 208/120 V. The 208 indicates the voltage between phases of the secondary windings.

Figure 19-29 shows the wiring for a three-phase **wye-to-wye** transformer connection. You can calculate output voltage in

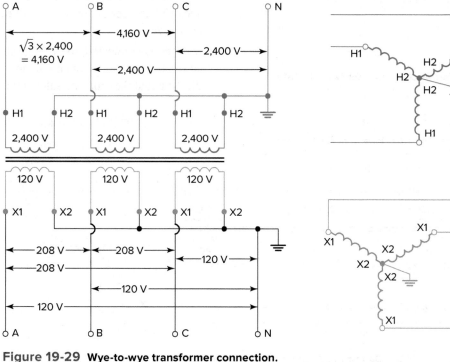
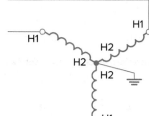
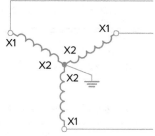

Figure 19-29 Wye-to-wye transformer connection.

the wye-to-wye transformer connections by applying the turns ratio rules for the single-phase transformer connections. The three-phase voltage, current, and power formulas associated with wye transformer connections can be expressed as follows:

$$I_{line} = I_{phase}$$

$$E_{line} = \sqrt{3} \times E_{phase} = 1.73 \times E_{phase}$$

$$E_{phase} = \frac{E_{line}}{\sqrt{3}} = \frac{E_{line}}{1.73}$$

$$kVA = \frac{\sqrt{3} \times E_{line} \times I_{Line}}{1,000}$$

EXAMPLE 19-10

Problem: For the wye-to-wye transformer connection shown in Figure 19-30, determine the following values:

a. Primary phase voltage.
b. Primary line voltage.
c. Secondary phase voltage.
d. Secondary line voltage.
e. Turns ratio for each phase transformer.

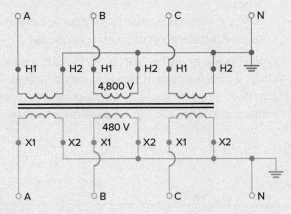

Figure 19-30 Transformer connection for example 19-10.

Solution:

a. $E_{phase} = 4,800$ volts

b. $E_{line} = E_{phase} \times 1.73$
   $= 4,800 \times 1.73$
   $= 8,304$ volts

c. $E_{phase} = 480$ volts

d. $E_{line} = E_{phase} \times 1.73$
   $= 480 \times 1.73$
   $= 830.4$ volts

e. Turns ratio $= \dfrac{E_P}{E_S}$
   $= \dfrac{4,800}{480}$
   $= 10:1$

Figure 19-31 shows the wiring for a three-phase **delta-to-delta** transformer connection, which is summarized as follows:

- The delta connection of the low-voltage secondary coils is similar to the primary coil connections with each winding end X2 connected to the beginning X1 of another secondary winding.

- Although the three secondary coils are connected to form a closed circuit, **no current** flows in the circuit at no load because the vector sum of three equal voltages 120 degrees out of phase with each other is always zero.

- If properly connected, the voltage should be zero across the last pair of open secondary coil leads, indicating that they may be safely connected together.

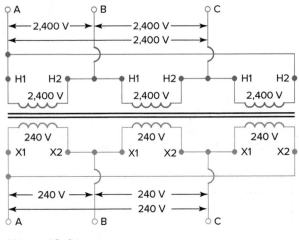
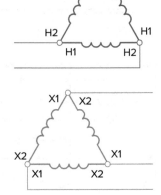

Figure 19-31 Delta-to-delta transformer connection.

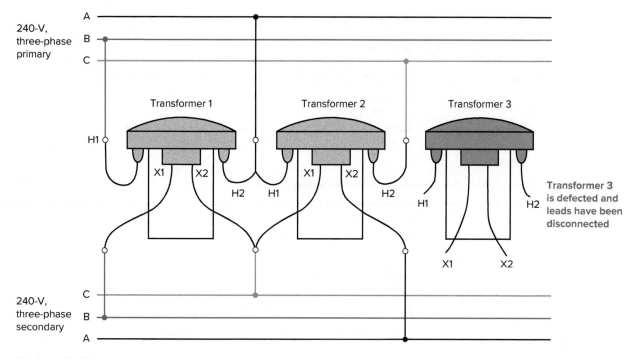

240-V, three-phase primary

A
B
C

Transformer 1   Transformer 2   Transformer 3

H1

X1   X2

H2   H1   X1   X2   H2   H1   H2

Transformer 3 is defected and leads have been disconnected

X1   X2

240-V, three-phase secondary

C
B
A

Figure 19-32 Open delta connection.

• In the delta connection, the line voltages are equal to the individual coil voltages. Unlike the wye connection, the delta connection does not provide a single common point. The three-phase voltage, current, and power formulas associated with a delta transformer can be expressed as follows:

$$E_{\text{line}} = E_{\text{phase}}$$

$$I_{\text{line}} = \sqrt{3} \times I_{\text{phase}} = 1.73 \times I_{\text{phase}}$$

$$I_{\text{phase}} = \frac{I_{\text{line}}}{\sqrt{3}} = \frac{I_{\text{line}}}{1.73}$$

$$\text{kVA} = \frac{\sqrt{3} \times E_{\text{line}} \times I_{\text{Line}}}{1,000}$$

In a three-phase delta-to-delta transformer bank, supplying a balanced three-phase load, each individual transformer shares one-third of the total load. If one of these transformers should fail, it can be taken out of service, and the two remaining transformers can supply three-phase power at a reduced power level. The resulting configuration obtained is shown in Figure 19-32 and is known as an **open delta** connection. Although the two remaining transformers are limited to about 57.7 percent of the total load, this configuration allows a circuit to remain powered during a failure of a transformer, albeit at a lower overall load factor.

EXAMPLE 19-11

Problem: For the delta-to-delta transformer connection shown in Figure 19-33, determine the following values:

a. Primary phase and line voltage.
b. Secondary phase and line voltage.
c. Turns ratio for each phase transformer.

Solution:

a. $E_{\text{phase}} = E_{\text{line}} = 138 \text{ kV}$

b. $E_{\text{phase}} = E_{\text{line}} = 4,160 \text{ V}$

c. $\text{Turns ratio} = \dfrac{E_P}{E_S}$

$$= \frac{138,000}{4,160}$$

$$= 33:1$$

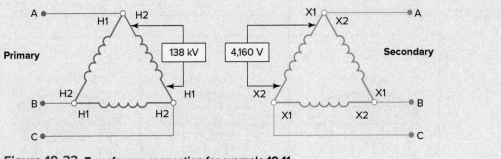

Figure 19-33 Transformer connection for example 19-11.

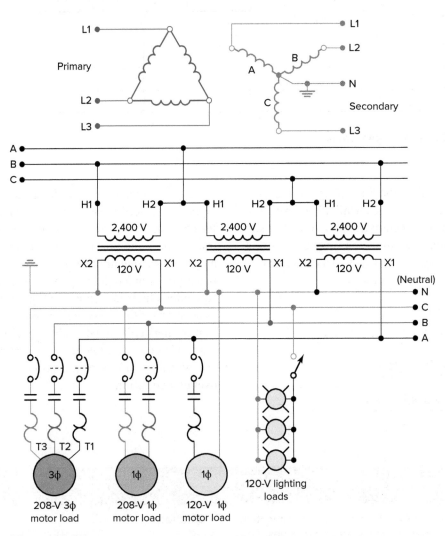

Figure 19-34 Delta-to-wye transformer connection.

The delta-to-wye three-phase transformer connection shown in Figure 19-34 is one of the most commonly used transformer connections and is summarized as follows:

- The **four-wire secondary** provides a neutral point for supplying line-to-neutral power to single-phase loads.

- The neutral point is also grounded for safety reasons.

- Three-phase loads are supplied at 208 V, while the voltage for single-phase loads is 208 V or 120 V.

- The loads connected to a transformer should be connected so that the transformer is **electrically**

balanced. Perfect electrical balance occurs when loads on a transformer are connected so that each phase of the transformer carries the same amount of current. Although perfect balance is rarely, if ever, achieved, it is something you should strive for.

The delta-to-wye connection is used for both step-up and step-down voltage applications. When used for **step-up voltage** transformation, the primary line voltage is stepped up by the transformer ratio and is *further increased by the factor of 1.73.* Since the secondary winding voltage is only 57.7 percent of the secondary output voltage, this also means that the insulation requirements of the secondary windings are reduced. This is particularly useful when the output voltage is very high.

EXAMPLE 19-12

Problem: For the delta-to-wye transformer connection shown in Figure 19-35, determine the following values:

 a. Rated primary line current.
 b. Rated primary phase current.
 c. Rated secondary line current.
 d. Rated secondary phase current.
 e. Current ratio of rated secondary to primary line current.

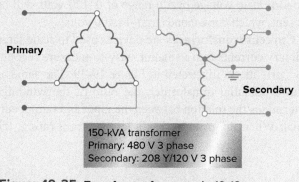

150-kVA transformer
Primary: 480 V 3 phase
Secondary: 208 Y/120 V 3 phase

Figure 19-35 Transformer for example 19-12.

Solution:

 a. Primary $I_{line} = \dfrac{kVA \times 1{,}000}{\sqrt{3} \times E_{line}}$

 $= \dfrac{150 \times 1{,}000}{1.73 \times 480}$

 $= 181\ A$

 b. Primary $I_{phase} = \dfrac{I_{line}}{\sqrt{3}}$

 $= \dfrac{181}{1.73}$

 $= 105\ A$

 c. Secondary $I_{line} = \dfrac{kVA \times 1{,}000}{\sqrt{3} \times E_{line}}$

 $= \dfrac{150 \times 1{,}000}{1.73 \times 208}$

 $= 417\ A$

 d. Secondary I_{phase} = Secondary I_{line}

 $= 417\ A$

 e. Current ratio $= \dfrac{\text{secondary } I_{line}}{\text{primary } I_{line}}$

 $= \dfrac{417}{105}$

 $= 3.97$ or approximately 4:1

Part 3 Review Questions

1. List three advantages of using a single three-phase transformer unit, as opposed to three separate single-phase transformers for the transformation of three-phase voltages.

2. The primary of a wye-connected distribution transformer is supplied by 13,800 volts. Calculate the voltage across each primary phase winding.

3. Answer each of the following with reference to the transformer connection shown in Figure 19-36.
    a. Identify the type of transformer connection shown.
    b. Determine the value of the primary line and phase voltage.
    c. Determine the value of the secondary line and phase voltage.
    d. Determine the value of the turns ratio for each phase winding.

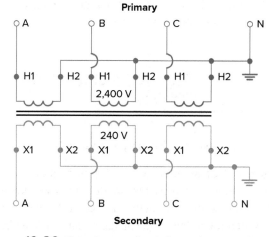

Figure 19-36 Transformer circuit for review question 3.

4. Three single-phase transformers are connected to form a three-phase transformer bank, as illustrated in Figure 19-37. Each single-phase transformer is rated for 25 kVA and operated with a primary voltage of 13.8 kV and a secondary voltage of 277 V.

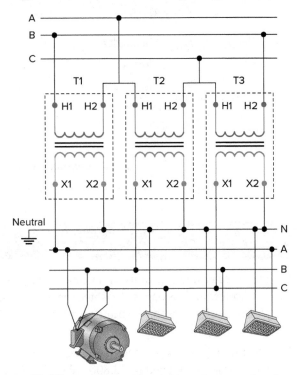

Figure 19-37 Transformer circuit for review question 4.

   a. Identify the type of connection shown.
   b. What is the total kVA rating of the transformer bank?
   c. What is the value of the single-phase voltage applied to each of the lighting loads?
   d. What is the value of the three-phase voltage applied to the motor load?
   e. What is the rated secondary phase current?
   f. What is the rated secondary line current?

5. How should you go about distributing loads on a three-phase, four-wire system?

PART 4 TRANSFORMER INSTALLATIONS

19.7 Instrument Transformers

Instrument transformers are small transformers used for measuring voltage and current in electric power systems and for power system protection and control. These transformers **step down** the large voltage or current of a circuit to isolate and protect measurement and control circuitry

Figure 19-38 Potential transformer.
©Flex-core

from the high currents or voltages present on the circuits being measured or controlled.

The **potential (voltage) transformer,** shown in Figure 19-38, can be used to supply a low voltage to voltmeters. The secondary low-voltage side of a potential transformer is usually wound for 120 volts, which makes it possible to use standard instruments of 120 volts. The primary side is designed to be connected in parallel with the circuit to be monitored. For example, it would not be safe or practical to measure a voltage of 14,400 volts directly. By inserting a voltage transformer in the circuit, a directly proportionate voltage over a range of 0–120 volts will be present, which corresponds to 0–14,400 volts.

Current transformers are devices used to scale large primary currents into a smaller, easy-to-measure secondary currents, as illustrated in Figure 19-39. The same as with a potential transformer, the ratio of the windings determines the relation between the input and output currents. When the primary has a large current rating, the

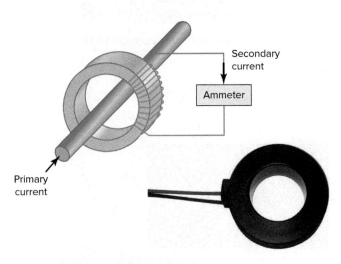

Figure 19-39 Current transformer.
Photo: ©Elkor Technologies Inc.

primary may consist of a straight conductor passing through the core. The secondary winding consisting of many turns is designed to produce 5 amperes when its rated current is flowing in the primary. In order to facilitate the production of standardized current devices, the secondary of a current transformer is always rated at *5 amps* regardless of its primary current rating. For example, assume you wish to measure 200 amps through a conductor using a current transformer. The secondary of the current transformer will produce a current which is proportionate to the current in the conductor over a range of 0–5 amps, which corresponds to 0–200 amps through the conductor.

A current transformer should *never be open-circuited while the main current is passing through the primary winding.* If the load is removed from the secondary winding while the main circuit current is flowing, the transformer acts to step-up the voltage to a dangerous level, due to the high turns ratio. For this reason, you have to place a short on the secondary winding before removing the secondary load while the main current is flowing through the primary winding.

19.8 Transformer Insulation Resistance

Transformer windings must have a sufficient **insulation resistance** to prevent the leakage of current to other windings or to the transformer case. A megohmmeter, commonly known by the trade name **megger,** is a type of ohmmeter used to measure very high values of resistance (megohm range) beyond the range of standard ohmmeters. Most ohmmeters, of the type found in multimeters, utilize a 9-V battery or less for resistance

Figure 19-40 Megger used to measure transformer insulation resistance.
©Fluke Corporation

measurements. To make an extremely high resistance test for insulation breakdown, a very high voltage is necessary. Megohmmeters have much higher voltage ratings. Some of the most common are designed to operate on one of the following values: 500 volts, 1,000 volts, and 10,000 volts. Figure 19-40 shows a megger used to measure transformer insulation resistance.

The circuit diagrams of Figure 19-41 illustrate how insulation tests of transformer windings are made using a megger. Both the resistance insulation to ground as well as between coil windings should be tested. When making the insulation to ground test, all windings should be grounded except the winding being tested. When making the insulation between coils test, be sure to test all possible coil winding combinations.

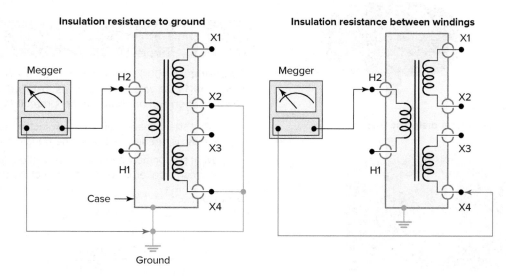

Figure 19-41 Insulation testing of transformer windings.

19.9 Cooling of Transformers

To maintain efficiency and life expectancy of a transformer, its cooling system needs to be operating at peak performance. For dry-type transformers, the ventilation system needs to be operating efficiently. For forced-air-cooled systems, the fan motors should be checked for proper lubrication and operation. Water-cooled systems must be tested for leaks and proper operation of pumps, pressure gauges, temperature gauges, and alarm systems.

When a liquid coolant. such as oil, is used its **dielectric strength** (measure of its electrical strength as an insulator) should be tested. *Water in the coolant will reduce its dielectric strength and the insulation quality.* In cases where the dielectric strength of the coolant is reduced significantly, conducting arcs may develop, causing short circuits when the transformer is energized. A standard oil dielectric test involves applying high voltage to a sample taken from the transformer and recording the voltage at which the oil breaks down, as illustrated in Figure 19-42.

Outdoor liquid-cooled transformers (Figure 19-43) usually use mineral oil, and liquid-cooled transformers for *indoor* use are filled with a synthetic liquid that is nonflammable and nonexplosive. Synthetic oil coolants must be handled with care as they sometimes cause skin irritations. One type, askarel-insulated transformers used in past years, contained polychlorinated biphenyls (PCBs), which are known to cause cancer. **Askarel** has been banned by the Environmental Protection Agency, and its use as a transformer coolant is being phased out. However, askarel coolants are still found throughout the electrical industry in older transformers, and direct contact with them should be avoided.

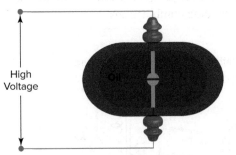

Figure 19-42 Transformer dielectric testing.

Figure 19-43 Liquid-cooled transformer.
©TheYok/iStock/Getty Images

Transformers can be installed either indoors or outdoors. Because of the potential hazards associated with some types of transformers, special installation requirements apply when transformers are installed indoors. A **transformer vault** is a structure or room in which power transformers and associated control and monitoring hardware are housed. They serve the following main purposes:

- Provide a means to isolate potentially hazardous electric components from unqualified personnel.
- Contain any fire or combustion that may occur as a result of a transformer malfunction.

19.10 Transformer Overcurrent Protection

Article 450 of the National Electrical Code (NEC) deals with requirements for transformers. Specifically, **450.3** covers **overcurrent protection** requirements for the transformer. Transformer overcurrent protection is required to protect the **primary windings** from short circuits and overloads and the **secondary windings** from overloads.

Rules for sizing overcurrent protection for a transformer operating at *more than 1,000 volts* are covered in Section 450.3(A) and Table 450.3(A) of the NEC and summarized as follows:

- When only primary protection is provided in a supervised location, the maximum current rating of the fuse is *250 percent* of the full-load current (FLA) of the primary.
- If a circuit breaker is used for protection, the maximum size is limited to *300 percent* of the full-load primary current.
- In both cases, when the calculated current value is not equal to a standard size of fuse or circuit breaker, the next larger size may be used.

EXAMPLE 19-13

Problem: A 400-kVA transformer, rated 7,200 volts primary and 1,200 volts secondary, is installed in a supervised location and has only primary protection.

 a. What size primary overcurrent protection device, using a fuse, is needed to protect the transformer?

 b. What size primary overcurrent protection device, using a circuit breaker, is needed to protect the transformer?

Solution:

a.
$$I_P = \frac{\text{kVA} \times 1,000}{E_P}$$
$$= \frac{400,000}{7,200}$$
$$= 55.55 \text{ amps}$$

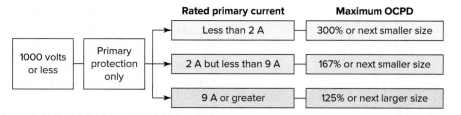

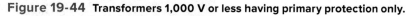

Figure 19-44 Transformers 1,000 V or less having primary protection only.

Maximum current rating of the fuse $= I_P \times 250\%$

$$= 55.55 \text{ amps} \times 2.5$$

$$= 138.88 \text{ A}$$

Next largest standard size: 138.88 A = 150 A

   b. Maximum current rating of the circuit breaker $= I_P \times 300\%$

$$= 55.55 \text{ amps} \times 3$$

$$= 166.65 \text{ A}$$

Next largest standard size: 166.65 A = 175 A

Rules for sizing an overcurrent protection device (OCPD) for a transformer operating at **1,000 volts or less** are covered in Section 450.3(B) and Table 450.3(B) of the code. If only primary protection is provided, the general rule is that the fuse or circuit breaker will not exceed **125 percent** of the full-load primary current. The exceptions made to this rule based on the maximum primary current are summarized in the chart shown in Figure 19-44.

EXAMPLE 19-14

Problem: What size primary overcurrent protection device (fuse or circuit breaker) is needed to protect the transformer circuit shown in Figure 19-45?

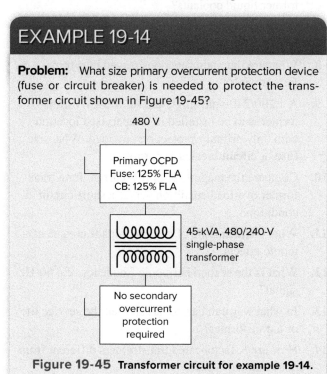

Figure 19-45 Transformer circuit for example 19-14.

Solution:

$$I_P = \frac{\text{kVA} \times 1,000}{E_P}$$

$$= \frac{45,000}{480}$$

$$= 93.75 \text{ amps}$$

Maximum primary overcurrent rating $= I_P \times 125\%$

$$= 93.75 \text{ amps} \times 1.25$$

$$= 117.2 \text{ A}$$

Therefore, use a 125-amp fuse or circuit breaker (the next highest fuse/fixed-trip circuit breaker size per NEC 240.6).

Transformer overcurrent may be classified as being due to overloads or short circuits. A transformer **overload** condition is said to exist when it is delivering from **one to six times** its normal current rating. In this case the current is confined to its normal path, and a temperature rise takes place in the transformer.

When a transformer **short-circuit** condition exists, current is not confined to the normal wiring channels to the load. Short-circuit currents can reach levels that are **hundreds of times** greater than the normal full-load operating current. Damage can be extensive if protective devices do not react in milliseconds to open the current path created. As discussed previously, the impedance rating of the transformer is used to calculate short-circuit current and to determine the interrupting current capacity required by the protective device.

19.11 Harmonics

A **linear load** is a load that does not distort the applied signal when operated within specifications. Linear loads include induction motors and resistive-type heating loads. **Nonlinear** loads include computers, variable-speed drives, solid-state motor starters, and solid-state lighting ballasts, which contain components that can **distort the applied** sine wave. A comparison of typical current waveforms for linear and nonlinear loads is shown in Figure 19-46.

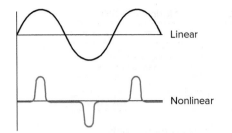

Figure 19-46 Linear and nonlinear current waveforms.

Nonlinear loads by nature demand current only during part of the cycle and as a result create harmonic effects. A **harmonic** is a signal which is a multiple of the fundamental frequency. For example, the third harmonic of a 60-Hz signal is a 180-Hz signal and the fifth harmonic is a 300-Hz signal. The specific harmonic frequencies produced and the amplitude of each signal vary with the circuit, and not all harmonics affect electrical distribution systems the same. Figure 19-47 illustrates how harmonics can cause an AC sine wave to become distorted.

Harmonics shorten the service life of a transformer by causing additional heat in transformer windings. In addition, selected harmonics cause additional heat in the neutral conductor. Because of these factors, many transformers need to be **derated** (oversized so that they operate at a fraction of their rated capacity) when used with nonlinear loads.

K-factor-rated transformers are transformers designed to operate with specific levels of nonlinear loading without derating. These transformers do not reduce the harmonics, but they are designed with windings and oversized neutrals that can handle the harmonic currents and have better ventilation to dissipate the heat more effectively.

Because the cost of manufacturing a transformer increases with the K-factor, transformers with different K-factors are available. **K-factor ratings** range between 1 and 50. The higher the K-factor, the more heat from harmonic currents the transformer is able to handle. Linear loads without harmonics, such as motors, incandescent lamps, and heating elements, have a K-factor of 1 (K-1). Circuits that include mostly nonlinear loads, such as computers and variable-frequency motor drives, have a K-factor of 20. The most common ratings for K-factor transformers are K-4, K-13, K-20, and K-30.

Part 4 Review Questions

1. What is the main function of an instrument transformer?

2. A potential instrument transformer has a rated primary voltage of 12,000 volts and a rated secondary voltage of 120 volts. What would the secondary voltage be when the primary voltage is 1,000 volts?

3. The current rating of the primary winding of a current transformer is 100 amperes, while its secondary rating is 5 amperes. An ammeter connected across the secondary indicates a current flow of 4 amperes. What is the value of the current flow in the primary?

4. For safety reasons the secondary circuit of an instrument current transformer should be closed whenever there is current in the primary circuit. Explain why.

5. A resistance test for insulation breakdown of a distribution transformer is to be made using a megger. Why would the use of a megger be preferred to that of a standard ohmmeter for this type of resistance measurement?

6. What is involved in the dielectric testing of transformer liquid coolants?

7. Which type of transformer liquid coolant has been banned because it contains PCBs?

8. What are the main functions served by a transformer vault?

9. A 1,200/240-volt, single-phase, 100-kVA transformer is to be installed in a supervised location with only primary protection provided. What size fuse or circuit breaker is permitted?

10. Compare the magnitude of overcurrent for a transformer overload condition versus a short-circuit condition.

11. What characteristic of nonlinear loads creates harmonic effects?

12. What is the second harmonic frequency of a 60-Hz signal?

13. In what way can harmonics shorten the service life of a transformer?

14. How are K-factor-rated transformers different from conventional types?

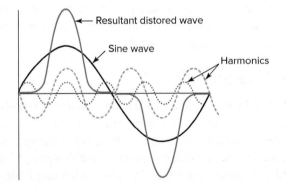

Figure 19-47 Harmonics can cause an AC sine wave to become distorted.

High voltage insulation tester
©Fluke Corporation

Electrical Installation and Maintenance

SECTION FOUR

CHAPTER TWENTY

Circuit Conductors and Wire Sizes

Conduit system
©Allied Tube and Conduit a part of Atkore International

LEARNING OUTCOMES

▶ Identify uses for different conductor forms.

▶ Properly select wire-insulating materials.

▶ Compare the AWG size and diameter of conductors.

▶ List the factors that determine a conductor's ampacity rating.

▶ Identify the factors that contribute to the resistance value of a conductor.

▶ Calculate line-voltage drop and line-power loss.

One of the most important parts of any electrical installation is the conductors that connect all the components. The conductors must be able to deliver the necessary energy on a continuous basis, without overheating or causing unacceptable voltage drops. In this chapter we study the different conductor forms and their application.

PART 1 ELECTRIC WIRE AND CABLE

20.1 Building Wire and Cable

Building wire or cable is used to deliver electricity to where it is needed. In general the National Electrical Code (NEC) recognizes three types of conductors: **Copper, aluminum,** and **copper-clad aluminum.** Unless specified, all conductors in the NEC are considered to be copper.

A **wire** is a single conductor and may be solid or stranded, as illustrated in Figure 20-1. A **solid wire** consists of a single conductor, while a **stranded wire** is made up of a number of single wires twisted together to form a single conductor. The purpose of stranding conductors is to provide increased flexibility. The choice between solid and stranded depends on the need for flexibility in handling and working with the wire. Smaller wires sizes are generally both stranded and solid, while larger wire sizes are normally stranded.

A **cable** consists of two or more wires insulated from each other and held together as a single assembly. Figure 20-2 shows a typical two-conductor (2/G) cable assembly consisting of one black wire, one white wire, and one ground wire. Note that the **ground wire** is never included in the cable conductor count.

20.2 Conductor Insulation

Insulated wire or cable is used for electrical isolation between conductors and from ground. This is accomplished by coating or wrapping them with various insulation materials having very high electric resistance. Bare conductors are available for **ground wires** only.

Insulation resistance refers to the resistance to current leakage through the insulation materials. An ordinary ohmmeter cannot be used for measuring the resistance of

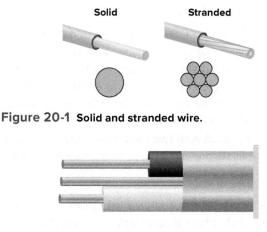

Figure 20-1 Solid and stranded wire.

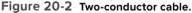

Figure 20-2 Two-conductor cable.

Figure 20-3 High-voltage insulation tester.
©Fluke Corporation

insulating material. To test adequately for insulation breakdown, it is necessary to use a much higher voltage than is furnished by an ohmmeter's battery. Insulation resistance can be measured with a high-voltage **megger** or **megohmmeter,** such as that shown in Figure 20-3, without damaging the insulation.

Factors that must be considered when selecting wire insulation include:

- Circuit voltages
- Surrounding temperature conditions
- Moisture
- Conductor flexibility
- Being resistant to fire, oil, or liquid fuels.

Thermoplastic is one of the most commonly used insulators. Regular thermoplastic is an excellent insulator, but it is sensitive to extremes of temperature. Thermoplastic insulation will soften and can melt if heated above its rated temperature and will stiffen at temperatures colder than 14°F (−10°C). Typical examples of thermoplastic insulation are types THHN, THHW, THW, THWN, and TW. Type TW thermoplastic is a weatherproof type, while type THW is both weatherproof and heat resistant.

Thermosetting insulation can withstand higher and lower temperatures. This insulation material will not soften, flow, or distort appreciably when subjected to heat and pressure. Typical examples of thermosetting insulation are types RHH, RHW, XHH, and XHHW.

Enameled wire (also called **magnet wire**) is a conductor with baked-on enamel film insulation, as shown in Figure 20-4. It is applied by repeatedly passing the bare wire through a solution of hot enamel to form a thin, tough coating that is a varnishlike plastic compound. It is called magnet wire because it is used for winding the coils in electromagnets, solenoids, transformers, motors, and generators.

Figure 20-4 Magnet wire.
©Theerawut_Chaiyatham/iStock/Getty Images

The National Electrical Code (NEC) requirements for conductor selection can be found in Chapter 3, Articles 320 through 340. The type of insulation applied and the wire size or gauge initially classify wires or conductors. The various types of insulation, in turn, are subdivided according to their maximum operating temperatures and the nature of their use.

The NEC has adopted various color code requirements for many of the conductors used in electrical distribution systems. The following are examples that apply to alternating current circuits:

- The grounded (identified) conductor must have an outer finish that is either continuous white or gray, or have an outer finish (not green) that has three continuous white stripes along the conductor's entire length.

- An ungrounded conductor must have an outer finish that is a color other than green, white, natural gray, or three continuous white stripes.

- *The NEC restricts use of the color green to equipment grounding and bonding conductors only.*

20.3 Cable Assemblies

A jacket or outer sheath may be applied over conductor insulation or cable core for mechanical, chemical, or electrical protection. A **cable assembly** is a flexible assembly containing two or more conductors grouped together within a common protective cover and used to connect individual components. The **NEC articles** describe exactly *where and how* the various types of cables should be used.

Nonmetallic-Sheathed Cable (Type NM)

Generally, nonmetallic-sheathed (NM) cable has two or more thermoplastic-insulated wires, one bare grounding wire, paper insulation surrounding the conductors, and a thermoplastic jacket or sheath. All nonmetallic-sheathed cable sold and used today is marked NM-B or NMC-B, as illustrated in Figure 20-5. The reason being, when originally developed type NM cable conductors had insulation rated for 60°C (140°F) which often became brittle due to overheating when NM cable was connected to

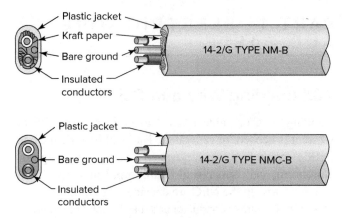

Figure 20-5 Nonmetallic-sheathed cable.

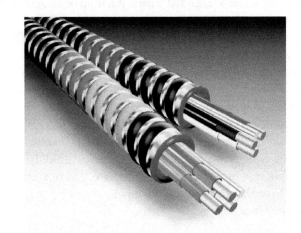

Figure 20-6 Armored cable (AC).
©AFC Cable Systems, Inc.

ceiling-mounted luminaires. **Type NM-B** is permitted in normally dry locations; **Type MNC-B** can be used in either dry or damp locations. This cable is a common wiring method for residential and commercial branch circuits. Because it provides minimal physical protection for the conductors, the installation restrictions are strict.

Armored Cable (Type AC)

Armored cable (type AC) contains insulated conductors individually wrapped in wax paper and enclosed in a spiral flexible **metal sheath** (either steel or aluminum; Figure 20-6). A bare bonding wire in continuous direct contact with the metal armor is placed inside the armor with the conductors. **Antishort bushings** (also called redheads or BX bushings) installed between the wire and the armored sheath prevent damage to the wire insulation.

Teck Cable (Type MC)

Teck cable is a type of metal-clad (MC) cable. All Teck cables come with a bare ground wire and uses aluminum interlocked armor (Figure 20-7). A unique feature of Teck

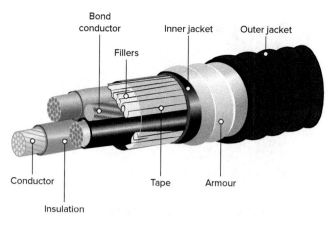

Figure 20-7 **Teck cable (type MC).**

Figure 20-8 **Service-entrance (SE) cable.**

style cables is the inner and outer PVC jacket. The **extra inner jacket** provides additional protection from external physical and chemical damage over typical aluminum interlocked products. One advantage of using Teck cable is that it avoids the need for using conduit or ducts. Because Teck cable has armor in its construction, it can be direct buried.

Service-Entrance Cable (Type SE)

Service-entrance (type SE) cable is a single or multiconductor assembly with or without an overall covering used for service-entrance wiring. Figure 20-8 shows a service-entrance cable used to convey power from the service drop to the meter base and from the meter base to the distribution panelboard. USE cable is SE cable identified for underground use and has a moisture-resistant covering.

20.4 Conduit Systems

Electrical **conduits** are defined in the NEC as a raceway of circular cross section with associated fittings, couplings, and connectors for the installation of electric conductors. Figure 20-9 shows a typical conduit system. There are many types of conduits; they can be either flexible or rigid. With or without options such as liquid-tight, or flame retardant, conduits are made from materials such as steel, aluminum, and polymers.

Advantages of conduit wiring include the following:

- Metallic conduits provide a high degree of **fire protection,** as well as the ability to safely contain overloaded or short-circuited conductors that could cause or contribute to a fire.

Figure 20-9 **Conduit system.**
©Allied Tube and Conduit a part of Atkore International

Figure 20-10 **Rigid metal conduit (RMC).**
©Allied Tube and Conduit a part of Atkore International

- The **rigidity** of conduit permits installation with fewer supports than the other types of wiring systems.
- Conduits can be sized to provide the easy installation of the additional conductors in the conduit.

Rigid Metal Conduit (RMC)

A rigid metal conduit (RMC) or thickwall conduit is a **threadable** raceway, generally made of steel or aluminum, as shown in Figure 20-10. This type of conduit provides the greatest amount of mechanical protection to conductors. Its use is permitted under all atmospheric conditions and in all types of occupancies as per the NEC articles. An additional benefit of using metal conduit is that the NEC recognizes a properly installed metal conduit system as an equipment grounding conductor.

Rigid Nonmetal Conduit (RNMC)

A rigid nonmetallic conduit (RNMC) is made of approved nonmetallic material that is resistant to moisture

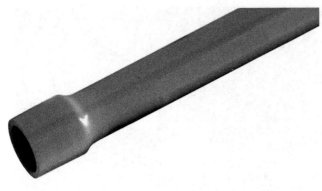

Figure 20-11 Rigid nonmetallic conduit (RNMC).
©Allied Tube and Conduit a part of Atkore International

Figure 20-13 Flexible metal conduit (FMC).
©Galco Industrial Electronics

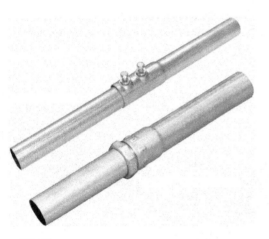

Figure 20-12 Electrical metallic tubing (EMT).
©Allied Tube and Conduit a part of Atkore International

Figure 20-14 Liquid-tight flexible metal conduit (LFMC).
©AFC Cable Systems, Inc.

and chemical atmospheres. Rigid PVC conduit, shown in Figure 20-11, protects conductors in the worst *corrosive locations.*

Electrical Metallic Tubing (EMT)

Electrical metallic tubing (EMT), or **thinwall** conduit, is made of lightweight steel tubing. Because of its lightweight construction, it is **threadless.** These features save much time and work when installing EMT. Lengths are connected together by using set screw or compression fittings, as illustrated in Figure 20-12.

Flexible Metal Conduit (FMC)

Flexible metal conduit (FMC) is made of an interlocking steel or aluminum strip (Figure 20-13), similar to armor cable. This type of conduit combines mechanical protection with maximum *flexibility* for nonhazardous locations. Flexible conduit is used in the installation of motors and/or machines with *vibrating* moving parts.

Liquid-Tight Flexible Metal Conduit (LFMC)

Liquid-tight flexible metal conduit (LFMC) is a raceway of circular cross section with an outer *liquid-tight, nonmetallic, sunlight-resistant jacket* over an inner helically wound metal strip (Figure 20-14). This type of conduit is excellent for use in damp locations, corrosive areas, or around machines where coolants, cutting, and/or lubricating liquids are likely to splash onto it.

20.5 Wire Sizes

The larger the diameter of a wire, the lower is its resistance and the greater its current-carrying capacity. If a wire carries more current than it is rated for, it will dangerously **overheat.**

The size of a solid wire is determined by its diameter and usually referred to by an equivalent gauge number, rather than by the actual diameter. **American wire gauge (AWG)** is the standardized wire gauge system used to represent the wire diameters. Table 20-1 shows the AWG gauge number

Table 20-1 AWG Gauge Number versus Diameter and Circular Mil Area		
AWG	**Diameter (in.)**	**Area (cmil)**
18	0.0403	1,624
16	0.0508	2,581
14	0.0641	4,109
12	0.0808	6,529
10	0.1019	10,380
8	0.1285	16,510
6	0.1620	26,250
4	0.2043	41,740
2	0.2576	66,370
1	0.2893	83,690
0 (1/0)	0.3249	105,560
00 (2/0)	0.3648	133,079
000 (3/0)	0.4096	167,772
0000 (4/0)	0.4600	211,600

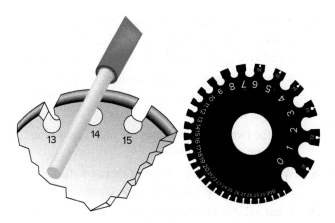

Figure 20-15 AWG wire gauge.

versus the diameter in inches and cross-sectional area in circular mils for some of the more common conductors. The table is for a single, solid, round conductor. Because there are also small gaps between the strands, a stranded wire will always have a slightly larger overall diameter than a solid wire with the same AWG. Note that the *larger* the gauge number, the *smaller* the actual diameter and area of the conductor.

In the AWG system, the cross-sectional *area* of a conductor is specified in **circular mils (cmil or CM).** A mil is one-thousandth of an inch. The cross-sectional area of a conductor is equal to the diameter (in mils) squared. For stranded wire, the size in circular mils is the total cross-sectional area and is equal to the area (in circular mils) of one strand multiplied by the number of strands.

EXAMPLE 20-1

Problem: Determine the cmil area of a 10 AWG solid, round conductor with a diameter of 0.1019 inch.

Solution:

$$\text{Diameter mils} = \text{Diameter inches} \times 1{,}000$$
$$= 0.1019 \times 1{,}000$$
$$= 101.9 \text{ mils}$$

$$\text{CM area} = d^2$$
$$= 101.9^2$$
$$= 10{,}384 \text{ cmils}$$

EXAMPLE 20-2

Problem: Find the diameter, in mils, of an 18 AWG wire with a cross-sectional area of 1,620 cmil.

Solution:

$$\text{cmil} = d^2$$
$$d = \sqrt{\text{cmil}}$$
$$= \sqrt{1{,}620}$$
$$= 40.25 \text{ mils}$$

The external diameter of a wire, including its insulation, has nothing to do with its wire size or current capacity. The AWG wire gauge, shown in Figure 20-15, can be used for determining the size of round wire, from 0 AWG to 36 AWG. To gauge the wire:

- The insulation is removed from one end.
- The bare end is inserted into the smallest *slot* in which it will fit without using force.
- The number stamped below the slot is the AWG of the wire.

Wire dimensions are specified in American wire gauge (AWG), circular mils (cmil), and thousands of circular mils (kcmil). For larger-wire sizes, greater than 4/0, the wire gauge system is typically abandoned for cross-sectional area measurement in thousands of circular mils **(kcmil),** with the k being used to denote a multiple of *thousand.*

Part 1 Review Questions

1. List the three types of conductors that are recognized by the NEC.
2. Why are larger-diameter wires generally stranded?

3. A cable assembly contains white, black, and red insulated wires and a bare ground wire. Would this be specified as a three- or four-conductor cable? Why?

4. Why is a megger, or megohmmeter, the preferred instrument for testing for the breakdown of electrical insulation?

5. List the five factors that must be considered when selecting a wire insulation.

6. What is the main advantage of thermosetting insulation over the thermoplastic type?

7. What color of insulation does the NEC restrict use to grounding and bonding conductors only?

8. Name three common types of cable assemblies.

9. Outline three advantages of a conduit wiring system.

10. What installation advantage is there when installing EMT versus RMC conduit?

11. What type of applications are FMC best suited for?

12. What is the circular mil area of an 2 AWG wire 257.6 mils in diameter? (Express your answer in cmil and kcmil units.)

13. What is the relationship between the AWG number and the diameter of the wire?

14. How are larger-size wires greater than 4/0 specified?

PART 2 CONDUCTOR AMPACITY AND LINE LOSSES

20.6 Conductor Ampacity

The **ampacity** rating of a conductor is the amount of current it can carry continuously under the conditions of use without overheating or exceeding its temperature rating. General requirements for conductor ampacity can be found in Article 310 of the National Electrical Code (NEC). The most prominent feature of Article 310 is its collection of ampacity tables. These tables set a maximum current value at which the conductor insulation shouldn't prematurely fail during normal use, under the conditions described in the tables. In general, conductor ampacity is determined by:

Material. Copper is a better conductor than aluminum, and so it can carry more current for a given gauge.

AWG wire size. The smaller the gauge number, the larger the cross-sectional area of the conductor and the more current it can carry.

Type of insulation. All insulated conductors have a maximum operating temperature at which the insulation of the conductor is not adversely affected. A conductor with more of a heat-resistant insulation will have a higher ampacity rating than one of equivalent size with a lower-insulator temperature rating.

Conductor length. The resistance of a wire increases as its length increases. The listed ampacities in the code tables assume that the length of the conductor will not increase the resistance of the circuit by a significant amount and that the full-rated voltage will be available at the end of the line. For long conductor runs, the wire diameter sizes must be increased above the required rated ampacity of the conductor to keep the amount of line voltage drop to an acceptable level.

Ambient air temperature. The higher the ambient temperature surrounding a conductor, the more difficult it is for the conductor to dissipate heat. When the ambient temperature is above 30°C (86°F), the ampacity rating of the conductor is reduced. NEC Table 310.15(B)(2)(a) gives the ampacity correction factors for situations where the ambient is expected to be higher or lower than 30°C (86°F). For example, the ampacity of a 1/0 AWG, aluminum, type THHN conductor when the ambient temperature is 100°F is found by taking the ampacity from the table and multiplying it by the appropriate correction factor. In this example, the ampacity would be 135 (amperes) × 0.91 (ambient temperature correction) = 122.85 amperes.

Installation conditions. Conductors that are run singly in free air (maximum typical air circulation) will have a higher ampacity rating than a similar conductor that is enclosed with other conductors in a cable or conduit. Adjacent load-carrying conductors affect operating temperature in two ways: the ambient temperature can be raised, and heat dissipation can be impeded. Table 310.15(B)(3)(a) contains the factors for more than three current-carrying conductors. This table states that when the number of conductors in a raceway or cable exceeds three, the ampacities are to be reduced by the appropriate percentage. For example, the ampacity of 12 No. 12 copper THHN conductors installed in one conduit can be found in Table 310-16. In this example, the ampacity (in the table) is 30 amperes and the derating factor for 12 conductors is 50 percent. Therefore, the ampacity would be 30 (amperes) × 0.5 (50%) = 15 amperes per conductor.

The National Electrical Code contains tables that list the ampacity for the approved types of conductor size, insulation, and operating conditions. These tables are a practical source of information that should be referred to for specific circuit installations.

20.7 Conductor Resistance

When a conductor carries a current, the conductor's resistance causes the conversion, into heat, of a portion of the electric energy being transmitted. The resistance of a wire increases as the *length* of the wire increases, as illustrated in Figure 20-16. The two wires are of the same material and diameter, but one is twice as long as the other. In this case the longer wire will have twice the resistance of the shorter wire. Tables indicating wire resistance generally specify the resistance of 1,000 feet of wire at a certain temperature.

EXAMPLE 20-3

Problem: The resistance of 1,000 feet of 2 AWG standard annealed copper wire is listed in a wire table as 0.1563 ohm per 1,000 ft at 20°C. What is the resistance of 250 ft of this wire?

Solution:

$$\text{Resistance} = \frac{\text{length (ft)}}{1,000} \times \text{resistance per } 1,000 \text{ ft}$$

$$= 250 \times \frac{0.1563}{1,000}$$

$$= 0.0391 \ \Omega$$

The larger the **diameter** of the conductor, the lower is its resistance. Large conductors allow more current flow with less voltage drop or loss. For example, a wire with a larger cross-sectional area than another wire has a lower resistance for a given length. For example, shown in Figure 20-17, the resistance for 1,000 feet of 10 AWG solid copper wire is 1.21 ohms, compared to 3.07 ohms per 1,000 feet for a smaller solid 14 AWG copper wire.

The resistance of a conductor varies with *temperature.* For copper and aluminum conductors, the higher the temperature, the higher the resistance. The two factors that determine the operating temperature of a conductor are the temperature of the surrounding air space (ambient temperature) and the amount of current flow through the conductor.

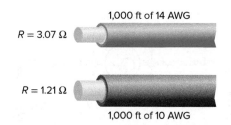

$R = 3.07 \ \Omega$ — 1,000 ft of 14 AWG

$R = 1.21 \ \Omega$ — 1,000 ft of 10 AWG

Figure 20-17 The larger the diameter, the lower the conductor resistance.

The *physical condition* of the conductor and/or connection will also affect resistance. A partially cut or nicked wire will act like smaller-diameter wire, with high resistance in the damaged area. Broken strands in the wire, poor splices, and loose or corroded connections also increase resistance.

Different **materials** have different atomic structures, which affect their ability to conduct electrons. Aluminum wire, because it is not as good a conductor as copper, has an ampacity approximately equal to that of copper wire two gauge sizes smaller. For example, 12 AWG aluminum wire has about the same ampacity as 14 AWG copper wire.

20.8 Line Voltage Drop and Power Loss

The resistance of the power source line conductors is normally *low* when compared to that of the load. In most instances, the conductors are treated as being ideal or perfect conductors of electricity. As a result, they are said to have zero resistance. In this case, the voltage value of the source is the same as that across the load, as illustrated in Figure 20-18.

In some circuits, the resistance of the conductors is important and must be taken into account. This is often the case where the load is located some **distance** from the voltage source—such as hundreds of feet or more. In this type of circuit, the voltage at the load can be significantly less than what appears at the energy source. When this is the case, the line voltage drop is the **difference** between the operating voltages of the source and that of the load, as

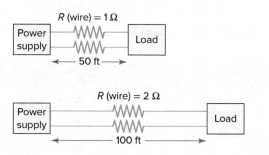

R (wire) = 1 Ω

Power supply — Load — 50 ft

R (wire) = 2 Ω

Power supply — Load — 100 ft

Figure 20-16 Conductor resistance increases with length.

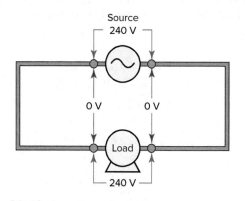

Source 240 V

0 V   0 V

Load

240 V

Figure 20-18 Zero line voltage drop.

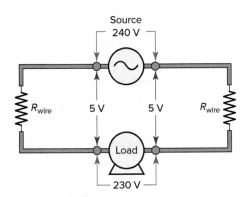

Figure 20-19 Determining line voltage drop.

illustrated in the circuit of Figure 20-19. In this example the supply voltage is 240 V, while the voltage at the load is 230 V, resulting in a line voltage drop of 10 V.

Excessive voltage drops result in wasted power and energy. Any voltage that drops between the supply and the equipment is lost to the equipment and in some applications can seriously affect the operation of the equipment. As an example, a substantial loss in voltage to a motor can cause reduction in motor horsepower, increased operating costs, increased motor heat buildup, and reduced motor life expectancy.

When sizing circuit conductors for extremely long circuit runs, the expected voltage drops are estimated before the installation. If necessary, wire diameter sizes are increased above the size required by ampacity to keep the line voltage loss within acceptable limits. While the NEC does not have requirements for voltage drop, it *recommends* limiting the voltage drop on branch circuit conductors to a maximum of *3 percent* and a total maximum of *5 percent* for feeder and branch circuit combined.

EXAMPLE 20-4

Problem: The single-phase voltage at the source and load of an operating branch circuit installation is measured and found to be 120 V and 118 V, respectively. Determine the amount and the percentage of the voltage drop.

Solution:

$$E_{VD} = E_{source} - E_{load}$$
$$= 120 \text{ V} - 118 \text{ V}$$
$$= 2 \text{ V}$$

$$\% \text{ voltage drop} = \frac{E_{source} - E_{load}}{E_{source}} \times 100$$
$$= \frac{2 \text{ V}}{120 \text{ V}} \times 100$$
$$= 1.67\%$$

The **voltage drop** across a wire is directly proportional to the resistance of the wire and the amount of current it is carrying. The voltage drop can be calculated according to Ohm's law as follows:

$$E_{VD} = I \times R_{wire}$$

EXAMPLE 20-5

Problem: The current flow through a DC circuit is calculated and found to be 11 amps. What would the expected line voltage drop be if the conductors supplying power to the load have a combined total resistance of 0.2 ohm?

Solution:

$$E_{VD} = I \times R_{wire}$$
$$= 11 \text{ A} \times 0.2 \text{ }\Omega$$
$$= 2.2 \text{ V}$$

Resistivity (K) is a constant, which is related to a wire's resistance per circular mil-foot (Figure 20-20). The resistivity of copper and aluminum wire is different, and both increase as the temperature increases. Table 20-2 lists the typical values of resistivity (K) for copper and aluminum at three different temperatures. The K values listed are approximate and will vary slightly, depending on the temperature and method used in determining K.

The **resistance** of a given length of a particular size of copper or aluminum wire can be calculated using the resistivity or K factor in the following wire resistance formula:

$$R = \frac{K \times L}{\text{cmil}}$$

where R = total resistance of the wire in ohms
K = resistivity in ohms per circular mil-foot (from tables)
L = length of the wire in feet
cmil = circular mil area of the wire (from tables)

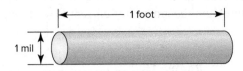

Figure 20-20 Circular mil-foot.

Table 20-2 Resistivity			
Resistivity K (ohms per circular mil-foot)			
	25°C	**50°C**	**75°C**
Copper	10.8	11.8	12.9
Aluminum	17.0	19.0	21.2

EXAMPLE 20-6

Problem: What is the resistance of 2,000-ft length of copper wire that has a cross-sectional area of 10,380 cmils? Assume a temperature of 75°C and a resistivity of 12.9 ohms per circular mil-foot.

Solution:

$$R = \frac{K \times L}{\text{cmil}}$$

$$= \frac{12.9 \times 2,000}{10,380}$$

$$= 2.48 \ \Omega$$

The wire resistance formula is transposed or rearranged to *determine K, L*, or cmil as follows:

$$K = \frac{R \times \text{cmil}}{L}$$

$$L = \frac{R \times \text{cmil}}{K}$$

$$\text{cmil} = \frac{K \times L}{R}$$

EXAMPLE 20-7

Problem: The resistance between the two ends of a length of 12 AWG copper wire (cross-sectional area of 6,530 cmils) is measured and found to be 0.8 ohm. Using a K of 12.9, calculate the approximate length of the wire.

Solution:

$$L = \frac{R \times \text{cmil}}{K}$$

$$= \frac{0.8 \times 6,530}{12.9}$$

$$= 405 \text{ feet}$$

At times it is necessary to compute voltage drop of an installation when the length, size of wire, and current are known. For **single-phase** systems the following formula is used:

$$E_{VD} \text{ (single phase)} = \frac{K \times I \times L \times 2}{\text{cmil}}$$

where E_{VD} = voltage drop in volts
K = resistivity in ohms per circular mil-foot (from tables)
L = length in feet from the beginning of the circuit to the load
cmil = circular mil area of the wire (from tables)

For **three-phase** systems a slightly different formula is used to calculate the voltage drop. The $K \times I \times L$ is multiplied by $\sqrt{3}$, or 1.73, instead of 2 as follows:

$$E_{VD} \text{ (three phase)} = \frac{K \times I \times L \times 1.73}{\text{cmil}}$$

EXAMPLE 20-8

Problem: A 240-V, single-phase circuit is being used to supply power to an electric water heater. The distance of the circuit from the panel to the load is 85 ft, the load current is 14 amperes, and the cross-section area of the 12 AWG wire is 6,530 cmils. Determine the approximate circuit voltage drop and the percentage voltage drop using a K of 12.9 ohms per circular mil-foot at 75°C.

Solution:

$$E_{VD} = \frac{K \times I \times L \times 2}{\text{cmil}}$$

$$= \frac{12.9 \times 14 \times 85 \times 2}{6,530}$$

$$= 4.70 \text{ V}$$

$$\% \text{ voltage drop} = \frac{E_{VD}}{E_{\text{source}}} \times 100$$

$$= \frac{4.70 \text{ V}}{240 \text{ V}} \times 100$$

$$= 1.96\%$$

The single-phase voltage drop formula is transposed or rearranged, as follows, to determine the *minimum cmil size conductor* that must be installed to remain below a specified amount of voltage drop value:

$$\text{cmil} = \frac{K \times I \times L \times 2}{E_{VD}}$$

EXAMPLE 20-9

Problem: Find the size of copper wire (use a K factor 12.9) required to carry a load of 45 amperes at 240 volts a distance of 500 feet with 2 percent voltage drop. Use Table 20-1 to determine the closest minimum size acceptable.

Solution:

$$E_{VD} = 240 \text{ V} \times 2\%$$

$$= 240 \times 0.02$$

$$= 4.8 \text{ V}$$

$$\text{cmil} = \frac{K \times I \times L \times 2}{E_{VD}}$$

$$= \frac{12.9 \times 45 \times 500 \times 2}{4.8}$$

$$= 120,938 \text{ cmils}$$

Referring to Table 20-1, it will be found that this size lies between 1/0 AWG and 2/0 AWG, so 2/0 AWG would be the wire size selected.

The single-phase voltage drop formula is transposed or rearranged, as follows, to determine the **maximum length (distance)** from the source to the load for a specified amount of voltage drop value:

$$L = \frac{cmil \times E_{VD}}{2 \times K \times I}$$

EXAMPLE 20-10

Problem: A 240-V single-phase circuit is to provide power to a load. Determine the maximum distance of the circuit from the power supply to the load if the conductor size is 6 AWG, the load current is 30 amperes, and the maximum voltage drop permitted is 1 percent. (Use a K factor of 12.9.)

Solution:

6 AWG area (from Table 20-1) = 26,250 cmils

$$E_{VD} = 240 \text{ V} \times 1\%$$
$$= 2.4 \text{ V}$$

$$L = \frac{cmil \times E_{VD}}{2 \times K \times I}$$
$$= \frac{26,250 \times 2.4}{2 \times 12.9 \times 30}$$
$$= 81.4 \text{ ft}$$

Current flow through a conductor also causes a **power loss,** in the form of heat, due to the conductor's resistance. Power loss in a wire is equal to the square of the current multiplied by the resistance of the wire:

$$P = I^2 \times R_{wire}$$

where
P = power in watts (W)
I = current in amperes (A)
R = resistance in ohms (Ω)

When the voltage drop of the circuit is known, the power loss can be more easily calculated using the equation

$$P = E_{VD} \times I_{line}$$

EXAMPLE 20-11

Problem: The total resistance of two 12 AWG copper conductors, 75 feet long, is 0.3 ohm (0.15 Ω for each conductor). The current of the circuit is 16 amperes. Calculate the total amount of power lost in the circuit conductors.

Solution:

$$P = I^2 \times R_{wire}$$
$$= 16^2 \times 0.15 \times 2$$
$$= 76.8 \text{ W}$$

Part 2 Review Questions

1. What does the ampacity rating of a conductor specify?

2. List the factors taken into consideration when determining the ampacity rating of a conductor.

3. Why is a copper conductor rated at a higher ampacity than an aluminum conductor of equivalent gauge size or diameter?

4. State the effect (increase or decrease) of each of the following on the resistance value of a circuit conductor:
   a. Increasing the length of the conductor.
   b. Decreasing the diameter of the conductor.
   c. Increasing the operating temperature of the conductor.
   d. Using the same-size aluminum conductor in place of a copper one.

5. a. What causes line voltage drop in a circuit?
   b. Under what condition is the line voltage drop considered to be zero?
   c. In what type of electrical installation must the resistance of the conductors be taken into account?

6. The NEC recommends that the voltage drop in a branch circuit should not exceed 3 percent of the supply voltage. Assuming the supply voltage feeding a load is 120 V and the voltage measured at the load is 118 V, determine:
   a. The amount of line voltage drop.
   b. The maximum line voltage drop acceptable based on a 3 percent maximum.
   c. The percentage voltage drop of this circuit.

7. Calculate the resistance of a 10 AWG (10,380 cmils) copper wire 300 ft long. Assume a temperature of 75°C and a resistivity of 12.9 ohms per circular mil-foot.

8. What is the approximate voltage drop on a 120-volt, single-phase circuit consisting of 14 AWG copper conductors (4,110 cmils) where the load is 5 amperes and the distance of the circuit from the panel to actual load is 60 ft? Use a K of 12.9 ohms per circular mil-foot.

9. A copper wire (with a K factor 12.9) is required to carry a load of 16 amps at 120 volts a distance of 130 feet with a voltage drop no greater than 3 percent.
   a. Calculate the value of the maximum allowable voltage drop.

b. Calculate the minimum cross-sectional area of the wire required.

c. Use Table 20-1 to determine the closest minimum size acceptable.

10. The resistance between the two conductor ends of a partly used 250-ft roll of 12 AWG copper wire (cross-sectional area of 6,530 cmils) is measured and found to be 0.25 ohm. Using a K of 12.9, calculate the approximate length of the cable remaining in the roll.

11. Determine the maximum distance a single-phase, 240-volt, 42-ampere load can be located from the panelboard so voltage drop does not exceed 3 percent. The circuit is to be wired with 8 AWG copper conductors, and a K of 12.9 is to be assumed.

12. **a.** What is the resistance of 500 ft of 10 AWG solid copper wire that is specified as 0.9989 ohm per 1,000 ft?

b. Calculate the line voltage drop if this overall length of wire is used in a circuit that draws 25 A of current.

c. Calculate the line power loss of the circuit.

13. According to the NEC, what is the maximum ambient temperature above which the value of the conductor ampacity is required to be reduced?

14. According to the NEC, what number of conductors in a raceway or cable must be exceeded before the value of the conductor ampacity is required to be reduced?

CHAPTER TWENTY-ONE

Fuses and Circuit Breakers

Domestic electrical distribution board, mounted on wall
©kanvag/123RF

LEARNING OUTCOMES

▶ Define the terms *overload* and *short circuit*.

▶ Compare the basic principle of operation of a fuse and a circuit breaker.

▶ State how fuses and circuit breakers are rated.

▶ Identify basic fuse types and typical applications.

▶ Test fuses and circuit breakers in and out of circuits.

▶ Understand how selective coordination of protective devices can prevent power blackouts.

Circuit **overcurrent protection** is a vital part of every electric circuit. Electric circuits can be damaged or even destroyed if their voltage and current levels exceed those for which they are designed. In general, fuses and circuit breakers are designed to protect personnel, conductors, and equipment. Both operate on the same principle: to interrupt or open the circuit as quickly as possible before damage can occur.

PART 1 CIRCUIT PROTECTION

21.1 Overloads and Short Circuits

An electric circuit is limited in the amount of current it can safely handle. Excessive current flow through a conductor will cause the conductor to heat up. The current capacity of the circuit is determined by the wire conductors used. An electric circuit is said to be **overloaded** when the amount of current flowing through it is more than its rated current capacity.

Overloads can occur in a home when the number of electrical loads connected into the same branch circuit exceeds its rated current capacity, as illustrated in Figure 21-1, and are summarized as follows:

- The branch circuit has a maximum rated capacity of 15 amps.
- The sum of the parallel load currents connected to it is 17 amps.
- Circuit is overloaded by 2 amps, and as a result the breaker trips.

Overloads can also be caused by defective equipment as well as overloaded equipment. For example a defective heater element, with lower than normal resistance, will cause the circuit current to increase above its normal operating value. Similarly, a motor can become overloaded by too heavy a load on the conveyor, as illustrated in Figure 21-2.

In an overloaded circuit the conductors are required to carry more current than they are safely rated to carry. Overloads normally range from *one to six times* the normal current levels. Harmless temporary surge currents that occur normally when motors are started up or transformers are energized cause **temporary overloads.** Such overload currents are of such brief duration that any temperature rise is trivial and has no harmful effect on the circuit components. **Continuous overloads,** which are sustained over a longer period of time, result in excessive heat being produced by the conductors, creating insulation deterioration and a potential fire hazard.

In general, the term **short circuit** refers to a circuit that is completed in the wrong way, allowing current to travel along a path where essentially *little or no resistance* is encountered. Figure 21-3 illustrates a short circuit across a battery voltage source which establishes a direct path from one side of the voltage source to the other without passing through a load. The total resistance is near zero as only the resistance of the wire and the battery will limit the value of the current flow.

When a short circuit occurs with voltage applied, the decrease in resistance results in a short-circuit current that can be **thousands of times** higher than normal operating current. A short circuit can develop when wire insulation deteriorates enough to expose bare conductors. This type of short circuit can occur in flexible cords, plugs, or appliances, as illustrated in Figure 21-4. The heat generated by

Figure 21-3 Short circuit.

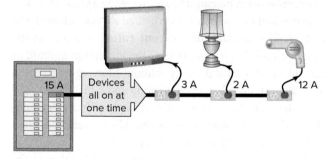

Figure 21-1 Overloaded circuit.

Figure 21-2 Overloaded motor.

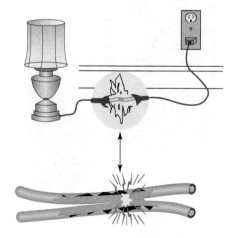

Figure 21-4 A short circuit can occur in a flexible cord.

this current will cause extensive damage to connected equipment and conductors. This *dangerous current must be interrupted immediately* when a short circuit occurs.

Bolted short circuit refers to a short circuit for which the main power supply line conductors become solidly connected together, resulting in the highest possible damaging fault current. This may occur from improper connections or metal objects being lodged between line conductors and is one of the most dangerous circuit faults.

Damaging short-circuit forces appear in the form of thermal (heat) and magnetic forces. **Thermal damage** is directly proportional to the product of the current and time that the current is allowed to flow. The **magnetic forces** between bus bars and other conductors may buckle and destroy bus bars and panels and cause conductors to be pulled out of their terminals.

21.2 Ratings of Protection Devices

The purpose of **overcurrent protection devices** is to protect electric circuits from the damage of too much current flow. Overcurrent protection can either be accomplished with a **fuse** or **circuit breaker.** Figure 21-5 shows a circuit breaker load center that provides overcurrent branch circuit protection. The functions that a protective device must be capable of performing include:

- Sense a short circuit or overload.
- Interrupt overcurrent conditions before damage is done to conductors and the other connected electrical components.
- Not open needlessly.
- Have no effect on normal circuit operation.

The resistance of a fuse or circuit breaker is very low and usually an insignificant part of the total circuit resistance.

Figure 21-5 Circuit breaker overcurrent protection load center.
Courtesy of Schneider Electric

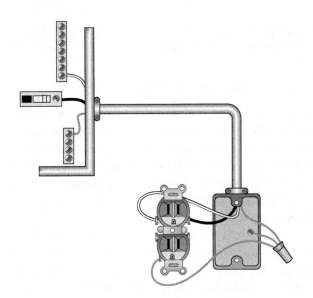

Figure 21-6 Connection of overcurrent protection device.

Under normal circuit operation it simply functions as a conductor. Fuses and circuit breakers are both connected in **series** with the circuit they protect. In general, these overcurrent devices must be installed at the point where the conductor being protected receives its power; as an example, at the beginning of a branch circuit, as illustrated in Figure 21-6.

Whenever wiring is forced to carry more current than it can safely handle, fuses will blow or circuit breakers will trip. These actions open the circuit, disconnecting the supply of electricity, but do not normally correct the problem. For this reason you should first try to locate and correct the problem *before* replacing a fuse or resetting a circuit breaker.

Fuses and circuit breakers are rated for both current and voltage. The **continuous-current rating** marked on the fuse or circuit breaker represents the maximum amount of current the device will carry without blowing or tripping open the circuit. The current rating must match the full-load current of the circuit as closely as possible. For example, undersized fuses blow easily, while oversized fuses may not provide enough protection.

The **voltage rating** of a fuse or circuit breaker is the highest voltage at which it is designed to safely interrupt the current. Specifically, the voltage rating determines the ability of the device to suppress the internal arcing that occurs when a current is opened under overcurrent or short-circuit conditions. The voltage rating must be at least equal to or greater than the circuit voltage. It can be higher but never lower. Low-voltage circuit breakers protect circuits using less than 1000 V of electricity.

The **interrupting-current rating** (also known as short-circuit rating) of a fuse or circuit breaker is the maximum current it can safely interrupt. If a fault current exceeds a level beyond the interrupting capacity of the protective device, the device may actually rupture, causing additional damage. The interrupting-current rating is many times greater than the continuous-current rating and should be far in excess of the maximum current the power source can deliver. Common interrupt ratings are 10,000 A, 50,000 A, and 100,000 A.

Current-limiting ability is a measure of how much current the protective device will *let through* the system. Current-limiting protective devices operate within less than one-half cycle. For example, a current-limiting fuse delivering a short-circuit current will start to melt within one-fourth cycle of the AC wave and clear the circuit within one-half cycle.

The **time-current characteristics or response time** of a protective device refers to the length of time it takes for the device to operate under fault current or overload conditions. Fast-acting-rated protective devices may respond to an overload in a fraction of a second, while standard types may take 1 to 30 seconds, depending on the amount of the current overload. For example, **fast-acting fuses** are very sensitive to increased current and used to protect exceptionally delicate electronic circuits that have a steady flow of current through them.

Part 1 Review Questions

1. Under what operating condition is an electric circuit considered to be overloaded?

2. Outline three possible scenarios that could cause a circuit to become overloaded.

3. Give two examples of harmless temporary circuit overloads that occur normally.

4. What is the effect on conductors when they are required to carry more current than they are safely rated to handle?

5. State the general definition for the term *short circuit*.

6. Compare the ways in which overload currents differ from short-circuit currents.

7. What makes a bolted short circuit the most dangerous type of short circuit?

8. In what two forms do damaging short-circuit forces appear?

9. List four functions of a protective device.

10. How are fuses and circuit breakers connected with respect to the circuits they protect?

11. How does the resistance of a fuse or circuit breaker compare with that of a load device?

12. In general, at what point in the circuit must overcurrent devices be installed?

13. Why should you try to locate and correct the problem before replacing a fuse or resetting a circuit breaker?

14. Explain what each of these fuse and circuit breaker ratings specify:
    a. Continuous-current rating.
    b. Voltage rating.
    c. Interrupting-current rating.
    d. Current-limiting ability.
    e. Time-current characteristics.

15. To what is the continuous-current rating of a fuse or circuit breaker matched?

16. Typically how long does it take a current-limiting fuse delivering a short circuit to clear the circuit?

PART 2   FUSES AND CIRCUIT BREAKERS

21.3 Types of Fuses

A **low-melting metal strip** or fusible link or links encapsulated and connected to contact terminals make up the fundamental parts of the basic fuse. When the current flow through this link is greater than the rating of the fuse, the metal strip will melt, opening the circuit.

Plug fuses are round fuses which screw into a base in the fuse holder to complete the circuit. A plug fuse contains a fuse element of soft wire or metal enclosed in glass housing, as illustrated in Figure 21-7. The strip of metal is designed to carry a given amount of electric current, such as 15 amps. If anything happens causing more current to flow in the circuit than the circuit and the fuse are designed to carry, the metal strip melts or burns out. This opens the circuit, stopping the flow of current and protecting the wiring.

Edison-base plug fuses feature metal threads similar to incandescent lamp bases. The Edison base can no longer be used, except in replacement of existing fuses. Type S fuses are tamper-resistant fuses designed to

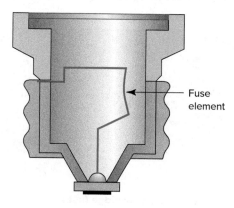

Fuse element

Figure 21-7 Plug fuse.

Fuse   Adapter

Figure 21-8 **Type S plug fuse and adapter.**

prevent switching fuse sizes in the same fuse base. These fuses use an adapter base and a fuse insert, such as illustrated in Figure 21-8. Once an adapter base is installed, fuses with a higher rating cannot be installed in that fuse box opening. Because of these tamper-resistant qualities, type S fuses are the only type allowed by the National Electrical Code for new installations.

Plug fuses have a maximum voltage rating of *125 V.* They are available in a number of common current sizes up to a maximum of *30 A.* They can be found in 120-V general house lighting and receptacle circuits. The current rating of the fuse is matched to the maximum-current rating of the circuit conductors. On the basis of this rating, an adapter of the proper size is inserted into the Edison-base fuse holder. The proper type S fuse is then screwed into the adapter.

The see-through glass body of the plug fuse is a great help for finding what caused a fuse to blow. If the glass front is black, it indicates that there has been a **short circuit.** If the glass front is clean and clear, it indicates that the circuit is **overloaded.**

Cartridge fuses operate exactly like plug-type fuses but are designed to carry much higher currents. The two basic types of cartridge fuses are the ferrule-contact type and the knife-blade type. Figure 21-9 shows the **ferrule**-type cartridge fuses, which are available in ampere ratings from 0 through *60 A.* Ferrule fuses are used in circuits up to *600 volts.*

The **knife-blade** cartridge fuse is used for circuit current ratings in *excess of 60 A.* The contact points of this

Figure 21-9 **Ferrule-type cartridge fuses.**
Courtesy of Cooper Bussmann

Figure 21-10 **Knife-blade fuse.**
Courtesy of Cooper Bussmann

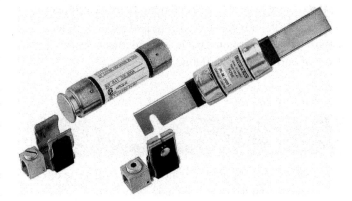

Figure 21-11 **Fuse rejection clips for current-limiting fuses.**
Courtesy of Cooper Bussmann

fuse are larger and more rugged, which allows it to handle higher current flows. Figure 21-10 shows a knife-blade fuse that is available in ampere ratings of *70 through 6,000 A.* The maximum voltage rating for knife-blade fuses is *600 volts.*

The NEC requires that fuse holders for current-limiting cartridge fuses be designed to reject non-current-limiting types of fuses. Figure 21-11 illustrates fuse rejection clips which accept only rejection-type fuses.

In broad terms, cartridge fuses are classified as one-time, renewable, dual-element, time-delay, current-limiting, or high-interrupting capacity. The **one-time cartridge fuse,** illustrated in Figure 21-12, consists of a fuse link enclosed in a tube of insulating filler material. The purpose

Figure 21-12 **One-time cartridge fuse.**

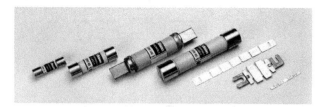

Figure 21-13 Renewable cartridge fuses and fuse links.
Courtesy of Cooper Bussmann

of the filler material is to suppress the arc when the fuse blows. These fuses have very little time delay, and their use is limited to short-circuit protection on circuits in which faults occur infrequently.

The **renewable cartridge fuses,** shown in Figure 21-13, are used to take advantage of lower replacement costs for the protection of mains and feeders in which faults occur frequently. Unlike the one-time cartridge fuse, these fuses contain a fuse link that can be *replaced* once blown. Although initially more expensive, this fuse will reduce maintenance costs over a long period of time.

Single-element fuses provide excellent short-circuit protection, but to accommodate temporary surges or transients, the fuse must be oversized. **Dual-element, time-delay fuses** provide protection from both short circuits and overloads by the use of two individual components on the same element (link). One element removes **overloads** and the other element removes **short circuits.** A typical dual-element, time-delay fuse is illustrated in Figure 21-14. The short-circuit element is a copper link with restrictive notches or segments. The overload thermal element portion is a spring-loaded device that opens the circuit when solder holding the spring in position melts. Dual-element, time-delay fuses are particularly advantageous for **motor installations.** This type of fuse can be sized closer to the motor running current and still carry starting currents; therefore, smaller switches and panels can be installed.

Cartridge fuses come in a wide range of types, sizes, and ratings. Underwriters Laboratories (UL) designates various classes as shown in Table 21-1.

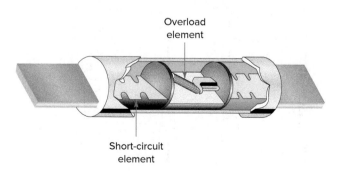

Figure 21-14 Dual-element, time-delay fuse.

Table 21-1 UL Fuse Classification

UL Class	Fuse Overload Characteristics
L	Time delay
RK1	Time delay Fast acting
RK5	Time delay
T	Fast acting
J	Time delay Fast acting
CC	Time delay Fast acting
CD	Time delay
G	Time delay
K5	Fast acting
H	Renewable fuse, fast acting

Figure 21-15 High-voltage fuse.
Courtesy of Cooper Bussmann

High-voltage fuses, such as the one shown in Figure 21-15, are rated for *over 600* volts and used to protect high-voltage utility power lines. They are specially constructed so that they will be safe for interruption of current at such high voltages and include expulsion, liquid, and solid-material types.

21.4 Testing Fuses

The **ohmmeter** is used to make an **out-of-circuit test** of a fuse, as illustrated in Figure 21-16. A good fuse should indicate a near zero resistance reading on the meter. An infinite reading on an analog ohmmeter indicates an open

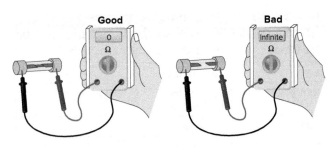

Figure 21-16 Ohmmeter fuse test.

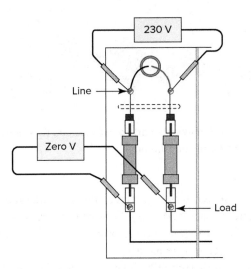

Figure 21-18 Zero volts on load side indicates one or both of the fuses are blown.

fuse. Depending on the manufacturer, digital ohmmeters may display an open infinite reading as 1 or OL (overload).

In the majority of electrical systems all **ungrounded** power line conductors must be installed with an overcurrent device connected in series. One overcurrent protection is required for low-voltage circuits, single-phase circuits of 120 volts or less, and all DC circuits, as illustrated in Figure 21-17. The neutral line wire in AC circuits and the negative line wire in DC circuits do not include overcurrent protection.

A **voltmeter** can be used to make an **in-circuit test** of a fuse. A properly operating fuse in a circuit with power applied should have close to **zero voltage** drop across it. The reason is that a fuse not blown is similar in resistance to a small piece of wire. Voltages are checked on the line and the load sides of the fuse, as illustrated in Figure 21-18. Full voltage on the line side and zero voltage on the load side indicate one or both of the fuses are blown. The operating voltage drop across the two terminals of a good fuse will be near zero because its resistance is normally low. If you read an *appreciable voltage* drop across the fuse, this means its resistance is high because it is burnt open.

21.5 Circuit Breakers

Circuit breakers are somewhat more sophisticated overcurrent devices than fuses and use a *mechanical device* to protect a circuit from short circuits and overloads. Like a fuse, the circuit breaker is connected in series with the circuit it protects. Circuit breakers are

rated in a manner similar to the one used for fuses. As with fuses, the ampere rating of a breaker must match the ampacity of the circuit it protects. Most low-voltage circuit breakers (less than 1000 volts) are housed in molded plastic cases, as illustrated in Figure 21-19.

Circuit breakers provide a **manual means** of energizing and deenergizing a circuit. In addition, circuit breakers provide automatic overcurrent protection of a circuit. A circuit breaker allows a circuit to be reactivated quickly after a short circuit or overload is cleared, unlike fuses, which must be replaced.

Circuit breakers use two principles of operation to protect the circuit: thermal and magnetic. **Thermal circuit breakers** consist of a heating element and mechanical latching mechanism. The heating element is usually a **bimetallic strip** that heats up when current flows through it. A simplified thermal circuit breaker is shown in Figure 21-20 and summarized as follows:

- The contacts and bimetallic strip are part of the current-carrying path.
- **Overload current** exceeding the breaker-overload rating heats the bimetallic strip.

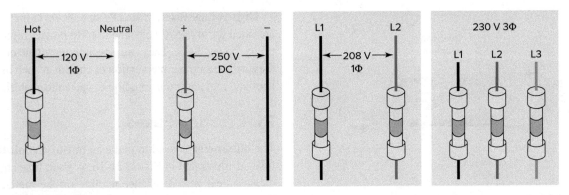

Figure 21-17 Conductors installed with an overcurrent device.

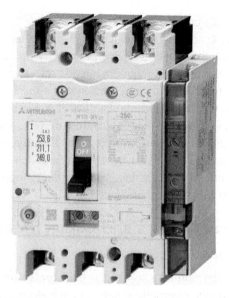

Figure 21-19 Molded case circuit breaker.
©Mitsubishi Electric

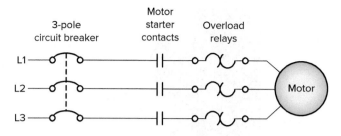

Figure 21-21 Magnetic instantaneous-trip circuit breaker.

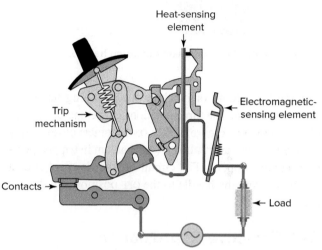

Figure 21-22 Thermal-magnetic circuit breaker.

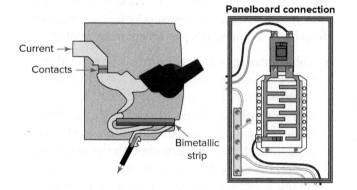

Figure 21-20 Thermal circuit breaker.

- The heated strip bends down and activates the spring-loaded trip mechanism to open the contacts.
- The time the bimetal needs to bend and trip the circuit varies inversely with the current.
- A thermal circuit breaker must cool off before you can reset it.
- Ambient temperatures affect the trip point so the breaker will require more current and take a longer time to trip in a very cold environment than it will in a hot environment.

Magnetic instantaneous-trip circuit breakers work on the principle of electromagnetism. The name comes from the electromagnet used to *sense short-circuit current.* Magnetic instantaneous-trip-only circuit breakers do not provide overload protection and are used on motor circuits for which overload protection is provided by a motor starter. In the schematic shown in Figure 21-21, a motor is

supplied through a three-pole circuit breaker, motor starter contacts, and separately supplied overload contacts. Heat generated from excessive current will cause the overload contacts to open, removing power from the motor.

Thermal-magnetic circuit breakers include both a magnetic-tripping function, for short-circuit protection, and a thermal-tripping function, for overload protection, as illustrated in Figure 12-22. Thermal-magnetic circuit breakers are also called inverse-time circuit breakers. As the alternative name **inverse-time** implies, the higher the overload, the shorter the time in which the circuit breaker will open. When an overload condition exists, the excess current will generate heat, which is detected by the bimetallic heat-sensing element. After a short period of time, dependent on the rating of the breaker and amount of overload, the breaker will trip, disconnecting the load from the voltage source. If a short circuit occurs, the electromagnetic sensor responds instantaneously to the fault current and disconnects the circuit.

The operating handle of a circuit breaker provides a manual means for energizing and deenergizing a circuit and must be capable of being reset after a fault condition has been cleared. All molded case circuit breakers are **trip free,** meaning that they cannot be prevented from

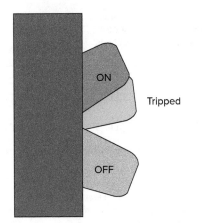

Figure 21-23 Circuit breaker operating handle.

tripping by holding or blocking the operating handle in the ON position. The three positions of the operating handle are shown in Figure 21-23: ON (contacts closed), OFF (contacts open), and TRIPPED (mechanism in tripped position, contacts open). The circuit breaker is *reset* after a trip by moving the handle to the OFF position and then to the ON position.

21.6 Selective Coordination of Protective Devices

Selective coordination refers to the installation of circuit protective devices in series such that when a fault occurs, only the device *nearest the fault* opens. The other devices remain closed, leaving other circuits unaffected. Figure 21-24 illustrates the operation of a selective coordination system. A short circuit has occurred in the circuit fed by branch circuit

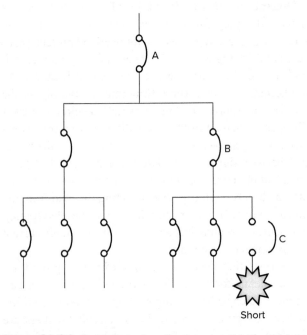

Figure 21-24 Operation of a selective coordination system.

breaker *C*. Power is interrupted to equipment supplied by circuit breaker *C* only. All other circuits remain unaffected.

Selective coordination is always more desirable than a system blackout. NEC requirements for selective coordination are mandatory for electrical systems such as elevator circuits where maximum reliability of power is critical. An analysis of overcurrent device time-current characteristics is required to ensure coordination. The circuit breakers are coordinated so that for any given fault value, the **tripping time** of each breaker is *greater* than tripping time for the next downstream breaker.

21.7 Load Centers and Circuit Breakers

Load Centers and Circuit Breakers

The main function of a residential **load center or panelboard** is to take electricity supplied by the electrical utility and distribute it throughout the home for lighting and appliances. In addition to providing terminating points for branch circuits, the distribution panel contains fuses or circuit breakers that provide overcurrent protection for the branch circuits.

Required manufacturer's panelboard rating include:

- **Voltage rating**—A panelboard is designed and intended for use only on a supply circuit involving two different potentials (e.g., 120/240 volts).
- **Current rating**—The current rating of a panelboard is the maximum continuous current that can be supplied through the main terminals. Unless the assembly, including the overcurrent device(s), is marked for use at 100 percent of its current rating, overcurrent protection devices should not be loaded continuously to more than 80 percent of their rating.
- **Short-circuit current rating**—A panelboard is required to be marked with the phrase "Short-Circuit Current Rating." This phrase indicates that (1) the overcurrent devices are capable of opening the circuit under fault conditions, and (2) the panelboard bus structure will withstand the magnetic forces generated by fault current passing through it.

Power to the load center is typically supplied by the **three-wire distribution system** shown in Figure 21-25. The primary utility voltage is in the kilovolt range. This voltage is stepped down to 240 V, which appears across the two outside leads of the transformer secondary winding. A center-tap wire on the transformer divides this voltage in half, giving 120 V between the center-tap connection and the outer leads. The two ungrounded wires are called the **hot** or **live wires** and have a black or red insulation covering. The center-tap wire is **grounded** (connected to earth) at the base of the transformer and is known as the **neutral wire.** The neutral wire is color coded with a **white** insulation covering. For

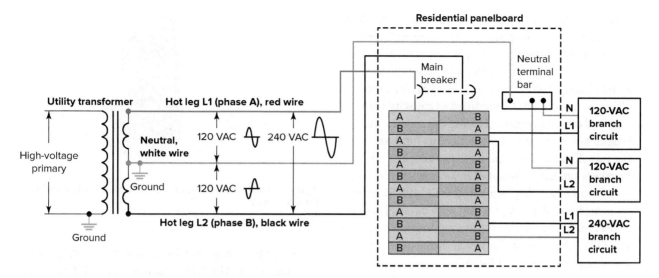

Figure 21-25 Three-wire distribution system.

safety purposes, the live wires are switch controlled and have fuses or circuit breakers connected in series with them. The neutral wire is grounded (connected to the earth) at the transformer and in the panelboard.

The NEC defines a **branch circuit** as the circuit conductors between the final overcurrent device protecting the circuit and the outlet(s). Typical residential branch circuits are shown in Figure 21-26. The wiring of a circuit breaker panelboard is basically the same as that of a fuse panelboard. A circuit breaker panelboard offers additional features such as tamper-proof current ratings, ease of resetting (instead of replacing), and absence of exposed live parts.

Most circuit breakers plug into their terminals. The plug-in mounting method is the main mounting method for residential circuit breakers. The **line side** is a clamp that clips onto the bus stab in the load center, while the load side terminal is the cable-out. Installation involves positioning the breaker in an anchor clip at one end, then pushing the connector clip into a snap-in hole at the other end. There are two types of residential breakers (Figure 21-27): single pole and double pole.

- **Single-pole** breakers used to supply multiple receptacle and lighting outlets are rated for 120 volts and 15 or 20 amps. They control standard lighting and outlet circuits, as well as some appliance circuits, in the home. These breakers themselves occupy a single slot in the load center.

- **Double-pole** breakers used to supply individual branch circuits used to large appliances such as cooking equipment and water heaters are rated for 240 volts and higher currents from 20 to 100 amperes. These breakers take up two slots in the load center. They are constructed so that both hot wires are protected and are opened and closed together.

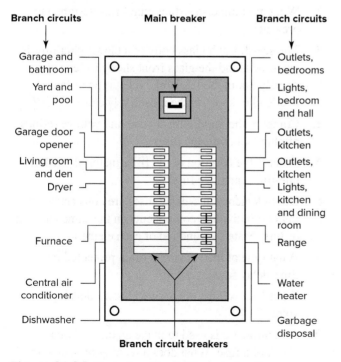

Figure 21-26 Typical residential branch circuits.

Figure 21-27 Single-pole and double-pole circuit breakers. Courtesy of Schneider Electric

- A **tandem circuit** breaker is a double circuit breaker that takes up the space of a single circuit breaker on a panelboard. Because tandem circuit breakers allow for two circuits to be installed on a panelboard in a one circuit breaker space, they're typically used after a panelboard has been filled to capacity with standard circuit breakers. The use of tandem circuit breakers is an acceptable practice, **as long as the panelboard is designed for tandem circuit breakers** and they're installed in locations within the panelboard where they're allowed. NEC 408.54 states, "A panelboard shall be provided with physical means to prevent the installation of more overcurrent devices than that number for which the panelboard was designed, rated, and listed."

Special application circuit breakers include:

- The **ground-fault circuit interrupter (GFCI)** circuit breaker (Figure 21-28) provides protection to people against overloads, short circuits, and ground faults. It senses hot wire-to-ground faults at 5 mA or higher, which is low enough to protect people. The NEC requires ground fault protection in locations where electricity can come into contact with water such as bathrooms, laundry rooms, kitchens, and swimming pools. GFCIs have test buttons, as do arc-fault circuit interrupters (AFCIs). Pushing the GFCI test button periodically is important to verify that the device is still providing the required protection.

- The **ground-fault equipment protector (GFEP)** circuit breaker is designed to protect equipment damage from ground faults. This circuit breaker senses hot wire to ground faults at 30 mA or higher. The NEC requires GFEP protection in applications such as fixed outdoor electric deicing and snow melting equipment and electric heat tracing and heat panels.

- **Arc-fault circuit interrupter (AFCI)** circuit breakers are designed to protect against unwanted or intermittent flow of current between two conductors

Figure 21-29 Dual-purpose arc-fault/ground-fault (AF/GF) circuit breaker.
©Eaton

that can damage nearby combustible material and start a fire. The NEC requires that AFCI protection for dwelling units must be provided for outlets and devices in specified required locations.

- **Dual-purpose arc-fault/ground-fault (AF/GF)** circuit breakers have been made available in response to changes in the NEC that now require AFCI protection in kitchens and laundry areas, as well as the historical requirement of ground-fault protection. The AF/GF circuit breaker interrupter, shown in Figure 21-29, offers both types of protection in a single unit.

Part 2 Review Questions

1. Explain how a fuse operates to protect a circuit.

2. What are tamper-resistant plug fuses designed to prevent?

3. The see-through glass body of a blown plug fuse is examined, and the glass front shows a clean clear open break in the fuse link. What type of fault, overload or short, is the most likely cause? Why?

4. Compare the maximum current and voltage ratings of plug and cartridge fuses.

5. What is the function of the filler material used in a one-time cartridge fuse?

6. Dual-element, time-delay fuses are constructed using two individual components on the same element or link. State the function of each element.

7. What type of load device is often protected by time-delay fuses?

8. What fuse voltage ranges are classified as high voltage?

9. An ohmmeter is used to make an out-of-circuit check of a fuse. What does a reading of near zero ohms indicate? Why?

Figure 21-28 Ground-fault circuit interrupter (GFCI) breaker.
©Eaton

10. A voltmeter is used to make an in-circuit test across a fuse. What does a reading of near zero volts indicate? Why?

11. What are the two main functions of a circuit breaker?

12. Compare the way the current is sensed in a thermal and magnetic instant-trip circuit breakers.

13. Thermal-magnetic circuit breakers are also known as inverse-time breakers. Why?

14. Name the three positions for the operating handle of a circuit breaker.

15. If an electrical system is selectively coordinated which breaker(s) should open when a fault occurs?

16. How are the time-current characteristics of protective devices applied in providing for selective coordination of an electrical system?

17. What are the two main functions of a residential load center or panelboard?

18. List three required manufacturer's panelboard ratings.

19. With reference to a three-wire residential distribution system:
    a. What is the value of the voltage between the two hot or live wires?
    b. What is the value of the voltage from the neutral wire to either of the two hot or live wires?
    c. Which wire(s) of the system in connected to ground?
    d. Which wire(s) of the system require overcurrent protection?

20. What is the NEC definition of a branch circuit?

21. What is the most common type of mounting method used for residential circuit breakers?

22. For what type of branch circuits are single-pole breakers required?

23. For what type of branch circuits are double-pole breakers required?

24. Under what conditions does the NEC allow tandem circuit breakers to be installed in a panelboard?

25. In what way does GFCI circuit breaker protection differ from GFEP circuit breaker protection?

CHAPTER TWENTY-TWO

Relays

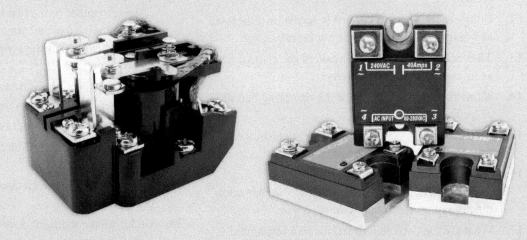

Electromagnetic and solid-state relays
©Mihancea Petru/123RF; ©Bancha Chuamuangpan/123RF

A relay is a device that is used to perform switching functions. The relay performs the same function as a switch, except that it is electrically operated instead of manually operated. Since relays are electrically operated, unlike traditional switches, they can be opened or closed from a remote location. In this chapter, you will learn about the different types of relays and their operating characteristics.

LEARNING OUTCOMES

▶ Compare electromagnetic and solid-state relays.

▶ Identify relay symbols used on schematic diagrams.

▶ Describe the different ways in which relays are used.

▶ Explain how relays are rated.

▶ Describe the operation of on-delay and off-delay timer relays.

▶ Explain the difference between a relay and contactor.

▶ Compare the operation of magnetic and solid-state contactors.

▶ Explain the operation of a one-shot timer.

▶ Have an understand of the operating principle of a protective relay.

PART 1 RELAY FUNDAMENTALS

22.1 Electromechanical Relay

Relays are switches that open and close circuits electromechanically or electronically. An **electromechanical relay,** shown in Figure 22-1, is essentially a remote-controlled switch. The relay turns a load circuit ON or OFF by energizing an **electromagnet,** which opens or closes **contacts** in the circuit.

Figure 22-2 illustrates the parts of a typical electromechanical relay which operates as follows:

- The relay consists of a coil, wound on an iron core, to form an electromagnet.

Figure 22-1 **Electromechanical relay.**
Courtesy of Omron Automation and Safety

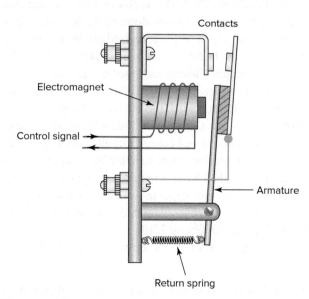

Figure 22-2 **Electromechanical relay operation.**

- When the coil is energized by a control signal, the core becomes magnetized and sets up a magnetic field that attracts the iron arm of the armature to it.
- As a result, the contacts on the armature close.
- When the current to the coil is switched off, the armature is spring-returned to its normal deenergized position and the contacts on the armature open.
- The coil and contacts are insulated from each other; therefore, under normal conditions, no electric circuit will exist between them.

A relay is made up of two circuits: the coil input or **control circuit** and the contact output or **load circuit,** as illustrated in Figure 22-3. The relay turns the load circuit on and off by operating the switch. Closing the switch energizes the electromagnet, which in turn closes the relay contacts to switch the load on.

A relay will usually have only one coil, but it may have any number of different contacts. Electromechanical relays contain both *stationary* and *moving* contacts, as illustrated in Figure 22-4, and are summarized as follows:

- The moving contacts are attached to the armature.
- Contacts are referred to as normally open (NO) and normally closed (NC).
- When the coil is energized, it produces an electromagnetic field.
- Action of this field, in turn, causes the armature to move, closing the NO contacts and opening the NC contacts.

Normally open contacts are open when the coil is deenergized and closed when the coil is energized. **Normally closed contacts** are closed when the coil is deenergized and open when the coil is energized. Each contact is normally drawn as it would appear with the coil *deenergized.*

A letter is used in most diagrams to designate the coil. For example the letter M frequently indicates a motor starter, while CR is used for control relays. The associated contacts will have the same identifying letters as illustrated in Figure 22-5.

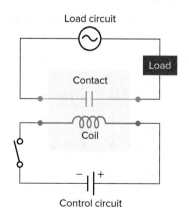

Figure 22-3 **Relay control and load circuits.**

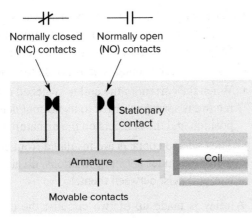

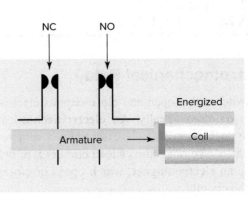

Figure 22-4 Relay NO and NC contacts.

Figure 22-5 Identification of relay coil and contacts.

22.2 Relay Applications

Relays can be used for **multiple-switching** applications. One relay coil/armature assembly is used to actuate more than one set of contacts. Those contacts may be normally open, normally closed, or any combination of the two. A simple example of this type of application is the relay control with two pilot lights illustrated in Figure 22-6. The operation of the circuit can be summarized as follows:

- With the switch open, coil CR1 is deenergized.
- The circuit to the green pilot light is completed through normally closed contact CR1-2, so this light will be on.
- At the same time, the circuit to the red pilot light is opened through normally open contact CR1-1, so this light will be off.
- With the switch closed, the coil is energized.

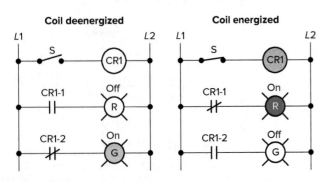

Figure 22-6 Relay multiple-switching application.

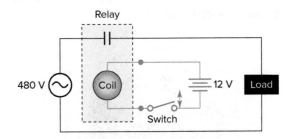

Figure 22-7 Relay control of a high-voltage circuit.

- The normally open contact CR1-1 closes to switch the red pilot light on.
- At the same time, the normally closed CR1-2 opens to switch the green pilot light off.

You can use a relay to control a **high-voltage load circuit with a low-voltage control circuit,** as illustrated in the circuit of Figure 22-7. This is possible because the coil and contacts of the relay are electrically insulated from each other. The relay's coil is energized by the low-voltage (*12-V*) source, while the contact interrupts the higher voltage (*480-V*) circuit. Closing and opening the switch energizes and deenergizes the coil. This, in turn, closes and opens the contacts to switch the load on and off.

You can also use a relay to control a high-current load circuit with a low-current control circuit. This is possible because the current that can be handled by the contacts can be much greater than what is required to operate the relay coil. Relay coils are capable of being controlled by low-current signals from integrated circuits and transistors, as illustrated in Figure 22-8.

The current control of the circuit is summarized as follows:

- The *2-mA* electronic control current signal switches the transistor on and off.

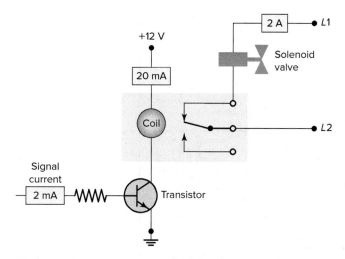

Figure 22-8 Relay control using a low-current signal.

- When the transistor is in the ON state, a current of **20 mA** is conducted through it to energize the relay coil.
- The NO contacts then close to operate the solenoid valve with *2 A (2,000 mA)* of current.

22.3 Relay Styles and Specifications

One popular style of relay is the general-purpose **ice cube relay,** so named because of its size and shape and the clear plastic enclosure surrounding the contacts. Figure 22-9 illustrates an eight-pin, plug-in-style ice cube relay. This relay contains two separate single-pole, double-throw contacts. Because the relay plugs into a socket, the wiring is connected to the socket, not the relay.

Relay options that aid in troubleshooting are also available. An ON/OFF **indicator** is installed to indicate the state (energized or deenergized) of the relay coil. A **manual override button,** mechanically connected to the contact assembly, may be used to move the contacts into their energized position for testing purposes. *Use caution*

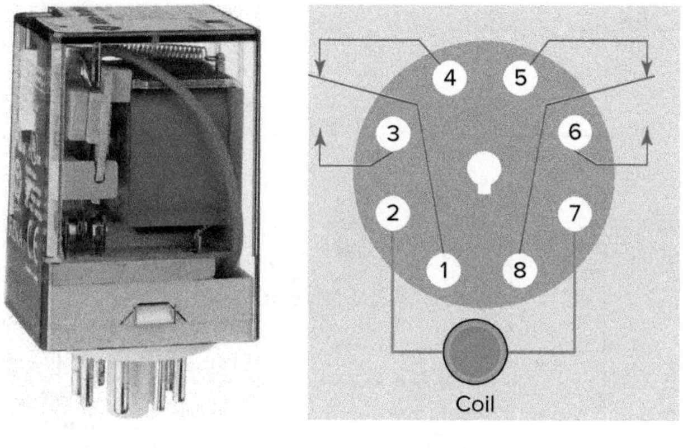

Figure 22-9 Plug-in-style ice cube relay.
Photo: Courtesy of Rockwell Automation, Inc.

when exercising this feature, as the circuit controlling the coil is bypassed and loads may be energized or deenergized without warning.

Relay coils and contacts have separate ratings. **Coils** are usually rated for:

- Type of operating current (DC or AC).
- Normal operating voltage or current.
- Permissible coil voltage variation (pickup and dropout).
- Coil resistance.
- Power consumption.

In general, relay **contacts** are rated in terms of the maximum amount of current the contacts are capable of handling at a specified voltage level and type (AC or DC). Relay contacts often have two ratings: AC and DC. For a given relay, the contact rating is **higher for AC currents** than for DC currents. Other considerations include the type of load (resistive or inductive), ON/OFF duty cycle and the environment to which the relay will be exposed.

Relays are available in a wide range of switching configurations. Figure 22-10 illustrates common relay contact switching arrangements. Like switch contacts, relay contacts are classified by their number of poles, throws, and breaks as follows:

- **Pole** is the number of switch contact sets.
- **Throw** is the number of conducting positions, single or double.
- **Break** designates the number of points in a set of contacts where the current will be interrupted during opening of the contacts. All relay contacts are constructed as single break or double break. Single-break contacts have lower current ratings because they break the current at only one point.

Part 1 Review Questions

1. Explain how an electromechanical relay operates.
2. Name the two circuits associated with a relay, and describe how they interact with each other.
3. Compare the operation of a normally open (NO) and normally closed (NC) relay contact.
4. Describe three common relay control applications.
5. Describe two common relay options used for troubleshooting purposes.
6. List six ways in which relay coils may be specified.
7. Define the terms *poles, throw,* and *break* as they apply to relay contact switching arrangements.

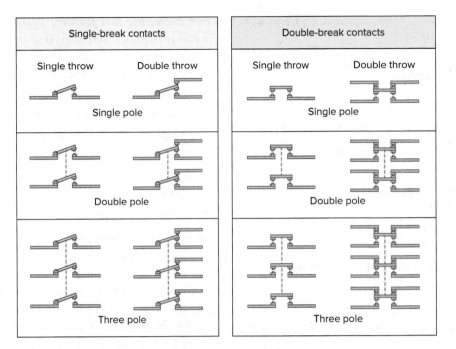

Figure 22-10 **Relay contact switching arrangements.**

PART 2 TYPES OF RELAYS

22.4 Solid-State Relays

Unlike electromechanical relays, **solid-state relays** do not operate with coils and contacts. Instead, solid-state relays use **semiconductor** switching devices such as transistors. These relays provide the advantages inherent in solid-state designs of no moving parts or contacts that can wear out. Solid-state relays are manufactured in a variety of configurations that including the hockey-puck type shown in Figure 22-11. A square or rectangle normally is used on the schematic to represent the relay along with the input and output connections.

Like electromechanical relays, solid-state relays provide **electrical isolation** between the input control circuit and the switched load circuit. A common method used to provide this electrical isolation is to have the input section illuminate a **light-emitting diode (LED)** that activates a photodetector device connected to the output section.

Solid-state relays are constructed with different main switching devices depending on the type of load being switched. Solid-state relays intended for use with DC loads use a power **transistor** connected to the load circuit, as shown in Figure 22-12. The operation of the circuit can be summarized as follows.

- Voltage applied to the control input turns the LED on.
- The photodetector connected to the transistor then turns the transistor on, allowing current flow to the load.
- The LED section of the relay acts like the **coil** of the electromechanical relay and requires a DC voltage for its operation.

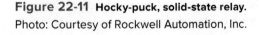

Figure 22-11 **Hocky-puck, solid-state relay.**
Photo: Courtesy of Rockwell Automation, Inc.

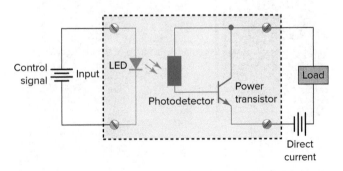

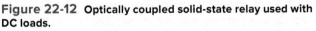

Figure 22-12 **Optically coupled solid-state relay used with DC loads.**

- The transistor section is equivalent to the **contacts** of the electromechanical relay.
- Because solid-state relays have no moving parts, their switching response time is many times *faster* than that of electromechanical relays. For this reason, when loads are to be switched continually and quickly, the solid-state relay is the relay of choice.

Advantages of electromechanical relays versus solid-state types, for specific application requirements, are listed as follows:

Electromechanical Relay Advantages
- Normally have multi-throw and multi-pole contact arrangements
- Contacts can switch AC or DC
- Low contact voltage drop, thus no heat sink is required
- Resistant to voltage transients
- No off-state leakage current through open contacts

Solid-State Relay Advantages
- No contacts to wear out
- No contact arcing to generate electromagnetic interference
- Resistant to shock and vibration because they have no moving parts
- Logic compatibility to programmable controllers, digital circuits, and computers
- Fast switching capability

22.5 Time-Delay Relays

Time-delay relays are control relays with a time delay built in. Their purpose is to *control an event based on time.* The difference between control relays and time-delay relays is when the output contacts open and close. With a control relay the contacts change state immediately when voltage is applied and removed from the coil. With on-delay timers, the contacts can open or close after some time delay. Typically, time-delay relays are initiated or triggered by one of two methods: application of input voltage or opening or closing of a trigger signal. Figure 22-13 shows a solid-state, 3-second, **on-delay timer** and associated relay connections. The timer is energized continuously with voltage applied to terminals L1 and L2. When the initiating contact closes, timing begins. At the 3-second time-out, the output contact closes.

There are two basic timing functions: on delay and off delay. An **on-delay timer** has a preset time period that must pass after the timer is switched on before any change of state of the timer contacts occurs. The operation of the **off-delay timer** is the exact opposite of that of the on-delay timer. When an off-delay timer is switched on, the timed contacts will change state immediately. When power is removed, however, there is a time delay before the off-delay timed contacts change to their normal deenergized positions.

Figure 22-14 shows the wiring diagram for the automatic pumping down of a sump using a level sensor switch and **plug-in, cube-type,** 15-second, off-delay timer. A solid-state timing circuit drives an internal electromechanical relay within the timer. The operation of the circuit can be summarized as follows:

- When the level rises to point A, the level sensor contact closes to energize the relay timer coil and close the NO contacts to the pump motor starter.
- This immediately turns the pump on to initiate the pumping action.
- When the height of the vessel level decreases, the sensor contacts open and timing begins.
- The pump continues to run and empty the tank for the 15-second length of the time-delay period.

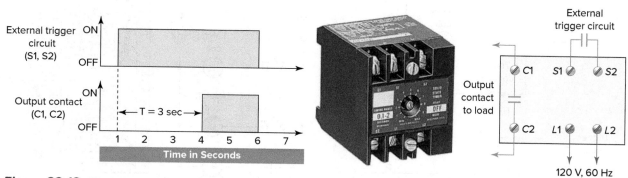

Figure 22-13 On-delay timer.
Photo: Courtesy of Rockwell Automation, Inc.

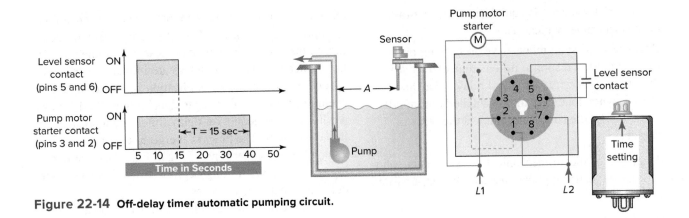

Figure 22-14 Off-delay timer automatic pumping circuit.

- At time-out the relay coil deenergizes, and the normally open relay contact reopens, turning the pump off.
- The timer has a built-in time adjustment potentiometer that is adjusted to empty the tank to a desired level before the pump shuts off.

A **one-shot timer** is used to shorten long trigger signals. Figure 22-15 shows a typical schematic and timing diagram for a one-shot, 1-second timer. When the trigger circuit contact is closed, the timer coil and output PL are energized and the timing begins. After the 1-second time delay is completed, the timer contact opens and the timing is reset. No matter how long the trigger circuit contact stays closed, the output PL remains on for just one second of time, then returns to its original off state.

22.6 Latching Relays

Mechanical latching relays use a **latch mechanism** to hold their contacts in their last set position until commanded to change state, usually by energizing a second coil. Figure 22-16

shows the circuit of a two-coil latching relay circuit, the operation of which is summarized as follows:

- The latch coil (L) requires only a single pulse of current to set the latch and hold the relay in the latched position.
- Similarly, the unlatch coil (U) is momentarily energized to disengage the mechanical latch and return the relay to the unlatched position.
- There is no deenergized position for the contacts of a latching relay. The contact is shown with the relay in the unlatched condition—that is, as if the unlatch coil were the last one energized.
- In the unlatched state, the circuit to the pilot light is open, so the pilot light is off.
- When the ON button is momentarily actuated, the latch coil is energized to set the relay to its latched position.
- The contacts close, completing the circuit to the pilot light, so the light is switched on.
- In cases of power loss, the relay will remain in its original latched or unlatched state when power is restored. This arrangement is sometimes referred to as a **memory relay**.

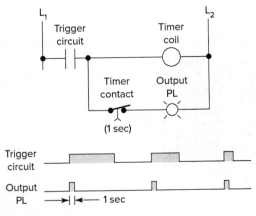

Figure 22-15 One-shot timer.

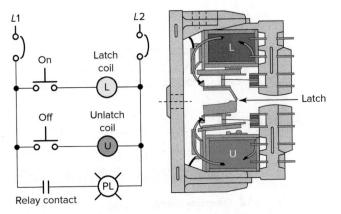

Figure 22-16 Two-coil latching relay circuit.

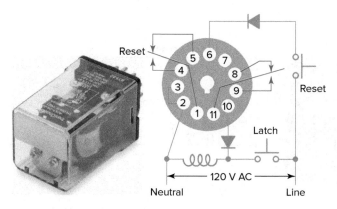

Figure 22-17 Single-coil magnetic latching relay.
Photo: ©Automation Direct

Magnetic latching relays are single-coil relays designed to be **polarity**-sensitive. Figure 22-17 shows a single-coil magnetic latching relay, the operation of which is summarized as follows:

- The direction of the current flow through the coil determines the position of the relay contacts.
- The double-pole, double-throw (DPDT) relay contacts are shown in the relay reset position.
- When the latch button is momentarily closed, a current pulse of proper polarity causes the contacts to change state.
- A **permanent magnet** is used to hold the contacts in the latch position without the need for continued power to the coil.
- When the RESET button is momentarily closed, a current pulse of opposite polarity causes the contacts to change state again
- Repeated pulses from the same input have no effect.

A common application for a latching relay involves **power failure.** Circuit continuity during power failures is often important in automatic processing equipment, where a sequence of operations must continue from the point of interruption after power is resumed rather than return to the beginning of the sequence. In applications similar to this, it is important not to have the relay control any devices that could create a safety hazard if they were to restart after a power interruption.

22.7 Protective Relays

The purpose of the **protective relay** is to detect abnormal or dangerous circuit conditions, ideally during their initial stage, and to either eliminate or significantly reduce damage to personnel and/or equipment. These relays are designed to initiate appropriate action in order to isolate the faulty portion of a circuit or system. Most of today's protective relays are microprocessor based and operate by

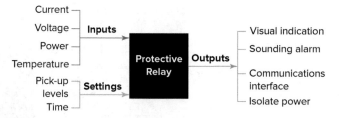

Figure 22-18 Protective relay.

receiving inputs, comparing them to set points, and providing outputs, as illustrated in Figure 22-18. The basic operating parameters for a protective relay are as follows:

- **Inputs** provide information from the system to make a decision. Input devices continuously monitor parameters such as current, voltage, resistance, power, and temperature to see if these parameters deviate from set limits. In some cases, the field input-sensing device is connected directly to the relay. In others, additional signal conditioning devices are needed to convert the measured parameters to a format that the relay can process. Instrument transformers are used to supply accurately scaled current and voltage quantities for measurement while insulating the relay from the high voltage and current of high-power systems.

- **Settings** allow the user to select the input levels and timing required for the relay to take the appropriate action. Setting's values are based on the detailed characteristics of the various elements of the circuit or system. The pickup setting is the value of actuating quantity (e.g., voltage or current) on which the relay initiates its operation. The time setting represents the elapsed time between the instant when actuating quantity exceeds the pickup value to the instant when the protective relay is activated.

- **Outputs** communicate the decision made by the protective relay. Typically, the relay will operate a contact of some sort to indicate that an input has surpassed a setting, or the relay can provide notification through visual feedback such as a meter or LED. Electronic microprocessor relay outputs have ability to communicate with a network or a programmable logic controller (PLC).

22.8 Contactors and Motor Starters

The **magnetic contactor** is similar in operation to the electromechanical relay. Generally, unlike relays, contactors are designed to make and break electric power circuit loads in excess of 15 A. Figure 22-19 shows a three-pole magnetic contactor. The action of the electromagnet, when

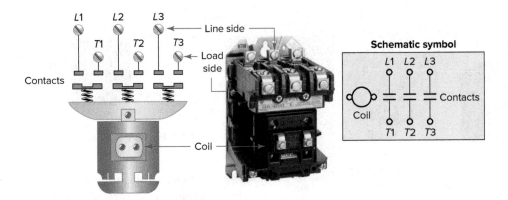

Figure 22-19 Three-pole magnetic contactor.

Photo: Courtesy of Rockwell Automation, Inc.

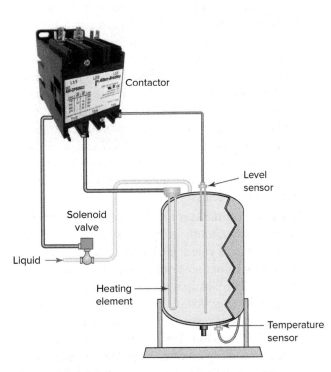

Figure 22-20 Contactor used in conjunction with a pilot device.

Photo: Courtesy of Rockwell Automation, Inc.

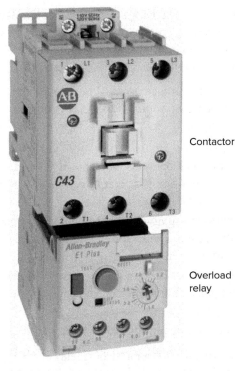

Figure 22-21 Motor starter consisting of a contactor and an overload relay.

Courtesy of Rockwell Automation, Inc.

the coil is energized, closes the sets of contacts. When the coil is deenergized, springs are used to assist in the opening of the contacts.

Contactors are used in conjunction with **pilot devices** to automatically control high-current loads. The pilot device, with limited current-handling capacity, is used to control current to the contactor coil, the contacts of which are used to switch heavier-load currents. Figure 22-20 illustrates a contactor used with pilot devices to control the temperature and liquid level of a tank. In this application the contactor coil connects with the level and temperature sensors to automatically open and close the power contacts to the solenoid and heating element loads.

The basic use for the magnetic contactor is for switching power in resistance heating elements, lighting, magnetic brakes, and heavy industrial solenoids. Contactors can also be used to switch motors if separate overload protection is supplied, as illustrated in Figure 22-21. In its most basic form, a **magnetic motor starter** is a contactor with an overload protective device, known as an overload relay (OL), physically and electrically attached.

Motor overload relays are a type of protective relay intended to protect motors against excessive heating due to prolonged motor current overloads. This relay's operation is designed to limit the duration of time that motor overload current can flow. Motor overload relay types include fixed

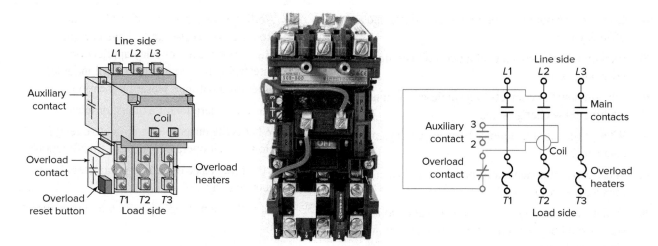

Figure 22-22 Magnetic motor starter.

Photo: Courtesy of Rockwell Automation, Inc.

bimetallic, interchangeable heater bimetallic, and electronic. The basic operating parameters for a motor overload (OL) relay consists of:

- An input that continuously monitors the motor current.
- The setting that sets the value of the motor overload current and the duration of time before the relay is activated.
- A normally closed (NC) output contact that, when activated, opens to deenergize the motor starter coil.

Selection of the overload relay is done using the manufacturer's table included with the magnetic motor starter. Normally magnetic starters come equipped with some *manufacturer-installed control wiring,* as illustrated in Figure 22-22.

Figure 22-23 shows a typical three-phase magnetic across-the-line AC starter schematic diagram. The circuit's operation can be summarized as follows:

- The control transformer is powered by two of the three phases. This transformer lowers the voltage to a more common value useful when adding lights, timers, or remote switches not rated for the higher voltages.
- When the START button is pressed, coil M energizes to close all M contacts. The M contacts in series with the motor close to complete the current path to the motor. These contacts are part of the **power circuit** and must be designed to handle the full-load current of the motor.
- Memory contact M (connected across the START button) also closes to seal in the coil circuit when the START button is released. This contact is part of the **control circuit;** as such, it is required to handle the small amount of current needed to energize the coil.

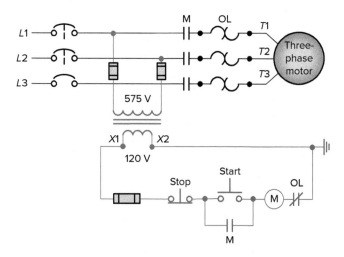

Figure 22-23 Three-phase magnetic across-the-line AC starter.

- The starter has three overload heaters, one in each phase. The normally closed (NC) relay contact OL opens automatically when an overload current is sensed on any phase to deenergize the M coil and stop the motor.
- The motor can be started or stopped from a number of locations by connecting additional START buttons in parallel and additional STOP buttons in series.

Part 2 Review Questions

1. What is the main advantage of solid-state relays over electromechanical types?

2. Explain how electrical isolation of the input and output sections of a solid-state relay can be accomplished.

3. Compare how output contacts are switched in a conventional control relay versus a time-delay relay.

4. Compare the operation of an on-delay timer and an off-delay timer.

5. Explain how the contacts of a two-coil latching relay change state.

6. Explain how the contacts of a single-coil magnetic latching relay change state.

7. What is the main difference between a relay and contactor?

8. What two components are combined to form a magnetic motor starter?

9. Which type of relay, electromechanical or solid-state, would be best suited for each of the following circuit applications?

a. Ability to switch either AC or DC loads.
b. Requires zero off-state leakage current.
c. High-speed, counter-relay application.
d. Interfacing to a computer control module.
e. Operating within a vibrating location.

10. Explain the operation of a one-shot relay.

11. a. What is the purpose of a protective relay?
b. How does it operate to achieve this function?

12. List the three basic operating parameters for a protective relay.

13. What is the monitored input for a motor overload relay?

CHAPTER TWENTY-THREE

Lighting Equipment

Incandescent, compact fluorescent, and LED lamps
©Takashi Honima/123RF

A light fixture, or luminaire, is an electrical device used to create artificial light and/or illumination by use of an electric lamp. The number of different types of lighting fixtures available staggers the imagination. This chapter provides an overview of the construction, operation, and installation of lighting equipment.

LEARNING OUTCOMES

▶ Understand the basics of incandescent lamps.
▶ Understand the basics of fluorescent lamps.
▶ Understand the basics of LED lamps.
▶ Develop a working knowledge of luminaire (lighting fixture) disconnects and thermal protection.
▶ Comprehend lighting control devices.
▶ Comprehend lighting control circuits.

23.1 Lamps

The **incandescent lamp** or lightbulb produces light by **heating** a tungsten filament wire to a high temperature until it glows. During the manufacture of an incandescent lamp, the air is removed from the glass bulb before it is sealed. This is done to prevent the oxidation of the filament. The filament would burn out quickly if oxygen were present within the bulb. The glass bulb is filled with inert gas, such as argon, resulting in a lower evaporation rate and longer filament life. The ends of the filament are brought out to a screw-type base, as illustrated in Figure 23-1, which makes replacement simple. The **lumen** is a measurement of the amount of light produced by a light source For incandescent bulbs, 10 percent of the energy they consume is used to create visible light, while the other 90 percent is wasted heat.

A **tungsten-halogen lamp** is an incandescent lamp with gases from the halogen family sealed inside the bulb, as shown in Figure 23-2. Its light output is similar to that of a regular incandescent bulb, but it uses up to 40 percent less power. The combination of the halogen gas and the tungsten filament produces a chemical reaction known as a **halogen cycle** which increases the lifetime of the filament and prevents darkening

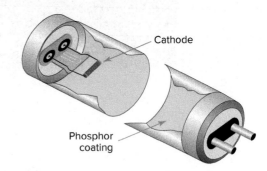

Figure 23-3 Fluorescent tube.

of the bulb by redepositing tungsten from the inside of the bulb back onto the filament. Halogen lamps provide an intense, white light in a very small casing and operate at **very high temperatures.** Do not touch the halogen bulb surface or inside reflectors with your bare hands. Oils from skin can lead to breakage or shorten the life of the lamp. Use clean gloves or lint-free cloth for installation and removal.

The **fluorescent lamp** belongs to a family of light sources known as **gas-discharge** lamps. Unlike incandescent lamps, they have no filament. Instead, current flows in an arc through mercury vapor between contacts called cathodes at each end of the tubular lamp, as illustrated in Figure 23-3. In operation, a very small amount of mercury mixes with inert gases to conduct the electric current. This in turn allows the phosphor coating on the glass tube to emit light. Compared with incandescent lamps, fluorescent lamps last longer and produce *less heat* while providing more light per watt.

Fluorescent lamps require an auxiliary component called a **ballast** (Figure 23-4) to operate. The ballast performs two functions:

- It produces a jolt of high voltage to vaporize the mercury inside the lamp and start the arc from one end to the other.
- Once the lamp is started, it limits current to the lower value needed for proper operation.

Compact fluorescent lamps come in a variety of shapes and styles, as illustrated in Figure 23-5. Some include built-in integral electronic ballasts and screw shells. Like all fluorescent lamps, they contain **mercury,** which complicates their disposal.

Figure 23-1 Incandescent lamp.

Figure 23-2 Tungsten-halogen lamp.
Courtesy of Osram Sylvania Inc.

Figure 23-4 Fluorescent lamp ballast.
©Standard Products, Inc.

Figure 23-5 Compact fluorescent lamps.
Courtesy of Osram Sylvania Inc.

Ballast

Figure 23-7 HID lamp fixture with ballast.
Courtesy of Lumetric Lighting Solutions, Inc.

Figure 23-8 LED lamp.
©Lumicrest LED Lighting Products

High-intensity discharge (HID) lamps, like fluorescent lamps, produce light by means of a gas-discharge arc in a tube. Unlike fluorescent lamps, these lamps operate at a much **higher current,** which flows through a much **shorter arc tube.** Compared with fluorescent and incandescent lamps, HID lamps deliver a greater proportion of their radiation in visible light as opposed to heat. There are three different types in common use: the mercury vapor lamp, the metal halide lamp (Figure 23-6), and the high-pressure sodium lamp. HID lamps provide the longest service life of any lighting type.

HID lamps can not be directly connected to an electric circuit; they must be operated with matching ballast, as illustrated in Figure 23-7. When first turned on, the lamp can take up to **10 minutes** to produce light, because the ballast needs time to establish the electric arc. If turned off, it must cool off before it will relight. The main advantages of high-intensity discharge lamps are high light efficiency, very long life, and high watt output from single fixtures. HID lighting systems are widely used in applications where high light levels are desired for large areas, such as stores, factories, and street lighting.

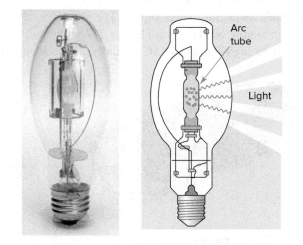

Arc tube

Light

Figure 23-6 High-intensity discharge (HID) lamp.
Photo: ©MemphisClay30/iStock/Getty Images

An **LED lamp** is a solid-state lamp that uses **light-emitting diodes (LEDs)** as the source of light. The light output of individual light-emitting diodes is small, compared to incandescent and compact fluorescent lamps, so multiple diodes are often used together, as shown on Figure 23-8. LEDs are low-voltage light sources, requiring a constant DC voltage or current to operate optimally. Individual LEDs used for illumination require **2 to 4 volts of direct current** (DC) power and several hundred **mA of current.** As LEDs are connected in series in an array, higher voltage is required. An **LED driver** is the power supply for an LED system, much like ballast is to a fluorescent or HID lighting system. For LEDs, 80 percent of the energy they consume is used to create visible light, while the other 20 percent is wasted heat.

One of the major characteristics of an LED is its color. The material used in the semiconducting element of an LED determines its color. An **RGB** LED can emit different colors by mixing the three basic colors red, green and blue, as illustrated in Figure 23-9. This type of LED consists of three separate LEDs red, green, and blue packed in a single case. By using pulse wide modulation (PWM) type of control, it is possible to mix different proportions of red, green, and blue in the same LED, producing nearly any color imaginable. A pulse width modulated signal

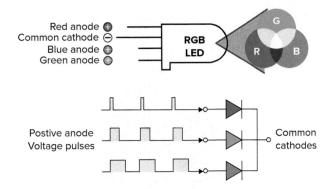

Figure 23-9 **RGB color LED.**

allows you to vary how much time the voltage to each anode is high (usually 5 V) at any time. You can change the proportion of time the signal is high compared to when it is low over a consistent time interval.

23.2 Wiring Luminaires

The National Electrical Code (NEC) defines a **luminaire** as a complete lighting unit consisting of a lamp or lamps together with the parts designed to distribute the light, to position and protect the lamps and ballast (where applicable), and to connect the lamps to the power supply.

The code allows ceiling outlet boxes to support luminaires (fixtures) weighing up to *50 lb,* as illustrated in Figure 23-10. Heavier luminaires must be supported by

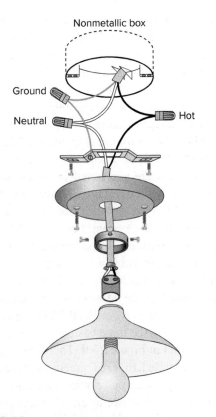

Figure 23-10 **Ceiling light fixture.**

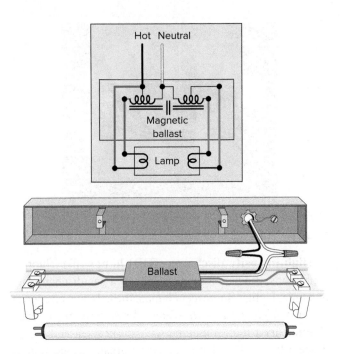

Figure 23-11 **Rapid start fluorescent fixture.**

outlet boxes listed for the weight to be supported or must be supported independently of the box.

A rapid-start fluorescent fixture is shown in Figure 23-11. This fixture contains a magnetic ballast circuit that utilizes continuous cathode heating, while the system is energized, to start and maintain lamp light output at efficient levels. Rapid-start ballasts may be either an electromagnetic, electronic, or a hybrid design.

There are two basic types of lamp ballasts: low-frequency magnetic ballasts and high-frequency electronic ballasts. The **magnetic ballast** uses a transformer to convert the input line voltage and current to the voltage and current required to start and operate the fluorescent lamps. Capacitors are added to assist lamp starting and power factor correction. But the output frequency is the same as the input frequency (60 Hz).

The **electronic ballast,** shown in Figure 23-12, uses an electronic switching power supply. The electronic circuit takes incoming 60 Hz power (**120 or 277 volts**) and converts it to high-frequency AC (**usually 20 to 40 kHz**).

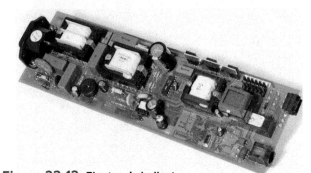

Figure 23-12 **Electronic ballast.**
Courtesy of MARLEX Engineering Inc.

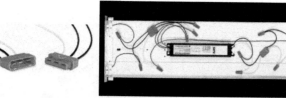

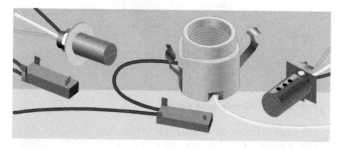

Figure 23-13 Luminaire disconnect.
©Thomas & Betts Corporation

Figure 23-14 Thermal protectors.

Electronic high-frequency ballasts increase lamp-ballast efficacy, leading to increased energy efficiency and lower operating costs.

The NEC requires all nonresidential fluorescent luminaires with ballasts to be equipped with a **luminaire disconnect** built into the fixture. This disconnect is required so that power to the fixture can be removed in order to service it without risk of exposure to dangerous live wires. Figure 23-13 shows the required installation of a luminaire disconnect between the power line and the ballast.

The NEC has special safety rules to prevent **overheating** of luminaires. The hot metal shells of incandescent luminaires, as well as overheated ballasts of fluorescent and HID luminaires, pose a potential fire threat if they contact flammable materials. Most fluorescent luminaires installed indoors are required to have integral **thermal protectors,** such as that shown in Figure 23-14, which automatically disconnect the power supply when their case temperature is excessive.

Part 1 Review Questions

1. How is light produced in an incandescent lamp?
2. Compare the makeup of tungsten-type incandescent lamps with that of tungsten-halogen types.
3. When handling a tungsten-halogen lamp, you should avoid touching the surface of the bulb with your bare hands. Why?
4. How is light produced in a fluorescent tubular lamp?
5. Which two functions are performed by the fluorescent lamp ballast?

6. How is light produced in a high-intensity discharge (HID) lamp?
7. What are the typical operating voltage and current levels for LED lamps?
8. What is an LED driver?
9. Which designation does the NEC use when referring to a complete lighting fixture?
10. What is the maximum weight of luminaires, allowed by the code, that can be supported by outlet boxes?
11. In what way does the magnetic fluorescent ballast differ from electronic types?
12. What is the function and purpose of a luminaire disconnect?
13. Why do certain types of luminaires require thermal protectors?
14. What is the lumen a measurement of?
15. What determines the color of the light emitted by an LED?

PART 2   LIGHTING CONTROL

23.3 Light Switches

The NEC lists requirements for AC-only and AC/DC general-use snap switches. These devices are also known as wall switches and toggle switches, and the important points for each are summarized as follows:

- Most switches used in lighting control wiring are *alternating current general-use snap switches,* similar to that shown in Figure 23-15. In addition to

Figure 23-15 General-use snap switch.
Courtesy of Legrand/Pass & Seymour

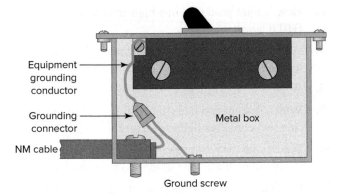

Figure 23-16 **Switch grounding.**

Figure 23-17 **Single-pole switch.**

voltage and current ratings, they are marked **AC only.** Most are dual-rated for *120 and 277 volts,* which allows them to be used on either 120-volt lighting circuits in homes or 277-volt lighting circuits in commercial construction. All snap switches that offer quiet operation are AC general-use snap switches.

- *Alternating current or direct current* general-use snap switches are typically of heavier construction than AC-only snap switches. AC/DC switches use springs to break their contacts quickly, which minimizes arcing when they are used to interrupt *DC circuits.*
- Snap switches designed for connection to either copper or aluminum conductors, are listed and marked **CO/ALR.**

The NEC requires that snap switches and similar control switches be effectively **grounded** and provide a means to ground metal faceplates, as illustrated in Figure 23-16. Switches are manufactured with equipment grounding terminals, normally a green hexagonal screw, for connection to the equipment grounding conductor. When a metal faceplate is installed on a switch box, it is grounded by connection to the metal yoke of the grounded snap switch by means of two no. 6-32 screws. Switches and other wiring devices are sold with small cardboard washers holding these screws in place. The washers should be removed before installing a metal faceplate, to ensure that it is grounded as required by the code.

A **single-pole** switch is used to control light(s) from *one location.* They have two positions indentified as ON and OFF, as illustrated in Figure 23-17. When this switch is pushed to the ON position, the circuit is completed between the two terminal points, allowing current to flow through the switch. In the OFF position, the contact between the two terminal points is broken, opening the internal-switch circuit. These switches are fastened to the outlet box so that the handle of the switch is moved up to turn the switch on and down to turn if off.

Three-way switches are used to control light(s) from *two locations.* Examples of this type of lighting control include hall or stairway lights and lights in a room with two entrances. The internal circuit of a three-way switch is such that it allows current to flow through the switch in either of its two positions. For this reason there are *no* ON/OFF marks on the operating handle. Three-way switches have three terminals that can be connected in two different positions, as illustrated in Figure 23-18. One terminal is called the **common** and is attached to the hot (ungrounded) conductor. The other two terminals, which are lighter in color, are called **traveler** terminals. Two conductors, called travelers, run from these terminals to the traveler terminals of a second three-way switch. The switched conductor runs from the darker common terminal of the second three-way switch to the outlet or other controlled load.

Four-way switches are used to control light(s) from *three or more locations.* One or more four-way switches are always used with a *pair* of three-way switches. Four-way switches have four terminals. The four-way switch, like the three-way switch, allows current to flow through the switch in either of its two positions. For this reason there are no ON/OFF marks on the operating handle.

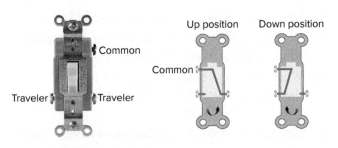

Figure 23-18 **Three-way switch.**
Photo: Courtesy of Cooper Wiring Devices

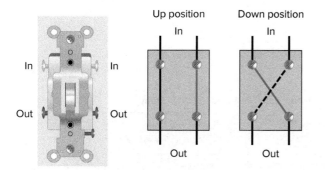

Figure 23-19 Four-way switch.

Figure 23-19 shows the internal switching of a four-way switch. The four-way terminals are labeled *in* and *out* (two of each). The mechanics of this switch are such that they either switch the travelers **straight through** or **crisscross** them.

23.4 Switch Wiring

General requirements that apply to all switch wiring are summarized as follows:

- Switching always takes place in the *ungrounded (hot) conductor.*
- Grounded conductors (neutrals) are *never permitted* to be switched.

- Single-pole switches aren't permitted to be installed *backward,* so that the controlled light(s) is ON when the switch handle indicates OFF.

Figure 23-20 shows the branch circuit wiring for two lamps controlled by a **single-pole switch.** The switch is connected in series with the hot ungrounded conductor, while the neutral ground is run directly to each lamp holder. Connecting the lamps in parallel provides the full 120 V across each lamp when the switch is closed. Connecting the two lamps in series would result in less than the normal 120-V operating voltage across each light. A second feature of the parallel connection of lamps is that they operate independently of each other so that if one burns out, the operation of the other is not affected.

Figure 23-21 shows the branch circuit wiring for one lamp that is controlled from two locations using **three-way switching.** The neutral white conductors in both switch boxes are spliced directly through to the lamp. The red and blue conductors that run between the two switch boxes are the **travelers.** The black conductor from the source is connected to the **common** terminal of the first switch and the black conductor going to the lamp is connected to the common terminal of the second switch. Whenever you operate either one of the three-way switches, the light changes its state—if it is ON, it turns off, and if it is OFF, it turns on.

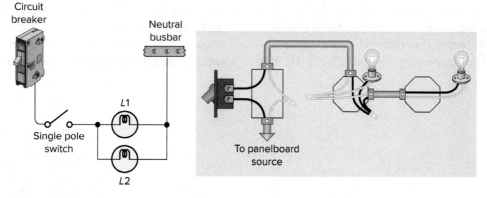

Figure 23-20 Two lamps controlled by a single-pole switch.

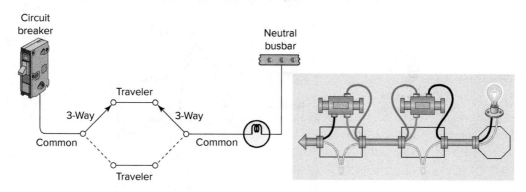

Figure 23-21 Three-way switching circuit.

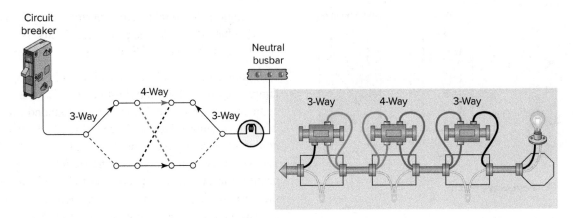

Figure 23-22 Lamp controlled from three locations.

Figure 23-22 shows the branch circuit wiring for one lamp that is controlled from *three* locations using a **four-way switch** in combination with 2 three-way switches. The four-way switch is connected to the **travelers** between the 2 three-way switches. When connected correctly, the actuation of any one of the switches will change the state of the light (turn the lamp either on or off). You must connect the travelers to the proper in and out *pairs* of terminals on the four-way switch; otherwise, the switching sequence will not operate properly. For more than three control locations, three-way switches are installed at the first two locations, and then a four-way switch is installed at each additional control location.

At times, it may be desirable to have a switch that controls one-half of a duplex receptacle. To do this, the link on the receptacle that connects the **live hot brass terminal screws** together is removed so that each half of the receptacle can operate independently (Figure 23-23). **Do not**

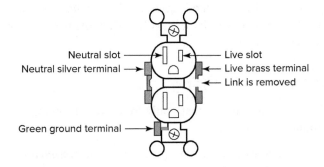

Figure 23-23 Split-wired receptacle.

break the link joining the silver-colored screw connections on the neutral side! These receptacles are often referred to as **split-wired** receptacles.

Figure 23-24 shows the schematic and wiring diagram for a **switch-controlled** receptacle. The switch controls the top

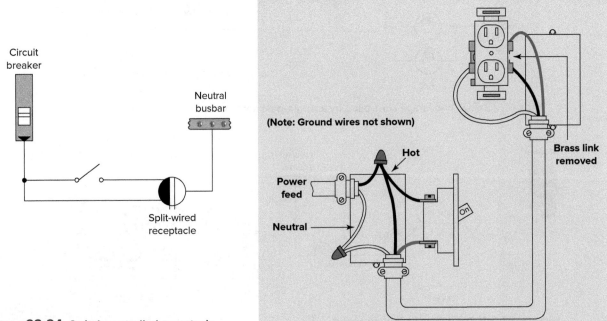

Figure 23-24 Switch controlled receptacle.

half of the receptacle, while the bottom half is live at all times. In a typical application, a portable lamp can be plugged into the top half and controlled by the switch, while at the same time, the other half is live at all times for use with appliances such a vacuum cleaners, radios, or televisions.

Other common types of lighting control devices include the following:

- **Dimmer switches** are used to control lighting level and can save energy. They operate by varying the power that goes to a lighting load. Dimmers work very well with incandescent bulbs, but for other types of bulbs, including compact fluorescent lamp (CFL) bulbs and light-emitting diode (LED) bulbs, check to see that they are **dimmable.**

- **Smart switches** allow you to control your home lighting using just your phone or tablet. Most come with an ON and OFF switch so you can immediately control it from its physical location as well as from your smartphone once you connect it to your Wi-Fi network.

- **Occupancy sensor switches** are motion detecting devices used to detect the presence of a person in a room or space in order to automatically turn lights on or off, using infrared, ultrasonic, microwave, or other technology.

- **Light sensor switches** are used to detect light. A dusk-to-dawn light sensor operates to switch lights on as night falls and off shortly after sunrise.

- **Programmable timer switches** allow you to program the operation of your lights. They often contain an LCD screen that shows time, day, and load status, and a manual override enables the light to be turned on/off without affecting the program.

Part 2 Review Questions

1. What is the most common dual-voltage rating for AC general-use snap switches?

2. How are switches designed for connection to either copper or aluminum identified?

3. What is the green hexagon screw terminal found on switches used for?

4. Explain how a single-pole switch is properly fastened to an outlet box.

5. What type of switches would be required to control a light from two positions?

6. What type of switches would be required to control a light from four positions?

7. Explain why there are no ON/OFF markings on the handles of a three-way or four-way switch.

8. Identify the three terminals associated with a three-way switch.

9. Identify the four terminals associated with a four-way switch.

10. Which conductor is never permitted to be switched?

11. What change must be made to a duplex receptacle in order to have it operate as a split-circuit receptacle?

12. State the type of lighting control you would use for each of the following lighting scenarios:
    a. Control lights left on in unoccupied spaces.
    b. Automatic control of lights in a parking lot.
    c. Wireless control of a lighting system.

CHAPTER TWENTY-FOUR

Electric Motors and Controls

Cutaway view of a electric induction motor
©Aaron Roeth Photography

An electric motor converts electric energy to mechanical energy by using interacting magnetic fields. Electric motors are used for a wide variety of residential, commercial, and industrial operations. This chapter deals with the operating principles of AC three-phase and single-phase motors along with the circuits used to control them. DC motor theory includes operating principles along with the circuits used to control them.

LEARNING OUTCOMES

▶ Understand the principle of operation of AC motors.

▶ Have a working knowledge of the construction, connection, and operating characteristics of different types of AC motors.

▶ Comprehend the operation of basic motor control circuits.

▶ Have a working knowledge of the construction, connection, and operating characteristics of different types of DC motors.

24.1 Rotating Magnetic Field

The operating principle of all motors is based on the fact that whenever a current-carrying conductor is placed inside a magnetic field, there will be mechanical force experienced by that conductor. A **rotating magnetic field** is key to the operation of all AC motors. The principle is simple. A magnetic field in the stator is made to rotate electrically around and around in a circle. Another magnetic field in the rotor is made to follow the rotation of this field pattern by being attracted and repelled by the stator field. Because the rotor is free to turn, it follows the rotating magnetic field in the stator.

Figure 24-1 illustrates the concept of a rotating magnetic field as it applies to the stator of a three-phase AC motor. The operation can be summarized as follows:

- Three sets of windings are placed *120 electrical degrees apart* with each set connected to one phase of the three-phase power supply.
- When three-phase current passes through the **stator windings,** a rotating magnetic field effect is produced that travels around the inside of the stator core.

- The **polarity** of the rotating magnetic field is shown at seven selected positions marked off at 60-degree intervals on the sine waves representing the current flowing in the three phases: A, B, and C.
- In the example shown, the magnetic field will rotate around the stator in a *clockwise* direction.
- Simply interchanging any two of the three-phase power input leads to the stator windings *reverses direction of rotation* of the magnetic field.

The speed of the rotating magnetic field varies *directly* with the **frequency** of the power supply and *inversely* with the number of **poles** constructed on the stator winding. This means the higher the frequency, the greater the speed, and the greater the number of poles, the slower the speed. Motors designed for 60 Hz use have synchronous speeds of 3,600, 1,800, 1,200, 900, 720, 600, 514, and 450 rpm. The synchronous speed of an AC motor can be calculated by the formula

$$S = \frac{120f}{P}$$

where S = synchronous speed in rpm
$\quad f$ = frequency, Hz, of the power supply
$\quad P$ = number of poles wound in each of the single-phase windings

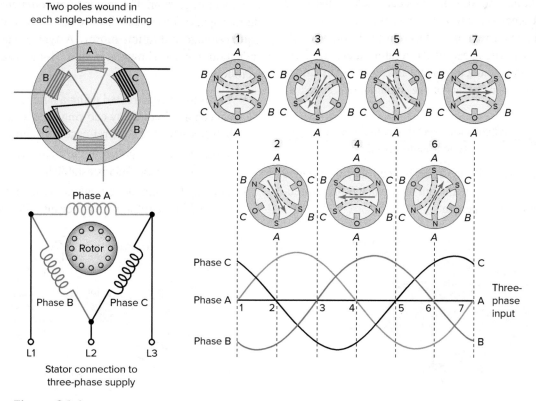

Figure 24-1 **Rotating magnetic field.**

EXAMPLE 24-1

Problem: Determine the synchronous speed of a four-pole AC motor connected to a 60-Hz electric supply.

Solution:

$$S = \frac{120f}{P}$$
$$= \frac{120 \times 60}{4}$$
$$= 1,800 \text{ rpm}$$

EXAMPLE 24-2

Problem: Determine the percentage slip of an induction motor having a synchronous speed of 1,800 rpm and a rated actual speed of 1,750 rpm.

Solution:

$$\text{Percent slip} = \frac{\text{synchronous speed} - \text{actual speed}}{\text{synchronous speed}} \times 100$$
$$= \frac{1,800 - 1,750}{1,800} \times 100$$
$$= 2.78\%$$

There are two ways to define AC motor speed, as illustrated in Figure 24-2. First is synchronous speed. The **synchronous speed** of an AC motor is the speed of the *stator's magnetic field rotation.* This is the motor's ideal theoretical, or mathematical, speed, since the rotor will always turn at a slightly slower rate. The other way motor speed is measured is called *actual speed.* This is the speed at which the shaft rotates. The nameplate of most AC motors lists the actual motor speed rather than the synchronous speed.

The rotor does not revolve at synchronous speed but tends to slip behind. Slip is what allows a motor to turn. If the rotor turned at the same speed at which the field rotates, there would be no **relative motion** between the rotor and the field and no voltage induced. Because the rotor slips with respect to the rotating magnetic field of the stator, voltage and current are induced in the rotor. The difference between the speed of the rotating magnetic field and the rotor in an induction motor is known as **slip** and is expressed as a percentage of the synchronous speed as follows:

$$\text{Percent slip} = \frac{\text{synchronous speed} - \text{actual speed}}{\text{synchronous speed}} \times 100$$

The slip increases with load and is necessary to produce useful **torque.** The usual amount of slip in a 60-Hz, three-phase motor is 2 or 3 percent.

24.2 Squirrel-Cage Induction Motor

The **AC induction motor** is by far the most commonly used motor. The induction motor is so named because *no external voltage* is applied to its rotor. Instead, the AC current in the stator *induces* a voltage across an air gap and into the rotor winding to produce rotor current and associated magnetic field. To visualize how the rotor rotates, a magnet mounted on a shaft can be substituted for the squirrel-cage rotor, as illustrated in Figure 24-3. The poles of the rotor are attracted by the poles of the stator, and a rotation is produced.

An induction motor rotor can be either a wound rotor or a squirrel-cage rotor. The majority of commercial and industrial applications usually involve the use of a **three-phase squirrel-cage induction motor.** A typical squirrel-cage induction motor is shown in Figure 24-4 and summarized as follows:

- The rotor is constructed using a number of single bars short-circuited by end rings and arranged in a hamster-wheel or squirrel-cage configuration.
- When voltage is applied to the stator winding, a rotating magnetic field is established.
- This rotating magnetic field causes a voltage to be induced in the rotor, which, because the rotor bars

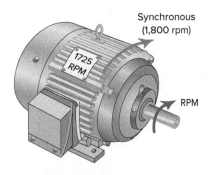

Figure 24-2 **Motor synchronous and actual speed.**

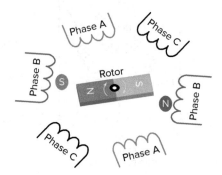

Figure 24-3 **Rotor rotation.**

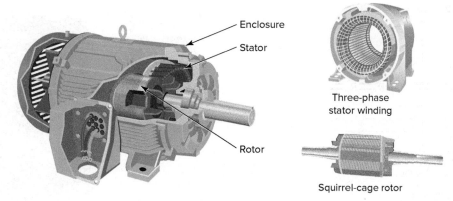

Figure 24-4 **Squirrel-cage induction motor.**

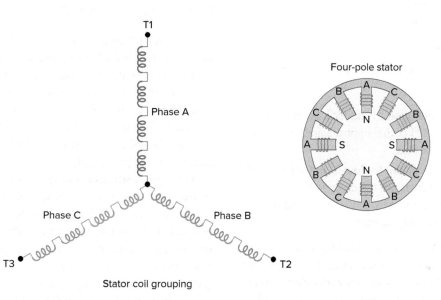

Stator coil grouping

Figure 24-5 **Four-pole, wye-connected stator windings.**

are essentially single-turn coils, causes currents to flow in the rotor bars.

- These rotor currents establish their own magnetic field, which interacts with the stator magnetic field to produce a torque.
- The resultant production of torque spins the rotor in the same direction as the rotation of the magnetic field produced by the stator.

A three-phase motor **stator winding** consists of three separate groups of coils, called **phases,** and designated A, B, and C. The phases are displaced from each other by 120 electrical degrees and contain the same number of coils connected for the same number of poles. **Poles** refer to a coil or group of coils wound to produce a unit of magnetic polarity. The number of poles a stator is wound for will always be an even number and refers to the total number of north and south poles per phase. Figure 24-5 shows a typical connection of coils for a **four-pole, three-phase, wye-connected** induction motor.

The *direction of rotation* of any three-phase motor can be changed by interchanging any *two of the three* main power lines to the motor stator. This causes the direction of the rotating magnetic field to reverse. Figure 24-6 shows the power circuit for reversing a three-phase motor. The forward contacts, F, when closed, connect L1, L2, and L3 to motor terminals *T1, T2, and T3,* respectively. The reverse contacts, R, when closed, connect L1, L2, and L3 to motor terminals *T3, T2, and T1,* respectively.

Loading of a squirrel-cage motor is similar to that of a transformer and can be summarized as follows:

- Both the transformer and induction motor involve changing flux linkages with respect to a primary (stator) winding and secondary (rotor) winding.
- The no-load motor current is low and similar to the exciting current in a transformer and is composed of a magnetizing component that creates the revolving flux.

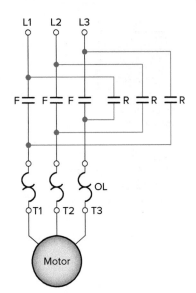

Figure 24-6 Power circuit for reversing a three-phase motor.

- When the induction motor is under load, the rotor current develops a flux that opposes and, therefore, weakens the stator flux.
- This allows more current to flow in the stator windings, just as an increase in the current in the secondary of a transformer results in a corresponding increase in the primary current.

The moment a motor is started, during the acceleration period, the motor draws a high inrush current. This starting current is also known as the **locked-rotor current.** Induction motors, started at rated voltage, have locked-rotor starting currents of up to **6 times** their nameplate full-load current. The locked-rotor current depends largely on the type of rotor bar design and can be determined from the **NEMA design code** letters listed on the nameplate.

Figure 24-7 **Multispeed squirrel-cage motors.**
Courtesy of Baldor Electric Co.

The squirrel-cage motor normally operates at essentially constant speed, close to the synchronous speed. A single-speed motor has one rated speed at which it runs when supplied with the nameplate voltage and frequency. A *multispeed* squirrel-cage motor, such as those shown in Figure 24-7, will run at more than one speed, depending on how the windings are connected to form a *different number of magnetic poles.*

Two-speed, single-winding induction motors are called **consequent pole** motors. These motors are wound for one speed, but when the winding is reconnected, the number of magnetic poles within the stator is doubled and the motor speed is reduced to one-half of the original speed, as shown in Figure 24-8. The low speed on a single-winding consequent pole motor is always *one-half of the higher speed.*

The **two-speed, two-winding** induction motor is made in such a manner that it is really two motors wound into one stator. One winding, when energized, gives one of the speeds. When the second winding is energized, the motor takes on the speed that is determined by the second winding,

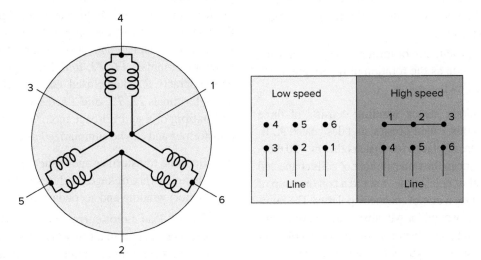

Figure 24-8 Two-speed, single-winding motor connections.

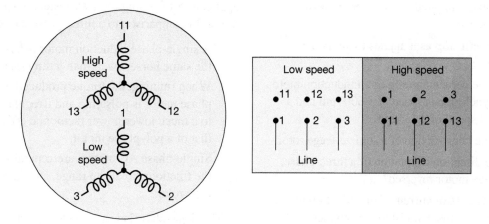

Figure 24-9 Two-speed, two-winding motor connections.

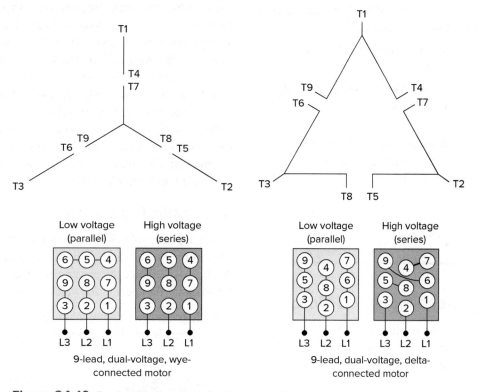

Figure 24-10 Dual-voltage motor winding connections.

as illustrated in Figure 24-9. The two-speed, two-winding motor can be used to get virtually any combination of normal motor speeds, and the two different speeds need not be related to each other by a 2:1 speed factor.

Single-speed AC induction motors are frequently supplied with multiple external leads for various voltage ratings in fixed-frequency applications. The multiple leads may be designed to allow either series or parallel reconnections, wye to delta reconnections, or combinations of these. Figure 24-10 shows typical connections for **dual-voltage** wye and delta series and parallel reconnections. *These types of reconnections should not be confused with the reconnection of multispeed, polyphase induction motors.* In the case of multispeed motors, the reconnection results in a motor with a different number of magnetic poles and therefore a different synchronous speed at a given frequency.

Part 1 Review Questions

1. Explain how a rotating magnetic field is established in a three-phase motor stator.

2. Calculate the synchronous speed of a six-pole AC motor operated from a standard 60-Hz voltage source.

3. Compare synchronous speed and actual speed of an AC motor.

4. How is voltage supplied to the rotor of an AC induction motor?

5. Define the term *slip* as it applies to an induction motor.

6. Calculate the percentage slip of an induction motor having a synchronous speed of 3,600 rpm and a rated actual speed of 3,435 rpm.

7. Describe the construction of a squirrel-cage rotor.

8. How is the direction of rotation of a three-phase, squirrel-cage motor reversed?

9. Why does the stator current of an induction motor increase when a mechanical load is applied to the rotor?

10. Compare the starting and full-load running currents of a typical AC induction motor.

11. What two factors determine the speed of a squirrel-cage induction motor?

12. What is the difference in construction between multispeed single-winding and separate-winding induction motors?

13. A two-speed motor requiring 1,750 rpm and 1,140 rpm is required for a motor installation. Would you require a single-winding or two-winding motor? Why?

PART 2 SINGLE-PHASE MOTORS

24.3 Single-Phase Induction Motor

Most home and business appliances operate on single-phase AC power. For this reason, **single-phase AC motors** are in widespread use. Figure 24-11 illustrates the application of a single-phase residential furnace blower motor. In comparison to a three-phase motor:

- A single-phase induction motor is larger in *size,* for the same horsepower, than a three-phase motor.

- When running, the torque produced by a single-phase motor is pulsating and irregular, contributing to a much lower power factor and efficiency than that of a polyphase motor.

- Single-phase AC motors are generally available in the fractional to 10-hp range, and all use a squirrel-cage rotor.

The single-phase induction motor operates on the principle of induction, just as does a three-phase motor. Unlike three-phase motors, they are *not self-starting.* Whereas a three-phase induction motor sets up a rotating field that can start the motor, a single-phase motor needs an *auxiliary means of starting.* Once a single-phase induction motor is running, it develops a rotating magnetic field. However, before the rotor begins to turn, the stator produces only a pulsating, stationary field.

A single-phase motor could be started by mechanically *spinning the rotor* and then quickly applying power. However, normally these motors use some sort of automatic starting. Single-phase induction motors are classified by their **start and run** characteristics.

24.4 Split-Phase Motor

A single-phase, split-phase induction motor uses a **squirrel-cage rotor** that is identical to that in a three-phase motor. Figure 24-12 shows the construction and wiring of a split-phase motor, the operation of which is summarized as follows:

- To produce a rotating magnetic field, the single-phase current is split by two windings, the main *running*

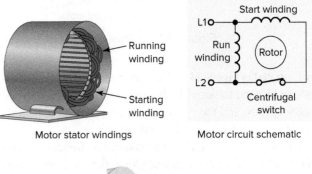

Motor stator windings   Motor circuit schematic

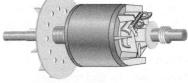

Squirrel-cage rotor

Figure 24-12 Split-phase induction motor.

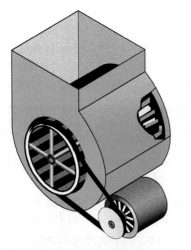

Figure 24-11 Single-phase residential furnace blower motor.

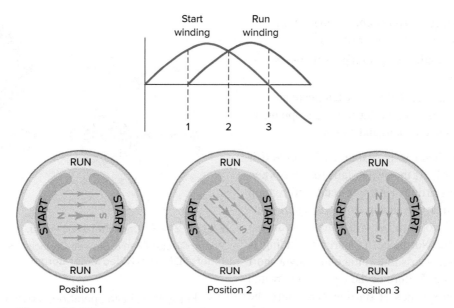

Figure 24-13 **Rotating magnetic field of a split-phase motor.**

winding and an auxiliary *starting* winding, which is displaced in the stator 90 mechanical degrees from the running winding.

- The starting winding is connected in series with a switch, centrifugally or electrically operated, to *disconnect* it when the starting speed reaches about 75 percent of full-load speed.

Phase displacement in a split-phase motor is accomplished by the difference in inductive reactance of the start and run windings, as well as the physical displacement of the windings in the stator. The **starting winding** is wound *on the top* of the stator slots with fewer turns of smaller-diameter wire. The **running winding** has many turns of large-diameter wire wound in the *bottom* of the stator slots that give it a higher inductive reactance than the starting winding.

The way in which the two windings of a split-phase motor produce a rotating magnetic field is illustrated in Figure 24-13 and can be summarized as follows:

- When AC line voltage is applied, the current in the starting winding *leads the current* in the running winding by approximately 45 electrical degrees.
- Since the **magnetism** produced by these currents *follows the same wave pattern,* the two sine waves can be thought of as the waveforms of the electromagnetism produced by the two windings.
- As the alternations in current (and magnetism) continue, the *positions of the north and south poles change* in what appears to be a clockwise rotation.
- At the same time the rotating field cuts the squirrel-cage conductors of the rotor and *induces a current* in them.

- The rotor current creates magnetic poles in the rotor, which *interact* with the poles of the stator rotating magnetic field to produce motor **torque.**

Once the motor is running, the starting winding must be *removed* from the circuit. Since the starting winding is of a smaller gauge size, continuous current through it would cause the winding to burn out. Either a mechanical centrifugal or electronic solid-state switch may be used to automatically disconnect the starting winding from the circuit.

The operation of a centrifugal-type switch for a split-phase motor is illustrated in Figure 24-14, and its operation can be summarized as follows:

- The switch consists of a centrifugal mechanism, which rotates on the **motor shaft** and interacts with a **fixed stationary** switch whose contacts are connected in series with the start winding.
- When the motor approaches its *normal operating speed,* centrifugal force overcomes the spring force,

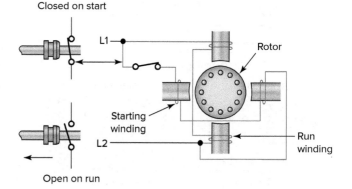

Figure 24-14 **Centrifugal switch operation.**

allowing the contacts to **open** and disconnect the starting winding from the power source.

- The motor then continues operating solely on its running winding.
- Motors using such a centrifugal switch make a distinct **clicking** noise when starting and stopping as the centrifugal switch opens and closes.

The centrifugal switch can be a source of trouble if it fails to operate properly. If the switch fails to close when the motor stops, the starting winding circuit will remain open. As a result, when the motor circuit is again energized, the motor will not turn but will simply produce a **low humming sound.** Normally the starting winding is designed for operation across line voltage for only a short interval during starting. Failure of the centrifugal switch to open within a few seconds of starting may cause the starting winding to char or burn out.

The split-phase induction motor is the simplest and most common type of single-phase motor. Its simple design makes it typically less expensive than other single-phase motor types. Split-phase motors are considered to have low or moderate starting torque. Reversing the leads to either the start or run windings, but not to both, changes the direction of rotation of a split-phase motor.

Dual-voltage, split-phase motors have leads that allow external connection for different line voltages. Figure 24-15 shows a NEMA-standard, single-phase motor with **dual-voltage run windings.**

- When the motor is operated at low voltage, the two run windings and the start winding are all connected in parallel.

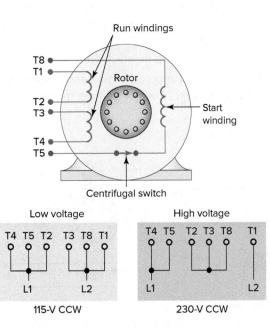

Figure 24-15 Dual-voltage, split-phase motor.

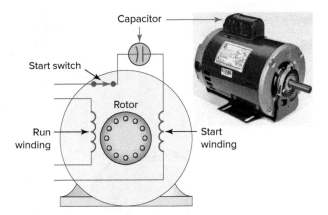

Figure 24-16 Capacitor-start motor.
Photo: ©Belted Fan & Blower, Capacitor Start, Open Drip proof (ODP) from Nidec Motor Corporation

- For high-voltage operation, the two run windings connect in series, and the start winding is connected in parallel with one of the run windings.

24.5 Single-Phase Capacitor Motor

The **capacitor-start motor,** illustrated in Figure 24-16, is a modified split-phase motor.

- A capacitor connected in series with the starting winding creates a phase shift of approximately 80 degrees between the starting and running winding.
- This is substantially higher than the 45 degrees of a standard split-phase motor and results in a higher **starting torque.**
- Capacitor-start motors provide more than double the starting torque with one-third less starting current than the split-phase motor.
- Like the split-phase motor, the capacitor-start motor also has a starting mechanism which disconnects not only the start winding but also the capacitor when the motor reaches about 75 percent of rated speed.

The capacitor-start motor application range is much wider because of higher starting torque and lower starting current. The capacitor can be a source of trouble if it becomes short-circuited or open-circuited. A **short-circuited** capacitor will cause an excessive amount of current to flow through the starting winding, while an **open capacitor** will cause the motor not to start.

The **permanent-capacitor** motor has a run-type capacitor permanently connected in series with the start winding. This makes the start winding an auxiliary winding once the motor reaches running speed. The run and auxiliary windings are identical in this type of motor, allowing the motor to be **reversed** by switching the capacitor from one winding to the other, as illustrated in Figure 24-17.

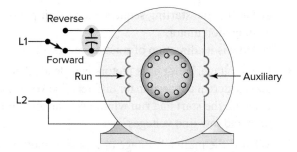

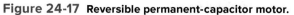

Figure 24-17 Reversible permanent-capacitor motor.

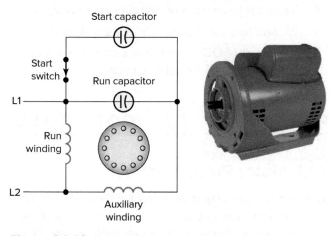

Figure 24-18 Capacitor-start/capacitor-run motor.
Photo: ©I. Pilon/Shutterstock

Single-phase motors run in the direction in which they are started, so whichever winding has the capacitor connected to it will control the direction.

The **capacitor-start/capacitor-run motor,** shown in Figure 24-18, uses both start and run capacitors.

- When the motor is started, the two capacitors are connected in parallel to produce a large amount of capacitance and starting torque.

- Once the motor is up to speed, the start switch *disconnects the start capacitor* from the circuit.

- The motor start capacitor is typically an **electrolytic** type, while the run capacitor is an **oil-filled** type. The electrolytic type offers a large amount of capacitance when compared to its oil-filled counterpart.

- It is important to note that these two capacitors **are not** interchangeable, as an electrolytic capacitor used in an AC circuit for more than a few seconds will overheat.

24.6 Shaded-Pole Motor

The **shaded-pole motor** has only one main winding and no start winding or switch, as illustrated in Figure 24-19.

- As in other induction motors, the rotating part is a squirrel-cage rotor.

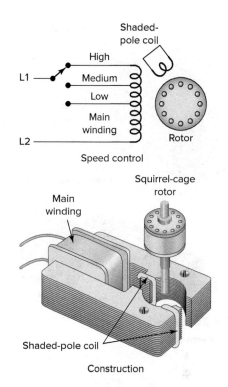

Figure 24-19 Shaded-pole motor.

- Starting is by means of a design that uses a continuous copper loop around a small portion of each motor pole.

- Currents in the copper loop delay the phase of magnetic flux in that part of the pole enough to provide a rotating field.

- This rotating field effect produces a *very low starting torque* compared to other classes of single-phase motors.

- Although direction of rotation is not normally reversible, some shaded-pole motors are wound with two main windings that reverse the direction of the field.

- Slip in the shaded-pole motor is not a problem, as the current in the stator is not controlled by a countervoltage determined by rotor speed, as in other types of single-phase motors.

- Speed can therefore be controlled merely by varying voltage, or through a multitap winding.

- Because of the weak starting torque, shaded-pole motors are built only in small sizes ranging from 1/20 to 1/6 hp. Applications for this type of motor include fans and blowers.

24.7 Universal Motor

The **universal motor,** shown in Figure 24-20, is constructed with series field windings, a wound armature winding, a commutator, and brushes. The series field and armature windings are connected in series with the power supply. Universal motors operate similar to **series DC motors.**

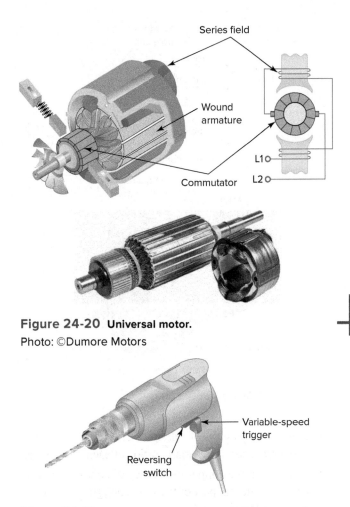

Series field

Wound armature

Commutator

L1

L2

Figure 24-20 Universal motor.
Photo: ©Dumore Motors

Variable-speed trigger

Reversing switch

Figure 24-21 Universal motor speed and direction controls.

Since the field winding and armature are connected in series, both the field winding and armature winding are energized when voltage is applied to the motor. Both windings produce magnetic fields which react to each other and cause the armature to rotate.

Although universal motors are designed to run on *AC or DC,* most are used for household appliances and portable hand tools that operate on single-phase AC power. The universal motor does not operate at a constant speed. The motor runs at a low speed with a heavy load applied and high speed with a light load. Both the *speed* and *direction* of rotation of a universal motor can be controlled, as illustrated in Figure 24-21. Reversing is accomplished by reversing the current flow through the armature with respect to the series field. Varying the voltage that is applied to the motor controls the speed.

Part 2 Review Questions

1. What is the major difference between the starting requirements for a three-phase and a single-phase induction motor?

2. **a.** Outline the starting sequence for a split-phase induction motor.
   b. How is its direction of rotation reversed?

3. Dual-voltage, split-phase motors have leads that allow external connection for different line voltages. How are the start and run windings connected for high and low line voltages?

4. What is the main advantage of capacitor motors over split-phase types?

5. Name the three types of capacitor motor designs.

6. Explain how the shaded-pole motor is started.

7. What type of DC motor is constructed in a manner similar to that of the universal motor?

PART 3   MOTOR CONTROLS

24.8 Motor Protection

Motor *protection* safeguards the motor, the supply system, and personnel from various upset conditions of the driven load, the supply system, or the motor itself. Figure 24-22 illustrates the different types of protection devices associated with a motor branch circuit.

Generally, motor branch circuit conductors that supply a single motor must have an ampacity of not less than *125 percent* of the motor's full-load current rating. This provision is based on the need to provide for a sustained running current that is greater than the rated full-load current and for protection of the conductors by the motor overload protective device set above the full-load current rating.

Overcurrent protection for motors and motor circuits is *different* from that for nonmotor loads. The most common method for providing overcurrent protection for nonmotor loads is to use a circuit breaker that combines overcurrent protection with short-circuit and ground-fault protection. However, this isn't usually the best choice for motors because they draw a large amount of current at initial start-up, usually around 6 times the normal full-load current of the motor. With rare exceptions, the best method for providing overcurrent protection for motors is to separate the overload protection devices from the short-circuit and ground-fault protection devices.

Motor overcurrent protection can be summarized as follows:

- **Short-circuit and ground-fault motor protection.** Branch and feeder fuses and circuit breakers protect motor circuits against the very high current of a short circuit or a ground fault. Fuses and circuit breakers

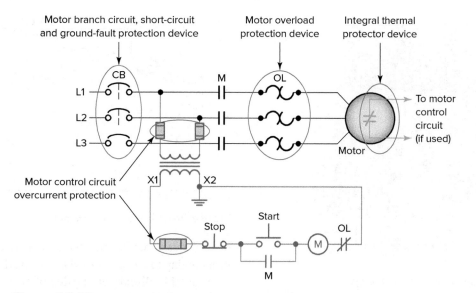

Figure 24-22 **Motor branch circuit protection.**

connected to motor circuits must be capable of ignoring the initial high inrush current and allow the motor to draw excessive current during start-up and acceleration.

- **Overload protection.** Overload devices are intended to protect motors, motor control apparatus, and motor branch circuit conductors against excessive heating due to motor overloads and failure to start. Motor overload may include conditions such as a motor operating with an excessive load or a motor operating with low line voltages or, on a three-phase motor, a loss of a phase. The motor overload devices are most often integrated into the motor starter.

A **thermal overload relay** uses a heater connected in series with the motor supply. If an overload occurs, the heat produced causes a set of contacts to open, interrupting the circuit. The **bimetallic type** of thermal overload relay, illustrated in Figure 24-23, uses a bimetallic strip made up of two pieces of dissimilar metal that are perma-

nently joined by lamination. The operation of the device can be summarized as follows:

- Heat causes the bimetallic strip to bend because the dissimilar metals expand and contract at different rates.
- Overload heating elements connected in series with the motor circuit heat the bimetal tripping elements in accordance to the motor load current.
- The movement/deflection of the bimetallic strip is used as a means of operating the trip mechanism and opening the normally closed overload contacts.

An **electronic overload relay,** such as that shown in Figure 24-24, senses motor current through a current transformer and the heating effect on the motor is computed. If an overload condition exists, the sensing circuit interrupts the power circuit. The tripping current can be adjusted to suit the particular application. Electronic overloads often perform additional protective functions, such as ground-fault and phase loss protection.

Integral thermal motor protection, such as that illustrated in Figure 24-25, pertains to a thermal protector

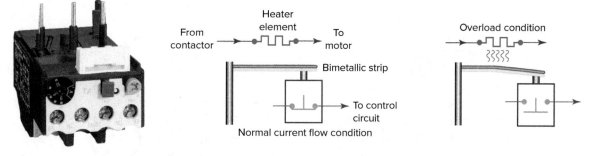

Figure 24-23 **Bimetallic thermal overload relay.**
Photo: Courtesy of Rockwell Automation, Inc.

Figure 24-24 Electronic overload relay.
Courtesy of Rockwell Automation, Inc.

Figure 24-25 Integral thermal motor protection.
©Bodine Electric Company

placed internally within a motor to protect the motor windings from overheating. Connected in series with the stator windings, the thermal overload protector opens the current in the coils if the winding gets too hot. Overload relays protect the motor by monitoring the motor current. They do not, however, monitor the actual amount of *heat generated within the winding.* Motors subject to such conditions as excessive starting cycles, high ambient motor temperatures,

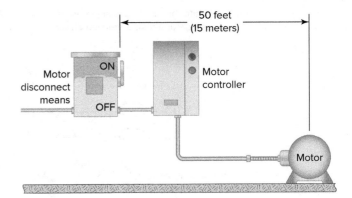

Figure 24-26 Motor disconnect.

or inadequate ventilation conditions may experience rapid heat buildup that is not sensed by the overload relay.

A suitable **disconnect device** of sufficient capacity is required within sight of the motor, in accordance with National Electrical Code requirements (no more than 50 feet), as illustrated in Figure 24-26. The purpose is to open the supply conductors to the motor, allowing personnel to work safely on the installation.

A **two-wire** control circuit provides what is known as **low-voltage release.** Low-voltage release interrupts the circuit when the supply voltage drops below a set value and automatically reestablishes the circuit when the supply voltage is restored. A **three-wire** control circuit provides what is known as **low-voltage protection.** Low-voltage protection also interrupts the circuit when the supply voltage drops below a set value. However, with low-voltage protection, the motor must be manually restarted upon resumption of normal supply voltage.

Basic two-wire and three-wire motor control circuits are shown in Figure 24-27. The operation of each is summarized as follows:

Two-Wire Circuit

• It consists of a normally open, maintained contact device that, when closed, energizes the coil of a magnetic motor starter, which, in turn, energizes the connected motor load.

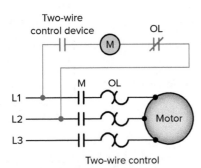
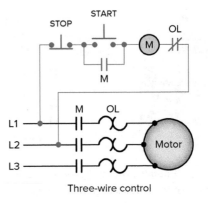

Figure 24-27 Two-wire versus three-wire motor control.

- In the event of a power failure, the magnetic motor starter will deenergize.
- Once power is restored, the magnetic motor starter will automatically reenergize, provided that none of the maintained contact devices have changed state.
- This is an advantage in applications such as refrigeration systems where you do not need someone to restart the equipment after a power failure.
- However, it can be extremely dangerous in applications where equipment starts automatically, placing the operator in danger.

Three-Wire Circuit

- It consists of a normally closed stop button (STOP), a normally open start button (START), a sealing contact (M), and the coil of a magnetic motor starter.
- When the normally open START button is pressed, the coil of the magnetic motor starter is energized. An auxiliary contact seals around the START button to provide a latched circuit.
- Pressing the normally closed STOP button deenergizes the circuit.
- In the event of a power failure, the magnetic motor starter will deenergize.
- Once power is restored the magnetic motor starter will not automatically reenergize. The operator must press the START button to initiate the sequence of operations once again.

24.9 Motor Starting

A full-voltage, or **across-the-line,** motor starter is designed to apply full line voltage to the motor upon starting. Figure 24-28

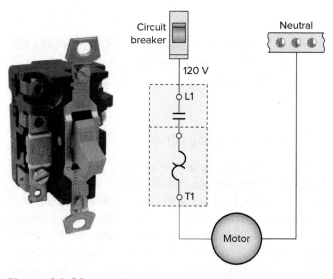

Figure 24-28 Fractional horsepower manual motor starter.
Photo: Courtesy of Schneider Electric

shows a single-pole, fractional horsepower manual motor starter consisting of a ***manually operated*** ON/OFF snap-action switch with overload protection, which works as follows:

- When the switch is moved to the ON or START position, the motor is connected directly across the line and in series with the starter contact and the thermal overload protection device.
- As current flows through the circuit, the temperature of the thermal overload rises, and at a predetermined overload temperature point, the device actuates to open the contact.
- When an overload is sensed, the starter handle automatically moves to the center position to signify that the contacts have opened because of overload and the motor is no longer operating.
- The starter contacts cannot be reclosed until the overload relay is reset manually.
- The starter is reset by moving the handle to the full off position after allowing about 2 minutes for the heater to cool.

Magnetic starters are used with larger motors or where **remote control** is desired. Figure 24-29 shows a typical **three-phase, across-the-line** magnetic starter, the operation of which is summarized as follows:

- The control transformer lowers the voltage to a more common value for control components not rated for the higher voltage.
- When the START button is pressed, coil M energizes to close all M contacts.
- The M contacts in series with the motor close to complete the current path to the motor. These contacts are part of the power circuit and must be designed to handle the full-load current of the motor.
- Memory contact M (connected across the START button) also closes to seal in the coil circuit when the START button is released. This contact is part of the control circuit; as such, it is required to handle the small amount of current needed to energize the coil.
- The starter has three overload heaters, one in each phase. The normally closed (NC) relay contact OL opens automatically when an overload current is sensed on any phase to deenergize the M coil and stop the motor.

Certain applications require a motor to operate in either direction. Interchanging any two leads to a three-phase induction motor will cause it to run in the reverse direction. **Reversing starters** are used to automatically accomplish this phase reversal. The power and control circuits for a

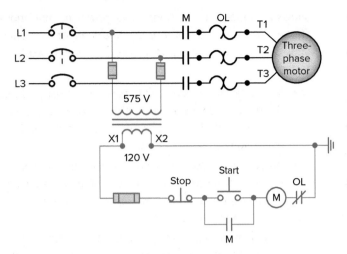

Figure 24-29 Three-phase, across-the-line magnetic starter.
Photo: ©Siemens Industry, Inc.

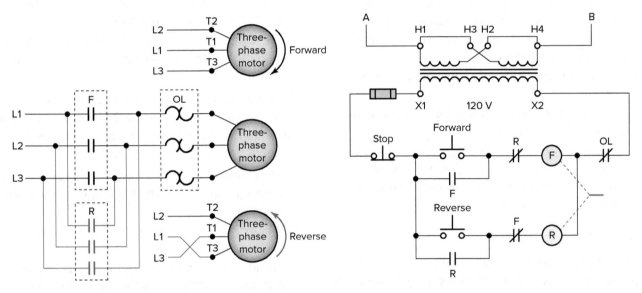

Figure 24-30 Magnetic full-voltage, three-phase reversing motor starter.

magnetic full-voltage, three-phase reversing motor starter are shown in Figure 24-30 and summarized as follows:

- The starter is constructed using 2 three-pole contactors with a single overload relay assembly.

- Power contacts (F) of the forward contactor, when closed, connect L1, L2, and L3 to motor terminals T1, T2, and T3, respectively.

- Power contacts (R) of the reverse contactor, when closed, connect Ll to motor terminal T3 and connect L3 to motor terminal T1, causing the motor to run in the opposite direction.

- When the motor is reversed, it is vital that *both contactors not be energized at the same time.* Activating both contactors would cause a short

circuit. Both mechanical and electrical interlocks are used to prevent the forward and reverse contactors from being activated at the same time.

24.10 Motor Stopping

The most common method of stopping a motor is to remove the supply voltage and allow the motor and load to coast to a stop. In some applications, however, the motor must be stopped more quickly or held in position by some sort of braking device. **Electrical braking** uses the windings of the motor to produce a retarding torque. The kinetic energy of the rotor and the load is dissipated as heat in the rotor bars of the motor.

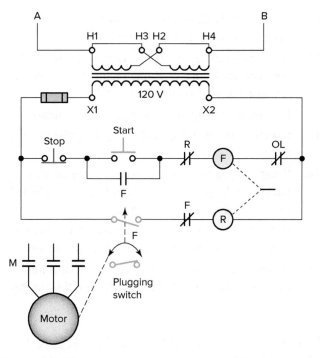

Figure 24-31 Forward-direction plugging circuit.

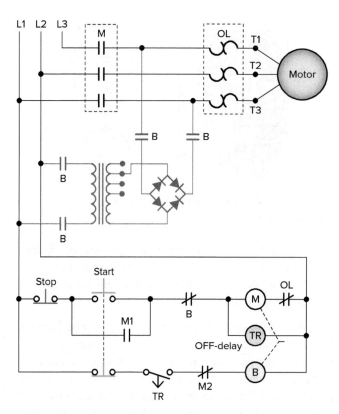

Figure 24-32 DC injection braking applied to an induction motor.

Two means of electrical braking are plugging and dynamic breaking. **Plugging** stops a polyphase motor quickly by connecting the motor for reverse rotation while the motor is still running in the forward direction. Figure 24-31 shows the control schematic for forward-direction plugging, the operation of which is summarized as follows:

- Pressing the START button closes and seals in the forward contactor. As a result, the motor rotates in the forward direction.
- The normally closed auxiliary contact F opens the circuit to the reverse contactor coil.
- The forward contact on the plugging switch closes.
- Pressing the STOP button deenergizes the forward contactor.
- The reverse contactor is energized, and the motor is plugged.
- The motor speed decreases to the setting of the plugging switch, at which point its forward contact opens and deenergizes the reverse contactor.
- This contactor is used only to stop the motor by using the plugging operation; it is not used to run the motor in reverse.

DC injection braking is achieved by removing the AC power supply from the motor and applying direct current to one of the stator phases. The circuit of Figure 24-32 is one example of how DC injection braking can be applied to a

three-phase AC induction motor. The operation of the circuit can be summarized as follows:

- The DC injection braking voltage is obtained from the bridge rectifier circuit, which changes the line voltage from AC to DC.
- Pressing the START button energizes starter coil M and off-delay timer coil TR.
- Normally open M1 auxiliary contact closes to maintain current to the starter coil, and normally closed M2 auxiliary contact opens to open the current path to braking coil B.
- Normally open off-delay timer contact TR remains closed at all times while the motor is operating.
- When the STOP button is pressed, starter coil M and off-delay timer coil TR are deenergized.
- Braking coil B becomes energized through the closed TR contact.
- All B contacts close to apply DC braking power to two phases of the motor stator winding.
- Coil B is deenergized after the timer contact times out. The timing contact is adjusted to remain closed until the motor comes to a complete stop.
- A transformer with tapped windings is used in this circuit to adjust the amount of braking torque applied to the motor.

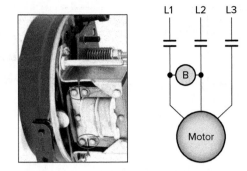

Figure 24-33 AC electromagnetic brake.
Photo: ©Warner Electric LLC

- The motor starter (M) and braking contactor (B) are mechanically and electrically interlocked so that the AC and DC supplies are not connected to the motor at the same time.

Unlike plugging or dynamic braking, **electromechanical friction brakes** can hold the motor shaft stationary after the motor has stopped. Most rely on friction in a drum or disc brake arrangement and are set with a spring and released by a solenoid. These motors are directly coupled to an AC electromagnetic brake, as shown in Figure 24-33. When the power source is turned off, the motor stops instantaneously and holds the load. Most come equipped with an external manual release device, which allows the driven load to be moved without energizing the motor.

Part 3 Review Questions

1. Why are motor branch circuit conductors required to have an ampacity not less than 125 percent of the motor full-load current?

2. Compare the method of providing branch circuit overcurrent protection used for nonmotor and motor loads.

3. List three things motor overload protection guards against.

4. Compare the operation of a thermal overload relay and that of an electronic overload relay.

5. What does integral thermal motor protection pertain to?

6. What is the basic NEC rule with regard to location of a motor disconnect device?

7. In what way is a full-voltage starter designed to start a motor?

8. Compare the way main contacts of a manual and magnetic starter are operated.

9. How is the direction of rotation of a three-phase induction motor reversed?

10. What is the basic component makeup of an electromagnetic reversing motor starter.

11. What undesirable circuit condition would result if both contactors in a reversing motor starter were to become energized at the same time?

12. Explain how plugging is applied to stop a motor.

13. Explain how DC injection braking is applied to stop a motor.

14. What braking function can an electromechanical friction brake perform that cannot be accomplished using plugging or dynamic braking?

15. A two-wire control circuit provides what is known as low-voltage release. Explain what this means.

16. A three-wire control circuit provides what is known as low-voltage protection. Explain what this means.

PART 4   DIRECT CURRENT MOTORS

A DC motor is an electric motor that runs on direct current (DC) electricity. Direct current motors are not used as much as alternating current (AC) types because all electric utility systems deliver alternating current. In general, they are more expensive than AC types, and maintenance of the brush/commutator assembly found on DC motors is significant compared to that of AC motor designs. For special applications, however, it is advantageous to transform the alternating current into direct current in order to use DC motors. DC motors are used for portable battery-operated devices or in large sizes to operate steel rolling mills, cranes, and elevators, which all require high torque and variable speed.

24.11 Permanent-Magnet DC Motor

Permanent-magnet (PM) DC motors use permanent magnets to supply the main field flux and electromagnets to provide the armature flux. Movement of the magnetic field of the armature is achieved by switching current between the armature coils. **Commutation** is the process of switching the field in the armature windings to produce constant torque in one direction, and the **commutator** is a device connected to the armature, which enables this switching of current. Figure 24-34 shows a simplified permanent-magnet motor, the operation of which is summarized as follows:

- Electric connection is made from the DC battery supply to the armature coils through the two brushes and commutator.

- This causes the armature to act as an electromagnet.

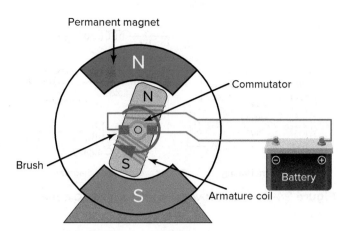

Figure 24-34 **Permanent-magnet DC motor.**

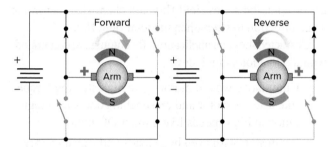

Figure 24-35 **Reversing the direction of rotation of a PM motor.**

- The armature poles are attracted to the permanent magnet field poles of opposite polarity, causing the armature to rotate in a clockwise direction.
- The commutator operates as a switch that reverses the polarity of the applied DC voltage as the armature coil turns past alternate north and south poles to maintain the relative polarity of the armature coils to ensure that torque continuously acts in the same direction.

The direction of rotation of a permanent-magnet DC motor is determined by the direction of the current flow through the armature. Reversing the polarity of the voltage applied to the armature will reverse the direction of rotation, as illustrated in Figure 24-35. Variable-speed control of a PM motor is accomplished by varying the value of the voltage applied to the armature. The speed of the motor varies directly with the amount of armature voltage applied. The higher the value of the armature voltage, the faster the motor will run.

24.12 Series DC Motor

Wound-field DC motors are usually classified as series wound, shunt wound, or compound wound. The connection for a **series-wound DC motor** is illustrated in Figure 24-36. A series-wound DC motor consists of a

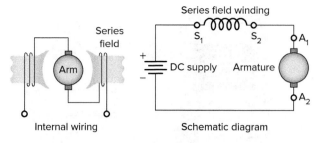

Figure 24-36 **Series-wound DC motor.**

series field winding (identified by the symbols S1 and S2) connected in series with the armature (identified by the symbols A1 and A2). Since the series field winding is connected in series with the armature, it will carry the same amount of current that passes through the armature. For this reason, the windings of the series field are made from heavy-gauge wire that is large enough to carry the full motor load current. Because of the large diameter of the series winding, the winding will have only a few turns of wire and a very low resistance value.

Characteristics and applications for series-wound DC motors include:

- High starting torque is ideal for starting very heavy mechanical loads, such as cranes, hoists, and elevators.
- Poor speed regulation as speed varies widely between no load and rated load.
- **Caution:** The no-load speed of a series motor can increase to the point of damaging the motor. For this reason, it should never be operated without a load of some type coupled to it.

24.13 Shunt DC Motor

The connection for a **shunt-wound DC motor** is illustrated in Figure 24-37. A shunt-wound DC motor consists of a shunt field (identified by the symbols F1 and F2) connected in parallel with the armature. This motor is called a shunt motor because the field is in parallel to, or "shunts," the armature. The shunt field winding is made up of many turns of small-gauge wire and has a much higher resistance and lower current flow compared to a series field winding.

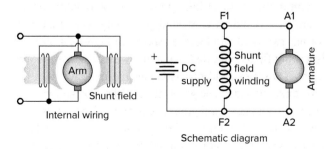

Figure 24-37 **Shunt-wound DC motor.**

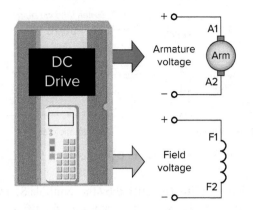

Figure 24-38 Separately excited DC motor and drive.

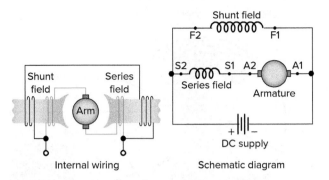

Figure 24-39 Cumulative compound-wound DC motor.

Characteristics and applications for shunt-wound DC motors include:

- It has a low starting torque.
- After the motor reaches full rpm, its torque is at its fullest potential.
- One of the main advantages of a shunt motor is its good speed regulation. It runs almost as fast fully loaded as it does with no load.
- Unlike series motor, the shunt motor will not accelerate to a high speed when no load is coupled to it.
- Shunt motors are particularly suitable for applications such as conveyors, where constant speed is desired and high starting torque is not needed.

A **separately excited DC motor** has the armature and field coils fed from separate supply sources. This type of motor has a field coil similar to that of a self-excited shunt motor. In the most common configuration, armature voltage control is used in conjunction with a constant or variable voltage field excitation. Figure 24-38 shows a typical **DC drive application** for a separately excited DC motor. An advantage to separately exciting the field is that the variable-speed DC drive can be used to provide independent control of the field and armature.

24.14 Compound DC Motor

A **compound-wound DC motor** is a combination of the shunt-wound and series-wound types. This type of DC motor has two field windings, as shown in Figure 24-39. One is a shunt field connected in parallel with the armature; the other is a series field that is connected in series with the armature. The shunt field gives this type of motor the constant-speed advantage of a regular shunt motor. The series field gives it the advantage of being able to develop a large torque when the motor is started under a heavy load. This motor is normally connected **cumulative-compound,** as

shown in Figure 24-39, so that the polarity of the shunt winding is such that it adds to the series fields. As such, under load, the series field flux and shunt field act in the same direction to strengthen the total field flux.

Characteristics and applications for cumulative compound-wound DC motors include:

- Cumulative wound motors give high starting torque like a series motor and reasonable good speed regulation at high speeds like a shunt DC motor.
- It can start with even huge loads and run smoothly after that.
- These motors are generally used where severe starting conditions are met and constant speed is required at the same time.

24.15 Direction of Rotation

The direction of rotation of a wound DC motor depends on the direction of the field and the direction of the current flow through the armature. If either the direction of the field current or the direction of the current flow through the armature of a wound DC motor is reversed, the rotation of the motor will reverse. If both of these two factors are reversed at the same time, however, the motor will continue rotating in the same direction.

For a series-wound DC motor, changing the polarity of either the armature or series field winding changes the direction of rotation. Figure 24-40 shows the power and control circuit schematics for a typical DC motor-reversing starter used to operate a series motor in the forward and reverse directions. In this application, reversing the polarity of the armature voltage changes the direction of rotation.

As in a DC series motor, the direction of rotation of a DC shunt and compound motor can be reversed by changing the polarity of either the armature winding or the field winding. Figure 24-41 shows the power circuit schematics for typical DC shunt and compound motor-reversing starters. The **industry standard** is to reverse the current through the armature while maintaining the current through the shunt

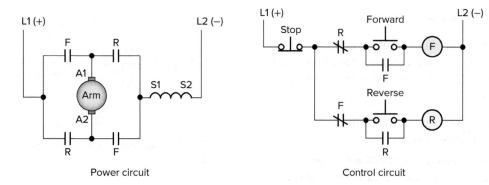

Power circuit   Control circuit

Figure 24-40 DC series motor-reversing starter.

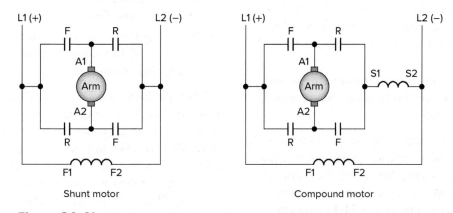

Shunt motor   Compound motor

Figure 24-41 DC shunt and compound motor reversing.

and series field in the same direction. For the compound-wound motor, this ensures a cumulative connection (both fields aiding) for either direction of rotation.

24.16 Motor Counter Electromotive Force (CEMF)

As the armature rotates in a DC motor, the armature coils cut the magnetic field of the stator and induce a voltage, or electromotive force (EMF), in these coils. This occurs in a motor as a by-product of motor rotation and is sometimes referred to as the generator action of a motor. Because this induced voltage opposes the applied terminal voltage, it is called **counter electromotive force,** or **CEMF.** Counter EMF is a form of resistance that opposes and limits the flow of armature current, as illustrated in Figure 24-42.

The overall effect of the CEMF is that this voltage will be subtracted from the terminal voltage of the motor so that the armature motor winding will see a smaller voltage potential. Counter EMF is directly proportional to the speed of the armature and the field strength. At the moment a motor starts, the armature is not rotating, so there is no CEMF generated in the armature. Full-line voltage is applied across the armature, and it draws a relatively large amount of current.

At this point, the only factor limiting current through the armature is the relatively low resistance of the wind-

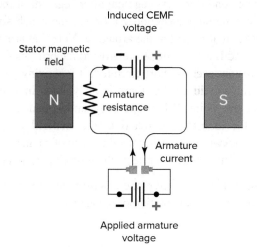

Figure 24-42 Motor counter electromotive force (CEMF).

ings. As the motor picks up speed, a counter electromotive force is generated in the armature, which opposes the applied terminal voltage and quickly reduces the amount of armature current.

When a motor reaches its full no-load speed, it is designed to be generating a CEMF nearly equal to the applied line voltage. Only enough current can flow to maintain this speed. When a load is applied to the motor, its speed will be decreased, which will reduce the CEMF, and more current will be drawn by the armature to drive the load.

24.17 Armature Reaction

In a DC motor, the commutator must reverse current through armature coils that left the influence of one field pole and are approaching the influence of an alternate field pole as follows:

- The width of a brush is made a little more than the width of a commutator segment.
- Whenever a brush spans two commutator segments, it short-circuits the two coils connected to these segments.
- The brushes are aligned so that they make contact with armature winding conductors that are moving parallel to the main field so that there is no voltage induced in them at this point, known as the **neutral plane.**
- Thus, the reversal of current directions in the two short-circuited coils can take place with the least amount of sparking.

The magnetic field produced by current flow through the motor armature conductors tends to distort and weaken the flux coming from the main field poles. Such distortion and field weakening of the stator field of the motor are known as **armature reaction.** Figure 24-43 shows the position of the neutral plane under no-load and loaded motor operating conditions. As segment after segment of the rotating commutator pass under a brush, the brush **short-circuits** coil after coil in the armature. Note that armature coils A and B are positioned relative to the brushes so that at the instant each is short-circuited, it is moving parallel to the main field so that there is no voltage induced in them at this point. When operating under loaded conditions, due to armature reaction, the **neutral plane** is shifted backward, opposing the direction of rotation. As a result, armature reaction affects the motor operation by:

- Shifting the neutral plane in a direction opposite to the direction of rotation of the armature.

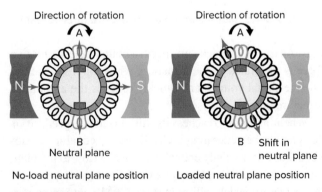

Figure 24-43 Position of the neutral plane under no-load and loaded motor operating conditions.

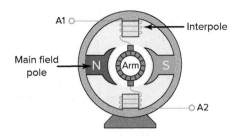

Figure 24-44 Interpoles positioned between the main field poles.

- Reducing motor torque as a result of the weakening of the magnetic field.
- Increasing the arcing at the brushes due to short-circuiting of the voltage being induced in the coils undergoing commutation.

When the load on a DC motor fluctuates, the neutral plane shifts back and forth between no-load and full-load positions. For small DC motors, the brushes are set in an intermediate position to produce acceptable commutation at all loads. In larger DC motors, **interpoles** (also called commutating poles) are placed between the main field poles, as illustrated in Figure 24-44, to minimize the effects of armature reaction. These narrow poles have a few turns of larger-gauge wire connected in series with the armature. The strength of the interpole field varies with the armature current. The magnetic field generated by the interpoles is designed to be equal to and opposite that produced by the armature reaction for all values of load current and improves commutation.

24.18 Motor Speed

Motor **speed regulation** is a measure of a motor's ability to maintain its speed from no load to full load without a change in the applied voltage to the armature or fields. Speed regulation is the ratio of the loss in speed, between no load and full load, to the full-load speed and is calculated as follows (the lower the percentage, the better the speed regulation):

$$= \frac{\text{no-load speed} - \text{full-load speed}}{\text{full-load speed}} \times 100$$

Motor **speed control** is one of the most useful features of a DC motor. DC motors are used in applications where variable speed and strong torque are required. The **base speed** of a DC motor is the speed the motor will operate at when full voltage is applied to both the armature and the field.

- When full voltage is applied to the field and reduced voltage is applied to the armature, the motor operates **below** base speed.

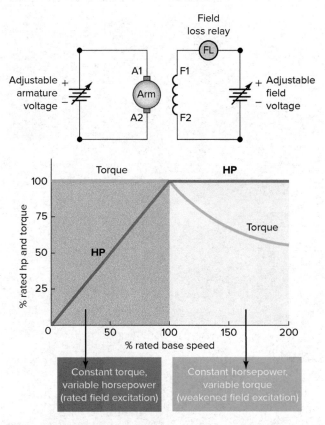

Figure 24-45 Coordinated armature and field voltage control DC motor.

- When full voltage is applied to the armature and reduced voltage is applied to the field, the motor operates **above** base speed.

Coordinated armature and field voltage control is used to maximize the speed range of DC motors, as illustrated in Figure 24-45.

- Initially, the motor is armature voltage-controlled for constant-torque, variable-horsepower operation up to base speed.
- The higher armature voltage, the higher the speed.
- Once base speed is reached, the field-weakening control is applied for constant-horsepower, variable-torque operation to the motor's maximum rated speed.
- The lower the field voltage and current, the higher the speed.
- **Should the motor suffer a loss of field excitation current while operating, the motor will immediately begin to accelerate to the top speed that the loading will allow. This can result in the motor virtually flying apart if it is lightly loaded. For this reason, some form of field loss protection must be provided in the motor control circuit that will automatically stop the motor in the event that current to the field circuit is lost or drops below a safe value.**

1. Give two reasons DC motors are seldom the first motor of choice for some applications.
2. What special types of processes may warrant the use of a DC motor?
3. Explain the function of the commutator in the operation of a DC motor.
4. **a.** How is the direction of rotation of a permanent-magnet DC motor changed?
   b. How is the speed of a permanent-magnet DC motor changed?
5. Why should a DC series motor not be operated without some sort of a load coupled to it?
6. What operating characteristic of a series DC motor makes it ideal choice for motor crane hoists?
7. Series DC motors are said to have very poor speed regulation. What does this mean?
8. Compare the makeup of a shunt field and series field winding.
9. What operating characteristic of a shunt DC motor makes it suitable for applications such as conveyors?
10. How are the series and shunt field windings of the compound-wound DC motor connected relative to the armature?
11. Compare the starting torque and speed regulation of a compound motor with that of the series and shunt motor.
12. In what manner is a cumulative-compound motor connected?
13. How can the direction of a wound DC motor be changed?
14. Explain how CEMF is produced in a DC motor armature.
15. **a.** What is motor armature reaction?
    b. Describe three effects that armature reaction has on the operation of a DC motor.
16. Explain how interpoles minimize the effects of armature reaction.
17. A DC shunt motor is operating with a measured no-load speed of 1,775 rpm. When full load is applied, the speed drops to 1,725 rpm. What is the percentage speed regulation?
18. **a.** How is the base speed of a DC motor defined?
    b. In what manner is the speed of a DC motor controlled below base speed?
    c. In what manner is the speed of a DC motor controlled above base speed?
19. Field loss protection must be provided for DC motors. Why?

CHAPTER TWENTY-FIVE

Electronic Controls

Microchips and electronics on a circuit board
©Gaertner/Alamy Stock photo

Electronic controls are used extensively throughout the electrical industry. Most electronic devices use semiconductor components to perform electronic control. This chapter presents a broad overview of diodes, transistors, thyristors, and integrated circuits (ICs) along with their application in electrical control circuits.

PART 1 ELECTRONIC COMPONENTS

25.1 Diodes

A **diode** is created by joining N- and P-type semiconductor materials together as illustrated in Figure 25-1. Where the materials come in contact with each other, a junction is formed. This device is referred to as a junction diode. Diode leads are identified as the **anode** lead (connected to the P-type material) and the **cathode** lead (connected to the N-type material).

The main operating characteristic of a diode is that it *allows* current in one direction and *blocks* current in the opposite direction. When a voltage is applied to a diode, it is referred to as a **bias voltage.** Figure 25-2 illustrates the two basic operating modes of a diode, forward bias and reverse bias, which can be summarized as follows:

- The diode will either allow or prevent current through the lamp, depending on the polarity of the applied voltage.
- When the polarity of the battery is such that electrons are allowed to flow through the diode, the diode is said to be **forward-biased.**
- When the polarity of the battery is such that electrons are not allowed to flow through the diode, the diode is said to be **reverse-biased.**

- A diode may be thought of like a switch: closed when forward-biased and open when reverse-biased.
- The arrow shown on the diode symbol points in the permitted direction of *conventional current flow (positive + to negative −).* This holds true for all semiconductor symbols possessing arrowheads.

The **light-emitting diode (LED)** is another type of diode device. An LED contains a PN junction that **emits light** when conducting current. Figure 25-3 illustrates a simple LED circuit, the operation of which is summarized as follows:

- When forward-biased, the energy of the electrons flowing through the resistance of the junction is converted directly to light energy.
- Because the LED is a diode, current will flow only when the LED is connected in forward bias.
- The LED is connected in series with a resistor that limits the voltage and current to the desired value.

LED displays come in a variety of packages for different applications. The most familiar is the *seven-segment LED display* for showing numbers, illustrated in Figure 25-4. By the correct segments being energized, the numbers 0 through 9 can be displayed.

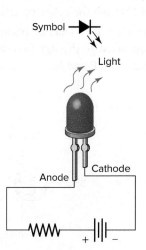

Figure 25-3 Light-emitting diode (LED).

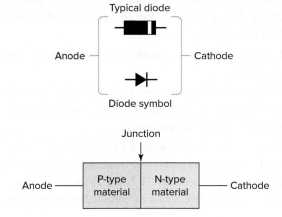

Figure 25-1 PN-junction diode.

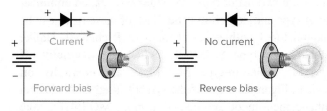

Figure 25-2 Diode forward and reverse biasing.

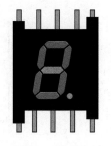

Figure 25-4 Seven-segment LED display.

25.2 Transistors

The **transistor** is a three-terminal semiconductor device primarily used to *amplify* a signal or *switch* a circuit on and off. This is accomplished by using a small amount of electricity to control a much larger supply of electricity. There are two general types of transistors: the bipolar junction transistor and the field-effect transistor.

The **bipolar junction transistor (BJT)** is a three-layer device constructed from two semiconductor diode junctions joined together, as illustrated in Figure 25-5, and summarized as follows:

- It consists of three sections of semiconductors: an **emitter** (E), a **base** (B) and a **collector** (C).
- The base region is very thin, so a small current in this region can be used to control a larger current flowing between the emitter and collector regions.
- The BJT is a **current amplifier** in that a small current flow from the base to the emitter results in a larger flow from the collector to the emitter.
- There are two types of BJT transistors, **NPN** and **PNP**, with different circuit symbols. The letters refer to the layers of semiconductor material used to make the transistor.
- NPN and PNP transistors operate in a similar manner, their biggest difference being the direction of current flow through the collector and emitter.
- Bipolar transistors are so named because the controlled current must go through two types of semiconductor material, P and N.

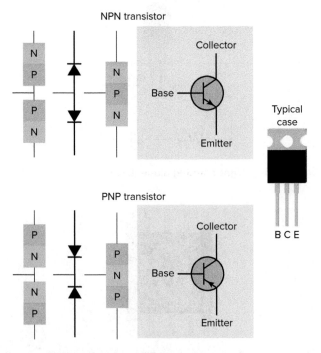

Figure 25-5 Bipolar junction transistor (BJT).

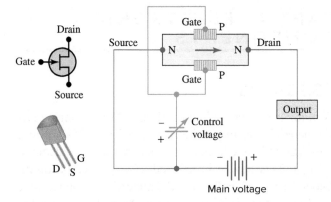

Figure 25-6 Junction field-effect transistor (JFET).

The bipolar junction transistor is a current-controlled device, while the **field-effect transistor (FET)** is a voltage-controlled device. The field-effect transistor uses basically *no input current.* Instead, output current flow is controlled by a varying **electric field.** The FET operates by the effects of an electric field on the flow of electrons through a *single* type of semiconductor material. This is why the FET is sometimes called a unipolar transistor.

Field-effect transistors can be divided into two main types: junction-gate types called **JFETs** and insulated-gate types called **MOSFETs.** A junction field-effect transistor (JFET) is shown in Figure 25-6, and its operation is summarized as follows:

- It is constructed with a bar of N-type material and a gate of P-type material. Because the material in the channel is N type, the device is called an **N-channel JFET.** The arrow indicates the direction of the electron flow through the device.
- JFETs have three connections, or leads: source, gate, and drain.
- The **gate** permits electrons to flow through or blocks their passage by creating or eliminating a channel between the source and drain.
- There are also **P-channel JFETs** that use P-type material for the channel and N-type material for the gate. The main difference between the N and P types is that the polarities of voltage they are connected to are opposite.

The gate of a MOSFET is *insulated* from the channel, resulting in much higher input impedance than the JFET. There are two major types of MOSFETs, called **enhancement type** and **depletion type.** Each of these types can be manufactured with an N channel or P channel. Unlike enhancement-type MOSFETs, which are **normally off** devices, depletion-type MOSFETs are **normally on** devices. Figure 25-7 shows the symbols used for N-channel enhancement-mode and depletion-mode MOSFETs. Note that, unlike the depletion-mode symbol, the line from the

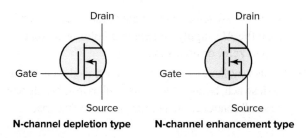

Figure 25-7 N-channel enhancement type and depletion type MOSFETs.

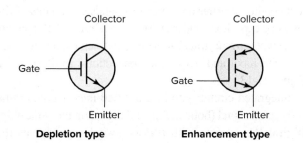

Figure 25-8 An N-channel, insulated-gate bipolar transistor (IGBT).

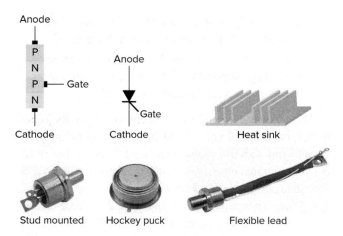

Figure 25-9 Silicon controlled rectifier (SCR).
Photo: Courtesy of Vishay Intertechnology, Inc.

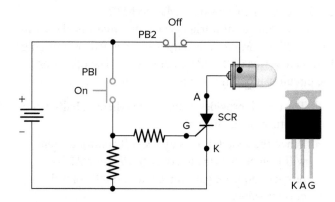

Figure 25-10 SCR operated from a DC source.

source to drain is **broken** for the enhancement-mode symbol. This implies the enhancement type MOSFET is normally off.

An **insulated-gate bipolar transistor (IGBT)** is a cross between a bipolar transistor and MOSFET. The IGBT has the output switching and conduction characteristics of a bipolar transistor but is voltage-controlled like a MOSFET. In general, this means it has the advantages of a high-current-handling capability of a bipolar with the ease of control of a MOSFET. Figure 25-8 shows the symbols used for N-channel, enhancement-type and depletion-type, insulated-gate bipolar transistors.

25.3 Thyristors

Thyristor is a generic term for a broad range of semiconductor components used as an **electronic switch.** Similar to a mechanical switch, thyristors have only two states: ON (conducting) and OFF (not conducting). Thyristors have no linear in-between state as transistors have. In addition to switching, they can also be used to adjust the amount of power applied to a load. Thyristors are mainly used with high currents and voltages. The silicon controlled rectifier and triac are the most frequently used thyristor devices.

Silicon controlled rectifiers (SCRs) are similar to diodes except for a third terminal, or gate, which controls, or turns on, the SCR. Basically, the SCR is a four-layer semiconductor device composed of an anode (A), cathode (K), and gate (G), as shown in Figure 25-9. Common SCR case styles include stud mounted, hockey puck, and flexible lead. SCRs function as switches to turn on or off small or large amounts of power. High-current SCRs that can handle load currents in

the thousands of amperes have provisions for some type of heat sink to dissipate the heat generated by the device.

In function, the SCR has much in common with a diode. Like the diode, it conducts current in only **one direction** when it is **forward-biased** from anode to cathode. It is unlike the diode because of the presence of a **gate** (G) lead, which is used to turn the device on. It requires a momentary positive voltage applied to the gate to switch it on. The schematic of an SCR switching circuit that is operated from a DC source is shown in Figure 25-10. The operation of the circuit can be summarized as follows:

- The anode of the SCR is connected so that it is positive with respect to the cathode (forward-biased).
- Momentarily closing pushbutton PB1 applies a positive voltage to the gate of the SCR, which switches the anode-to-cathode circuit into conduction, thus turning the lamp on.
- Once the SCR is on, it stays on, even after the gate voltage is removed.
- The only way to turn the SCR off is to reduce the anode-cathode current to zero by removing the source voltage from the anode-cathode circuit.

- Momentarily pressing pushbutton PB2 opens the anode-to-cathode circuit to switch the lamp off.
- When operated from an alternating current source, the change of polarity of the source voltage causes the device to automatically switch off.

The **triac** is essentially equivalent to two SCRs joined in reverse parallel (paralleled but with the polarity reversed) and with their gates connected together. The result is a bidirectional electronic switch that can be used to provide load current during **both halves** of an AC supply voltage. Triac connections, shown in Figure 25-11, are labeled main terminal 1 (MT1), main terminal 2 (MT2), and gate (G). The leads are designated this way since the triac acts like two opposing diodes when it is turned on and neither lead always acts like a cathode or an anode. Gate current is used to control current from MT1 to MT2.

The schematic of a triac switching circuit is shown in Figure 25-12. Maximum output is obtained by utilizing both half-waves of the AC input voltage. The operation of the circuit can be summarized as follows:

- The circuit provides random (anywhere in half-cycle) fast turn on of AC loads.
- When the switch is closed, a small control current will trigger the triac to conduct. Resistor R1 is provided to limit gate current to a small control current value.

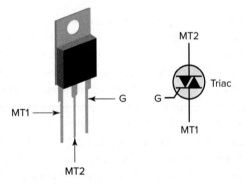

Figure 25-11 Triac.

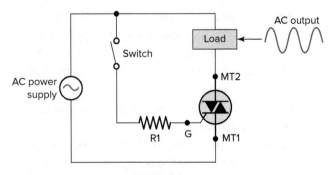

Figure 25-12 Triac switching circuit.

- When the switch is then opened, the triac turns off when the AC supply voltage and holding current drop to zero, or reverse polarity.
- In this way, large currents can be controlled even with a small switch, because the switch will have to handle only the small control current needed to turn on the triac.

25.4 Integrated Circuits (ICs)

An **integrated circuit (IC),** sometimes called a **chip,** is a semiconductor wafer on which thousands or millions of tiny resistors, capacitors, and transistors are fabricated. ICs provide a *complete circuit function* in one small semiconductor package with input and output pin connections, as illustrated in Figure 25-13.

Integrated circuits can be classified into analog, digital, and mixed signal (both analog and digital on the same chip). **Digital** ICs operate with ON/OFF switch-type signals that have only two different states, called low (logic 0) and high (logic 1). **Analog** ICs contain amplifying-type circuitry and signals capable of an *unlimited number of states.*

The analog and digital processes can be seen in a simple comparison between a light dimmer and a light switch. A light dimmer involves an analog process, which varies the intensity of light from off to fully on. The operation of a standard light switch, on the other hand, involves a digital process; the switch can be operated only to turn the light off or on.

An **operational amplifier (op-amp)** is a commonly used analog IC that can perform a wide variety of processing tasks, including amplifying weak signals from sensors. Typically, an op-amp has two inputs called + and − (or VIN+ and VIN−) and a single output. Negative feedback is used to control the gain of the overall op amp circuit.

The symbol and an inverting op-amp circuit is shown in Figure 25-14. The circuit operation is summarized as follows:

- Resistors values R_{in} and R_{fb} set the voltage gain of the amplifier.
- Resistor R_{in} is called the input resistor.
- Resistor R_{fb} is called the feedback resistor.

Figure 25-13 Integrated circuit (IC).

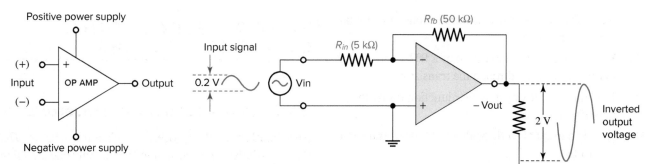

Figure 25-14 **Op-amp inverting amplifier.**

- The ratio of the resistance value of R_{fb} to that of the resistance R_{in} sets the voltage gain as follows:

$$\text{Op-amp voltage gain} = \frac{R_{fb}}{R_{in}}$$

$$= \frac{50 \text{ k}\Omega}{5 \text{ k}\Omega}$$

$$= 10$$

- According to the calculated gain (10), applying an AC signal of 0.2 volt peak-to-peak will produce an inverted amplified signal of 2 volts peak-to-peak at the output.
- You can set the gain by simply using different values for R_{in} and R_{fb}.

Electrostatic discharge (ESD) is defined as the transfer of electrostatic charges between bodies at different potential caused by *direct contact* or an *induced* electrostatic field, as illustrated in Figure 25-15. All integrated circuits are sensitive to electrostatic discharge to some degree. If static discharge occurs at a sufficient magnitude, some damage or degradation (the IC is weakened and will often fail later) will usually occur. This damage is mainly due to the current flowing through ICs during discharge or by high voltage penetrating the insulation.

Precautions that should be taken when working on integrated circuits include:

- Never handle sensitive ICs by their leads.
- Keep your work area clean, especially of common plastics.

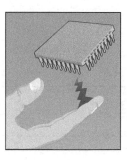

Figure 25-15 **Electrostatic discharge (ESD).**

Figure 25-16 **Antistatic wrist strap band.**
©Andrew_Howe/iStock/Getty Images

- Handle printed circuit boards by their outside corners.
- Always transport and store sensitive ICs and control boards in antistatic packaging.
- Wear an antistatic wrist strap band that connects you to ground, as shown in Figure 25-16, when working on deenergized circuits.

Part 1 Review Questions

1. What is the operating characteristic of a diode?
2. Compare the forward-bias and reverse-bias operating modes of a diode.
3. In what direction does the arrowhead on the symbol of the diode point?
4. Under what operating condition does an LED emit light?
5. Explain how energized segments are selected for display on a seven-segment LED.
6. What two main functions do transistors perform?

7. Name the three semiconductor sections of a bipolar junction transistor.

8. Which semiconductor section of a BJT controls the main current flow through the transistor.

9. Explain in what way a bipolar junction transistor amplifies current.

10. How is current flow through a field-effect transistor controlled?

11. **a.** Name the three leads found on a junction field-effect transistor.
   b. To which two leads is the input control voltage connected?

12. Why are field-effect transistors said to be unipolar?

13. In what way is the gate of a MOSFET constructed different from that of a JFET?

14. In what way is the operation of a depletion-mode MOSFET different from that of an enhancement-mode MOSFET?

15. In what way is the insulated-gate bipolar transistor operating characteristics similar to BJT and MOSFET types?

16. In what way is the operation of a thyristor similar to that of a mechanical switch?

17. What are the similarities and differences between an SCR and a diode?

18. Compare the way in which the control of an SCR is different when operated from an AC supply than when operated from a DC supply.

19. SCRs are unidirectional devices, while triacs are bidirectional. What does this mean as far as the operation of each concerned?

20. Describe the makeup of an integrated circuit.

21. Compare the type of signals used to operate analog and digital IC circuitry.

22. In what way can an electrostatic discharge do damage to an integrated circuit?

23. For the op-amp circuit shown, calculate the voltage gain (see Figure 25-17).

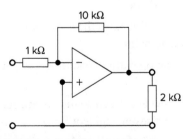

Figure 25-17 Circuit for problem 23.

PART 2 ELECTRONIC CONTROL DEVICES

25.5 Diode Rectifiers and Inverters

Rectification is the process of changing **AC to DC.** Because **diodes** allow current to flow in only one direction, they can be used as rectifiers. Figure 25-18 shows a **half-wave rectifier** circuit, the operation of which is summarized as follows:

- The AC input is applied to the primary of the transformer; the secondary voltage supplies the rectifier and load.
- During the positive half-cycle of the AC input wave, the anode side of the diode is positive.
- The diode is then forward-biased, allowing it to conduct a current to the load. Because the diode acts as a closed switch during this time, the positive half-cycle of the AC waveform is developed across the load.
- During the negative half-cycle of the AC input wave, the anode side of the diode is negative.
- The diode is now reverse-biased; as a result, no current can flow through it. The diode acts as an open switch during this time, so no voltage is produced across the load.
- Thus, applying an AC voltage to the circuit produces a pulsating DC voltage across the load.

The half-wave rectifier makes use of only half of the AC input wave. A less pulsating and greater average direct current can be produced by rectifying both half-cycles of the AC input wave, as illustrated in Figure 25-19, using a full-wave bridge rectifier. Such a rectifier circuit is known as a full wave. The bridge

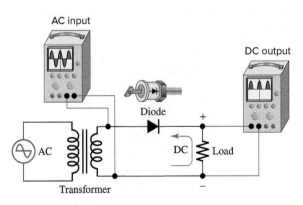

Figure 25-18 Half-wave rectifier circuit.

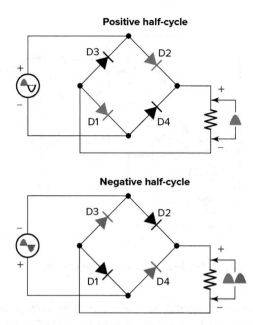

Positive half-cycle

D3   D2

D1   D4

Negative half-cycle

D3   D2

D1   D4

Figure 25-19 Full-wave rectifier circuit.

rectifier makes use of four diodes and its operation is summarized as follows:

- During the positive half-cycle, the anodes of Dl and D2 are positive (forward-biased), whereas the anodes of D3 and D4 are negative (reverse-biased).
- Electron flow is from the negative side of the line, through D1, to the load, then through D2, and back to the other side of the line.
- During the next half-cycle, the polarity of the AC line voltage reverses.
- As a result, diodes D3 and D4 become forward-biased. Electron flow is now from the negative side of the line through D3, to the load, then through D4, and back to the other side of the line.
- Note that during this half-cycle, the current flows through the load in the same direction, producing a full-wave pulsating direct current.

An inverter performs the opposite function of a rectifier. **Inverters** convert direct current (DC) to alternating current (AC) by using electronic circuits. Typical application include:

- Converting a 12-V battery voltage into conventional household 120-VAC voltage.
- Conversion of the variable DC output of a solar panel into a clean sinusoidal 50- or 60-Hz AC current that is then applied directly to the commercial electric grid or to a local, off-grid electrical network.
- Part of an uninterruptible power supply (UPS), such as that shown in Figure 25-20, used to deliver backup power to computers.

Figure 25-20 Inverter example: an uninterruptible power supply.

©Gitanna/iStock/Getty Images

25.6 Transistor Switching and Amplification

When a **transistor** is used as a *switch,* it has only two operating states, ON and OFF. Figure 25-21 shows an example of a transistor switching circuit. The operation of the circuit is summarized as follows:

- A low-power transistor is used to switch the current for the relay's coil.
- With the proximity sensor switch open, *no base or collector current* flows, so the transistor is switched off.
- The relay coil will be deenergized, and voltage to the load will be switched off by the normally open relay contacts.
- When the transistor is in the OFF state, the collector current is zero, the voltage drop across the collector and emitter is 12 V, and the voltage across the relay coil is almost zero or negligible.
- The proximity sensor switch, on closing, establishes a small base current that drives the collector fully on.

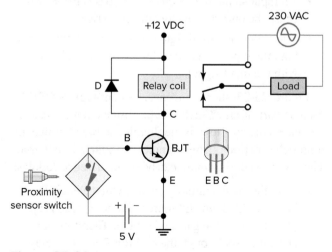

Figure 25-21 Transistor switching circuit.

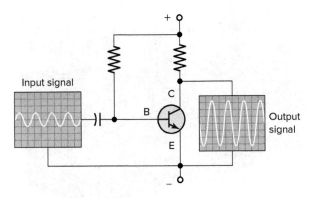

Figure 25-22 Transistor amplifier circuit.

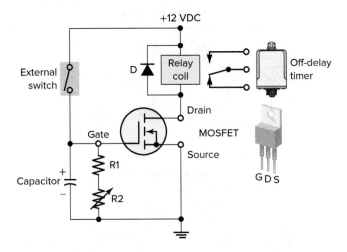

Figure 25-23 MOSFET off-delay timer circuit.

- As a result, the relay coil is energized, and its normally open contacts close to switch on the load.
- When the transistor is in the ON state, collector current is at its maximum value and the voltage across the collector and emitter drops to near zero while that across the relay coil increases to approximately 12 V.
- Inductive loads, such as the coils of relays and solenoids, produce a high transient voltage at turnoff. The diode is connected in reverse bias to suppress this countervoltage (a voltage in reverse polarity) when the coil goes from the ON to OFF state from becoming high enough to damage the transistor.

In general terms, an **amplifier** is a device that takes a signal and increases its amplitude. Figure 25-22 shows an example of a transistor amplifier circuit. The operation of the circuit is summarized as follows:

- The input signal is injected between the base and emitter.
- The output voltage is taken from between the collector and emitter.
- An input signal that increases the base current will cause the output collector voltage to fall.
- An input signal that decreases the base current will cause the output collector voltage to rise.
- The collector output signal is greater in amplitude than the input signal and 180 degrees out of phase with the input signal.

Figure 25-23 shows an enhancement-type MOSFET used as part of an **off-delay cube timer** circuit. Because the gate current flow is negligible, a broad range of time-delay periods, from minutes to hours, is possible. The operation of the circuit can be summarized as follows:

- With the switch initially open, a voltage is applied between the drain and source, but no voltage is applied between the gate and source. Therefore, no current flows through the MOSFET, and the relay coil will be deenergized.

- Closing the switch results in a positive voltage being applied to the gate, which triggers the MOSFET into conduction to energize the relay coil and switch the state of its contacts.
- At the same time the capacitor is charged to 12 VDC.
- The circuit remains in this state with the relay coil energized as long as the switch remains closed.
- When the switch is opened, the timing action begins.
- The positive gate circuit to the 12-V source is opened.
- The stored positive charge of the capacitor keeps the MOSFET switched on.
- The capacitor begins to discharge its stored energy through R1 and R2 while still maintaining a positive voltage at the gate.
- The MOSFET and relay coil continue to conduct a current for as long as it takes the capacitor to discharge.
- The discharge rate, and thus the off-delay timing period, is adjusted by varying the resistance of R2. Increasing the resistance will slow the rate of discharge and increase the timing period. Decreasing the resistance will have the opposite effect.

25.7 Thyristor Control Applications

Silicon controlled rectifiers (SCRs) are used in reduced-voltage motor starters to reduce the amount of voltage delivered to a motor on starting. Figure 25-24 shows a *three-phase, SCR reduced-voltage starter* power circuit made up of two contactors: a start contactor (C1) and a run contactor (C2). The operation of the circuit can be summarized as follows:

- For this application the SCR delivers a varying amount of power to the motor by means of *phase angle control.*

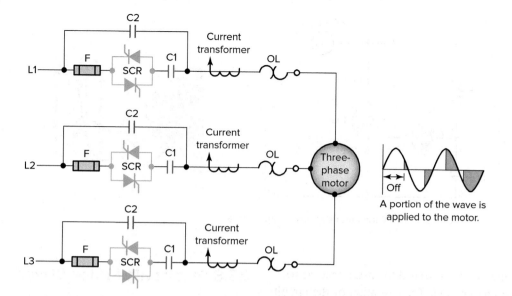

Figure 25-24 Three-phase, SCR reduced-voltage starter power circuit.

- When the motor is first started, the start contacts (C1) close, and reduced voltage is applied to the motor through the *reverse-parallel*-connected SCRs.
- Triggering of the SCRs is controlled by control circuits that chop the applied sine-wave system power so that only a *portion* of the wave is applied to the motor.
- The voltage is then automatically increased until the motor is at full-line voltage.
- At this point the run contacts (C2) close, and the motor is connected directly across the line and runs with full power applied to the motor terminals.

The **triac** is a bidirectional electronic switch, which means it can conduct current in either direction. This operating characteristic makes the triac an ideal component for switching AC power loads. Figure 25-25 illustrates how a triac is used to switch the AC voltage to control the ON/OFF state of a lamp. The output module of a programmable logic controller (PLC) serves as the link between the PLC's microprocessor and field load devices. An optical isolator separates the output signal from the PLC processor circuit from the field load devices. The operation of the circuit can be summarized as follows:

- As part of its normal operation, the processor sets the outputs on or off according to the logic program.
- When the processor calls for the lamp to be on, a small voltage is applied across the LED of the optical isolator.
- The LED emits light, which switches the phototransistor into conduction.
- This in turn switches the triac into conduction to turn on the lamp.

In addition to switching, a triac can be used to *vary* the amount of power supplied to an AC load, as illustrated in Figure 25-26. When used for this type of application, a

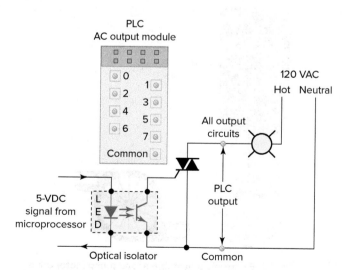

Figure 25-25 Triac switching of a PLC output module.

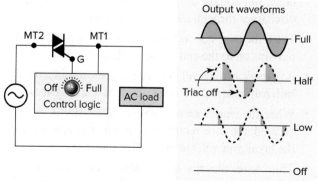

Figure 25-26 Triac variable-power circuit.

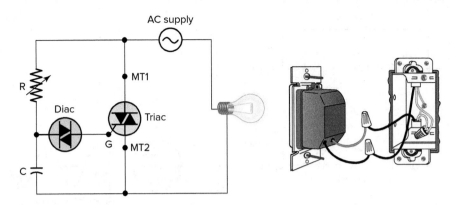

Figure 25-27 Triac incandescent lamp dimmer.

control triggering circuit is needed to ensure that the triac conducts at the proper time. The operation of the circuit can be summarized as follows:

- The trigger circuit controls the point on the AC waveform at which the triac is switched on. It proportionally turns on a percentage of each power line half-cycle.
- The resulting waveform is still alternating current, but the average current value is adjustable.
- Since the trigger can cause it to trigger current in either direction, it is an efficient power controller from essentially zero to full power.

Most lamp dimmer switches are manufactured with a triac as the power control device. A simplified triac incandescent lamp dimmer circuit is shown in Figure 25-27. The operation of the circuit can be summarized as follows:

- The **diac** is a two-terminal device that behaves like two diodes connected in opposite directions. Current flows through the diac whenever the voltage across it reaches its rated **breakover voltage.**
- With the variable resistor set to its lowest value, the capacitor will charge rapidly at the beginning of each half-cycle of the AC voltage.
- When the voltage across the capacitor reaches the breakover voltage of the diac, the capacitor voltage discharges through the gate of the triac.
- Thus, the triac conducts early in each half-cycle and remains on to the end of each half-cycle.
- As a result, current will flow through the lamp for most of each half-cycle and produce maximum lamp brightness.
- When the resistance of the variable resistor is increased, the time required to charge the capacitor to the breakover voltage of the diac increases.
- This causes the triac to fire *later* in each half-cycle. So, the amount of time current flows through the lamp is reduced, and less light is emitted.

25.8 Electronic Motor Drives

The primary function of an **electronic motor drive** is to control the speed, torque, direction, and resulting horsepower of a motor. Unlike constant-speed systems, the adjustable-speed drive permits the selection of an infinite number of speeds within its operating range. As an example, the use of adjustable-speed electronic drives in pump and fan systems can greatly increase their efficiency. For example, in systems that use throttles or dampers to interrupt the flow as a means of control dissipate useless energy. Running a system this way is like driving a car with the accelerator pressed to the floor while controlling speed with the brake. An electronic adjustable-speed drive, such as that shown in Figure 25-28, allows precise control of motor output with significant savings in the power required to handle the load.

Induction motors, the workhorses of industry, rotate at a fixed speed that is determined in part by the frequency of the supply voltage. The preferred method of speed control for induction motors is to alter the **frequency** of the supply

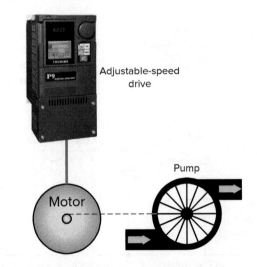

Figure 25-28 Electronic adjustable-speed pump motor drive.
Photo: ©Toshiba International Corporation

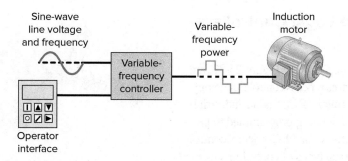

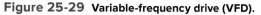

Figure 25-29 **Variable-frequency drive (VFD).**

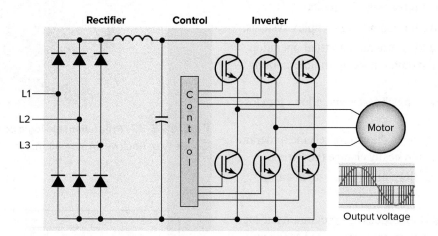

Figure 25-30 **The three sections of a variable-frequency drive.**

voltage. Since the basis of the drive's operation is to vary the frequency to the motor in order to vary the speed, the most common name for the system is the **variable-frequency drive (VFD).** VFDs convert the fixed-frequency supply voltage to a continuously variable frequency, as illustrated in Figure 25-29, thereby allowing adjustable motor speed.

Figure 25-30 shows a simplified diagram of the three sections of a variable-frequency drive. The operation of each section is summarized as follows:

- **Rectifier section.** The full-wave, three-phase diode rectifier converts the 60-Hz power from a standard utility supply to either fixed or adjustable DC voltage.
- **Inverter section.** Electronic switches—power transistors or thyristors—switch the rectified DC on and off, and produce a current or voltage waveform at the desired new frequency.
- **Control section.** An electronic circuit receives feedback information from the driven motor and adjusts the output voltage or frequency to the selected values. Usually, the output voltage is regulated to produce a constant ratio of voltage to frequency (V/Hz). Controllers may incorporate many complex control functions.

Converting DC to variable-frequency AC is accomplished using an inverter. Most currently available inverters use pulse

width modulation (PWM), illustrated in Figure 25-31, because the output current waveform closely approximates a sine wave. The operation of a PWM inverter is summarized as follows:

- Power semiconductors switch DC voltage at high speed, producing a series of short-duration pulses of constant amplitude.
- Output voltage is varied by changing the width and polarity of the switched pulses.
- Output frequency is adjusted by changing the switching cycle time.
- The resulting current in an inductive motor simulates a sine wave of the desired output frequency.

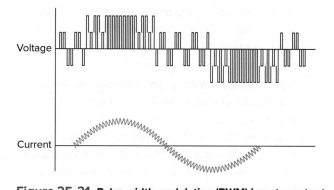

Figure 25-31 **Pulse width modulation (PWM) inverter output.**

25.9 Programmable Logic Controllers (PLCs)

Programmable logic controllers, shown in Figure 25-32, are now the most widely used industrial process control technology. A **programmable logic controller (PLC)** is an industrial-grade computer that is capable of being programmed to perform control functions. Benefits of a PLC system control compared with other conventional types of control include:

- It eliminates much of the hard-wiring associated with conventional relay control circuits.
- It is cost-effective for controlling complex systems.
- When running a PLC program, a visual operation can be seen on a monitor, making troubleshooting simpler.
- A PLC is relatively much faster and allows for closer process tolerances.
- It allows the changing of the control scheme (program) without having to physically change the wiring.
- Computational abilities allow more sophisticated control.

Figure 25-33 shows the sections of a programmable logic controller. Its basic functions are summarized as follows:

- **Power supply.** The power supply of a PLC system converts the available line voltage into low-voltage DC required for the internal circuitry.
- **Processing unit.** The central processing unit (CPU), also called a processor, and associated memory form the intelligence of a PLC system. The CPU evaluates the status of inputs, outputs, and other data as it executes a stored program. The CPU then sends signals to update the status of the outputs.
- **Input section.** The input section receives signals from input field devices, such as switches and sensors, and converts them to logic signals that can be used by the CPU.
- **Output section.** This section controls the system by operating output field devices, such as motor starters, contactors, solenoids, and the like. They convert control signals from the CPU into digital or analog values that can be used to control various load devices.
- **Programming device.** A programming device is used to enter or change the PLC's program or to monitor or change stored values. Once entered, the program is downloaded and stored in the PLC's memory. A personal computer is the most commonly used programming device and communicates with the PLC via a communications port.

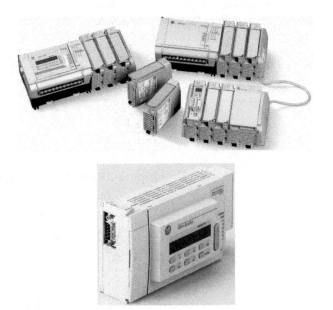

Figure 25-32 Programmable logic controllers (PLCs).
Courtesy of Rockwell Automation, Inc.

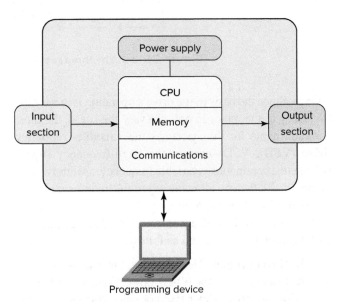
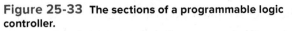

Figure 25-33 The sections of a programmable logic controller.

The programmable logic controller **program** consists of a series of instructions that direct the PLC to execute actions. A programming language provides rules for combining the instructions so that they produce the desired actions. **Relay ladder logic** is the most common programming language used with PLCs. Its origin is based on electromechanical relay control. The ladder logic program graphically represents rungs of contacts, coils, and special instruction blocks.

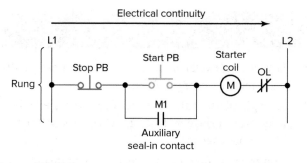

Figure 25-34 Hardwired motor START/STOP circuit.

Instruction	Symbol	State
XIC Examine if closed	⊣ ⊢	If the input device is **open** the instruction is **false** If the input device is **closed** the instruction is **true**
XIO Examine if open	⊣/⊢	If the input device is **open** the instruction is **true** If the input device is **closed** the instruction is **false**
OTE Output energize	⊣( )⊢	If the rung has logic **continuity** the output is **energized** If the rung does **not** have logic continuity the output is **deenergized**

Figure 25-35 Basic PLC instructions.

Figure 25-34 shows the traditional electrical diagram for a hardwired motor START/STOP circuit, the operation of which is summarized as follows:

- The rung is said to have electrical continuity whenever a current path is established between L1 and L2.
- Pressing the START pushbutton results in electrical continuity to energize the starter coil (M) and close the seal-in contact (M1).
- After the START pushbutton is released, electrical continuity is maintained by the seal-in contact.
- When the STOP pushbutton is pressed, electrical continuity is lost, and the starter coil deenergizes and contact M1 opens.

On an electrical diagram, the symbols represent real-world devices. In the electrical diagram, the electrical states of the devices are described as being OPEN/CLOSED or OFF/ON. In the ladder logic program, instructions are either FALSE/TRUE or binary 0/1. The three basic PLC instructions, as illustrated in Figure 25-35, are:

XIC Examine if closed instruction

XIO Examine if open instruction

OTE Output energize instruction

Each of the input and output connection points on a PLC has an **address** associated with it. This address will indicate which PLC input is connected to which input device and which PLC output will drive which output device. Addressing formats vary between PLC manufacturers. PLCs with fixed input and outputs typically have all their input and output locations predefined.

Figure 25-36 illustrates how the basic PLC instructions are applied in a **programmed** START/STOP motor circuit. The operation of the program can be summarized as follows:

- The normally closed STOP pushbutton is closed, making the stop instruction (I1) true.
- Closing the START pushbutton makes the start instruction (I2) true and establishes logical continuity of the rung.
- Rung logic continuity energizes the motor starter coil.
- The starter auxiliary contact M1 closes, making its instruction (I3) true.
- After the START pushbutton is released, electrical continuity is maintained by the true I3 instruction.

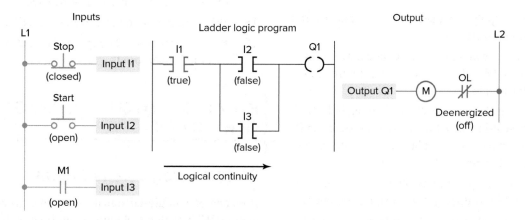

Figure 25-36 Programmed motor START/STOP circuit.

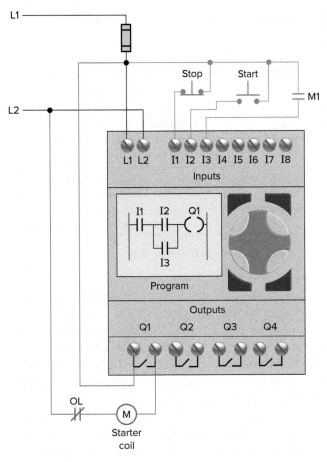

Figure 25-37 PLC wiring diagram for motor START/STOP circuit.

1. Give a brief explanation of how a single-phase, half-wave rectifier changes AC to DC.

2. A single-phase, half-wave rectifier is replaced with a full-wave bridge rectifier. In what ways will the DC output change?

3. Compare the function of a rectifier with that of an inverter.

4. Compare the operation of transistor switching and amplifying devices.

5. What operating characteristic of a MOSFET is utilized to provide long time-delay periods for electronic timers?

6. What type of SCR control is utilized to provide a varying amount of power for a three-phase, reduced-voltage starter?

7. What operating characteristic of a triac makes it an ideal electronic switch for switching AC power loads?

8. Explain how a diac is utilized to control power in a triac lamp dimmer circuit.

9. What is the primary function of an electronic motor drive?

10. Give a brief explanation of how an adjustable-speed electronic motor drive can increase the electrical efficiency of a motor pumping system.

11. List the three major sections of an electronic variable-frequency drive, and state the main function performed by each.

12. What is a programmable logic controller (PLC)?

13. When compared to conventional hardwired control systems, PLC systems are much easier to troubleshoot. Why?

14. List the five major parts of a PLC system, and state the main function performed by each.

15. What are the three basic PLC instructions?

16. By what means are the input and output connections to a PLC indentified?

PART 3 DIGITAL LOGIC CIRCUITS

25.10 Analog versus Digital

Both analog and digital signals are used to transmit information. With analog technology, information is translated into electric pulses of varying amplitude. In digital

Figure 25-37 illustrates typical PLC wiring, with input/output notation, designed to implement the motor START/STOP control. A fixed controller with eight predefined fixed inputs (I1 to I8) and four predefined fixed relay outputs (Q1 to Q4) is used to control and monitor the motor START/STOP operation. The wiring is completed as follows:

- The power supply is connected to terminals L1 and L2 of the controller.
- Q1 normally open output relay contact, M starter coil, and the OL relay contact are hard-wired in series with L1 and L2.
- The STOP pushbutton, START pushbutton, and M1 auxiliary seal-in contact inputs are connected to I1, I2, and I3, respectively, while the motor starter coil is connected to output Q1.
- The ladder logic program is entered using the front keypad and LCD display or a personal computer.
- Power is applied, and the PLC is placed in the run mode to operate the system.

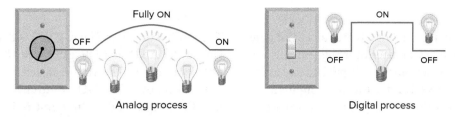

Figure 25-38 Analog versus digital process.

technology, translation of information is representative by two distinct amplitudes. The difference between analog and digital processes can be seen in a simple comparison between a light dimmer and a light switch circuit.

- A lamp dimmer involves an **analog** process, which varies the intensity of light from fully OFF to fully ON over a range of brightness (Figure 25-38).

- The operation of a lamp switch involves a **digital** process, meaning that the switch can be operated only to turn the light fully OFF or fully ON.

Digital logic is the foundation of many electronic devices and control systems. All digital equipment operates on the **binary principle.** The term *binary principle* refers to the idea that some things can be thought of as existing in **only one of two states.** These states are 1 and 0. The 1 and 0 can represent ON or OFF, open or closed, true or false, high or low, or any other two conditions. The key to the speed and accuracy with which binary information can be processed is that there are only two states, each of which is distinctly different. There is no in-between state, so when information is processed the outcome is either yes or no.

25.11 Logic Gates

The **logic gate** is an elementary building block of a digital circuit. A logic gate may have a number of inputs but only one output that is activated by particular combinations of input conditions. The two-state binary concept, applied to gates, is the basis for making decisions. The operations performed by digital equipment are based on three fundamental logic functions: AND, OR, and NOT. Each gate

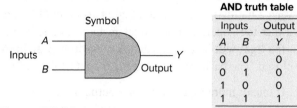

Figure 25-39 AND gate.

AND truth table		
Inputs		Output
A	B	Y
0	0	0
0	1	0
1	0	0
1	1	1

function has a **rule** that will determine outcome and a **symbol** that represents the operation.

AND Gate

The **symbol** drawn in Figure 25-39 is that of an **AND gate.** An AND gate is a device with two or more inputs and one output. The **rule** that determines its output is stated as follows: **The AND gate output is 1 only if all inputs are 1.** Logic gate truth tables show each possible input to the gate or circuit and the resultant output depending upon the combination of the input(s). Since logic gates are digital ICs (integrated circuits), their input and output signals can be in only one of two possible digital states—that is, logic 0 or logic 1. Thus, the logic state of the output of a logic gate depends on the logic states of each of its individual inputs.

The AND logic gate operates similarly to control devices connected in **series.** Figure 25-40 shows the comparison between a hardwired and digital logic two-input AND gate. Note that they operate in a similar fashion. For each circuit, the light will be on only when both **switch A and switch B** are closed. The basic rules that apply to an AND gate are:

- If all inputs are 1, the output will be 1.
- If any input is 0, the output will be 0.

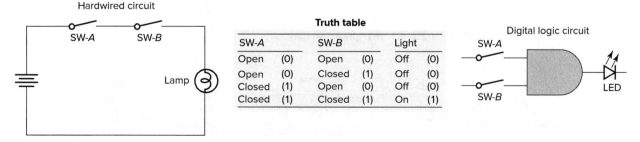

Truth table						
SW-A		SW-B		Light	
Open	(0)	Open	(0)	Off	(0)
Open	(0)	Closed	(1)	Off	(0)
Closed	(1)	Open	(0)	Off	(0)
Closed	(1)	Closed	(1)	On	(1)

Figure 25-40 Hardwired versus digital logic two-input AND gate.

OR Gate

The **symbol** drawn in Figure 25-41 is that of an **OR gate.** An OR gate can have any number of inputs but only one output. The **rule** that determines its output is stated as follows: **The OR gate output is 1 if one or more inputs are 1.** The truth table shows the resulting output Y from each possible input combination.

The OR logic gate operates similarly to control devices connected in **parallel.** Figure 25-42 shows the comparison between a hardwired and digital logic two-input OR gate. Note that they operate in a similar fashion. For each circuit, the light will be on if switch **A or switch B or both** are closed. The basic rules that apply to an OR gate are:

- If one or more inputs are 1, the output is 1.
- If all inputs are 0, the output will be 0.

NOT Gate

The **symbol** drawn in Figure 25-43 is that of a **NOT gate.** Unlike the AND and OR gates, the NOT gate can have **only one input.** The rule that determines its output is stated as follows: **The NOT output is 1 if the input is 0, and the output is 0 if the input is 1.** The result of the NOT operation is always the inverse of the input, and the NOT gate is, therefore, called an **inverter.** The NOT gate is often depicted by using a **bar** across the top of the letter, indicating an inverted output. The small **circle** at the output of the inverter is referred to as a bubble and indicates that an inversion of the logical function has taken place.

Figure 25-44 shows an example of the inverting action of the NOT gate as applied to the operation of a normally open pushbutton input. When the pushbutton is not pressed (representing the absence of a signal, or 0) the output LED will be on (1). When the pushbutton is pressed (representing the presence of a signal, or 1) the output LED will be off (0). The basic rules that apply to an NOT (inverter) gate are:

- If the input is 1, the output is 0.
- If the input is 0, the output will be 1.

An AND gate with an inverted output is called a **NAND** gate. The NAND gate operates as an AND gate followed by a NOT gate. The NOT symbol placed at the output of an

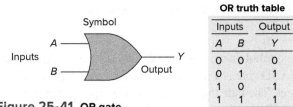

OR truth table

Inputs		Output
A	B	Y
0	0	0
0	1	1
1	0	1
1	1	1

Figure 25-41 OR gate.

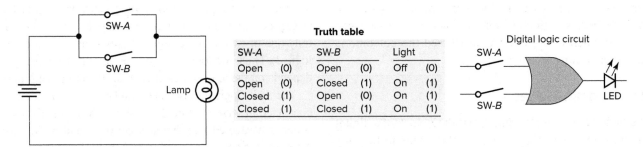

Figure 25-42 Hardwired versus digital logic two-input OR gate.

Truth table

SW-A		SW-B		Light	
Open	(0)	Open	(0)	Off	(0)
Open	(0)	Closed	(1)	On	(1)
Closed	(1)	Open	(0)	On	(1)
Closed	(1)	Closed	(1)	On	(1)

NOT truth table

A	NOT A
0	1
1	0

Figure 25-43 NOT (inverter) gate.

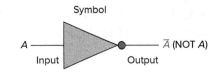

Truth table

Pushbutton		LED	
Not pressed	(0)	On	(1)
Pressed	(1)	Off	(0)

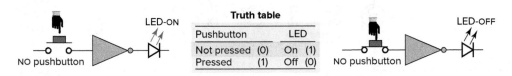

Figure 25-44 Inverting action of the NOT gate.

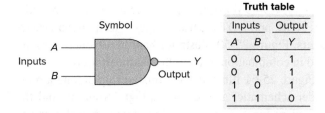

Figure 25-45 NAND gate.

Truth table

Inputs		Output
A	B	Y
0	0	1
0	1	1
1	0	1
1	1	0

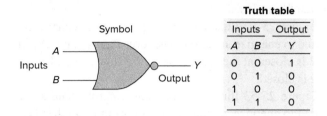

Figure 25-46 NOR gate.

Truth table

Inputs		Output
A	B	Y
0	0	1
0	1	0
1	0	0
1	1	0

AND gate would invert the normal output result. The NAND gate symbol and truth table are shown in Figure 25-45. The output of all NAND gates is high if any of the inputs are low. The basic rules that apply to an NAND gate are:

- If all inputs are 1, the output will be 0.
- If any input is 0, the output will be 1.

An OR gate with an inverted output is called a **NOR** gate. The NOR gate operates as an OR gate followed by a NOT gate. The NOT symbol placed at the output of an OR gate would invert the normal output result. The NOR gate symbol and truth table are shown in Figure 25-46. The output of all NOR gates are low if any of the inputs are high. The basic rules that apply to a NOR gate are:

- If all inputs are 0, the output will be 1.
- If any input is 1, the output will be 0.

The Exclusive-OR (XOR) Gate

Another gate that is similar to an OR gate is the **exclusive-OR (XOR) gate.** The XOR gate symbol and truth table are shown in Figure 25-47. The output of this circuit is HIGH only when one input or the other is HIGH, but **not both.** The exclusive-OR gate is commonly used for

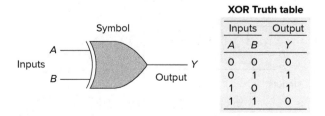

Figure 25-47 XOR gate.

XOR Truth table

Inputs		Output
A	B	Y
0	0	0
0	1	1
1	0	1
1	1	0

the comparison of two binary numbers. The basic rules that apply to an XOR gate are:

- If both inputs are different, the output will be 1.
- If both input are the same, the output will be 0.

TTL and CMOS Logic Gates

TTL stands for **transistor-transistor logic** and is a classification of integrated circuits. The name is derived from the use of two bipolar junction transistors or (BJTs) in the design of each logic gate. **CMOS (complementary metal oxide semiconductor)** is also another classification of ICs that uses field-effect transistors in the design.

TTL gates operate on a nominal power supply voltage of 5 volts. Ideally, a TTL high signal (1) would be 5 volts, and a TTL low signal (0) zero volts. Acceptable input signal voltages range from **0 volts to 0.8 volts** for a low logic state, and **2 volts to 5 volts** for a high logic state. In a similar manner, acceptable output signal voltages range from **0 volts to 0.4 volts** for a low logic state, and **2.4 volts to 5 volts** for a high logic state.

MOS gate circuits have input and output signal specifications that are quite different from TTL. For a CMOS gate operating at a power supply voltage of 5 volts, the acceptable input signal voltages range from **0 volts to 1.0 volts** for a low logic state and **3.5 volts to 5 volts** for a high logic state. Acceptable output signal voltages range from **0 volts to 0.1 volts** for a low logic state, and **4.9 volts to 5 volts** for a high logic state.

25.12 Combination Logic Circuits

Combinational logic circuits allow complex circuits to be made. A typical combinational logic circuit has multiple inputs and one or more outputs. For every combination of signals at the input terminals, there is a definite combination of signals at the output terminals. In many combination logic circuits, sensor switches act as the input devices. The combination logic circuit examines its inputs and then makes a decision based on them. The output is determined by:

- The binary state of the gate inputs.
- The types of logic gate circuits used.
- The manner in which the logic gates are interconnected.

Boolean equations are often used in the simplification of combination logic circuits. By translating a logic circuit's function into symbolic (Boolean) form and applying certain algebraic rules to the resulting equation, the operations are simplified. Boolean equations as related to AND, OR, and NOT gates are shown in Figure 25-48 and summarized as follows:

- Inputs are represented by capital letters A, B, C, and so on, the output is represented by capital Y.
- The dot (·), or no symbol, represents the AND gate operation.
- An addition sign (+) represents the OR gate operation.
- A bar over the letter $\overline{A}$ represents the NOT gate operation.

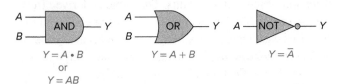

Figure 25-48 Boolean equations as related to AND, OR, and NOT gates.

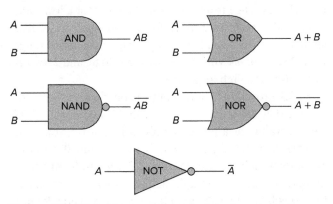

Figure 25-49 Logic gates used singly to form logical statements.

Figure 25-49 illustrates how logic gates AND, NAND, OR, NOR, and NOT are used singly to form **logical statements.** Figure 25-50 illustrates how basic logic gates are used in combination to form **Boolean equations.**

Figure 25-51 illustrates the relationship among a relay ladder schematic, a PLC ladder logic program, and the equivalent logic gate circuit. This logic control scenario requires that the output be energized whenever limit switch LS1 or LS2 is closed at the same time as the pressure switch PS is closed.

Figure 25-52 illustrates the logic control scenario required to energize an output whenever limit switch LS1 or LS2 is closed at the same time as the flow switch FS1 or FS2 is closed.

Figure 25-53 illustrates the logic control scenario required to energize an output whenever limit switches LS1 and LS2 are both closed or limit switch LS3 is closed.

Figure 25-54 illustrates the logic control scenario required to energize an output whenever limit switches LS1 and LS2 are both closed or whenever limit switches LS3 and LS4 are both closed.

Figure 25-55 illustrates the logic control scenario required to energize an output whenever limit switches LS1 is closed and the normally closed pushbutton PB is not activated.

Figure 25-56 illustrates the logic control scenario required for a pushbutton interlocking circuit. The output of this circuit is energized only when pushbutton A or B is activated, but not both.

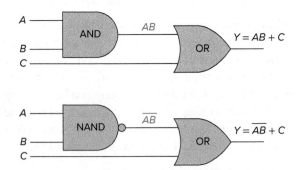
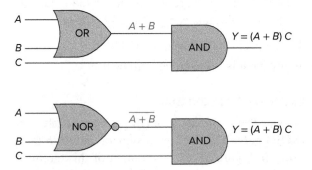

Figure 25-50 Logic gates used in combination to form Boolean equations.

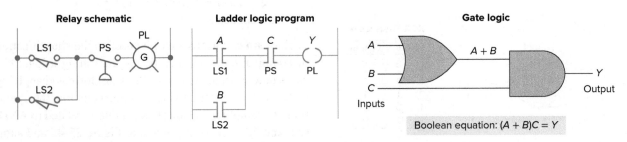

Figure 25-51 Two limit switches connected in parallel and in series with a pressure switch.

Relay schematic
Ladder logic program
Gate logic

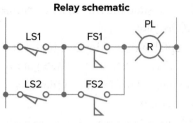
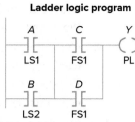
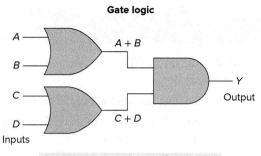

Boolean equation: $(A + B)(C + D) = Y$

Figure 25-52 Two parallel connected limit switches in series with two parallel connected flow switches.

Relay schematic
Ladder logic program
Gate logic

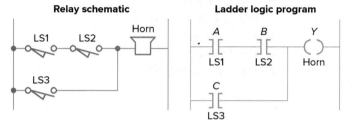
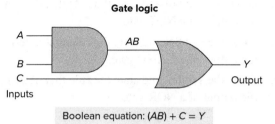

Boolean equation: $(AB) + C = Y$

Figure 25-53 Two series connected limit switches in parallel with a third other limit switch.

Relay schematic
Ladder logic program
Gate logic

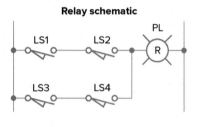
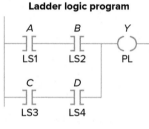
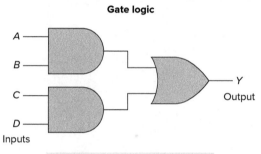

Boolean equation: $(AB) + (CD) = Y$

Figure 25-54 Two series connected limit switches in parallel with two other limit switches that are connected in series.

Relay schematic
Ladder logic program
Gate logic

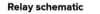

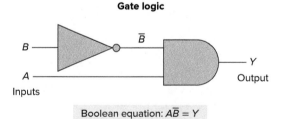

Boolean equation: $A\overline{B} = Y$

Figure 25-55 Normally open limit switch in series with normally closed pushbutton.

Relay schematic
Ladder logic program
Gate logic

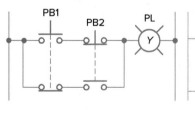
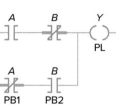
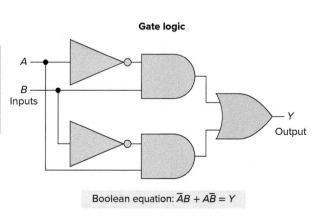

Boolean equation: $\overline{A}B + A\overline{B} = Y$

Figure 25-56 Implementation of an exclusive-OR (XOR) function using NOT, AND, and OR gates.

1. What is the major difference between an analog and digital signal?
2. What is the fundamental basis of the binary principle?
3. Draw the symbol and state the rule that determines the output of an AND gate.
4. Draw the symbol and state the rule that determines the output of a OR gate.
5. Draw the symbol and state the rule that determines the output of a NOT gate.
6. Draw the symbol and state the rule that determines the output of a NAND gate.
7. Draw the symbol and state the rule that determines the output of a NOR gate.
8. Draw the symbol and state the rule that determines the output of a XOR gate.
9. Compare the acceptable digital input signal voltages range for a TTL gate versus a CMOS gate IC.
10. List three factors that determine the output of a combination logic circuit.
11. Develop a combination logic gate circuit for each of the following Boolean equations using AND, OR, and NOT gates:

   a. $Y = ABC + D$
   b. $Y = AB + CD$
   c. $Y = (A + B)(\overline{C} + D)$
   d. $Y = \overline{A}(B + CD)$
   e. $Y = \overline{A}B + C$
   f. $Y = (ABC + D)(E\overline{F})$

Electricity for the Trades, 3/e
Multisim Simulation Lab Manual

Table of Contents

Section One   Fundamentals of Electricity

—(continued)

	Topic	Lab Assignment	File
Chapter 6	Measuring Voltage, Current, and Resistance	- Determining polarity	CH 06_01
		- Circuit continuity	CH 06_02
		- Test lead connections	CH 06_03
		- Ground-referenced voltages	CH 06_04
		- Multimeter troubleshooting (1)	CH 06_05
		- Multimeter troubleshooting (2)	CH 06_06
		- Power supply measurements (1)	CH 06_07
		- Power supply measurements (2)	CH 06_08
		- Series circuit measurements (1)	Ch 06_09
		- Current measurement	CH 06_10
		- Resistance measurement	CH 06_11
		- Component troubleshooting	CH 06_12
		- Branch circuit current measurement	Ch 06_13
		- Series circuit measurements (2)	CH 06_14
		- Parallel circuit measurements	CH 06_15
Chapter 7	Ohm's Law	- Calculating current (1)	CH 07_01
		- Current/resistance relationship	CH 07_02
		- Calculating voltage (1)	CH 07_03
		- Calculating resistance (1)	CH 07_04
		- Current/voltage relationship	CH 07_05
		- Calculating power	CH 07_06
		- Calculating voltage (2)	CH 07_07
		- Calculating current (2)	CH 07_08
		- Calculating resistance (2)	CH 07_09
		- Volts/amps/watts relationship	CH 07_10
		- Ohm's law troubleshooting	CH 07_11
Chapter 8	Resistors	- Variable resistor	CH 08_01
		- Rheostat	CH 08_02
		- Potentiometer	CH 08_03
		- Series resistors	CH 08_04
		- Voltage-divider	CH 08_05
		- Series/parallel resistors (1)	CH 08_06
		- Current divider	CH 08_07
		- Series/parallel resistors (2)	CH 08_08
		- Series/parallel resistors (3)	CH 08_09
		- Series/parallel resistors troubleshooting	CH 08_10
Chapter 9	Electricity and Magnetism	- Step-down transformer	CH 09_01
		- Control relay	CH 09_02
Chapter 10	Electric Power and Energy	- Calculating power (1)	CH 10_01
		- Power/current calculation (1)	CH 10_02
		- Power/resistance calculation	CH 10_03
		- Power/current calculation (2)	CH 10_04
		- Calculating power (2)	CH 10_05
		- Calculating energy	CH 10_06

—(continued)

	Topic	Lab Assignment	File
Chapter 11	Solving the DC Series Circuit	- Circuit characteristics	CH 11_01
		- Calculating E, I, and R (1)	CH 11_02
		- Calculating E, I, and R (2)	CH 11_03
		- Circuit analysis (1)	CH 11_04
		- Circuit analysis (2)	CH 11_05
		- Voltage measurement	CH 11_06
		- Aiding and opposing voltages	CH 11_07
		- Voltage source resistance	CH 11_08
		- Series conductor resistance	CH 11_09
		- Circuit troubleshooting (1)	CH 11_10
		- Circuit troubleshooting (2)	CH 11_11
		- Circuit troubleshooting (3)	CH 11_12
		- Circuit troubleshooting (4)	CH 11_13
		- Oven control circuit	CH 11_14
Chapter 12	Solving the DC Parallel Circuit	- Circuit characteristics	CH 12_01
		- Calculating E, I, and R (1)	CH 12_02
		- Calculating E, I, and R (2)	CH 12_03
		- Circuit analysis (1)	CH 12_04
		- Circuit analysis (2)	CH 12_05
		- Voltage polarity	CH 12_06
		- Circuit troubleshooting (1)	CH 12_07
		- Circuit troubleshooting (2)	CH 12_08
		- Circuit troubleshooting (3)	CH 12_09
		- Circuit solving (1)	CH 12_10
		- Circuit solving (2)	CH 12_11
		- Circuit solving (3)	CH 12_12
		- Circuit solving (4)	CH 12_13
Chapter 13	Solving the DC Series-Parallel Circuit	- Circuit characteristics	CH 13_01
		- Kirchhoff's laws	CH 13_02
		- Circuit analysis	CH 13_03
		- Circuit troubleshooting (1)	CH 13_04
		- Circuit troubleshooting (2)	CH 13_05
		- Circuit troubleshooting (3)	CH 13_06
		- Three-wire circuit characteristics	CH 13_07
		- Three-wire circuit analysis (1)	CH 13_08
		- Three-wire circuit analysis (2)	CH 13_09
		- Three-wire circuit troubleshooting	CH 13_10
Chapter 14	Network Theorems	- Practical voltage source	CH 14_01
		- Practical current source	CH 14_02
		- Superposition theorem (1)	CH 14_03
		- Superposition theorem (2)	CH 14_04
		- Superposition theorem (3)	CH 14_05
		- Superposition theorem (4)	CH 14_06
		- Superposition theorem (5)	CH 14_07
		- Thevenin's theorem (1)	CH 14_08
		- Thevenin's theorem (2)	CH 14_09
		- Thevenin's theorem (3)	CH 14_10
		- Norton theorem (1)	CH 14_11
		- Norton theorem (2)	CH 14_12

—(continued)

	Topic	Lab Assignment	File
Chapter 15	Alternating Current Fundamentals	- Time period	CH 15_01
		- Peak voltage	CH 15_02
		- RMS voltage	CH 15_03
		- Bridge rectifier	CH 15_04
		- Distribution transformer	CH 15_05
		- Three-phase system (1)	CH 15_06
		- Three-phase system (2)	CH 15_07
		- AC series resistive circuit	CH 15_08
		- AC parallel resistive circuit	CH 15_09
		- Three-phase resistive circuit (1)	CH 15_10
		- Three-phase resistive circuit (2)	CH 15_11
Chapter 16	Inductance and Capacitance	- L/R time constant	CH 16_01
		- Inductance DC/AC circuits	CH 16_02
		- Inductance/current flow	CH 16_03
		- Inductance/frequency	CH 16_04
		- Inductive reactance	CH 16_05
		- Inductors series connected	CH 16_06
		- Inductors parallel connected	CH 16_07
		- Inductor/phase relationship	CH 16_08
		- Inductor/wattage	CH 16_09
		- Calculating inductive reactance	CH 16_10
		- Inductors series connected X_L	CH 16_11
		- Inductors parallel connected X_L	CH 16_12
		- Inductive volt/amps reactive	CH 16_13
		- Capacitor charge/discharge	CH 16_14
		- RC time constant	CH 16_15
		- Capacitance DC/AC circuits	CH 16_16
		- Capacitance/current flow	CH 16_17
		- Capacitance/frequency	CH 16_18
		- Capacitive reactance	CH 16_19
		- Capacitors series connected	CH 16_20
		- Capacitors parallel connected	CH 16_21
		- Capacitor/phase relationship	CH 16_22
		- LC parallel circuit	CH 16_23
		- Capacitor/wattage	CH 16_24
		- Capacitive reactance	CH 16_25
		- Capacitive volt/amps reactive	CH 16_26
Chapter 17	Resistive, Inductive, Capacitive (RLC) Series Circuits	- R circuit	CH 17_01
		- RL circuit	CH 17_02
		- Adding and opposing voltages	CH 17_03
		- Triangular vector addition	CH 17_04
		- Parallelogram vector addition	CH 17_05
		- Pythagorean theorem	CH 17_06
		- RL circuit impedance	CH 17_07
		- RL voltage and current	CH 17_08

—(continued)

Topic	Lab Assignment	File
	- *RL* power	CH 17_09
	- *RL* circuit analysis	CH 17_10
	- *RC* circuit impedance	CH 17_11
	- *RC* voltage and current	CH 17_12
	- *RC* power	CH 17_13
	- *RC* circuit analysis	CH 17_14
	- *RLC* circuit waveforms	CH 17_15
	- *RLC* circuit voltages	CH 17_16
	- *RLC* circuit analysis	CH 17_17
	- *RLC* power measurement	CH 17_18
	- *RLC* power calculations	CH 17_19
	- *RLC* circuit analysis	CH 17_20
	- *RLC* series resonant circuit	CH 17_21
	- *RLC* resonant circuit analysis	CH 17_22
	- *RLC* resonant frequency calculation	CH 17_23
	- *RLC* resonant frequency Response	CH 17_24
Chapter 18	Resistive, Inductive, Capacitive (*RLC*) Parallel Circuits	
	- *RL* circuit impedance	CH 18_01
	- *RL* voltage and current	CH 18_02
	- *RL* circuit impedance	CH 18_03
	- *RL* power	CH 18_04
	- *RL* circuit analysis	CH 18_05
	- *RL* power calculations	CH 18_06
	- *RC* voltage and current	CH 18_07
	- *RC* power	CH 18_08
	- *RC* circuit analysis	CH 18_09
	- *RC* current and power	CH 18_10
	- *RC* circuit analysis	CH 18_11
	- *LC* current calculations	CH 18_12
	- *LC* impedance	CH 18_13
	- *LC* power calculations	CH 18_14
	- *RLC* voltage and current	CH 18_15
	- *RLC* current calculations	CH 18_16
	- *RLC* impedance	CH 18_17
	- *RLC* power measurements	CH 18_18
	- *RLC* parallel resonant circuit	CH 18_19
	- Motor load monitoring	CH 18_20
	- Inductive load monitoring	CH 18_21
	- Motor power factor	CH 18_22
	- Motor PF correction	CH 18_23
Chapter 19	Transformers	- Turns ratio	CH 19_01
	- Turns ratio/voltage	CH 19_02
	- Turns ratio/current	CH 19_03
	- Circuit analysis	CH 19_04
	- Polarity test circuit	CH 19_05
	- Dual voltage transformer	CH 19_06
	- Three-phase transformer analysis	CH 19_07
	- Three-phase transformer system	CH 19_08
	- Delta-to-wye transformer	CH 19_09

—(continued)

Section Four   Electrical Installation and Maintenance

Countervoltage, inductance and, 195
CPU. *See* Central processing unit
Crane electromagnets, 91
Cumulative-compound DC motors, 356
Current (*I*), 29. *See also specific types*
   in AC resistive circuit, 189–191
   in balanced three-wire circuit, 154
   capacitor phase shift and, 209
   in DC parallel circuits, 130–133
   in DC series-parallel circuit, 144–152
   defined, 3
   in delta-connected configuration, 189
   electricity transmission and, 102
   electromagnetism and, 87
   as flow, 18
   for HID, 331
   inductance and, 196
   in magnetic circuit, 90
   measurement, 51–52
   Ohm's law and, 34, 62–63
   in parallel circuits, 42
   in parallel *RL* circuits, 243
   in parallel *RLC* circuits, 256–258
   power and, 64–65
   rating, for load centers, 314
   in series *RC* circuits, 225–226
   in series *RL* circuits, 219
   in series *RLC* circuits, 232
   short-circuits and, 307
   in superposition theorem, 161–165
   in Thevenin's theorem, 167, 168–169
   transformers, 288–289
   in transformers, 271, 272–275, 280
   in unbalanced three-wire circuit, 154
   voltage and, 28
Current amplifier, for BJT, 362
Current divider circuits, 77–78
Current electricity, 23–24
Current flow
   in AC, 177
   to capacitors, 202, 206
   in conductors, 304
   diodes and, 361
   in electric circuits, 35
   inductive reactance and, 198–199
   magnetic field and, 88
   in Norton's theorem, 171
   in parallel circuits, 127
   in parallel *RL* circuits, 244
   in resonant series circuits, 239
   in series circuit, 113, 114
   in series *LC* circuits, 231
   in series-parallel circuits, 140–141
Current probe, 51
Current source, 160
   in Norton's theorem, 170
   in superposition theorem, 160
Current-limiting ability, for overcurrent
   protection devices, 309
Cycle, of AC, 181
Cylindrical alternator rotor, 180

D

DC (direct current), 111–123
   AC and, 177–178, 183
   AC to DC rectification, 366–367
   from batteries, 25

electric power of, 105, 107
generators, 93, 178
for LED lamps, 331
for light switches, 334
for magnets, 82
for PV, 100
relays for, 321
solving for, 112
symbols, 38
for universal motors, 348
voltage, 23, 26, 49
voltmeters for, 184
DC circuits
   electrons in, 178
   inductive reactance in, 197
   power for, 190
DC injection braking, for motor
   stopping, 353
DC motors, 354–359
   armature reaction in, 358
   cemf for, 357
   direction of rotation for, 356–357
   speed control for, 358–359
DC parallel circuits, 130–133
   batteries for, 134
   polarity of, 133–134
   solving, 144–155
   troubleshooting, 134–135
DC resistance, inductance and, 196
DC series circuit, 116–120
   polarity, 119–120
   troubleshooting, 122–123
DC series-parallel circuits
   current in, 144–152
   polarity in, 152
   power in, 144–152
   resistance in, 144–152
DC to AC inverters, 25, 367
Deep-cycle battery, 134
Delta-connected configuration, 189
   for single-speed AC induction
     motors, 343
   for three-phase transformers, 283
Delta-to-delta connection, for three-phase
   transformers, 284–285
Delta-to-wye connection, for three-phase
   transformers, 287
Depletion type MOSFETs, 362
Derating, of transformers, 292
Diacs, 370
Dielectric
   capacitance and, 201, 202, 203
   for charged capacitors, 202
   leakage current, 210–211
   strength, 290
Digital
   versus analog, 374–375
   analog-to-digital converter circuit, 47
   direct current, 177
   integrated circuits, 364
Digital multimeter (DMM), 48, 49
   for resistance, 53, 54
   specifications, 55–56
   wireless display for, 51, 52
Dimmer switches, 337
   for incandescent lamp, 370
Diodes, 361. *See also* Light-emitting diode

of semiconductors, 18
test, DMM, 56
Direct current. *See* DC
Direction, of vectors, 213
Discharge, of capacitors, 202, 205,
   206, 209
Disconnect device, for motors, 350
DMM. *See* Digital multimeter
Doping, 18
Dot notation, for transformer polarity, 279
Double-pole, double throw (DPDT), 325
Double-pole circuit breakers, 315
DPDT. *See* Double-pole, double throw
Dry-type transformers, 278
Dual-element, time-delay fuses, 311
Dual-purpose arc-fault/ground-fault
   (AF/GF), 316
Dual-voltage
   run windings, in single-phase induction
     motors, 346
   transformers, 280–281
   wye connections, for single-speed AC
     induction motors, 343

E

E. See Voltage
Eddy current, 178
   losses, 276
   in transformers, 272
Edison-base plug fuses, 309–310
Effective value, AC, 183
Efficiency, of transformers, 276–277
Electric circuit diagrams, 34, 38–40
   resistors on, 74
Electric circuits, 33–34. *See also specific*
   types
   breadboarding for, 43–44
   computer simulation software for, 43–44
   current flow in, 35
   linear element in, 66
   nonlinear element in, 66
   symbols for, 38
Electric energy, 104
   from batteries, 25
   electric power and, 32
   from heat, 26
Electric field, of FET, 362
Electric generator. *See also* Alternators;
   Generators
   mechanical-magnetic energy and,
     26–27
Electric power (P), 32
   calculating, 104–107
   measurement, 107
Electric service, 186
Electric shock, 3
   GFCI and, 11
   static, 21
Electrical continuity, testing, 19
Electrical degrees, 182
   120 degrees apart, 186–187
Electrical isolation, solid-state relays
   for, 322
Electrical metallic tubing (EMT), 298
Electrical motor braking, 352–353
Electrical noise, 104
Electrically balanced loads, 286–287